Geneti[illegible]
Bact[illegible] [illegible]ilen[illegible]

Genetics of Bacterial Virulence

CHARLES J. DORMAN
Department of Biochemistry, University of Dundee

OXFORD
BLACKWELL SCIENTIFIC PUBLICATIONS
LONDON EDINBURGH BOSTON
MELBOURNE PARIS BERLIN VIENNA

Editorial Offices:
Osney Mead, Oxford OX2 0EL
25 John Street, London WC1N 2BL
23 Ainslie Place, Edinburgh EH3 6AJ
238 Main Street, Cambridge
Massachusetts 02142, USA
54 University Street, Carlton
Victoria 3053, Australia

Other Editorial Offices:
Librairie Arnette SA
1, rue de Lille
75007 Paris
France

Blackwell Wissenschafts-Verlag GmbH
Düsseldorfer Str. 38
D-10707 Berlin
Germany

Blackwell MZV
Feldgasse 13
A-1238 Wien
Austria

First published 1994

Set by Semantic Graphics, Singapore
Printed and bound in Great Britain
at the Alden Press, Oxford

DISTRIBUTORS

Marston Book Services Ltd
PO Box 87
Oxford OX2 0DT
(*Orders*: Tel: 0865 791155
Fax: 0865 791927
Telex: 837515)

USA
Blackwell Scientific Publications, Inc.
238 Main Street
Cambridge, MA 02142
(*Orders*: Tel: 800 759-6102
617 876-7000)

Canada
Oxford University Press
70 Wynford Drive
Don Mills
Ontario M3C 1J9
(*Orders*: Tel: 416 441-2941)

Australia
Blackwell Scientific Publications Pty Ltd
54 University Street
Carlton, Victoria 3053
(*Orders*: Tel: 03 347-5552)

A catalogue record for this title
is available from the British Library

ISBN 0-632-03662-1

Library of Congress
Cataloging in Publication Data

Dorman, Charles J.
Genetics of Bacterial Virulence/
by Charles J. Dorman.
p. cm.
Includes bibliographical references
and index.
ISBN 0-632-03662-1
1. Virulence (Microbiology)
2. Bacterial genetics. I. Title.
QR175.D67 1994
616′.014—dc20

Contents

Preface

This book is being written during a period which sees the emergence of a more complete understanding of how it is that bacteria function as integrated cells. Unifying themes are particularly well developed in the area of gene regulation, the subject of this book. In this field, decades of research have yielded a rich harvest of data and it has become possible recently to describe some of the means by which the expression of different genes is coordinately regulated in response to changes in the environment. This has involved an appreciation of the complexities of multifactorial regulatory systems and the fundamental contributions made by genome structure itself. An exciting corollary to this is the possibility of applying these insights in order to obtain an understanding of how bacterial genes involved in disease processes are controlled. This work attempts to describe the outcome of this application.

It is not the purpose of this book to describe particular bacterial pathogens and their specific interactions with their hosts in detail. Instead, it is aimed at those interested in understanding how virulence genes are regulated and how such regulatory circuits are integrated with other control networks governing gene expression in the cell. Of necessity, background material is provided on the major pathogenic processes to which virulence genes contribute. Therefore, some information on familiar themes such as attachment, adhesion, invasion, intracellular survival and spread, iron acquisition, host defence avoidance et cetera is included. However, highly detailed descriptions of these processes have already been given in other publications and are not repeated here.

This volume presents, for the first time, a review of the gene regulatory circuits that underly these bacterial pathogenic processes. It attempts to place each regulatory system in the context of existing information on bacterial gene regulation. To that end, background information is provided on fundamental topics such as transcriptional and post-transcriptional regulation, general and site-specific recombination, transposition, replicon structure (primarily chromosomes and plasmids) and genome structure. The important topic of coordinated regulation of gene expression in response to the environment is covered at length.

I have not attempted to list exhaustively the multitude of genetic systems which have now been characterized at the molecular level in bacterial pathogens but have chosen those examples that are the best understood and are the most suitable for the purpose of illustrating how coordinated gene regulatory mechanisms work. Consequently I apologize to those colleagues whose contributions have not been included. Coverage of regulation at 'post-nucleic acid stages' of expression has been kept to a minimum, so topics concerned with secretion of or post-translational modification of gene products are not dealt with in detail.

The inspiration for this book comes from associations with many workers in the fields of prokaryotic genetics, biochemistry and medical microbiology spanning over 15 years. I am indebted particularly to John McEvoy, Tim Foster, Gordon Dougan, Bill Shaw, Chris Higgins, David Lilley and Philip Cohen. I also acknowledge the steadfast support given to me by Niamh Ní Bhriain over the past decade, both inside and outside research.

Introduction

Ultimately, the composition of a cell is determined by the expression of its genetic material. Knowledge of how expression of the genetic material is controlled is essential to an understanding of how the cell develops and functions. Bacterial cells, the subject of this book, have a remarkable capacity to reinvent themselves as they endeavour to adapt to changing environmental conditions. As simple, unicellular organisms, they have to make the most of their small, haploid complement of just a few thousand genes to ensure that they are not overwhelmed by (often sudden) variations in their surroundings. To do this, they have evolved elaborate control mechanisms to turn genes on or off as required, ranging from relatively straightforward transcription activator or repressor proteins to systems that act at the level of the structure of the genetic material itself.

From the point of view of environmental adaptation, pathogenic bacteria infecting their host face a considerable challenge. Like the pathogen, the host has undergone a great deal of environmental adaptation and will have evolved traits designed specifically to hinder infecting organisms. If this was not the case, there would be no host to infect. Thus, at a fundamental level, the host–pathogen interaction represents a form of biological politics in which two contending parties seek to reconcile their competing interests. If the host prevails, the infection fails; if the bacterium wins an outright victory, the host suffers disease and the bacterium may find itself without a host. Clearly, there are no calculations on either side about how much to resist or to persist. Each does its utmost to prevail and the outcome is a reflection of the relative strengths of the opponents. The success of the microbe will be determined by its genetic fitness for the task, by the presence or absence of competing organisms, by the state of the host and by the physical and chemical nature of each environmental niche it must colonize during the process. The success of the host in resisting attack will be influenced by such all-pervading factors as nutrition and genetically encoded traits. Illness can render a normally resistant host susceptible to infection, since a host already compromised (as, for example, with an immunodeficient individual) is less likely to withstand successfully a bacterial infection than is an uncompromised individual. Furthermore, those factors that render a particular host a 'soft target' may allow the microbe to do too much damage, resulting in loss of the host and a need to begin again. These observations are essential to the identification of those host groups most likely to suffer grievously from infection. It is not an accident that life-threatening microbial infections are so commonly a feature of the lives of those humans who are the most socially and economically deprived.

In considering the development of pathogenic processes and their regulation, it may be pertinent to pose the (rather anthropocentric) question 'what is it that bacteria want out of life?' The answer is probably that they desire to divide and pass on their genetic information to their descendants and to do so in an environment that poses the minimum of risk to the enterprise. Those bacteria that elect to 'come in from the cold' and associate with a host require specialized factors: adhesins will enable them to attach securely to host surfaces; capsules can hide and protect the bacterium from the host defences; invasion factors will admit them to a privileged intracellular world; specialized enzymes allow them to disable host defences or to move from one cellular compartment to another or from cell to cell; toxins may assist colonization at several stages of infection and may liberate nutrients from lysed host cells; specialized chelators help the bacteria to win access to the abundant yet normally unavailable iron sequestered within transferrin, lactoferrin and related compounds; some pathogens even possess the means to mimic host features, thus evading the immune system. All of these abilities are genetically encoded within the bacterium and many of the genes are subject to regulation in response to cues picked up from the environment or are controlled by quasiprogrammed mutation events which alter the profile of their expression across the population of infecting bacteria. It is the purpose of this book to examine the means by which such control is exercised.

Bacteria possess the ability to gather information about the physical and chemical nature of their environments. Recently, insights have been gained into the mechanisms by which they gather this information and transduce it to the cellular machinery charged with mounting an appropriate response. In many cases, that response involves a change in the pattern of gene expression in the cell. Since the environment is a multicomponent complex, the overall response must be one that deals with all of the stimuli received. For this reason, it is no surprise to learn that the bacteria appear to have 'networked' their response systems, to borrow a jargon term from another field of information technology, that of computers. So successful is this approach to environmental adaptation that it appears to have been recapitulated in Gram-positive and Gram-negative bacteria of many genera. Presumably, in terms of the life-or-death issue of adaptation, it is a system that patently 'works'. Since human beings are frequently disadvantaged because of the success of these bacterial coping strategies, it is certainly in our interests to become intimately familiar with how they function.

The issue of environmental response through modulation of gene expression is relevant to the field of medical microbiology because it has become apparent that the regulatory networks used for general adaptation are also important during infection. Furthermore, many of the specific regulatory mechanisms governing 'virulence gene' expression work in the same way as their housekeeping counterparts and frequently work cooperatively with pleiotropic regulators first identified as controlling gene expression in the free-living situation.

In terms of the development of a philosophy of pathogenicity, this realization is largely contemporaneous with the recognition that infection can be regarded as an ecological process. Just as bacteria living in the natural environment must cope with an ever-changing (and frequently hostile) environment, so it is with bacteria as they infect a human, animal or plant host. Therefore, it is not surprising to discover that the gene regulatory mechanisms which aid the bacterium in the wild can assist during infection, nor is it unexpected that the control systems which operate specifically to regulate virulence gene expression work in a manner directly analogous to those concerned with the regulation of housekeeping genes.

From the point of view of the researcher into gene regulatory mechanisms, this realization comes as a great benefit. This is because the molecular detail of an impressive array of gene control mechanisms has already been worked out in well-understood prokaryotes such as *Escherichia coli* and *Salmonella typhimurium* (both of which, incidently, occur in highly pathogenic forms). Thus, in many ways, the hard work has already been done for the medical microbiologist, or it is being done at the moment. Microbial genetics, biochemistry and biophysics have all contributed to a profound understanding of how it is that information concerning the physical and chemical make-up of the outside world is obtained by bacteria and transmitted to the nucleoid where it promotes changes in gene expression. The literature contains a vast store of information on the role of protein phosphorylation and methylation in information transduction, including high-resolution crystal structure and nuclear magnetic resonance data on participating molecules. Much is known about protein–DNA interactions and how these lead to modifications of the major processes of DNA. Knowledge of the structure of DNA has advanced impressively since the elucidation of its double-helical nature in 1953 and the importance of DNA structural transitions to processes such as transcription is widely appreciated. The *in vivo* significance of non-B-DNA structures such as A-DNA and Z-DNA and cruciform structures is becoming apparent. The many roles of different manifestations of DNA tertiary structure, such as bends, loops and supercoils, in processes such as gene expression are already generally accepted. Knowledge of the nature of the bacterial nucleoid, which traditionally has lagged behind that of its eukaryotic equivalent, has increased dramatically. It is now understood that the prokaryotes, like the eukaryotes, possess low-molecular-weight, basic proteins capable of compacting DNA and that these play both an architectural role in the nucleoid and regulatory roles in controlling gene expression and other reactions of DNA.

There is a growing appreciation of the forces at work within the bacterial genome that dictate its stability. The existence of extrachromosomal elements such as plasmids has been established for a considerable period of time but we continue to learn more about their roles in the reassortment of the bacterial genome. Recent demonstrations of the ability of self-transmissible plasmids to transfer themselves from bacteria to eukaryotes have profound implications for our understanding of the size of the pool of genetic material from which bacteria might acquire novel characteristics. Genetic information contributing to virulence can be carried on self-transmissible plasmids and these can be stored in non-pathogenic species to be donated later to virulent strains, amplifying their ability to cause harm to the host. In this way, non-pathogenic bacteria can collaborate with pathogens by acting as reservoirs of genes potentially harmful to the host.

Plasmids have evolved many successful strategies to ensure that their host does not lose them. The most blatant is a form of molecular blackmail in which the plasmid manufactures a cytotoxic compound to which only it has the antidote. Thus, should the cell 'lose' the plasmid, the unstable antidote soon breaks down and the toxic compound kills the 'unfaithful' cell. In this way, only plasmid-containing cells survive in the population. This may be regarded as an example of replicon self-interest working as a force to maintain genome stability.

In terms of disseminating genetic information, bacterial viruses may serve functions analogous to those of plasmids and there are several examples of phage carrying virulence factors such as toxin genes. Shuttling of genetic information between replicons (such as the chromosome and plasmids) can be achieved by mobile genetic elements (the transposons) or by more conventional recombination mechanisms. Some transposable elements are very sophisticated indeed and even have the ability to organize their own cell-to-cell transfer. An extremely significant consequence of this intercellular shuttling of genetic information is the dissemination of genetic information rendering cells resistant to antimicrobial agents, including antibiotics. This already represents a major public health issue and one that will continue to grow in importance as the usefulness of more and more drugs becomes diminished in the treatment of infectious diseases caused by bacteria. Other significant insights into genome reassortment come from the discovery of repetitive DNA sequences which serve as substrates for the recombination machinery of the cell, allowing portions of the genome to be amplified or deleted, often in response to environmental cues.

Gene movement between cells has obvious implications for the spread of information contributing to virulence in the bacterial population. However, gene movement between different sites in the genome of the same cell may be significant in terms of the facility of gene expression. Recent insights into the dynamic nature of the transcritionally active genome suggest that not all sites may be equivalent from the point of view of transcription. Not only are different regions of the bacterial chromosome thought to be supercoiled to different extents, the level and type of supercoiling in those domains can be modulated significantly by the transcriptional activities of the resident genes. Transcription complexes passing along the DNA template produce a phenomenon equivalent to the exertion of a topological 'Doppler effect', inducing waves of positive supercoils ahead and waves of negative supercoils behind. These can impinge on the activities occurring within neighbouring sequences and this has forced us to reassess our models of how the genome 'works' as an integrated unit. It also introduces yet another level of regulation of DNA processes.

The complexity of the contribution of DNA supercoiling to gene expression increases when one is forced to take account of the fact that it is environmentally responsive. Thus, changes in parameters such as temperature, osmolarity or growth phase result in changes in the level of supercoiling and these have the potential to be exploited in the regulation of the transcription pattern of the cell. Given the general importance of DNA structure to the expression of all genes and the impact that changes in supercoiling have on DNA tertiary structure, this feature of the genome has the potential to function as a general regulator of transcription. It can also exert an influence on DNA replication, on recombination and on transposition, permitting it to have widespread effects on the cell cycle and on genome stability. Supercoiling may also affect differentially chromosomal and plasmid sequences, resulting in changes in plasmid stability and/or in changes in plasmid-linked gene expression which may be more or less exaggerated than those seen on the chromosome. It is possible that some genes which work on plasmids may cease to function if moved to the chromosome (and vice versa) and this could permit the pathogen to store genetic information in a cryptic state until the cell encounters conditions where those genes confer a selective advantage. At this point, any cells in the population that have activated the genes will be selected for and proceed to cause infection. There is evidence that *E. coli* keeps some genetic information in such a cryptic form and the β-glucoside transport system is an example of this phenomenon. It has been reasoned that since some β-glucosides generate toxic products when metabolized, the population maintains just a few cells in a condition in which they can use these molecules as a carbon source. If β-glucosides are the sole carbon source available and those cells with active *bgl* operons transport and metabolize them, they will have a selective advantage over the Bgl^- majority, provided they have not encountered the toxic variety. By playing Russian roulette with the environment, the bacterial population risks a few members in the interests of gaining an advantage under the correct circumstances.

The topic of phase variation, in which more or less specific changes to the nucleotide sequence of the affected gene result in a switch between expression states, has attracted considerable attention. Mechanisms of phase variation can involve site-specific recombination events in which specialized recombinases acting at specific sites bring about a novel joint in the DNA, for example as a result of an inversion, which alters gene expression. Another mechanism involves the deletion or amplification of bases within homopolymeric sequences, resulting in transcriptional or post-transriptional changes in gene expression. Base modification (for example, through methylation) provides a further method of generating a quasistable change in the expression of particular genes. In most cases, the procedure is reversible, giving the cell the possibility of reverting to its original expression profile at some later time. From the point of view of a pathogenic bacterium, these mechanisms provide an ideal method of changing the structure of the cell at points where it may become vulnerable to immune recognition, such as the bacterial cell surface. Not surprisingly, many of the genes are subject to phase-variable control code for cell surface components such as outer membrane proteins or adhesins.

It can be seen that distinct but complementary mechanisms of gene control operate in bacteria. At one level, gene expression is directly under the control of the environment. At another level, gene expression is subject to more or less stochastic processes which confer on the cell potential advantages. If these are selected for by the current environment, this cell will be in a strong position to colonize its niche. These stochastic processes include both the phase-variable gene expression systems and conventional, random mutation events.

In recent years a body of opinion has grown up which holds that not all mutation events in bacteria are necessarily random but, instead can be directed by environmental conditions (Cairns *et al.*, 1988; reviewed in Foster, 1992). Thus, the possibility of non-darwinian, environmentally directed evolution has added yet another level of complexity to the mechanisms concerned with the expression of genetic information. The means by which such apparently directed mutations arise and become 'fixed' has provoked a great deal of debate but a common requirement appears to be a crisis in the life of the bacterium which must be resolved by the acquisition of a specific piece of genetic information. In the experimental systems used, this is usually an ability to use a particular sugar as a carbon source. However, one could easily imagine that a need to elaborate a particular form of 'virulence' factor could also be demanded in order to resolve an environmental crisis during infection.

An attractive possibility by which the desired mutation could be acquired envisages 'dead' bacterial cells (i.e. those no longer capable of dividing) as mutation 'factories' (Higgins, 1992). Here, an environmentally triggered acceleration in the mutation rate occurs, changing the structure of the genome at a rate up to 1000 times faster than in dividing cells. (This is not so fanciful when one recalls that bacteria do indeed possess inducible mutator systems.) Next, the altered genetic information is exported to cells still capable of dividing. This could be done by any of the well-documented mechanisms of gene transfer that bacteria are known to possess (plasmid-mediated transfer, bacteriophage-mediated transfer or transfer of DNA to naturally transformable cells). Dividing cells which acquire the urgently needed allele will then thrive and go on to colonize the niche successfully. In this way the mutation becomes fixed in the successful portion of the population and no non-darwinian processes need to be invoked. The novel component of the theory concerns the period of accelerated evolution that occurs under stress, which corresponds with the idea of 'punctuated gradualism', invoked by some palaeontologists to explain the sudden appearance over a short span of evolutionary time of new species in the fossil record without apparent intermediate forms (reviewed in Gould, 1989).

In terms of an improved ability to combat bacterial pathogens, how does more detailed information about how they organize their genetic affairs assist us? By knowing how bacteria interrogate their environment, process the information obtained, transmit it to the genome and elicit a change in gene expression, we can identify novel targets for intervention in these processes. One possibility is to use antimicrobial agents to inhibit the stimulus–response system. This suffers from the same long-term disadvantage as any other strategy based on such agents — the emergence of resistant strains. There is also the possibility (impossible to quantify) that the resistance mechanism may involve a change in the structure of the target component of the system which not only confers resistance but improves its performance. Thus, prolonged use of the drug could result in the selection of pathogens that are both resistant and more virulent.

Another option is to use detailed knowledge of how the regulatory processes operate to design, by rational means, strains disabled in important aspects of their environmental adaptation machinery. These organisms might be expected to misinterpret their surroundings, to activate the wrong genes at the wrong time and generally to mount an inappropriate environmental response. This approach is really a development of the successful methods based on auxotrophy used by Stocker and others to produce attenuated mutant strains of *Salmonella* for use as vaccines (Hoiseth & Stocker, 1981). The difference is that in this case, the intention is not to produce auxotrophy, but rather a form of microbial psychosis.

The chapters in this book deal in turn with (i) the mechanisms and the philosophy of bacterial virulence; (ii) the structure of the bacterial genome, indicating how its components influence gene expression and how its integrity is maintained despite the considerable forces at work which attempt to reassort it; (iii) the role of bacterial plasmids in gene expression and dissemination; (iv) gene rearrangements and how these are achieved; (v) how gene rearrangements influence specifically the expression of virulence factors; (vi) the many mechanisms involved in the regulation of gene expression both at the level of transcription and in the post-transcriptional phase; (vii) how these are applied to control the expression virulence genes; (viii) the means by which virulence gene expression is regulated in a networked manner. In this way the reader is taken through the material in a step-by-step manner arriving finally at a description of the mechanisms by which virulence gene expression is coordinately regulated.

1 Bacterial Virulence: Philosophy and Mechanisms

Introduction

This book deals with the mechanisms by which bacterial cells adapt to changes in their environment by altering the expression of their genes. In particular, it seeks to illustrate how these adaptive mechanisms are used by disease-causing organisms to control the expression of the virulence factors which render them pathogenic. Before proceeding to a detailed discussion of the control mechanisms, the nature of infection and the virulence factors required for it will be considered.

Bacterial infectivity

Prokaryotic cells are capable of existing in different types of relationships with host organisms. Some of these interactions result in no harm to the host or are even beneficial, as in the case of the normal gut flora of mammals. Other interactions can result in host damage of different degrees of severity. The extent of the damage is a function both of the health of the host (an already debilitated host will be much more susceptible to damage than a healthy individual) and of the virulence of the bacterium. Bacterial virulence can be enhanced by the use of specialized virulence factors. Examples of these include toxins which damage host cells and tissues, capsules which mask the bacterium from the host defences, enzymes which degrade specifically host defence molecules such as immunoglobulins, colonization factors (such as adhesins) which enable the bacterium to attach strongly to host tissues and invasins which allow the bacterium to enter host cells. In extreme cases, the bacterium can even insert part of its own genome into that of the host, altering the genetic activity of the host in a manner that favours the pathogen. This behaviour is similar to that displayed by viruses and is seen in some bacteria that infect and cause a form of cancer in plants.

All of these pathogenic processes are genetically determined, which means that the virulent organisms possess genes that code for them, thus distinguishing themselves from other bacteria which do not have these genes. Non-virulent bacteria can still live in association with a host but do not have the facility to cause disease easily; these are referred to as commensal organisms. However, the distinction is not as simple as one between 'armed' and 'unarmed' bacteria. It is possible for a bacterium to possess DNA sequences capable of coding for virulent traits without expressing them. Thus the organism is superficially harmless (phenotypically commensal) but is genetically virulent. Understanding how the latent virulence traits come to be expressed is partly a matter of studying gene regulation. It is also possible for a normally harmless bacterium to cause a disease if it encounters a host which is disabled in an appropriate way. Such a bacterium is classified as an opportunistic pathogen. This organism may lack obvious virulence traits such as toxins but it still causes a disease in the host. To do this, it must adapt to the environment at the site of infection. Again, this usually involves a change in the gene expression profile of the bacterium. Regulation of genes involved in environmental adaptation implies an ability on the part of the bacterium to interpret its surroundings. As will be seen in this book, many genes are regulated in response to specific environmental cues. These are chemical signals and physical signals which provide information about the presence of toxic substances, about temperature, oxygen availability, osmolarity, nutrient levels, pH and other parameters. The acquisition of this information allows the organism to construct a picture of its surroundings. It is then able to respond in an appropriate manner by modulating its gene expression programme. In a dedicated pathogen, the virulence genes form a part of that programme.

Before going on to describe in later chapters the transcriptional and post-transcriptional regulatory mechanisms and the DNA rearrangement events that determine the expression of bacterial virulence factors, it is necessary to discuss these factors and their utility to the pathogen.

Bacterial virulence factors

These are usually thought of as the specialized features (adhesins, capsules, toxins, iron chelators, etc.) that the infecting organism uses to gain access to its niche on or within the host, to colonize that niche, to escape the host defences and to exclude competing organisms. There is a wide variety of virulence factors in use in the bacterial world and knowledge of these is uneven from organism to organism. Similarly, knowledge of the process each participates in is frequently more complete in some systems than others. In the case of no single pathogen has all the detail of infectivity been worked out at the molecular level.

In this book it will not be possible to review exhaustively the enormous literature on bacterial virulence factors. These may be classified broadly by the functions they carry out. In this chapter, each stage of infection will be considered in turn and the requisite virulence factor introduced, with a suitable example where a well-understood one exists. It must be remembered that not every pathogen carries out every type of interaction with the host. For example, *Neisseria gonorrhoeae* lives exclusively in association with its human host and does not have to adapt to a free-living or a vector-associated phase of growth; *Vibrio cholera* is an intestinal pathogen but is non-invasive and so does not require the elaborate internalization apparatus used by invasive bacteria. Thus, each pathogen presents us with a unique natural history. Nevertheless, comparative analyses show that pathogens and their activities are characterized by recurring themes. These will be seen later in terms of the infection process and (more to the point) elsewhere in this volume in terms of the mechanisms by which gene expression is regulated.

Contact between the bacterium and its site of infection represents a preliminary stage in infection. Bacteria possess specialized factors required for attachment to the host, and these are usually referred to as adhesins. The need for such factors is self evident, since they permit a stable association with the host to be established. Adhesins also contribute to cell-to-cell interactions between infecting bacteria. These interactions promote the formation of microcolonies which appear to be necessary for future stages in the infection, perhaps because a minimum number of bacteria must cooperate in order to bring about a particular event such as invasion.

A great deal of research has been carried out into the nature of bacterial adhesion and this is reflected in a vast literature on the subject. As usual, not all systems are appreciated in the same detail. Adhesin biology represents an area of bacterial virulence where the issue of gene regulation has been studied in great detail. This is supported by an impressive amount of information about the structure and function of the gene products, at least in a limited number of systems (particularly those from *E. coli*). The Pap (pyelonephritis-associated pilus) system of uropathogenic *E. coli* strains are particularly well understood (for a review see Tennent *et al.*, 1990). These strains cause urinary tract infections, contributing to pyelonephritis and cystitis, and are important bacterial pathogens. The *pap* operon codes for an impressive array of proteins, only some of which are directly concerned with adhesion. The rest are there to regulate the transcription of the operon, to ensure that the molecules required to build the pilus get to the surface of the cell in a correctly folded form, to control the growth of the pilus so that it is neither too short nor too long, and to support the adhesin, so that it is presented to its receptor on the surface of the host cell in a productive manner (for a review see Hultgren *et al.*, 1991). In addition to the pilus type of adhesin, bacteria can express other types. These are usually surface proteins (such as outer membrane proteins in Gram-negative species) and their interactions with the host have generally not been worked out in detail.

Further degrees of intimacy in the host–pathogen interaction are possible. Some bacteria possess the means to enter host cells and replicate there. This requires specialized determinants called invasion factors (Falkow, 1991). These frequently make use of host signalling systems to promote the internalization of the bacterium by cells not normally proficient for phagocytosis (such as epithelia). This process requires significant rearrangements of the host cell surface, including cytoskeletal components (Bliska & Falkow, 1993; Rosenshine & Finlay, 1993). Invasion factors have been studied in a number of pathogens (including eukaryotic pathogens) and tend to be given names that reflect their contribution to an aggressive process. Examples of these include 'invasin' encoded by *Yersinia* species, 'internalin' in the Gram-positive pathogen *Listeria monocytogenes* and 'penetrin' in the eukaryotic parasite *Trypanosome cruzi* (Gaillard *et al.*, 1991; Isberg & Falkow, 1985; Ortega-Barria & Pereira, 1991). A further example from *Yersinia enterocolitica* is the Ail protein (Miller & Falkow, 1988). The interaction between invasin and the host cell has been characterized in detail at the molecular level and the host receptor identified (Isberg & Leong, 1990). In the case of internalin, penetrin and Ail, the host receptor has yet to be identified. The docking of *Yersinia* with its host is an interesting example of molecular mimicry. The receptor is a host surface protein belonging to the β1 chain integrin group. The integrins bind to extracellular matrix proteins such as collagen, fibronectin or laminin that contain an Arg-Gly-Asp (RGD) sequence. They are involved in cell-to-cell contacts, in transducing extracellular signals and in attaching the cytoskeleton to the cell membrane (reviewed in Hynes, 1992). Thus, by using an integrin as its receptor, the bacterium is subverting a pre-existing system for cell-to-cell interaction within the host.

The internalized bacterium is confined within a phagocytic vacuole. Some have the means to escape from the vacuole and to colonize the host cell cytoplasm. Examples of such bacteria include *Listeria monocytogenes* and *Shigella flexneri* (Cossart *et al.*, 1989; Sansonetti *et al.*, 1986). Subsequent steps in invasion require other bacterially encoded functions which can rearrange the host cytoskeleton and exploit host cell structures for intracellular and intercellular spread of the bacteria. *Shigella flexneri*, the cause of bacillary dysentery, is non-motile. Yet by using its ability to bring about polymerization of host cell actin, it can move at great velocity across the cytoplasm (reviewed in Sansonetti, 1992). *S. flexneri* is a Gram-negative enteric bacterium. Intriguingly, the Gram-positive intracellular pathogen, *Listeria monocytogenes* uses a similar mechanism to promote its own intracellular and intercellular spread (Dabiri *et al.*, 1990; Portnoy *et al.*, 1992). This illustrates the phenomenon of mechanistic convergence so often encountered in biology, in which distantly related organisms use quite similar systems to achieve similar ends. This convergence will be seen to be particularly relevant to the topic of gene regulation later in this book.

Part of the pathogenic process involves avoiding destruction by the host defences. For example, by adopting an intracellular life-style, *Listeria monocytogenes* avoids contact with the defences, which is consistent with the observation that antibody contributes little to host immunity to this organism (Portnoy *et al.*, 1992). Pathogens can also hide from the host defences in shelters of their own making. An example of this is the synthesis by bacteria of capsules composed of extracellular polysaccharide. These contribute to virulence in both animal and plant pathogens (Jann & Jann, 1990; Roberts & Coleman, 1991). It is possible for a bacterium to disguise itself as part of the host by placing appropriate molecules at its outer surface. The K1 capsule of pathogenic *Escherichia coli* is a very poor immunogen and has antiphagocytic properties. Thus it affords a considerable degree of protection to the bacterium. This poor immunogenicity has been rationalized by pointing to the similarity between the capsule and host glycoproteins and glycolipids. The capsule is a linear homopolymer of α-2,8-linked sialic acid, which is also a component of these eukaryotic cell surface molecules. So, by using a structure composed of sialic acid, the bacterium appears as 'self' to the human immune system. Sialic acid capsules are found on other bacteria too, including meningococci and the group B streptococci (Edwards *et al.*, 1982; Silver & Vimr, 1990). This is an interesting example of molecular mimicry and shows how a pathogenic microorganism is able to disguise itself in a manner that confounds the host defences. Strategies based on such subterfuges are commonplace in nature at the macromolecular level and so it is not very surprising to discover such mechanisms being used for molecular recognition. In the case of the *E. coli* K1 capsule, there is specific identity in structure between the polysialic moiety and N-CAM (the neural cell adhesion molecule), which is a glycoprotein expressed in the membranes of neuronal cells and required for cell–cell adhesion. This structural identity is probably responsible for the immunotolerance of K1-encapsulated *E. coli* by the host. As far as the immune system can tell, these bacteria are *part* of the host.

If bacteria cannot avoid attack by immunoglobulin molecules, they can sometimes synthesize specific enzymes to destroy them. Secretory IgA1 antibodies are associated with the mucosal surfaces of the body and it is across these surfaces that pathogenic bacteria enter host tissues. Some (such as the meningitis-causing pathogens *Haemophilus influenzae, Neisseria meningitidis, Streptococcus pneumoniae* and the urinary tract pathogen *Proteus*) produce proteases capable of cleaving host secretory IgA1. While all these enzymes perform a similar task, they fall into structurally distinct classes. Some are serine proteases (Neisseriae, *Haemophilus influenzae*), some are metalloproteases (*Streptococcus sanguis*) while others are thiol-activated proteases (*Bacteroides* [now *Prevotella*] *melaninogenica*). Once again, we see bacteria using a diverse group of molecules to perform a common task.

Sometimes engulfment by 'professional' phagocytes is unavoidable. If engulfed, some bacteria have the ability to survive and even to multiply within this hostile environment. For example, Salmonellae can synthesize protective factors to survive within phagolysosomes where they are subjected to conditions of oxidative stress, low pH and attack by antimicrobial defence peptides. In such an inhospitable environment, it is essential that the bacterium activates its own defences quickly and on cue. This response is achieved through the familiar circuit of environmental sensing and gene activation. Interestingly, a central regulator of this response, the PhoQ/PhoP system (discussed in Chapter 8) is also involved in controlling several housekeeping genes in response to environmental stimuli. This is a good example of control circuits for virulence gene regulation being integrated with the general regulatory networks of the cell. It is a theme that this book will return to again and again.

Toxins represent a class of virulence determinant that has received a great deal of attention. These are produced by some pathogens to carry out specific tasks. For example, an invasive microorganism such as *Listeria monocytogenes* may use a pore-forming toxin (listeriolysin O) to degrade the membrane of its phagocytic vacuole, allowing it to escape into the host cytoplasm (Cossart *et al.*, 1989). A non-invasive bacterium, such as *Vibrio cholerae* may inflict tissue damage on its host with a toxin as part of its pathogenic process. This may release otherwise scarce nutrients which the bacterium can exploit. The molecular detail of cholera toxin structure and function has been elucidated. It has an A_1B_5 subunit structure and the B subunit is needed to bind to the surface of the host cell, where it recognizes and binds to a specific receptor. This is a carbohydrate receptor called ganglioside G_{M1}. The A subunit is the business end of the toxin and it is this which is cleaved and enters the host cell. The host target is adenylate cyclase (reviewed in Wren, 1992). Following cleavage, the enzymatically active A1 fragment remains attached to A2 by a disulphide bridge which must be reduced to allow release of A1. Once in the cell, A1 catalyses the ADP-ribosylation of an arginine residue on the regulatory component of adenylate cyclase. This induces increased synthesis of cAMP in the cell which is believed to result in the electrolyte and water loss characteristic of the diarrhoea caused by this toxin. The subject of bacterial protein toxins is very large and is beyond the scope of this book. However, these molecules also show strongly convergent tendencies in terms of their structure and function (see Alouf & Freer, 1991). Thus, the basic format of the cholera toxin is recapitulated in related protein toxins from other bacteria such as *E. coli.*

Infecting bacteria are forced to use the host as a source of iron. Unfortunately for them, free iron is rarely available. Most of it is intracellular within haeme, ferritin and other molecules. Extracellular iron, which should be the most immediately useful to the bacterium, is usually sequestered by iron-binding glycoproteins such as transferrin (in serum) and lactoferrin (associated with secretions and with polymorphonuclear leukocyte granules). These circumstances call for a strong response from the bacterium. There is a large body of evidence showing that pathogenicity is attenuated significantly if the infecting organism cannot fulfil its iron requirements (in a recent review, Payne & Lawlor (1990) list 17 species of bacterial pathogen whose virulence or lethality in animal models is enhanced by exogenous iron). Bacteria can compete successfully for this valuable resource by synthesizing and secreting efficient chelators of their own. They also possess uptake systems to recover the iron–chelator complexes from the external mileau. These bacterial systems have such a high affinity for iron that they can take it away from host chelators (Payne & Lawlor, 1990; Otto *et al.*, 1992;). Alternatively, some bacteria have developed the means to capture host iron-chelator complexes directly and use these as a source of iron. For example, *Neisseria* species can use transferrin-bound iron directly and *Yersinia pestis* can use haeme (Perry & Brubaker, 1979; Mickelsen & Sparling, 1981); *Listeria monocytogenes* is able to remove iron from transferrin (Cowart & Foster, 1985).

As is the case with the protein toxins, the literature on bacterial iron acquisition systems is vast. While iron is undoubtedly in short supply in the host, bacteria face the same problems in most other niches too. Thus, non-pathogens also require an efficient iron-scavenging apparatus. That in use in *E. coli* K-12 has received a great deal of attention. Here, genes coding for siderophore synthesis and for the surface receptors required for iron–siderophore uptake occur on the chromosome and are subject to coordinated regulation in response to iron levels (described in Chapter 8). This organism synthesizes a siderophore called enterobactin (also called enterochelin) whose genetics and biochemistry have been studied in great detail (for a review see Neilands, 1990). An analogous system occurs in *Salmonella* species. Invasive strains of *E. coli* synthesize a siderophore called aerobactin. Genes for aerobactin synthesis are usually carried on plasmids but are subject to the same regulatory signals as the chromosomally encoded genes of the enterochelin biosynthetic pathway. This is a further example of the control of virulence gene expression coming under the umbrella of housekeeping gene regulators.

Neat classification of virulence factors is not always possible. For example, the ability of the plant pathogen *Agrobacterium tumefaciens* to send part of its own genome into the nucleus of the host is a high form of parasitism more often associated with viruses than with bacteria. These prokaryotic genes possess eukaryotic expression signals and when expressed reprogramme the plant to produce substances which benefit the infecting microorganism. A further problem in classification concerns factors which contribute to both housekeeping and virulence functions in the bacterium. An environmental sensing system which is essential to adaptation in the free-living state may also control the expression of factors required directly for virulence (such as the PhoQ/PhoP system already referred to in this chapter). Genetic studies in which mutations in such systems are found to have pleiotropic effects on pathogenicity point to the widespread occurrence of such overlaps. Presumably, these findings merely reflect the integrated nature of the bacterial cell and the fact that the virulence systems are often incorporated into regulatory circuits that also govern expression of 'housekeeping' functions (reviewed in Dorman & Ní Bhriain, 1992a).

Coordination of the pathogenic process

The purpose of this brief (and necessarily incomplete) survey of bacterial virulence factors has been to introduce some of the key concepts associated with different stages of the pathogenic process. Despite the brevity of the survey, certain important themes can be detected. One of these concerns the ability of bacteria to carry out a particular function using a variety of molecular structures. Thus, not all adhesins are pili and not all invasins are homologues of the classical invasion factor of *Yersiniae*. Another theme is that the bacteria have the ability to subvert existing host functions in order to achieve their own aims. Thus, they can use mechanisms concerned with host cell-to-cell contact to attach themselves to the surfaces of eukaryotic cell; they can use systems concerned with reordering the cytoskeleton in order to promote their own internalization within host cells; they can expolit the self-/non-self-discriminatory ability of the immune system to disguise themselves as host components. Another feature of pathogens is their ability to use overtly aggressive methods alongside these much more subtle mechanisms in order to prevail. Hence the use of enzymes that destroy antibody molecules or toxins that degrade host cell components such as lipid bilayers.

The use of these virulence factors represents a significant metabolic investment on the part of the bacterium and it is not surprising to discover that their expression is often tightly regulated. The regulatory mechanism frequently (but by no means invariably) involves the control of gene expression at the level of transcription. The systems used to activate or repress gene expression are usually operated in response to environmental cues. It is as though the bacterium 'sees' its world as a series of environmental stimuli. This picture is composed of physical components (temperature, pH, osmotic pressure, etc.) and chemical components (e.g. carbon, nitrogen or phosphate availability, the presence of a particular carbohydrate). Certain combinations of environmental parameters indicate a free-living situation and others a host-associated existence. Each of these broad categories can be further subdivided, e.g. 'free-living and aquatic' or 'host-associated and attached to a mucosal surface'. In this way, the gene expression profile of the cell seen in any environment is determined largely by the signals that that environment sends to the bacterium. This is true of all bacteria, but in the case of the virulent organisms those genes required for infection are coregulated with the 'standard' ones. As was discussed in the introduction, frequently gene regulatory systems controlling virulence gene expression also control 'housekeeping' functions i.e., the sorts of genes one might expect to find in *E. coli* K-12. This is particularly true for genes subject to multilayered control (and many virulence genes are). Here, the virulence gene may possess a dedicated regulator of its own but also share a second regulator with other genes. This sharing of regulators underlies coordinated control. Such control is bound to be crucial in the expression of the complex phenotypes characteristic of bacterial pathogens. Specific functions are required for each stage of infection and they must be expressed on cue. Similarly the expression of other functions may be incompatible with success at a particular step. For example, pilus expression is important for invasion of eukaryotic cells by Neisseriae but this may be antagonized by the presence of a capsule (Virji *et al.*, 1991). Hence the overriding need to coordinate gene expression in order to get the process right.

It is also important to consider the role of the bacterial population as well as the individual cells in an infection. This is because the expression profile of the population may not be identical in each cell within a particular niche. There may be influences from microenvironmental conditions. For example, cells within a microcolony may receive different signals from those at its surface. Another important contributor to population variability in gene expression is mutation. This stochastic process is at work within the infecting bacteria and forms a background of random genetic events upon which the more regular gene control systems operate. Its importance to virulence factor expression is illustrated by the large number of systems in which mutation plays a quasiregulatory role. Frequently, the genes affected code for surface components which may be immunogenic and one can imagine that varying the expression of these can assist in circumventing the host defences. These random events can either switch the gene on or off (phase variation) or subtly alter the structure of the gene product (antigenic variation). They can involve a change at a single base pair or more extensive changes. These can be deletions or amplifications which change the reading frame of the gene or they can involve recombination events which move coding sequences between silent sites and expression sites on the genome. The latter operation can occur by means of the homologous recombination machinery of the cell or can be brought about by a site-specific system. There are many variations on these themes and several examples will be discussed later in this volume.

The aim of the remainder of this book is to show the reader interested in the control of virulence factor expression how this is achieved in bacterial cells at the level of the gene. To assist with the understanding of unfamiliar concepts, a great deal of background material is included. Thus, a discussion of site-specific recombination in relation to adhesin expression is preceded by a section on the nature of this type of recombination and how it differs from general recombination. Similarly, before dealing with the contributions of mobile genetic elements to the expression of virulence factors, these elements are themselves explained. The same is true of all of the major methods by which transcription in bacteria is currently known to be regulated. This has to be so since most of them are also used to govern transcription of virulence genes!

2 The Bacterial Genome: The Chromosome

Introduction

This book is concerned with how genes in bacterial pathogens are regulated. Before attempting a discussion of any particular gene and its control mechanisms, it is necessary to describe the environment in which bacterial genes are maintained. For this reason, this chapter opens with a description of the bacterial genome. This consists of all the genetic material in the cell, and includes the chromosome and any plasmids and bacteriophage (bacterial viruses) that may be present. Genes that contribute to pathogenesis have been detected on all these molecules. The present chapter focuses on the bacterial chromosome and describes the structure and topological state of the DNA and its interactions with other components of the nucleoid (both catalytic and structural). It also reviews the major processes of DNA, which include replication, transcription and recombination (general, site-specific and transpositional). In the case of DNA replication, the means by which this activity is coordinated with the remainder of the bacterial cell cycle is also described.

The bacterial chromosome

The chromosome represents the major DNA molecule in the bacterial nucleoid and it will be discussed here. Plasmids and phage are dealt with in later chapters. Most knowledge of the bacterial chromosome has come from studies of just a few organisms. *Escherichia coli* is understood in the most detail and the founding role played in molecular genetics by this Gram-negative, rod-shaped, enteric bacterium is widely recognized. Although the strain upon which most research has been carried out, *E. coli* K-12, is not a bacterial pathogen, the basic principles gleaned from its study appear to hold true in most bacteria, even those greatly separated from it in evolutionary terms.

The bacterial chromosome is a circular molecule

The chromosome of *E. coli* K-12 is a covalently closed circular molecule of double-stranded DNA of about 4.67 million base pairs (bp) (Kohara *et al.*, 1987; Krawiec & Riley, 1990; Rudd, 1992). If we imagine an idealized gene to be one kilobase pair (kb) long, this chromosome has the capacity to hold about 4700 genes laid end to end on one strand of the DNA duplex. By 1990, some 1403 *E. coli* genes had been identified and mapped, with varying degrees of accuracy

(Bachmann, 1990). While it is clear that not all the DNA in the *E. coli* chromosome forms part of genes, these figures indicate that even in this, the most-studied of prokaryotes, much remains to be done. Most other bacteria also posses circular chromosomes. Exceptions include the spirochaete *Borrelia*, which has a linear chromosome (Baril *et al.*, 1989; Ferdows & Barbour, 1989).

Bacterial DNA

The base composition of bacterial DNA varies between species (something that has been widely exploited as a diagnostic tool in bacterial classification). However, the following remarks about bacterial DNA structure are applicable to all eubacterial species. Deoxyribonucleic acid in eubacterial cells is overwhelmingly in the form of a right-handed B-DNA duplex (Drlica & Riley, 1990), although unusual conformations such as left-handed DNA segments and cruciform structures can exist *in vivo* (Palecek, 1991). Non-standard DNA can be detected in *Bacillus subtilis* during spore formation. Here, the conformation of the DNA is altered from the usual B form to the form known as A-DNA through the binding of small acid-soluble proteins as the spore forms (Mohr *et al.*, 1991b). Thus, despite these exceptions, DNA in bacteria may be thought of as being of the B form as illustrated in standard biochemistry textbooks.

Bacterial DNA is negatively supercoiled

To appreciate fully the dynamic situation that obtains when genes are expressed, it is necessary to consider briefly the physical properties of DNA *in vivo*. This is a difficult subject but will reward careful study by helping the reader to visualize events taking place at the molecular level during processes such as transcription.

Most of the DNA in the bacterial cell is in the form of closed loops or covalently closed circles which, from the point of view of DNA secondary structure, are topologically equivalent. The DNA duplex in these loops and circles is maintained in an 'underwound' state, and this imparts torsional tension to the molecule (Fig. 2.1; Box 2.1). This tension may promote strand separation or a distortion of the DNA helical axis in which the duplex coils about itself. This coiling of the already coiled DNA duplex is referred to as 'supercoiling' and when DNA supercoils in the opposite sense to the right-handed coiling of the duplex, it is said to be 'negatively supercoiled' or 'underwound' (Fig. 2.1).

DNA that is 'overwound' (i.e. contains additional helical turns in the same sense as the right-handed B-DNA duplex) produces right-handed supercoils and is referred to as 'positively supercoiled' or 'overwound' (Fig. 2.1; Box 2.1). The tendency of negatively supercoiled DNA to melt favours thermodynamically those reactions of DNA that depend on strand separation to proceed. These include recombination, replication and transcription. Consequently, factors influencing the degree to which DNA is negatively supercoiled should also affect the efficiency of these processes.

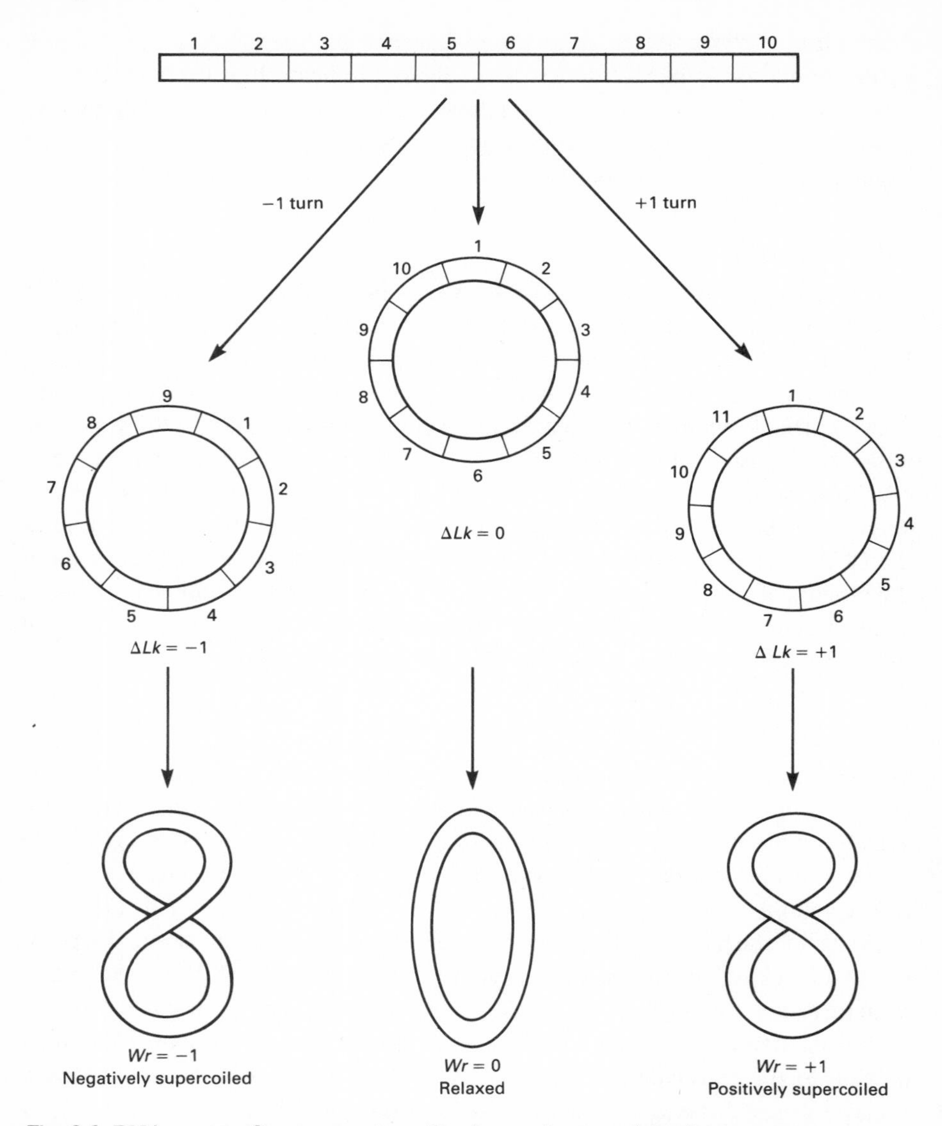

Fig. 2.1 DNA supercoiling parameters. The bar at the top of the figure represents a linear piece of DNA with 10 helical turns. By unwinding the linear molecule by one turn (−1) and ligating, the circular product on the left is generated. This molecule has a linking deficit of 1, i.e. $\Delta Lk = -1$. By overwinding the linear molecule by one helical turn and then ligating, the product on the right is generated. This has an enhanced linking value of 1, i.e. $\Delta Lk = +1$. By ligating the linear molecule without underwinding or overwinding, the relaxed molecule in the centre is generated ($\Delta Lk = 0$). The underwound molecule on the left can adopt a negatively supercoiled form, as shown at the bottom of the figure. Here, the change in linking number has been converted to a change in writhe (Wr = −1). The overwound circular molecule on the right can adopt a positively supercoiled conformation in which the excess in linking number is converted to writhe ($Wr = 1$). The relaxed molecule in the centre remains in its thermodynamically most favourable form, with $\Delta Lk = Wr = 0$.

Box 2.1 DNA supercoiling

Any covalently closed DNA molecule can be described by a characteristic set of values for the topological parameter known as the linking number (*Lk*) and the geometrical parameters known as twist (*Tw*) and writhe (*Wr*). *Lk* is a measurement of the number of times one DNA strand winds completely around the other in the duplex. *Tw* is a measurement of the coiling of the DNA strand about the duplex axis and is a description of how the DNA strands wind. *Wr* is a measurement of the writhing of the helical axis and approximates to the intuitive notion of supercoiling (*Wr* is not equivalent to the number of superhelical turns because it is a geometric parameter whose value is dependent upon the position from which the molecule is observed). The relationships of these parameters are given by the expression:

$$Lk = Tw + Wr$$

Lk is necessarily an integer with a positive value. DNA molecules with identical nucleotide sequences which differ from one another in *Lk* are known as topological isomers, or topoisomers. By convention, the linking number of the relaxed (non-supercoiled) form of a DNA molecule is designated Lk^{0}. Changes in *Lk* can be achieved only by breakage of one or both of the DNA strands followed by strand passage through the gap and religation (Fig. 2.1). Reductions in *Lk* are achieved by underwinding the duplex and increases in *Lk* result from overwinding the duplex. The change in *Lk* is described by the relationship:

$$\Delta Lk = Lk - Lk^{0}$$

For reductions in *Lk*, ΔLk has a negative sign and the resulting underwound molecule will be negatively supercoiled. The consequences of changes in Lk for the other parameters is given by the equation:

$$\Delta Lk = \Delta Tw + Wr$$

showing that variations in *Lk* are partitioned between *Tw* and *Wr*. The specific linking difference, or superhelix density, (σ) is a size-independent parameter. The value of σ is calculated from:

$$\sigma = \Delta Lk/Lk^{0}$$

For naturally occurring DNA molecules purified free of protein, σ is typically close to -0.06.

Supercoiled molecules possess energy as a consequence of their topological state. This energy is available to do thermodynamic 'work'. The free energy of supercoiling (ΔG_{sc}) is related in a quadratic manner to the change in linking number thus:

$$\Delta G_{sc} = (K.RT/N)\Delta Lk^2$$

where K is a proportionality constant equal to 1050 for DNA molecules greater than 2 kb, R is the gas constant and T the absolute temperature (see Box 2.1 for definition of ΔLk). This relationship tells us that relatively small changes in linking number can result in significant adjustments in the free energy of supercoiling (for reviews see Cozzarelli *et al.*, 1990; Fisher, 1984; Lilley, 1986b; Smith, 1981; Vinograd *et al.*, 1965).

Topoisomerases regulate DNA supercoiling

If changes in supercoiling can affect DNA-dependent processes such as transcription, we need to appreciate those factors that influence supercoiling. Supercoiling can be varied catalytically by special enzymes called topoisomerases. Several of these have been purified to homogeneity and the mechanisms by which they act on DNA are understood in detail. In all cases they alter the linking number of DNA (Box 2.1) by strand breakage and reunion mechanisms. As usual, the picture is clearest in *E. coli* and this bacterium possesses at least four DNA topoisomerases. These are topoisomerase I, DNA gyrase (or topoisomerase II), topoisomerase III and topoisomerase IV. The biological significance of only topoisomerase I and gyrase is appreciated in any depth (Table 2.1).

DNA gyrase is the topoisomerase that introduces negative supercoils into DNA and it does so via a type II mechanism (Fig. 2.2). Type II enzymes change the linking number of topologically closed DNA molecules in steps of two (Cozzarelli, 1980a). They are also found in eukaryotes but DNA gyrase has the unique property of being able to introduce *negative* supercoils, reducing the linking number in steps of two in the process (Gellert *et al.*, 1976; Yanagida & Sternglanz, 1990) It derives the energy required to do this from ATP, which means that DNA supercoiling levels are indirectly modulated by the size of the cellular ATP pool (Drlica, 1984, 1990; Hsieh *et al.*, 1991a,b). In principle, this relationship could provide a link between DNA topology and the physiology of the cell.

In *E. coli*, the activity of DNA gyrase is balanced by the countervailing influence of DNA topoisomerase I. This is a type I topoisomerase which relaxes DNA by removing negative supercoils, increasing the linking number of the

Table 2.1 DNA topoisomerases in *Escherichia coli*

Protein	Subunit structure	Subunit molecular weight (kDa)	Topoisomerase function	Gene(s)
Topo I	α	105	Type 1 topoisomerase	*topA*
Gyrase (topo II)	$\alpha_2\beta_2$	97 (α) 90 (β)	Type 2 topoisomerase	*gyrA* (α) *gyrB* (β)
Topo III	α	73.2	Type 1 topoisomerase	*topB*
Topo IV	$\alpha_2\beta_2$	75 (α) 66.8 (β)	Type 2 topoisomerase	*parC* (α) *parE* (β)

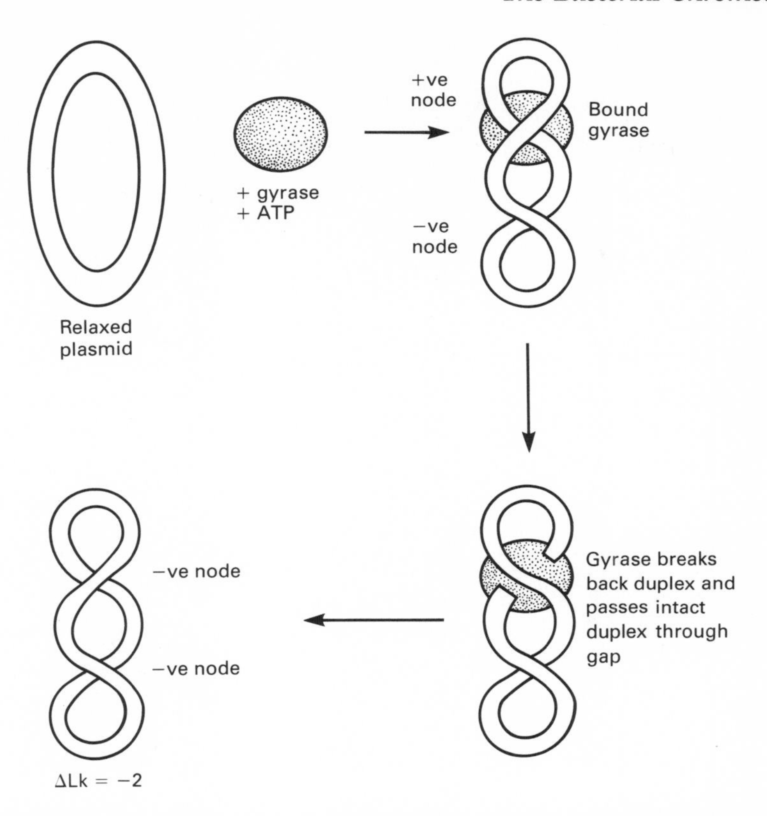

Fig. 2.2 Sign inversion model of gyrase activity. In the presence of ATP, DNA gyrase introduces a negative supercoil into a relaxed plasmid. On binding to the DNA gyrase introduces a positive node at the binding site and a compensating negative node elsewhere on the molecule. Gyrase then converts the positive node to a negative node by strand cleavage, followed by passage of the intact duplex through the gap and religation. The product of the reaction is a supercoiled molecule with two negative nodes, i.e. the plasmid has undergone a change in linking number of −2.

molecule in steps of one (DiNardo *et al.*, 1982; Pruss *et al.*, 1982). This 'swivelase' does not consume ATP during the reaction; energy stored in the supercoiled DNA molecule permits relaxation to proceed once topoisomerase I has made a single-stranded break in the duplex (Cozzarelli, 1980b; Wang, 1985; 1987) (Fig. 2.3).

The amounts of DNA gyrase and DNA topoisomerase I in the cell are controlled by DNA supercoiling at the level of transcription of their respective genes. DNA topoisomerase I is a monomeric enzyme and the gene that codes for it, *topA*, is activated transcriptionally by elevated levels of DNA supercoiling (Tse-Dinh, 1985; Tse-Dinh & Beran, 1988). Regulation of *topA* transcription is complex and involves more than one promoter, one of which is regulated by the heat shock response (Tse-Dinh & Beran, 1988; Lesley *et al.*, 1990).

DNA gyrase is made up of four subunits, two copies of the A protein encoded by *gyrA*, and two copies of the B protein encoded by *gyrB*. The promoters of *gyrA*

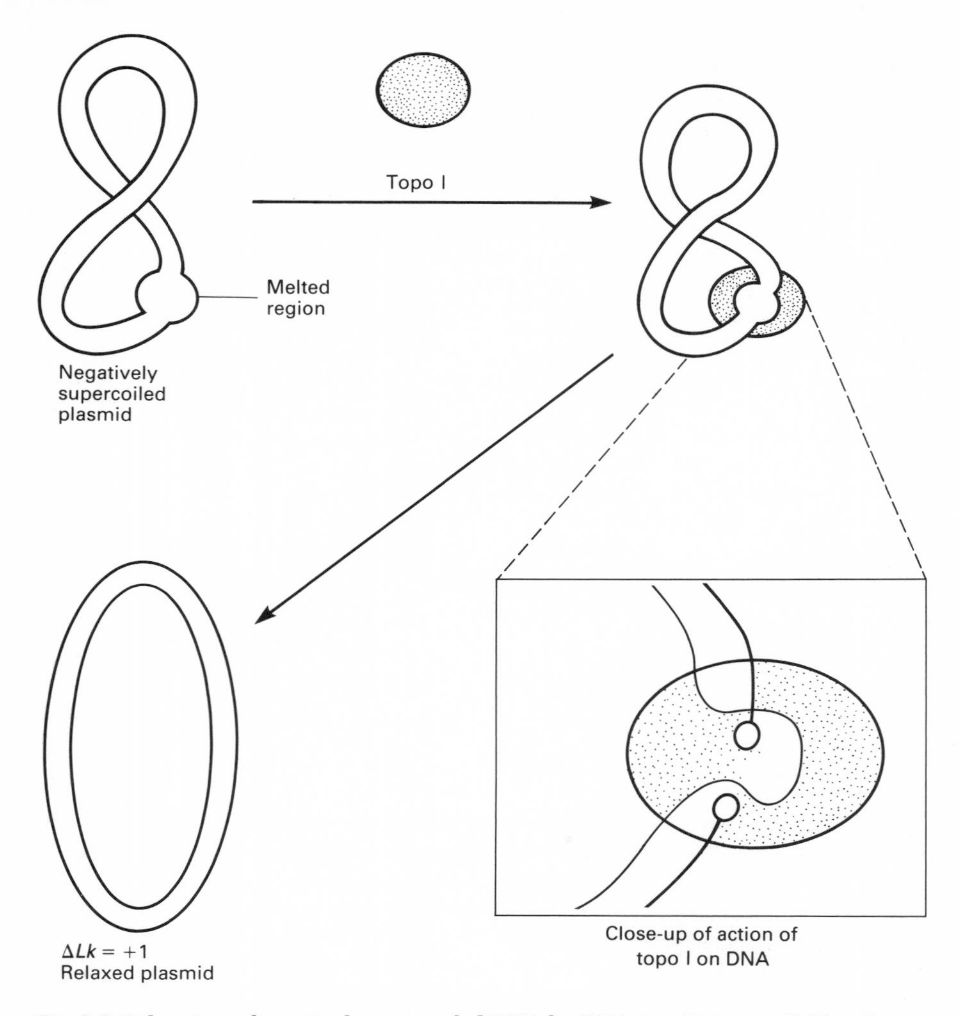

Fig. 2.3 Relaxation of negatively supercoiled DNA by DNA topoisomerase I. Negative supercoiling has resulted in an area of melted duplex in a plasmid. DNA topoisomerase I binds at this position and cleaves one strand of the duplex (see close-up view). The protein remains bound to the cleaved DNA via its active site tyrosine residue. Using a swivelase action, topoisomerase I passes the intact DNA strand through the gap, relaxing the plasmid in steps of one. The product has undergone a change in linking number of +1.

and *gyrB* are activated by a decline in DNA supercoiling levels (Menzel & Gellert, 1983; 1987). This points to an interesting mechanisms by which DNA supercoiling can be regulated. As negative supercoiling levels become excessive, the *topA* promoter is preferentially activated; if DNA becomes too relaxed, the *gyr* promoters are activated. In this way, supercoiling levels are thought to be kept at a value appropriate for cellular survival (Fig. 2.4). This type of regulation is called a homoeostatic balance.

Further evidence in support of homoeostasis in supercoiling control comes

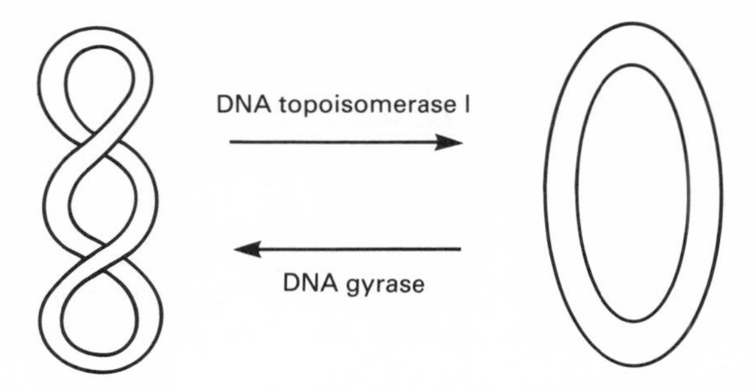

Fig. 2.4 Homeostatic regulation of DNA supercoiling. The homoeostatic model of DNA supercoiling regulation envisages a situation in which the level of supercoiling is maintained at values close to the optimum for cellular survival through the countervailing activities of DNA gyrase and DNA topisomerase I. The former introduces negative supercoils when these are in deficit while the latter prevents the DNA from becoming oversupercoiled. The promoter of the *topA* gene is activated by increases in supercoiling while the promoters of the *gyrA* and *gyrB* genes are activated by a loss of supercoils. Thus, the concentrations of the topoisomerases in the cell are themselves controlled at the level of transcription by supercoiling.

from genetic studies with mutants deficient in *topA*, the gene for topoisomerase I. Mutations in *topA* result in elevated levels of supercoiling in cellular DNA, presumably because gyrase has an unrestricted freedom to supercoil DNA. One might anticipate that this is deleterious for the cell, and *topA* mutants are certainly less viable than their wild-type parents. However, several independent studies have shown that *topA* mutants acquire additional mutations that compensate for the loss of *topA.* Many of these compensatory mutations map to the *gyr* genes and they restore the level of supercoiling to that of wild-type cells. Others map to the locus that contains the genes coding for topoisomerase IV. In these compensated strains, the mutant is found to have acquired additional copies of the topoisomerase IV genes through a gene amplification event. Since topoisomerase IV is a type II enzyme with an ability to relax DNA, it is thought that extra copies of its genes in the cell compensate for the loss of the relaxing activity normally provided by topoisomerase I (DiNardo *et al.*, 1982; Dorman *et al.*, 1989b; Pruss *et al.*, 1982; Raji *et al.*, 1985).

All these data are consistent with an inability on the part of the cell to tolerate deviations from the wild-type level of DNA supercoiling and with a consequent need to acquire compensatory mutations in order to reset supercoiling to near-wild-type levels if *topA* is lost. This is in keeping with the existence of a homoeostatic control circuit concerned with maintaining the *in vivo* level of DNA supercoiling at a value appropriate for the survival of the cell (Menzel & Gellert, 1983).

Macromolecular processes change the supercoiling of the DNA template

DNA is subject to several macromolecular processes such as transcription and replication which require strand separation. One might expect such processes to be influenced by the degree of negative supercoiling in the DNA since this parameter is known to favour strand separation. Recently, it has begun to be appreciated that these macromolecular processes themselves have the ability to influence DNA supercoiling. This makes for a more complicated situation and calls for a reconsideration of the roles of the topoisomerases. We must now recognize that not only are topoisomerases needed to control supercoiling levels, they are also required to reset them following disturbance by processes such as transcription.

Many experiments have demonstrated that when a gene is transcribed by RNA polymerase, the associated unwinding of the DNA duplex introduces a domain of relaxed (or even positively supercoiled DNA) ahead of the transcription complex and negative supercoils in its wake (Figueroa & Bossi, 1988; Liu & Wang, 1987; Pruss & Drlica, 1986, 1989; Tsao *et al.*, 1989; Wu *et al.*, 1988; Fig. 2.5). These differentially supercoiled domains become targets for topoisomerases (Fig. 2.6). Gyrase will remove positive supercoils by the same mechanism by which it introduces negative ones into relaxed DNA. DNA

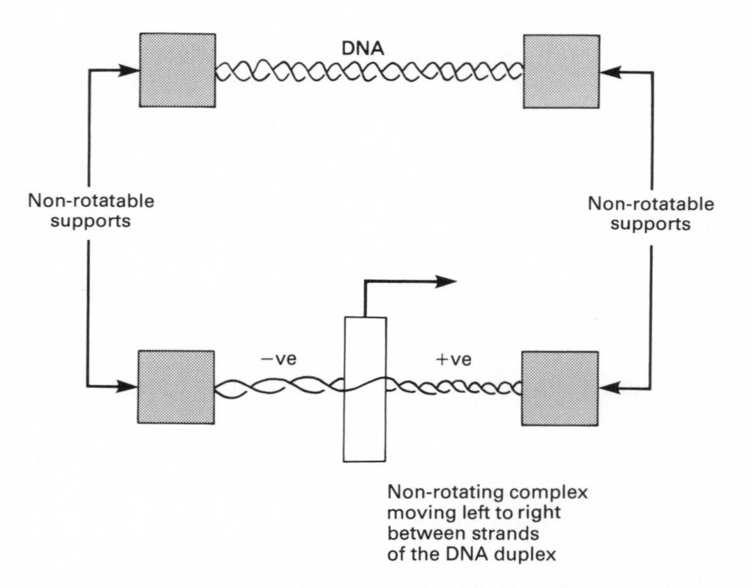

Fig. 2.5 Differential supercoiling of a DNA template during transcription. A length of duplex DNA is shown connected to two non-rotatable supports. If a body, such as a transcription complex, moves through the duplex, the template becomes partitioned into a domain of underwound (i.e. negatively supercoiled) DNA behind the complex and a domain of relaxed or even positively supercoiled DNA (as shown here) ahead of the complex. The only ways to resolve this situation are: (i) for the DNA to rotate, which it cannot; (ii) for the transcription complex to rotate, which it can only do with difficulty due to the torsional drag of the transcribing proteins, the attached mRNA and the translational machinery with its nascent polypepties; and (iii) for DNA topoisomerase I to relax the domain of negative supercoils and DNA gyrase to relax the domain of positive supercoils. These topoisomerase-catalysed events probably represent the principal way in which this situation is resolved *in vivo*.

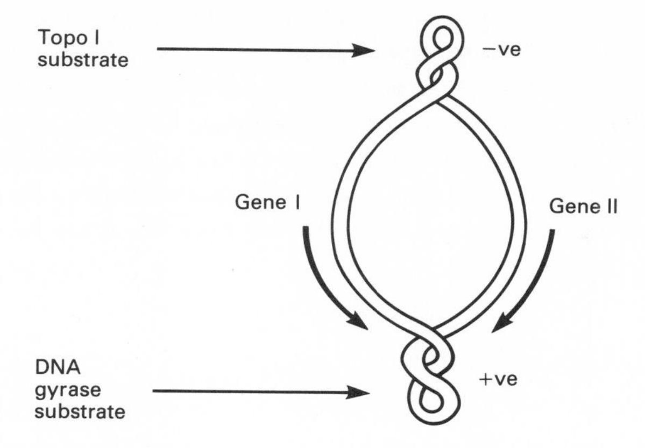

Fig. 2.6 Twin supercoiled domains generated by divergent transcription units. A plasmid containing two divergently transcribed genes (I and II) becomes partitioned into two differentially supercoiled domains; that behind the moving transcription complexes is negatively supercoiled while that ahead of the complexes is positively supercoiled. The negatively supercoiled domain is a substrate for DNA topoisomerase I while the positively supercoiled domain is a substrate for DNA gyrase.

topoisomerase I will relax the highly negatively supercoiled domain. It is assumed that under normal conditions, the topoisomerases of the cell deal with these overwound and underwound domains, quickly resetting the level of supercoiling to that of the DNA before transcription. Nevertheless, the appearance of highly negatively supercoiled domains and positively supercoiled domains on the chromosome could have important consequences for the expression of neighbouring genes, whose promoters may find this topological disturbance either inhibitory or stimulatory. Several lines of evidence demonstrate the potential of this topological disturbance not only to influence transcription of adjacent genes but also to drive structural transitions such as the formation of Z-DNA. If this is so, then one might expect that the passage of other macromolecules through the DNA might produce analogous effects. In this way, DNA polymerase and DNA recombination complexes may exert such effects during DNA replication and recombination, respectively (Honigberg & Radding, 1988; Wang & Liu, 1990; Wang *et al.*, 1990).

In this chapter, it is shown that bacterial DNA is found almost invariably in the form of negatively supercoiled B-DNA. In this section it is shown that positive supercoils may occur transiently in most eubacteria in response to events like transcription. They can also be introduced experimentally by incapacitation of DNA gyrase through antibiotic treatment (Lockshon & Morris, 1983).

While positive supercoiling of DNA is the exception rather than the rule in most bacteria, this is not so in some exotic species. For example, positively supercoiled DNA may be a normal feature of some thermophilic archaebacteria (Nadal *et al.*, 1986). In fact, some of these organisms have even been found to possess a topoisomerase that specifically introduces positive supercoils into DNA and that can also remove negative supercoils. This enzyme has been given the appropriate name of 'reverse gyrase' (Kikuchi & Asai, 1984; Kikuchi, 1990). Reverse gyrase is also present in some thermophilic eubacteria (Bouthier de la Tour *et al.*, 1991). Thus, some bacteria are capable of adapting to extreme environments by making profound adjustments to the structure of their DNA. In the next section it will be shown that much more modest changes to DNA structure form a common part of the bacterium's repertoire of responses to environmental change.

DNA supercoiling responds to changes in growth conditions

It is now recognized that certain environmental stresses experienced by bacteria result in alterations in the topology of DNA and that these have important consequences for the major processes of DNA. Specifically, it has been discovered that changes in growth phase, nutrient availability, osmolarity, temperature and the aerobic/anaerobic switch (in facultative anaerobes) produce fluctuations in the linking number of DNA and that these have important consequences for the control of prokaryotic transcription (Balke & Gralla, 1987; Dixon *et al.*, 1988; Dorman *et al.*, 1988, 1990; Goldstein & Drlica, 1984; Higgins *et al.*, 1988a; Ní Bhriain *et al.*, 1989; Whitehall *et al.*, 1992; Yamamoto & Droffner, 1985). Experiments using plasmids as reporters of DNA supercoiling have shown that when *E. coli* cells are grown anaerobically or exposed to osmotic stress, their linking number decreases, which is consistent with an increase in negative supercoiling of plasmid DNA (Dorman *et al.*, 1988; Higgins *et al.*, 1988a) (Fig. 2.7). Changes in growth temperature produce shifts in

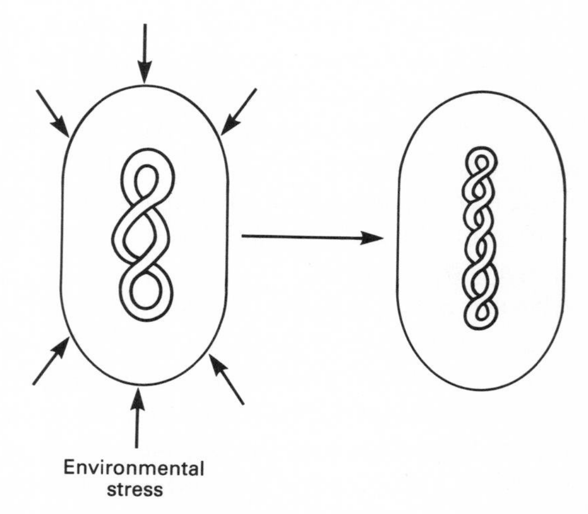

Fig. 2.7 Environmental stress can alter DNA supercoiling. Changes in environmental parameters (such as growth medium osmolarity) can result in a change in the level of supercoiling in bacterial DNA.

plasmid DNA supercoiling that vary with the species studied (Dorman *et al.*, 1990; Goldstein & Drlica, 1984). In addition to effects on transcription, environmentally wrought changes in DNA supercoiling modulate other DNA structural transitions such as cruciform extrusion (Dayn *et al.*, 1991; McClellan *et al.*, 1990).

All this points to an attractive mechanism by which changes in the environment could alter simultaneously the transcription of many genes in the cell. The plasmid studies show clearly that DNA linking numbers are modulated significantly and the responses of many promoters to supercoiling are well established. In addition, there is evidence that the effects detectable on plasmids reflect changes on the chromosome. Some workers have been able to measure changes in the level of supercoiling of the nucleoid as a whole and shown that trends detected on plasmids hold true for the chromosome. In other studies, supercoiling-sensitive promoters on plasmids were found to be equally sensitive when studied on the chromosome. These results give us confidence that the experimental systems in use tell us something useful about events on the chromosome. However, important questions do remain to be answered about how fluctuations in particular environmental parameters can affect DNA supercoiling, whether on plasmids *or* on the chromosome.

It must be admitted that the mechanism by which environmental change influences DNA supercoiling remains largely unknown and it is unclear if the major effect due to the linking number change is one of altered DNA twist or writhe (Wang & Syvanen, 1992; see Box 2.1 for definitions). It is possible that changes in intracellular ion concentrations or the levels of nucleoid-associated histone-like proteins may play a role (these low-molecular-weight, DNA-organizing proteins are discussed in detail later). For example, potassium ions are rapidly accumulated by enteric bacteria following osmotic stress and ions such as potassium alter the topology of DNA directly (Anderson & Bauer, 1978; Sutherland *et al.*, 1986). These observations are consistent with effects on DNA topology due to simple changes in intracellular ion concentrations. The interaction of nucleoid proteins with the DNA can also be modulated by changes in electrostatic charge due to alterations in the ionic environment resulting from environmental disturbance (Drlica & Rouvière-Yaniv, 1987).

An attractive possibility is that changes in the environment alter topoisomerase activity. For example, it has been shown that osmotic stress and anaerobic growth increase the intracellular ATP/ADP ratio in *E. coli* with a concomitant increase in the activity of the ATP-dependent DNA gyrase (Hsieh *et al.*, 1991a;b). Other possibilities include post-translational modification of topoisomerases by processes such as phosphorylation catalysed by putative protein kinases (and associated antagonistically acting protein phosphatases) whose activities are environmentally regulated. The post-translational modification model could apply equally to the structural nucleoid-associated proteins. Much more research is needed to discover the true mechanism(s).

The nucleoid

For a long time, knowledge of this structure lagged behind that of eukaryotic cells. Recently, there has been a great improvement in our understanding of the nature of the bacterial nucleoid, founded principally on work carried out with *E. coli.* The bacterial nucleoid consists of the chromosome and its associated macromolecules. As pointed out previously, the chromosome is in a negatively supercoiled state. This assists with the storage of the molecule within the bacterial cell since a supercoiled circle is much more compact than a relaxed one. Attempts at quantifying the level of *in vivo* chromosomal DNA supercoiling have produced values inconsistent with a model in which all the DNA is subject to torsional stress. In other words, the chromosome appears to be much less supercoiled than it ought to be. While it is accepted that bacterial DNA *is* supercoiled, experimental evidence based on several independent approaches indicates that about half of the supercoils are constrained by association with proteins in structures analogous to eukaryotic chromatin (Bliska & Cozzarelli, 1987; Lilley, 1986a; Pettijohn, 1982; Pettijohn & Pfenninger, 1980). This is supported by the observation that when bacterial or bacteriophage DNA is released from partially disrupted cells by gentle lysis, it can be seen as delicate, beaded fibres similar to those of eukaryotic chromatin, which break down rapidly, leaving just naked, supercoiled DNA (Griffith, 1976) (Fig. 2.8). So, the missing superhelical tension can be accounted for when one allows for association of the chromosome with structural proteins within a macromolecular complex akin to chromatin.

Given the large size of the chromosome and the relatively small volume of the bacterial cell, it is not surprising to find that the DNA in the nucleoid is highly concentrated, being present at 10–35 mg/ml (Kellenberger, 1990). Microscopic studies have indicated that the nucleoid of *E. coli* is confined to a ribosome-free zone in the cell and has a lobed shape with a cleft surface (Drlica, 1987; Hobot *et al.*, 1985). Gene expression appears to take place exclusively at the surface of the nucleoid. This is consistent with data indicating that this is where the ribosomes are and where translation must take place. If gene expression really is excluded from the interior of the nucleoid, this would appear to pose problems in terms of gene accession, since genes to be expressed need to be located within the bulk nucleoid, brought to the surface, transcribed and then returned. How the cell manages its gene retrieval, expression and storage system is not understood. Given the rapidity with which bacteria adapt to new environments by altering their gene expression programme, this system must be highly efficient. Studies of all bacteria (including pathogens) will benefit enormously from a more complete understanding of the workings of the nucleoid.

Nucleoid-associated proteins

In the last section, the bacterial chromosome was described as part of the nucleoid, a macromolecular complex which includes protein. Much research

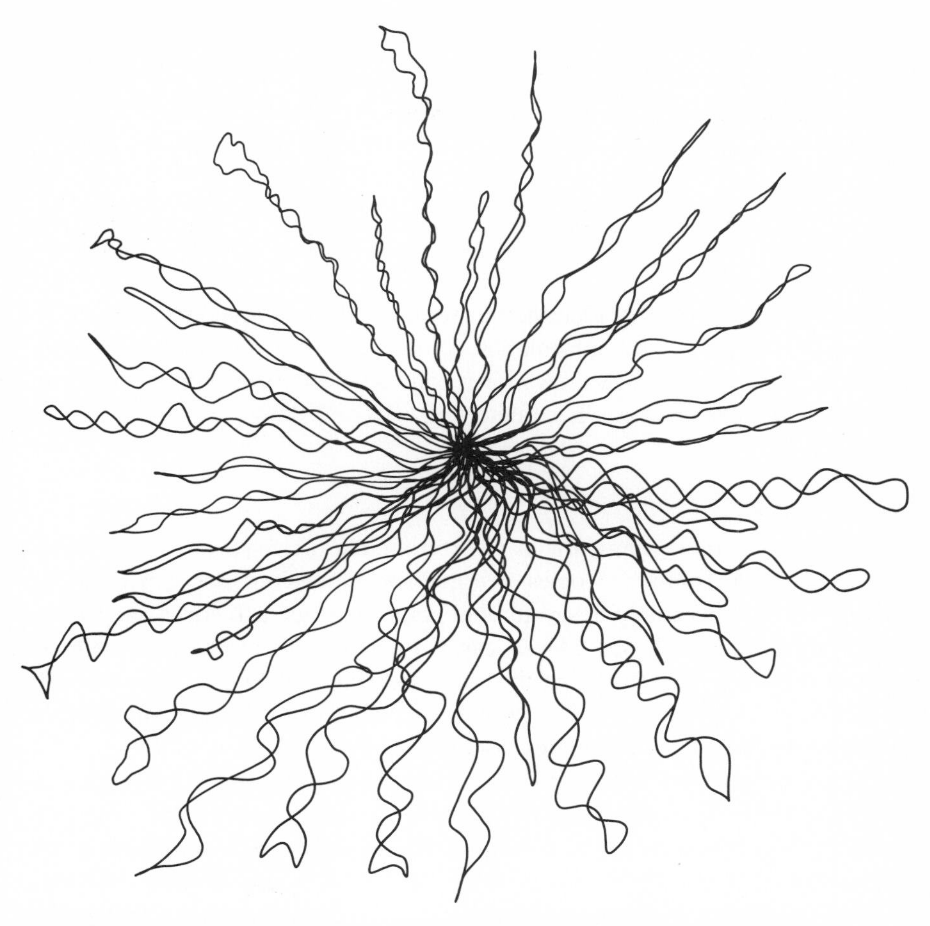

Fig. 2.8 The bacterial nucleoid contains a chromosome folded into topologically independent domains, each of which could be supercoiled to different levels. However, there is little experimental evidence to support different supercoiling of chromosomal domains (see text).

has been carried out into the biochemical nature of the proteins associated with the bacterial nucleoid and into the possibility that they may organize bacterial DNA into structures similar to eukaryotic chromatin (Drlica & Rouvière-Yaniv, 1987; Pettijohn, 1988; Schmid, 1990). Several of these have now been characterized biochemically in some detail and these are discussed next (Table 2.2). This is a rapidly moving field and our knowledge about nucleoid structure is constantly improving. However, from time to time erroneous impressions have been made. For example, not all of the proteins originally classified as being nucleoid-associated are still so-regarded. One of them, called H-protein or H2A-like protein, is now recognized as being a ribosomal protein (Bruckner & Cox, 1989) and not a nucleoid-associated histone-like protein, as originally thought (Hübscher *et al.*, 1980). The molecules discussed in the following

Table 2.2 Nucleoid-associated proteins in *Escherichia coli*

Protein	Subunit structure	Subunit molecular weight (kDa)	Function	Gene(s)
FIS	α_2	11.2	DNA bending	*fis*
HNS	α_2	15.4	DNA curve binding	*hns*
HU	$\alpha\beta$	9.5 (α), 9.5 (β)	DNA wrapping	*hupB* (α), *hupA* (β)
IHF	$\alpha\beta$	11.2 (α), 10.5 (β)	DNA bending	*himA* (α), *himD* (β)

sections have been reliably identified in terms of their interactions with DNA. Once again, the bulk of the information concerning them is derived from work with *E. coli* but where studies have been conducted in other genera, similar (if not identical) molecules have been detected.

HU

Regarded as the most abundant of the histone-like proteins in *E. coli*, HU is present in 20 000 dimeric copies per genome equivalent per exponentially growing cell. It is necessary to say at what point in the growth curve the protein measurement was made since levels can vary. This seems to be true for most nucleoid proteins, suggesting that the structure of the nucleoid is dynamic. HU is also known as HLPII and HB2 (Pettijohn, 1982). Thus, like other proteins in this class, it suffers from a complex nomenclature.

HU is a heterodimer of 9500 kDa subunits, is basic and wraps DNA without displaying overt sequence specificity. However, it does display some structural specificity (see later). Its physical properties and amino acid composition are reminiscent of eukaryotic histone proteins. HU has the ability to wrap DNA into particles resembling nucleosomes *in vitro* and it has been pointed out that it may take 8–10 HU dimers in association with 275–290 bp of DNA to form such a nucleosome (Broyles & Pettijohn, 1986; Rouvière-Yaniv *et al.*, 1979). It can mediate very tight DNA curvature, allowing DNA sequences as short as 99 bp to form circles (Hodges-Garcia *et al.*, 1989) and it binds preferentially to kinked or bent DNA (Pontiggia *et al.*, 1993). Thus, a major biological property of HU may be to impart flexibility to DNA in order to facilitate the interaction of other proteins with the DNA. Early work suggested that HU was not distributed throughout the nucleoid but associated with the metabolically active nucleoid surface (Dürrenberger *et al.*, 1988; Kellenberger, 1990). It is now known that HU is evenly distributed through the nucleoid in living *E. coli* cells, although its role as a facilitator of heterologous protein binding to DNA may cause it to be found preferentially at the transcriptionally active nucleoid surface under certain circumstances (Shellman & Pettijohn, 1991). Its structure is highly conserved and HU-like proteins have now been isolated from a wide range of bacteria, including *Bacillus stearothermophilus*, bacteriophage SPO1 from *Bacillus subtilis, Clostridium pasteurianum, Pseudomonas aeruginosa, Rhizobium meliloti, Salmonella typhimurium* and *Thermoplasma acidophilum* (reviewed in Drlica & Rouvière-Yaniv, 1987).

A role for HU in assisting DNA processes is supported by genetic evidence. Cells harbouring mutations in both genes required for HU synthesis are almost inviable and have an abnormal cell cycle. Mutants with an imbalance in the HU subunit ratio display a similarly aberrant phenotype. *In vitro* studies have shown HU to be required for chromosome replication, transposition of transposon Tn*10* and phage Mu and the site-specific recombination event that regulates expression of flagellae in *Salmonella typhimurium* (reviewed in Rouvière-Yaniv *et al.*, 1990). It is also known to be the main four-way junction-binding protein of *E. coli*, indicating that it plays a role in homologous recombination (Pontiggia *et al.*, 1993). HU is clearly an important protein which makes many significant contributions to the smooth running of events within the nucleoid.

Integration host factor

Integration host factor (IHF) is a close relative of HU and is a member of the histone-like protein family. IHF is so-called because it was identified initially as playing an accessory role in the integration and excision of bacteriophage λ in *E. coli* K-12. Early estimates indicated that IHF was a relatively rare protein in the cell (Drlica & Rouvière-Yaniv, 1987). More recent evidence suggests that it may be very abundant, with between 20 000 and 100 000 copies (Ditto, M. & Weisberg, R., cited in Freundlich *et al.*, 1992). Unlike HU, IHF is very fastidious in terms of the DNA sequences to which it will bind. This allows it to participate in specialized tasks in a specific manner.

IHF is a heterodimeric protein with physical characteristics similar to those of HU and its subunits are encoded by two unlinked genes (Drlica & Rouvière-Yaniv, 1987). Genetic and *in vitro* studies have demonstrated that IHF contributes to a wide variety of cellular functions, including the control of transcription and site-specific recombination (reviewed in Friedman, 1988a; Freundlich *et al.*, 1992). The manner in which IHF binds to DNA is unusual in that it uses two-stranded β-ribbons to dock with the minor groove of the B-DNA helix (Nash, 1990) (Fig. 2.9). Interactions with the major groove are much more common and the protein structures which contact the DNA are frequently in the form of structures such as 'helix-turn-helix' motifs. In these respects IHF mimics the eukaryotic transcription factor, TFIID, whose DNA-binding domain is similar to that of IHF (Lee *et al.*, 1991; Nash & Granston, 1991). Once bound, IHF exerts a dramatic effect upon the DNA, bending the helix by an angle of up to 140° (Kosturko *et al.*, 1989). Interestingly, TFIID binding also results in DNA bending (Horikoshi *et al.*, 1992).

The influence of IHF on the shape of DNA plays an important role in bringing together distant sites and so increasing the probability of reactions that

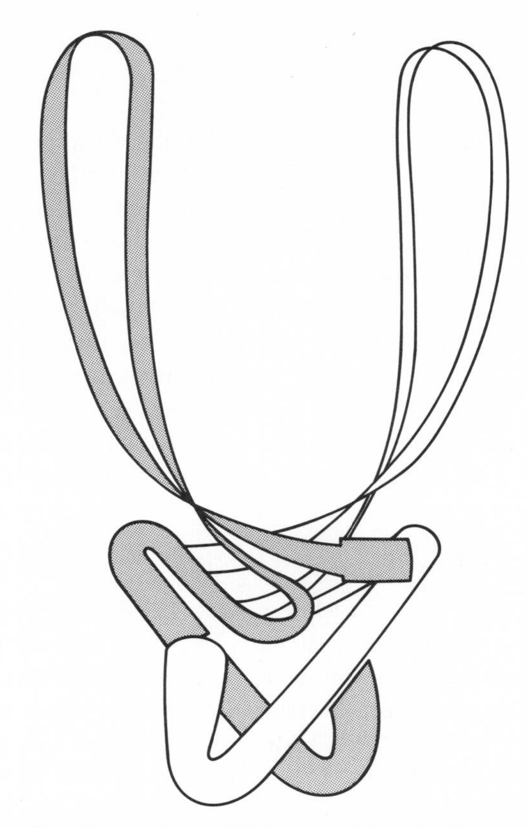

Fig. 2.9 Integration host factor. A simplified representation of an integration host factor (IHF) heterodimer is shown. This is based on an analysis of HU protein family members by Drlica & Rouvière-Yamiv (1987). The extended arms are thought to interact with the minor groove of DNA at the protein binding site. The consensus binding site is C/TAANNNNTTGATA/T, where N = A, C, G or T.

depend on meetings between these sites or proteins bound at these sites. Such long-range interactions include the formation of synapses in site-specific recombination where the sequences which recombine may lie at well-separated sites on the same DNA molecule. Alternatively, bending can help transcription initiation by bringing transcription activators bound to distant enhancer sequences into contact with promoter-bound RNA polymerase in order to catalyse the formation of an open transcription complex.

Thus, IHF may be regarded as a specialized member of the HU family of general DNA binding proteins. In some studies, the cellular concentration of IHF has been shown to peak at the interface between exponential growth and stationary phase, making IHF-dependent processes more likely to proceed at this point in the growth curve than at any other (Fig. 2.10). Conversely, IHF-inhibited processes are less likely to occur at this point in the growth cycle (Bushman *et al.*, 1985). Thus, IHF may also serve as a link between those DNA processes that depend on it and cellular physiology.

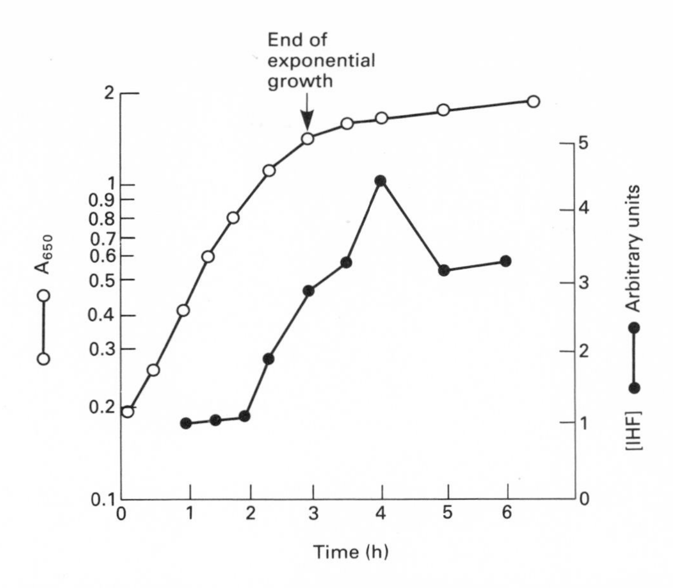

Fig. 2.10 Cellular levels of integration host factor (IHF) vary with growth phase. During early exponential phase, cells contain little IHF. As the cells move from exponential to stationary phase, the IHF levels peak. Based on the data of Bushman *et al.* (1985).

H-NS (H1)

Protein H-NS (or H1) is another major component of the *E. coli* nucleoid. It is a neutral protein with a molecular weight of 15 500. The cell contains about 20 000 copies of H-NS (H1). A 1984 study indicated that H-NS (H1) exists in three isoforms called H1a, H1b, and H1c with the H1a isoform accumulating in stationary phase and inducing significant compaction into DNA, equivalent to that seen in nucleosome cores (Spasskey *et al.*, 1984). These three isoforms were distinguished by isoelectric focusing. The molecular basis of the difference between them is unknown. *In vitro* studies have looked at the potential of H-NS (H1) to influence transcription. It was found that H1a impedes open complex formation by RNA polymerase at the *lacL8* UV5 promoter although it does not interfere with the binding of RNA polymerase to the promoter (Spasskey *et al.*, 1984). It has also been shown to be a silencer of transcription in several other systems (Göransson *et al.*, 1990; Jordi *et al.*, 1992). Thus, H-NS (H1) has the capacity to influence transcription negatively. These effects have been shown to result from specific interactions with DNA and not simply from a general binding in the vicinity of the affected promoter. This is because H-NS (H1) can affect differentially transcription from two promoters located on the same plasmid (Ueguchi & Mizuno, 1993). If this protein silenced transcription generally, both promoters would have been expected to be negatively affected. Given that there is a degree of specificity in its interactions with DNA, H-NS must have

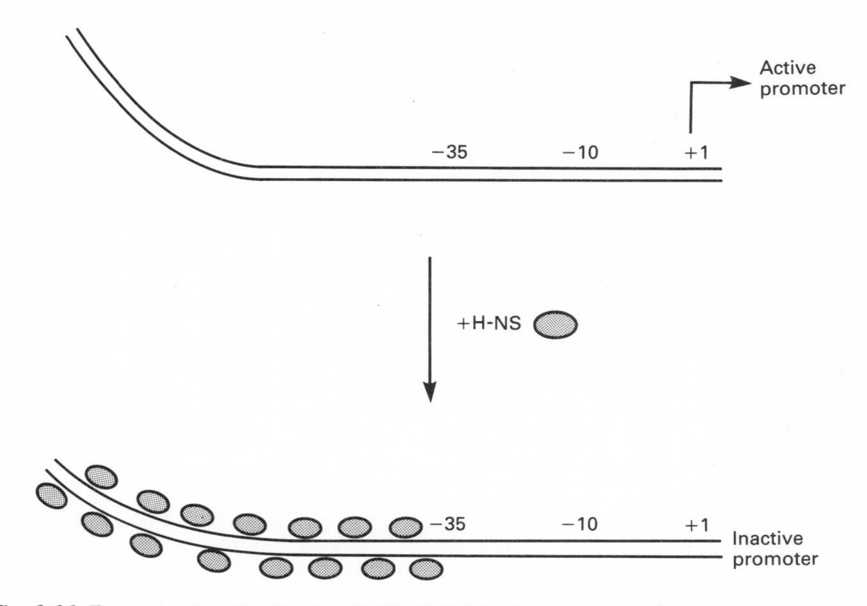

Fig. 2.11 Promoter inactivation by H-NS. H-NS is thought to bind preferentially to curved DNA sequences, often found upstream of strong promoters. By binding to such sequences, the protein 'silences' the promoter, possibly by altering the topology of the DNA.

some means of recognizing the sequence to which it will bind. Some workers have proposed a consensus DNA sequence for H-NS binding (Rimsky & Spassky, 1990) but most studies indicate that H-NS binds preferentially to sequences that are curved (Dersch *et al.*, 1993; Owen-Hughes *et al.*, 1992; Yamada *et al.*, 1991) (Fig. 2.11).

H-NS (H1) is encoded for the *osmZ* locus in *E. coli* and *Salmonella typhimurium* (reviewed in Higgins *et al.*, 1990). By general agreement, this important gene has been renamed *hns*. The *hns* gene is negatively autoregulated and is also under growth phase control (Dersch *et al.*, 1993). H-NS accumulation in stationary phase is a consequence of increased transcription of the *hns* gene (Dersch *et al.*, 1993). This is surprising since one might expect the higher levels of H-NS found in stationary phase cells to repress *hns* transcription. Perhaps changes in the topology of the DNA to which H-NS binds could account for the loss of negative autoregulation as the culture enters stationary phase. Alternatively, a change in the conformation or isoform of the protein might explain this effect.

Mutations in *hns* are highly pleiotropic and affect the expression of several genes, including genes involved in bacterial virulence (see Chapter 8). H-NS also influences both site-specific and general (homologous) recombination systems (Dri *et al.*, 1992; Higgins *et al.*, 1988a). Some mutations in *hns* change the linking number of plasmids used as reporters of *in vivo* DNA supercoiling, although the extent and direction of the linking number change depends upon the allele of *hns* (Higgins *et al.*, 1990). Furthermore, the level of the H-NS protein in the cell appears to be critical for the maintenance of genome integrity; too much (Barr *et al.*, 1992) or too little (Lejeune & Danchin, 1990) can result in an increased rate of mutation.

FIS

FIS is a 11 240 Da site-specific DNA-binding protein which acts as a homodimer (reviewed in Finkel & Johnson, 1992). It possesses a helix-turn-helix motif similar to that seen in one of the major classes of DNA-binding proteins. This is in contrast to the minor groove docking ribbons of HU and IHF, discussed earlier. FIS has homology to NtrC, a transcription activator of σ^{54}-dependent promoters (Johnson *et al.*, 1988; Koch *et al.*, 1988). NtrC is an enhancer-binding protein which activates transcription through a mechanism that relies on DNA looping. It is discussed in detail in Chapter 7. Several proteins required to regulate specific virulence genes in several species of bacteria show NtrC homology. That FIS has some sequence homology to NtrC does not mean that it activates transcription by a similar mechanism. It simply shows that both proteins interact with DNA in a similar manner. What happens after binding differs significantly for each protein.

FIS was discovered originally as a factor required to stimulate site-specific inversion systems catalysed by recombinases of the invertase family and derives its name from this function (Factor for Inversion Stimulation). It binds to 'enhancer' sequences in the Hin flagellar phase variation system of *Salmonella typhimurium* and to the recombinational enhancer of the Gin system of bacteriophage Mu (see Chapter 4) (Johnson & Simon, 1985; Johnson *et al.*, 1986b; Kahmann *et al.*, 1985; Koch & Kahmann, 1986). It should be noted that this enhancer-binding activity is not related to that displayed by NtrC.

FIS bends DNA by about 95° on binding and this is probably important in its biological function (Thompson & Landy, 1988). DNA bending has come to be recognized as a very important feature of many regulatory systems governing not just transcription but also recombination. It is a way of bringing distant sites on the same DNA molecule close together and is important in controlling the expression of many virulence factors. DNA bending is considered in more detail in Chapter 6.

FIS is also involved in modulating Int-Xis-catalysed excision of bacteriophage λ prophage from the *E. coli* chromosome under conditions in which the Xis 'excisionase' is present in low concentrations (see section on Bacteriophage λ integration and excision in Chapter 4) (Thompson & Landy, 1989). The λ recombination system has been invaluable in working out the functions of many DNA-binding proteins known now to contribute to the control of systems important in virulence. Thus FIS has the ability to participate in the organization of higher-order nucleoprotein structures in several recombination systems. In addition, FIS acts as a transcription activator in *E. coli*, binding to upstream activator sequences (UAS) in the 5′ region of genes coding for stable RNA species (rRNA and tRNA) and facilitating the binding of RNA polymerase to the promoter (Nilsson *et al.*, 1990). Work with the *rrnB* P_1 promoter indicates that the role of FIS is most likely to involve a direct interaction with RNA polymerase rather than simply a bending of DNA which brings distantly bound factors into closer proximity with the promoter (Gosink *et al.*, 1993). How generally applicable this observation is to other FIS-dependent promoters is currently unknown. In addition to influencing transcription and site-specific recombination, FIS is involved in the control of transposition. Specifically, FIS regulates the transposition of transposon Tn5 and of insertion sequence IS*50* (Weinreich & Reznikoff, 1992).

Like IHF, FIS availability varies with growth phase (Fig. 2.12). Unlike IHF, FIS levels are high in growing cells but almost undetectable in stationary phase cells (this is the opposite of IHF; Thompson *et al.*, 1987). This could enable FIS to link control of transcription, site-specific recombination and transposition events to cellular physiology (Nilsson *et al.*, 1990; Thompson *et al.*, 1987). Presumably, FIS availability is a reflection, at least in part, of the expression of the gene that

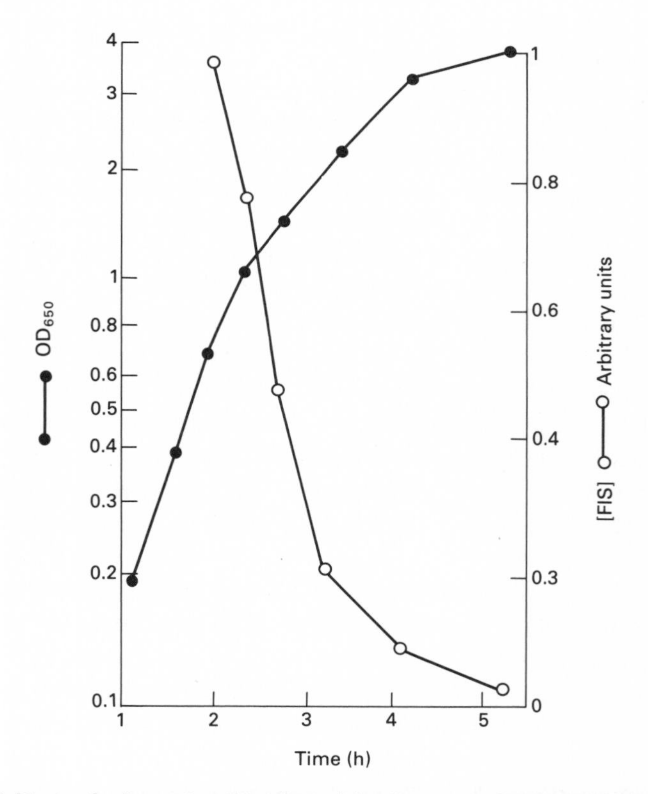

Fig. 2.12 FIS (Factor for Inversion Stimulation) levels vary with growth phase. During early exponential phase FIS levels are high, but then decline rapidly. Based on the data of Nilsson *et al.* (1990).

codes for it (*fis*). The regulation of transcription of the *fis* gene has turned out to be complex. It is known to be both negatively autoregulated (i.e. FIS turns *fis* transcription off) and to be subject to control by the stringent response (Ninnemann *et al.*, 1992). The stringent response is associated intimately with the control of stable RNA expression and those who are interested are directed to the description in Chapter 6.

HLPI

HLPI, coded for by the *firA* gene, is a 17 kDa histone-like protein thought to be associated with the outer membrane of enteric bacteria (Hirvas *et al.*, 1990). It is one of the least well understood histone-like proteins in *E. coli.* Its cellular location seems unlikely for a protein with a role in chromatin organization. It might reflect an association of the origin of replication of the bacterial chromosome with the outer membrane (Hendrickson *et al.*, 1982; Kusano *et al.*, 1984). This issue is discussed later under the cell cycle (see 'Partitioning of daughter nucleoids').

Chromosomal domains and gene expression

Having discussed the nature of the bacterial chromosome and its association with the other components of prokaryotic chromatin, it is necessary to consider those DNA topological aspects of the chromosome that might influence gene expression. This requires an appreciation of the possibility of 'geographical' influences on gene expression.

Physical-chemical studies have indicated that the chromosome of growing *E. coli* cells, doubling every 30 min, is segregated into 43 ± 10 domains of supercoiling per genome equivalent of DNA or 120 ± 30 domains per nucleoid (Sinden & Pettijohn, 1981; Worcel & Burgi, 1972). Each domain would consist of about 100 kb, on average. In support of this, electron microscopic analysis of chromosomes released from bacterial cells following gentle lysis reveals that the unbroken DNA molecule is folded into between 12 and 80 supercoiled loops (Delius & Worcel, 1973; 1974) (Fig. 2.8). The topological independence of the domains means that DNA supercoiling in one loop may be released without affecting supercoiling in the others. Taken together, these observations suggest that different domains of the chromosome may be supercoiled to different degrees with important consequences for DNA processes taking place within them. Thus, a gene in one domain may experience a different DNA topological environment to a gene in another and this could have important implications for promoter function. Once again, we are faced with an attractive idea. Is there evidence in its support?

Beckwith *et al* (1966) measured expression from the *lac* promoter inserted at 11 different sites on the *E. coli* chromosome as part of an integrated F′ *lac* episome. Nine strains gave results consistent with positional effects imposed by gene dosage acting as a function of distance from *oriC*, the origin of replication (see the next section for a discussion of gene dosage as a function of position with respect to the origin of chromosomal replication). However, insertions of the *lac* promoter at two of the sites gave results that were inconsistent with effects due to distance from *oriC* alone. Significantly, transcription from the *lac* promoter has since been shown to be sensitive to changes in DNA supercoiling (Borowiec & Gralla, 1987; Sanzey, 1979). These data may have indicated differences in transcription due to the location of *lac* within chromosomal domains whose DNA was supercoiled to levels dissimilar to those found at the usual chomosomal location of this operon. This would fit with the idea of differentially supercoiled domains.

Independent studies by a second research group which placed *lac* at 17 differerent sites on the *E. coli* chromosome produced no data that could not be explained by gene dosage alone (Masters *et al.*, 1985). Similar results were recorded by Schmid & Roth (1987) using a different promoter in another bacterium. Here, the *Salmonella typhimurium hisGDCB* genes and their promoter were translocated to 16 different sites on the chromosome of that organism and no effects on expression were detected that could not be explained by gene dosage. This result is particularly significant given that, like *lac*, *hisG* possesses a promoter with a pronounced sensitivity to DNA supercoiling *in vitro* (reviewed in Winkler, 1987). On the other hand, the *his* data are complicated by the fact that *in vivo*, the primary sensor of changes in DNA supercoiling is not the *hisG* promoter itself but the unlinked *hisR* gene which codes for $tRNA^{His}$, and regulates *his* transcription via the *his* attenuator (Figueroa *et al.*, 1991; O'Byrne *et al.*, 1992; Rudd & Menzel, 1987). In the Schmid & Roth experiments, the *hisR* gene would have remained at its native chromosomal location in all cases. The overall impression given by these results is that evidence in favour of position effects on transcription due to differential supercoiling of the chromosome is far from overwhelming. Nevertheless, it remains a nice idea and will undoubtedly attract further investigation.

Chromosomal replication and gene dosage

The gene dosage effects discussed above arise from the way in which the chromosome is replicated (Fig. 2.13). There is only one origin of replication, designated *oriC* and this is located at 84 min on the *E. coli* genetic map (Bachmann, 1990). The chromosome is replicated by a bidirectional mechanism, with replication forks moving away from *oriC* in opposite directions and ultimately overlapping in the termination region (*Ter*) between 23 and 36 min (Bachmann, 1990, François *et al.*, 1990; Hidaka *et al.*, 1991). Consequently, those genes located closest to the origin of replication will double in copy number before those near to the *Ter* region. If the cell is growing rapidly, further rounds of chromosome replication may initiate before cell division, permitting four copies of *oriC*-proximal genes to exist in the cell as against one copy of a *Ter*–located gene. This may have important consequences for the siting of genes

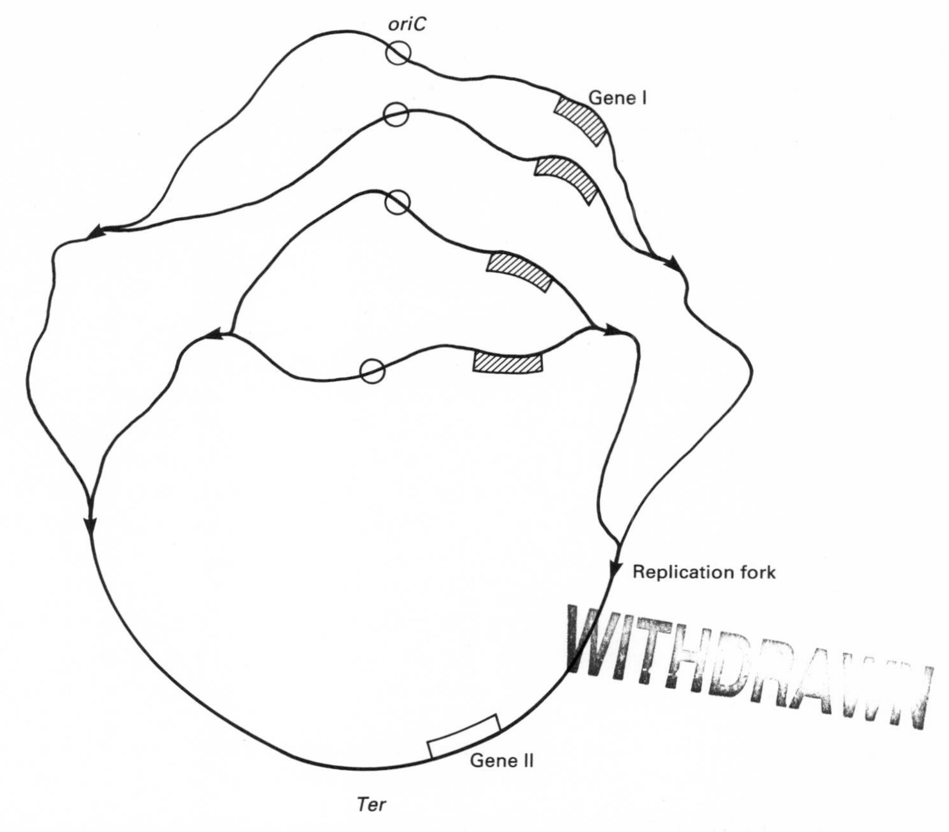

Fig. 2.13 Gene dosage as a function of location on the chromosome. The *Escherichia coli* chromosome is shown undergoing sequential rounds of replication within a rapidly growing cell. Two genes (I and II) are affected differentially by these events. Gene I, which is proximal to the origin of replication (*oriC*), is present in four copies whereas Gene II, which is proximal to the terminus of replication (*Ter*), is present in just one copy.

on the chromosome. If this is true, it may provide a strong influence for the conservation of the genetic map as seen, for example, with the related enteric bacteria *Escherichia coli* and *Salmonella typhimurium*. Although these bacteria are believed to have diverged over 100 million years ago, their genetic maps are remarkably conserved in terms of the order of the genes (Bachmann, 1990; Ochman & Wilson, 1987; Sanderson & Roth, 1988).

Colliding polymerases

The chromosome is a very busy place with many activities occurring simultaneously. Thus, it presents an interesting problem in 'traffic management'. It is possible that the bi-directional mode of chromosomal replication may exert an influence on gene expression due to the orientation of the transcription unit with respect to *oriC* (Fig. 2.14). It has been proposed that if a gene is orientated against the direction of DNA replication, DNA polymerase may collide with

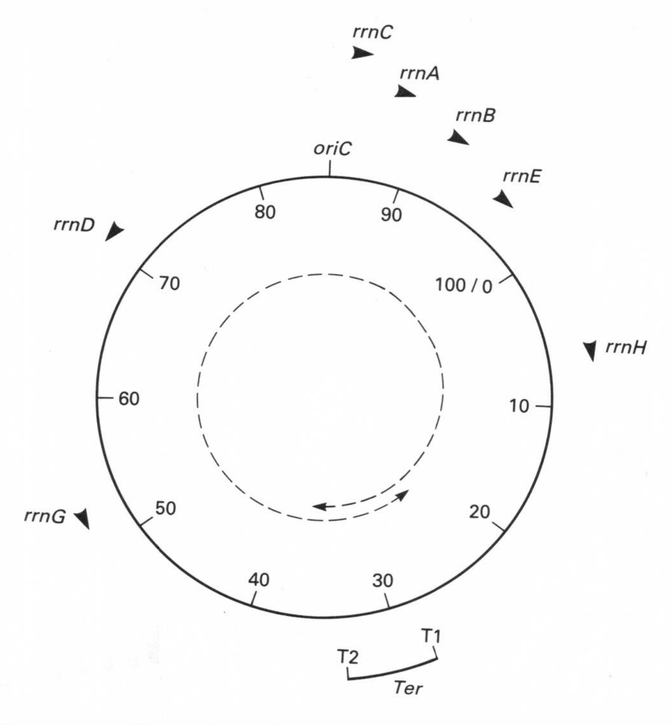

Fig. 2.14 Polymerase traffic on the *Escherichia coli* chromosome. The solid circle represents the *E. coli* chromosome divided into 100 map units. The inner dashed curves represent the movements of the replication forks from *oriC* to *Ter*. The large arrows around the circumference represent the seven ribosomal RNA operons, which are major transcription units. These arrows are aligned with the direction of replication fork movement, which is thought to be significant in preventing collisions between DNA and RNA polymerases.

RNA polymerase while the latter is in the act of transcription. Orientation in the same direction as replication fork movement will lead to an alignment of polymerases with a delay in the passage of the faster-moving DNA polymerase resulting, rather than a collision. Support for this model comes from an analysis of the direction of orientation of transcription units on the *E. coli* genetic map; those driven by the strongest promoters tend to be aligned with the direction of chromosome replication. Genes with weaker promoters appear to be aligned in either direction at random (Brewer, 1988; 1990). Genes within the *Ter* (replication termination) region cannot win because the replication forks overlap there (François *et al.*, 1990). However, most of the strongest promoters in the cell are found outside this area.

The major processes of DNA

These are the processes by which DNA is replicated, by which the genetic information is transcribed into message and by which the genome is reassorted through general or site-specific recombination and transposition. All these activities influence profoundly the expression of genetic information and their consideration is important to an understanding of how bacteria seek to regulate this process. The following sections review briefly these topics and indicate sources of more detailed information for those who want it.

1 Chromosome replication and the bacterial cell cycle

The replication of the circular chromosome is one of three major events that make up the bacterial cell cycle. The other two are the segregation of the daughter nucleoids and the division of the daughter cells (Fig. 2.15). These are now regarded as three independent processes which are coordinated rather than three aspects of one highly complex process (Nördstrom *et al.*, 1991). Genetic and biochemical analyses have helped identify many of the key participants in these processes but the mechanisms by which the processes are coordinated in time and space are still not fully understood (de Boer *et al.*, 1990; Donachie & Robinson, 1987; Helmstetter, 1987).

Cell division

The division of an *E. coli* cell into two daughter cells is accomplished by the laying down of a division septum at the centre of the parent cell followed by division of that cell. If the nucleoid has been partitioned to either side of the septum before division, both daughter cells will be viable. The poles of the newly divided cells possess division septum-like characteristics and in certain mutants, asymmetric division can occur at these to produce cells without chromosomes known as 'minicells'.

The predominant regulator of division appears to be the FtsZ protein; on achieving a critical intracellular level, FtsZ will initiate septation (Fig. 2.16). Once initiated, other division functions (such as SulA, SfiC, etc.) complete it. FtsZ-initiated septation will occur at the central division site and at the polar sites. The decision to divide at the central septum and not at the poles is influenced by the products of the *minC, minD* and *minE* genes. MinCD is an inhibitor of septation and this is given topological specificity by the MinE protein, that is to say, MinE influences the MinCD inhibitor to prevent septation at the poles but not at the central site. The interplay between these factors and FtsZ ensures that division occurs at the central division site (Fig. 2.16) (readers wanting a more detailed treatment are referred to deBoer *et al.*, 1990; Lutkenhaus, 1990; Nordström *et al.*, 1991).

Chromosomal DNA replication

In *Escherichia coli*, chromosomal DNA replication is a complicated process involving the participation of more than 30 proteins (Kornberg & Baker, 1992; Marians, 1992). It is initiated at a unique site, *oriC*, located at 84 min on the

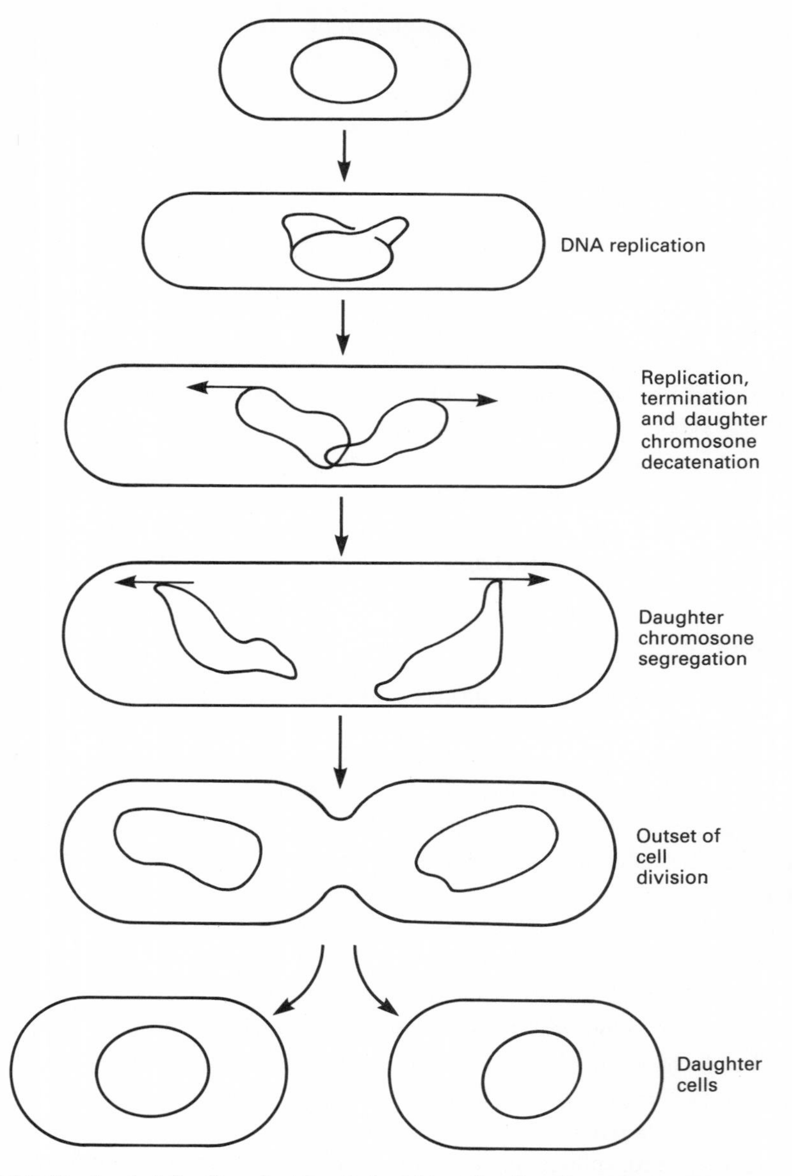

Fig. 2.15 The bacterial cell cycle. The *Escherichia coli* cell cycle consists of three independent but coordinated events. These are chromosome replication, daughter chromosome segregation and cell division.

E. coli genetic map (Bachmann, 1990). This site is composed of four 9 bp binding sites for DnaA (known as the 'R repeats'), three 13 bp direct repeats with which DnaA interacts once it has bound at the 9 bp R repeats and 5′-GATC-3′ sequences which are targets for the *dam* methylase (Zyskind, 1990) (Fig. 2.17). Transcription of *dam* is regulated by DnaA, illustrating how tightly integrated

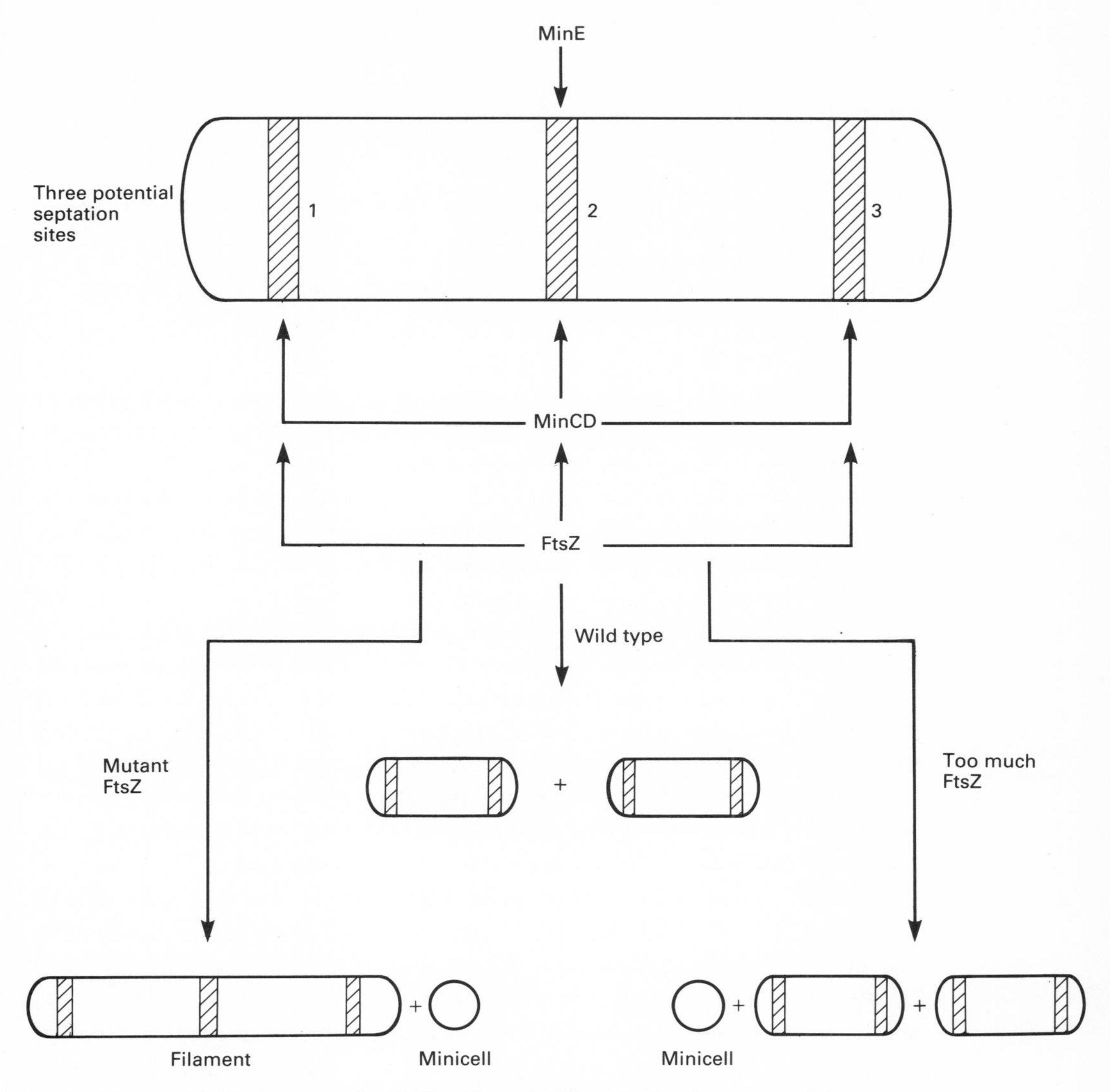

Fig. 2.16 Control of septation. The *Escherichia coli* cell possesses three potential septation sites and requires a mechanism to ensure that the central one is used in preference to the polar ones. This mechanism depends on the relative abundances of the Min and FtsZ proteins. Aberrant levels of FtsZ can result in filamentation (failure to divide) and minicell production (division at polar septation sites).

the replication system is. The methylation state of these sequences (present in 11 copies) controls binding of *oriC* to the bacterial outer membrane, an important feature in the timing of reinitiation of replication (see later). Dam sites are methylated on both strands under normal circumstances (Barras & Marinus, 1989). However, behind the moving DNA replication fork, the sites are hemi-methylated since the newly synthesized DNA has not yet been modified.

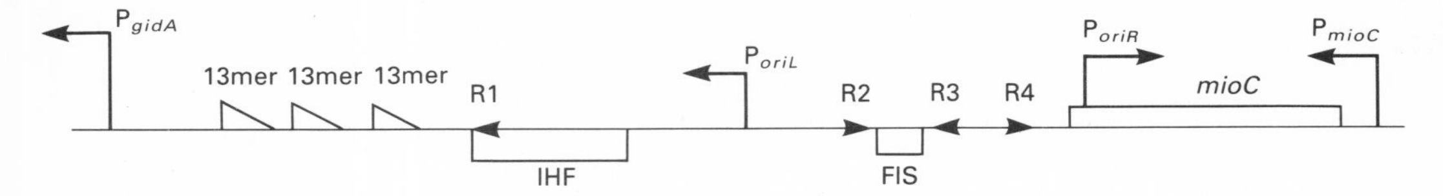

Fig. 2.17 Structure of the origin of replication of the *Escherichia coli* chromosome. A 1 kb sequence consisting of a minimal origin of replication (*oriC*). The locations of binding sites for the histone-like proteins FIS and IHF are indicated, together with DnaA 'boxes' (R1, R2, etc.) and 13 bp directly repeated sequences. Various promoters (P) in the *oriC* region are also shown.

This provides the cellular DNA surveillance machinery with a mechanism to discriminate between new and old DNA and is used by DNA repair enzymes to correct mismatches that arise during replication.

Dam methylation has been proposed as a control element in chromosome partitioning, with fully methylated *oriC* sequences being cytosolic and hemimethylated *oriC* being attached to the membrane. After 20 min at 30°, the *oriC* sites become fully methylated and detach from the membranes (Ogden *et al.*, 1988). Although it has been thought that this membrane anchoring might help in partitioning the daughter chromosomes to the daughter cells, *dam* mutants do not appear to be significantly impaired in chromosome partitioning (Vinella *et al.*, 1992). The true significance of the membrane attachment appears to be that it prevents binding of the newly synthesized *oriC* site by DnaA (Landoulsi *et al.*, 1990). Thus, the methylated state of the DNA and its attachment to the membrane is a device for timing the initiation of chromosome replication rather than assuring the fidelity of daughter chromosome partitioning.

To initiate replication, DnaA must bind ATP. In the first step, 20–40 monomers of DnaA bind at *oriC* R repeats. In the presence of histone-like protein HU and provided DnaA is bound to ATP, an open complex forms to facilitate the formation of the prepriming complex. This assembles upon addition of DnaB and DnaC and is accompanied by additional DNA unwinding. DnaB binds in the region of the 13 bp repeats and its helicase activity, assisted by DNA gyrase, unwinds the template for subsequent priming and replication. The transcription of nearby genes (especially *mioC*) may modulate the initiation of replication (reviewed in Zyskind, 1990; see also Chapter 6).

Other factors influencing the initiation of DNA replication include the methylation state of the *oriC* 5′-GATC-3′ sequences and the level of supercoiling of the DNA. Supercoiling assists the unwinding of DNA at *oriC* at the onset of replication and DnaA prefers to bind to supercoiled rather than relaxed DNA (Fuller & Kornberg, 1983; Kowalski & Eddy, 1989; Louarn *et al.*, 1984). DnaG primase is required to produce RNA primers needed for DNA synthesis from dNTPs by DNA polymerase III. Replication proceeds bidirectionally from *oriC* via a theta (Θ) intermediate and is terminated in a region of the *E. coli* chromosome directly opposite the origin known as *Ter* (McMacken *et al.*, 1987) (Fig. 2.18). The termination region extends over 450 kb and contains six symmetrically disposed

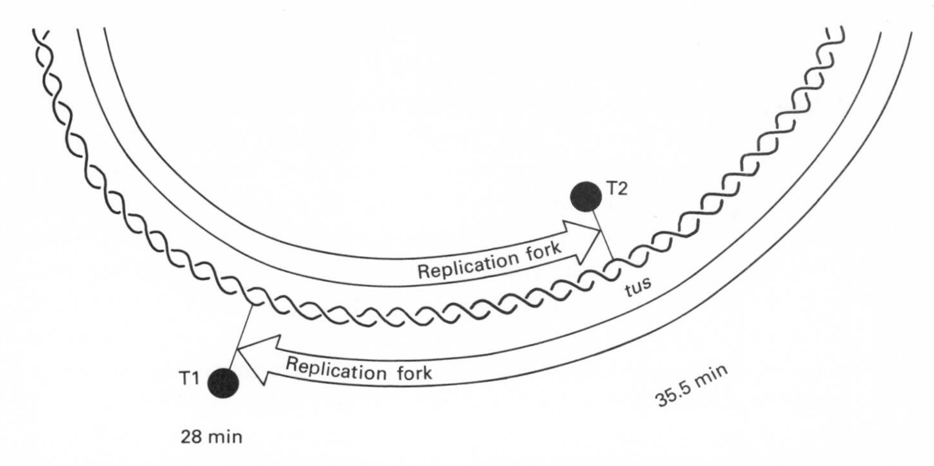

Fig. 2.18 Simplified diagram of the replication termination region of the *Escherichia coli* chromosome. In-coming replication forks (indicated by arrows) are trapped by the unidirectionally active termination sites (T). The gene coding for the termination protein Tus is located close to T2.

cis-acting sites concerned with trapping the replication forks. Replication forks may enter the traps but they are not permitted to leave again. Two primary sites, T1 and T2, are involved and these function in a polar manner. T1 inhibits anticlockwise movement while allowing clockwise forks to pass; T2 permits passage of anticlockwise forks while inhibiting clockwise fork movement. Operation of the T1 and T2 termination sites depends on the presence of the 36 kDa Tus (the termination utilization substance) protein. Tus acts in a polar fashion to inhibit the unwinding activity of the DNA helicases associated with the approaching replication fork. The gene coding for Tus, *tus*, is located close to T2 (reviewed in Kuempel *et al.*, 1990) (Fig. 2.18).

Partitioning of daughter nucleoids

The products of a round of chromosome replication are two interlinked (catenated) daughter chromosomes. In this condition, partitioning into daughter cells is not possible. The cell decatenates the chromosomes using DNA topoisomerases. DNA gyrase plays an important role in decatenation, but other enzymes may also be required (Hussain *et al.*, 1987; Kato *et al.*, 1989). One of these is thought to be topoisomerase III, the product of the *topB* gene of *E. coli* (DiGate & Marians, 1989). This enzyme is a type I topoisomerase enzyme with potent DNA decatenation activity (DiGate & Marians, 1988; Srivenugopal *et al.*, 1984). Whether this decatenating activity is relevant to chromosome segregation *in vivo* is unclear since *topB* null mutants of *E. coli* are viable (DiGate & Marians, 1989).

Further decatenating activity is thought to be provided by a novel type II topoisomerase, topoisomerase IV, which has been discovered in *E. coli* and *Salmonella typhimurium*. This is encoded by two genes, *parC* and *parE*, previously identified as being required for normal chromosome partitioning (Adams *et al.*,

1992; Kato *et al.*, 1988; 1990b). Interestingly, when *E. coli* mutants deficient in *topA*, the gene for topoisomerase I, are subjected to environmental stress, they amplify the region of the chromosome containing the topoisomerase IV genes, increasing their copy number. This amplification event is RecA-dependent (Dorman *et al.*, 1989b; Kato *et al.*, 1990b; see section on Genome Rearrangements and Gene Dosage in Chapter 4) and suggests that the activities of the different topoisomerases in the cell may be partially redundant, with one being able to take over some of the functions of another under some circumstances. Apart from its role as a decatenase, topoisomerase IV has an ATP-dependent type II DNA-relaxing activity, and this may be the activity that compensates for the loss of topoisomerase I in *topA* null mutants. The ParC protein is associated with the cytoplasmic membrane, indicating that topoisomerase IV is a membrane-bound topoisomerase (Hiraga, 1992). The cell appears to divide its decatenation functions principally between gyrase and topoisomerase IV. Gyrase deals with catenanes arising from tangling and recombination while topoisomerase IV unlinks those arising from DNA replication (Adams *et al.*, 1992).

A site-specific recombination system appears to be required for efficient resolution of chromosome dimers. The resolution site is called *dif* (for deletion induced filamentation) and it lies within the *Ter* region, which is concerned with the termination of chromosome replication. The *dif* sequence displays 28 bp of homology with the *cer* region of plasmid ColE1, the site of action of the XerC site-specific recombinase which resolves ColE1 multimers (Chapter 3). The *xerC* gene is located at 3700 kbp on the *E. coli* chromosome and its product acts on both the *cer* and the *dif* sites (Blakely *et al.*, 1991; Clerget, 1991; Kuempel *et al.*, 1991). Mutants with *dif* deletions produce anucleate cells and are derepressed for the SOS system. The Dif phenotype is suppressed in strains deficient in homologous recombination, suggesting that the XerC/*dif* system is required to resolve chromosomal multimers generated by homologous recombination between daughter chromosomes (Blakely *et al.*, 1991).

The *E. coli ruvB* gene also seems to be involved in chromosome partitioning. This SOS regulon gene is primarily involved in homologous recombination (see later) and its role in chromosome partitioning is thought to be secondary (Hiraga, 1992). However, *recA* mutants of *E. coli* have been shown to be deficient in chromosome segregation such that 10% of *recA* cells contain no DNA following cell division. Thus, this component of the SOS response appears to contribute in an important manner to chromosome partitioning (Zyskind *et al.*, 1992). It has been proposed that RecA may form a complex with replication forks stalled at Tus-bound termination sites; inactivation of the *tus* gene removes the need for RecA in constitutive stable replication mutants (Dasgupta, S., Bernander, R. & Nordström, K. cited in Zyskind *et al.*, 1992).

Searches for genes concerned with the movement of daughter chromosomes at cell division have led to the discovery of so-called *muk* mutants (from the Japanese word *mukaku*, which means 'anucleate') (Hiraga, 1992). Mutations in the *mukA* gene result in abnormally high levels of anucleate cells in a bacterial population growing in minimal glucose medium (Hiraga *et al.*, 1989). The *mukA* gene is allelic with *tolC*, a gene coding for an outer membrane protein. Mutations in *tolC* are highly pleiotropic, resulting in a reduction in growth rate, alterations in the expression of the outer membrane porin protein OmpF, haemolysin secretion, tolerance to colicin E1, increased sensitivity to sodium dodecyl sulphate, to basic dyes, to some antibiotics and to deoxycholate (Webster, 1991). How TolC contributes to chromosome partitioning is not understood.

The *mukB* gene codes for an enormous protein of almost 177 kDa. This hydrophilic protein consists of five structural domains (Fig. 2.19). The aminoterminal domain (Domain I) is predicted to be globular and contains a consensus sequence for ATP binding. Domains II and IV are predicted to be α-helical and

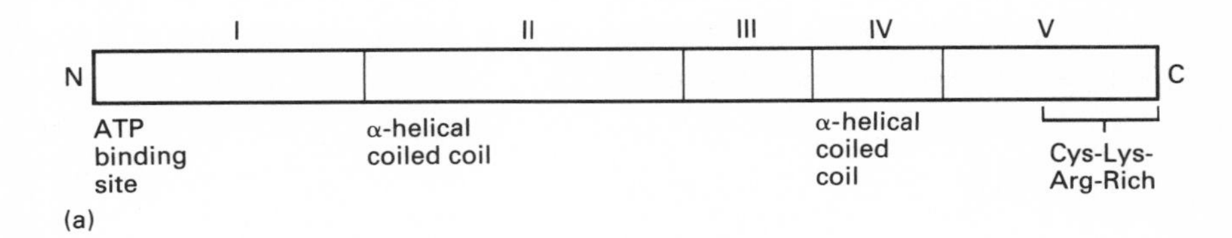

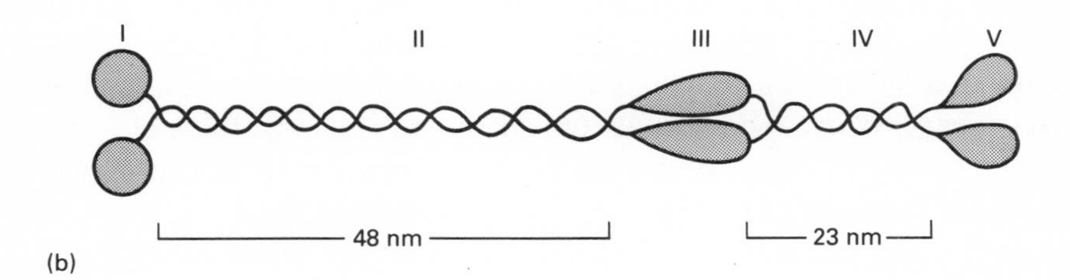

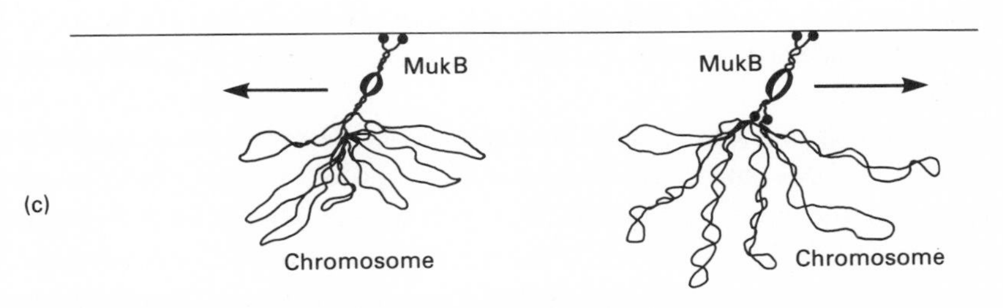

Fig. 2.19 Structure and function of chromosome partition protein MukB. (a) The domain structure of the MukB protein. (b) A predicted structure for a mukB homodimer, with the locations of the five major domains illustrated. (c) The function of MukB in chromosome partitioning. Here, MukB connects the daughter chromosomes to a putative filament (perhaps a protein polymer) for segregation to the daughter cells. These filaments perform a role analogous to spindle fibres in eukaryotic mitosis.

to display a coiled-coil structure. Domain III appears to be globular, as is the carboxyterminal domain (Domain V) (reviewed in Hiraga, 1992). Domain V contains three potential zinc-finger-like structures and this portion of the protein is also rich in the basic amino acids arginine and lysine. It has been proposed that the carboxyterminal domain may be involved in interactions with DNA (or with other proteins). The structure of MukB makes it a good candidate for a chemomechanical or force-generating enzyme, the first to be described in bacteria. By analogy with kinesin heavy chain and myosin heavy chain in eukaryotes, MukB is thought to form a homodimer in which the head region with its ATP-binding site would form the motor portion of the protein. Parts of the domains showing coiled-coil structure show weak homology to analogous regions of myosin heavy chain and rat dynamin. The contribution of MukB to chromosome partitioning has been proposed to involve an interaction with the DNA and with a prokaryotic cytoskeleton (Fig. 2.19). As yet, no definitive evidence for such an interaction is available.

Pleiotropic roles of cell cycle regulators

It is useful to bear in mind that factors with important roles in chromosome replication often have additional functions in the cell. Frequently, these functions are revealed by genetic analyses in which mutations in the genes coding for replication factors are found to have pleiotropic effects. For example, the DnaA protein which initiates replication of the chromosome (in the presence of ATP and the histone-like protein, HU) is also needed to replicate some plasmids and bacteriophage, is involved in certain transposition events (e.g. those of transposon Tn5) and is a negative regulator of expression of several genes, including its own, *dnaA* (Bramhill & Kornberg, 1988; Georgopoulos, 1989).

2 Transcription

From the standpoint of gene regulation, transcription is one of the most important processes of DNA. Its salient features will be described briefly here. As usual, readers seeking a more detailed discussion are referred to the literature cited.

The first step in transcription involves a period of promoter searching and recognition by RNA polymerase holoenzyme ($\alpha_2\beta\beta'\sigma$). Transcription initiation requires that the multicomponent RNA polymerase (R) makes a productive interaction with a free promoter sequence (P) in the DNA of the gene to be expressed. The result is the formation of a (reversible) closed complex. The next step is an isomerization of the closed complex (RP_c) to an irreversible open complex (RP_o) through local melting of the DNA duplex. In the presence of the four transcription substrates (nucleoside triphosphates or NTPs), the open complex is converted to the initial transcribing complex (ITC). Here, polymerase is stably anchored at the promoter, catalytically generating nested oligomers up to nine nucleotides in length in repeated, reiterative acts of abortive initiation. The sequence involved in the abortive initiations is called the initial transcribed sequence (ITS) (Krummel & Chamberlain, 1989). This plays an important role in the transition between abortive cycling and transcript elongation.

Polymerase next moves from the promoter sequence into the region to be transcribed, a step called promoter clearance. This ends the cyclic abortive reaction and involves the release of sigma factor, the relinquishing of promoter-anchoring contacts, cessation of abortive initiation and the commencement of processive elongation by the core RNA polymerase ($\alpha_2\beta\beta'$). These stages may be summarized as follows:

$$R + P \overset{K_B}{\longleftrightarrow} RP_c \overset{k_f}{\longrightarrow} RP_o \longrightarrow RNA$$

In the first step, RP_c is in rapid equilibrium with P, characterized by an association constant, K_B The forward rate constant for the second step is k_f. This step is generally irreversible. The values of these constants are an indication of the strength of particular promoters. An important determinant of promoter strength is the sequence of bases in the areas of recognition and contact by RNA polymerase. For σ^{70} programmed polymerase, a consensus promoter sequence has been deduced (Fig. 2.20; Chapter 6) and its crucial components are the −10 and −35 boxes and the length and the sequence of the spacer region between these. In general, the more closely a promoter matches the idealized consensus sequence, the more powerful a transcription initiation signal it will provide. Thus, mutations that reduce homology to the consensus weaken promoter strength (Auble & deHaseth, 1988; Borowiec & Gralla, 1987). The issues of transcription regulation and promoter strength are developed more fully in Chapter 6.

3 General (homologous) recombination

General recombination involves a genetic exchange between two homologous DNA sequences located either on the same molecule or on different molecules (for reviews see Smith, 1988; Weinstock, 1987). It is termed 'general' because it can occur between any two places on the participating molecules with sufficient homology to initiate the reaction. General recombination provides an essential mechanism for the reassortment of individual genomes or for the incorporation of new genetic material from an external source. It is fundamental to the processes of general transduction (in which bacteriophage transfer bacterial DNA between cells; Chapter 4), conjugal transfer of DNA (mediated by self-transmissible or mobilizable plasmids; Chapter 3) and those types of transforma-

−35 −10

TTGACA — 17 bp — TATAAT — +1

Fig. 2.20 Consensus sequence for σ^{70}-dependent promoter.

tion reaction that result in the physical combining of the foreign DNA with that of the recipient cell.

The process of general recombination may be reduced to one in which the 3′-OH end of a single-stranded DNA (ssDNA) molecule enters a double-stranded DNA (dsDNA) duplex and, if a region of homology is located, the ssDNA forms a heteroduplex with the dsDNA. This heteroduplex is then processed into a complete recombinant molecule. This involves movement of a heteroduplex joint along the recipient duplex in a process known as 'branch migration' (Fig. 2.21). As the invading strand is fed into the recipient molecule, it displaces its native strand counterpart, producing a 'D-loop', so-called because of its shape (Fig. 2.21). If the invading ssDNA is merely a single-stranded leader for a duplex, then once branch migration is underway, the displaced strand from the recipient molecule can find and pair with its complement in the donor molecule. This leads to reciprocal exchange between the duplexes (Fig. 2.21). The extent of reciprocal exchange is determined by the duration of the branch migration stage (Fig. 2.21). The junction which forms the site of strand exchange between duplex molecules, is termed a Holliday structure, named after its proposer, Robin Holliday (1964). If the participating duplexes are separated and rotated through 180° the result is a square-planar form called a Chi structure (after the letter of the Greek alphabet whose shape it resembles). Recombination is terminated by the nicking of DNA strands by endonuclease. This resolves the junction and yields the recombination products. Nicks in the DNA are then sealed by the enzyme DNA ligase (Fig. 2.21).

Much of the molecular detail of general recombination has been worked out in *E. coli* where there is considerable evidence for distinct pathways (reviewed in Mahajan, 1988). These are the RecBCD pathway, which is the main pathway in wild type cells, the RecE pathway, which operates in *recBC sbcA* mutants and the RecF pathway, which functions in *recBC sbcB* mutants (Clark, 1973). All three pathways depend on RecA. The RecBCD pathway is the major pathway of recombination between chromosomal DNA and homologous linear DNA. The RecF pathway is active in the recombination of plasmid molecules and in the repair of ultraviolet radiation damage.

Recently, it has been discovered that Holliday junctions are processed in *E. coli* by the products of the *ruvA*, *ruvB* and *ruvC* genes (reviewed in Taylor, 1992). RuvA and RuvB move the Holliday junction in an ATP-dependent reaction. The junction is then resolved by the RuvC protein which cleaves the junction. The *ruvA* and *ruvB* genes are organized as an operon whose transcription is under negative control by the LexA repressor, i.e. this operon is activated by the SOS response. Since this response is called upon in cells that have undergone DNA damage, this arrangement makes intuitive sense. Since *ruv* mutations have only a minor effect on the efficiency of RecBCD recombination (for example, following conjugal transfer of DNA), there must be another method of processing Holliday junctions. Another protein, RecG, appears to be involved here. Its gene, *recG*, forms part of a complex operon which includes the genes *spoT* (the bifunctional

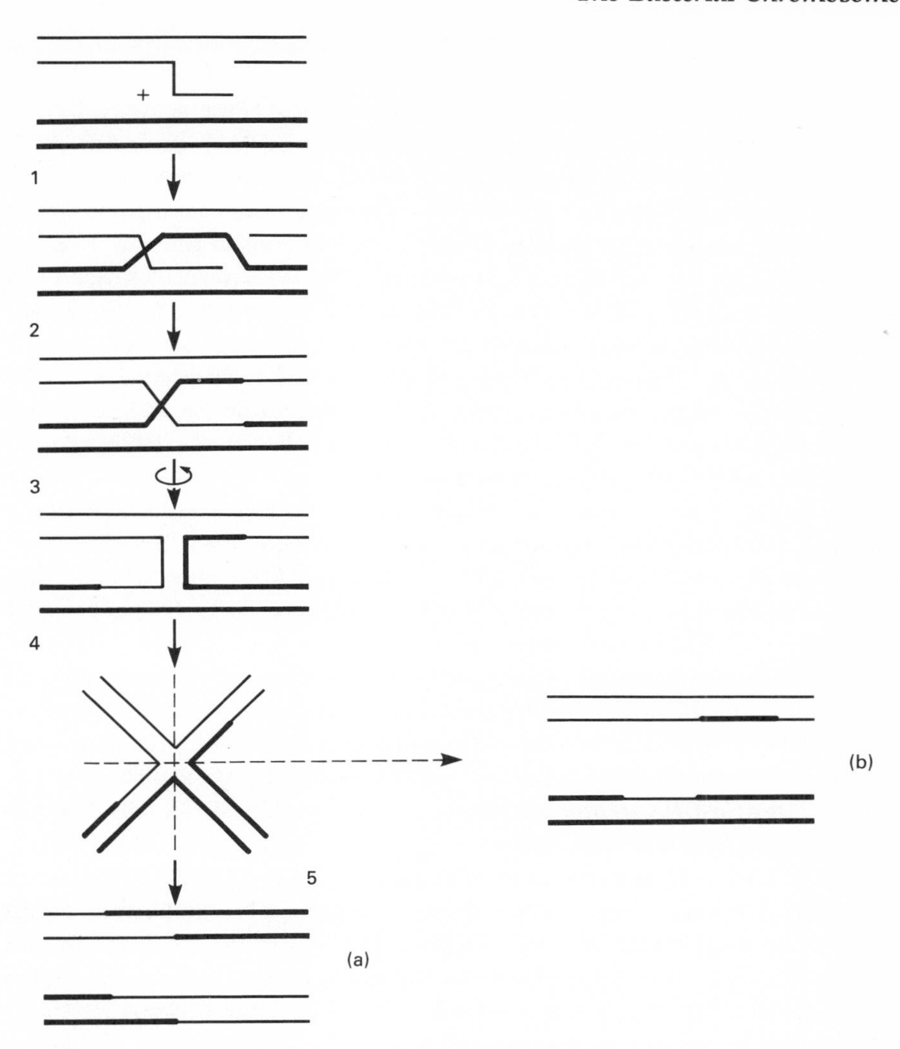

Fig. 2.21 General (homologous) recombination. Step 1, single-stranded DNA coated with SSB protein and RecA synapses with an homologous sequence to form a D-loop. Step 2, the D-loop is nicked and base-pairing permits formation of the Holliday junction; junction movement (branch migration) occurs. The junction is symmetrical (hence steps 3 and 4). Step 5, the junction is resolved resulting in either a 'splice' (a) or a 'patch' (b).

ppGpp synthetase II and ppGpp hydrolase of the Stringent Response, see Chapter 7) and the gene coding for the ω subunit of RNA polymerase (Lloyd & Sharples, 1991). (Thus, *E. coli* (and other bacteria) exploit operons as a way of coregulating the expression of factors contributing to distinct but fundamentally important processes.) Recently, the abundant nucleoid-associated protein HU, described earlier in this chapter, has been found to bind to the junction

(Pontiggia *et al.*, 1993). This shows that local DNA topology is crucial to general recombination, a not altogether unexpected finding.

RecA is a 37.5 kDa protein with DNA-dependent ATPase activity. It binds avidly and cooperatively to single-stranded DNA (ssDNA), promotes ATP-dependent assimilation of the ssDNA into homologous double stranded DNA (dsDNA) and unwinds dsDNA processively. It promotes base pairing and strand exchange between dsDNA molecules with 3′-ssDNA ends to generate Holliday junctions. It can also pair (anneal) complementary ssDNA molecules. RecA can act as an ATP-dependent co-protease, cleaving λ cI repressor (see Bacteriophage λ integration and excision, Chapter 4) and the LexA repressor of the SOS pathway (Cox & Lehman, 1987). Homologues to *recA* have been detected in *Bacillus subtilis* (where the gene is called *recE*) (Lovett & Roberts, 1985), *Erwinia carotovora* (Keener *et al.*, 1984), *Proteus mirabilis* (West *et al.*, 1983), *Proteus vulgaris* (Keener *et al.*, 1984), *Pseudomonas aeruginosa* (Ohman *et al.*, 1985), *Neisseria gonorrhoeae* (Koomey & Falkow, 1987), *Rhizobium meliloti* (Better & Helinski, 1983), *Salmonella typhimurium* (Pierre & Paoletti, 1983), *Shigella flexneri* (Keener *et al.*, 1984) and *Vibrio cholerae* (Goldberg & Mekalanos, 1986a). This suggests that homologous recombination probably proceeds by a broadly similar mechanism in all these organisms.

RecBCD (exonuclease V) is composed of the products of the *recB*, *recC* and *recD* genes. The B, C and D components are 140, 130 and 65 kDa proteins, respectively. This enzyme is a dsDNA-dependent ATPase with ATP-dependent dsDNA and ssDNA exonuclease and ssDNA endonuclease activities. It can unwind linear dsDNA to produce ssDNA or dsDNA with ssDNA tails. RecBCD interacts with Chi sequences in DNA.

Chi sites are 8 bp stretches of DNA (sequence: 5′-GCTGGTGG-3′) that occur in the *E. coli* genome every 5–10 kb and stimulate general recombination events up to 10 kb away (Smith & Stahl, 1985). As RecBCD moves along the DNA it unwinds the helix, and if it approaches a Chi site from the 3′ end, it will cut the upper strand of the DNA 4–6 bp from the Chi site (Fig. 2.22) (Taylor *et al.*, 1985). This action is thought to provide RecA with the 3′-OH ssDNA needed to promote recombination (Mahajan, 1988; Smith, 1988).

E. coli RecE (exonuclease VIII) is a 140 kDa protein that degrades dsDNA processively from the 5′ end to produce long 3′ tails. This exonuclease activity is ATP-independent. The function of RecF remains obscure and its contributions to general recombination have been inferred mainly from indirect genetic evidence (Blander *et al.*, 1984). It has been proposed that RecF may assist RecA by making ssDNA available to it (Madiraju *et al.*, 1988). SSB is a 19.5 kDa protein that preferentially binds to ssDNA in a cooperative manner. In addition to its role in recombination, SSB is also required for DNA replication. It helps RecA to bind ssDNA and thus promotes RecA-catalysed reactions (Cox & Lehman, 1987; Mahajan, 1988). Mutations in *sbcA* of the cryptic Rac (Recombination Activation) prophage lead to synthesis of exonuclease VIII (the *recE* gene product) which can promote general recombination in the absence of

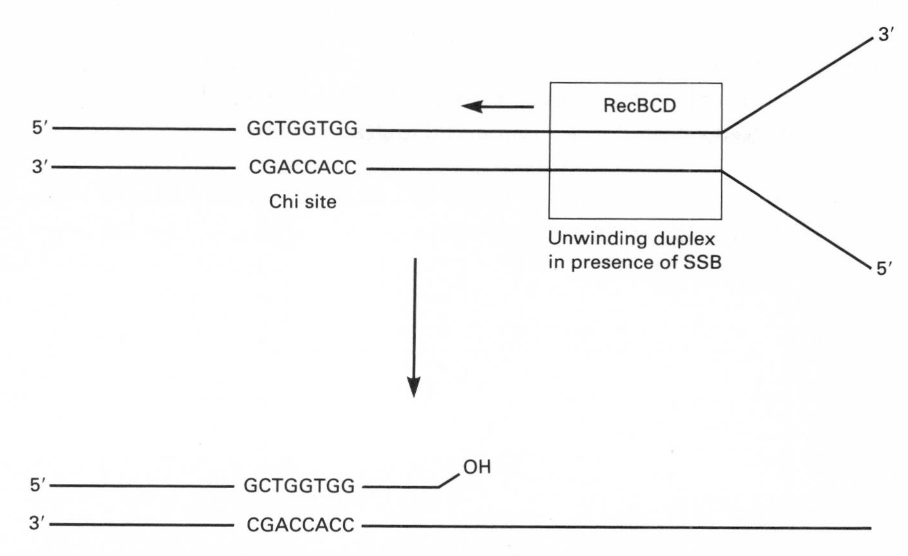

Fig. 2.22 Processing of DNA by RecBCD at a chi site. RecBCD moves towards the Chi site, unwinding the DNA duplex in the presence of SSB protein. The upper strand is cleaved 3′ to the chi site, generating the free 3′-OH needed for recombination.

RecBCD (Bachmann, 1990; Smith, 1988). The *sbcB* gene codes for exonuclease I which degrades ssDNA to mononucleotides in the 3′-to-5′ direction. Mutations in *sbcB* suppress recombination deficiency associated with a loss of RecBCD (exonuclease V) activity (Philips *et al.*, 1988; Smith, 1988). It is now recognized that *recBCD* suppression by mutations in *sbcB* also requires mutations in another gene, *sbcC*, the nature of whose product is not understood (Lloyd & Buckman, 1985; Smith, 1988).

4 Site-specific recombination and transposition

Site-specific recombination differs from homologous (or general) recombination in a number of respects. While homologous recombination involves substrate DNAs which share considerable homology with exchange taking place anywhere within that region of homology, the recombining sequences in site-specific reactions are usually short and the reaction occurs at a single, specific site. Homologous recombination is promoted by the proteins of the homologous recombination pathway whereas site-specific recombination involves specific recombinases and cofactors and is independent of the general recombination pathway. Site-specific recombination may be sub-divided into transposition and 'conservative' site-specific recombination (Figs 2.23, 2.24 & 2.25). These processes are mechanistically distinct and generate markedly different end products. Nevertheless, both bring about major rearrangements in the genomes of the cells in which they occur.

Transposition involves mobile DNA elements which have no sequence homology with their sites of insertion. This form of recombination is non-reciprocal and strand exchange is accompanied by DNA replication, although the amount of replication is small in the case of 'non-replicative' transposition (confusingly also referred to as 'conservative' transposition) (Figs 2.23 & 2.24). In the conservative site-specific recombination systems, recombination involves specific sequences in the reacting DNAs which usually display at least a short region of homology. Here, recombination is reciprocal and involves precise

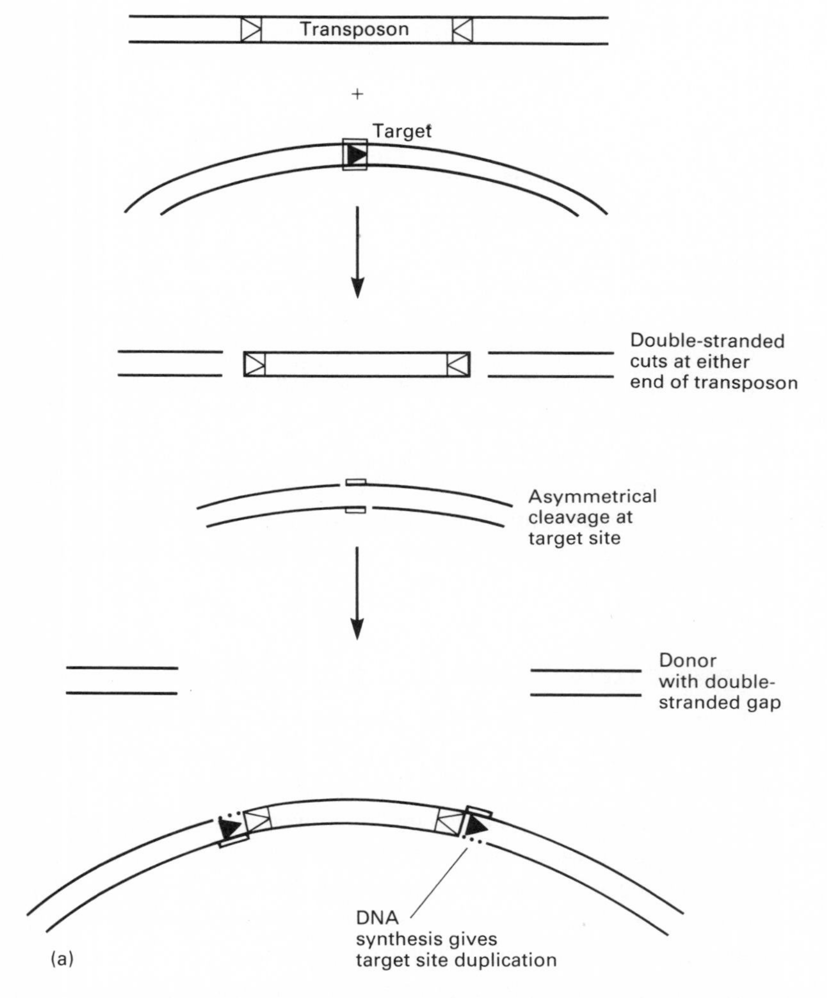

Fig. 2.23 Transposition pathways. (a) The conservative (or non-replicative) pathway. Double-stranded cuts are made at the termini of the transposon and an asymmetrical cleavage is made at the target site. The transposon inserts at the target site where a small amount of DNA replication duplicates the site. The donor molecule contains a double-stranded cleavage. If this is not repaired, the molecule will be lost.

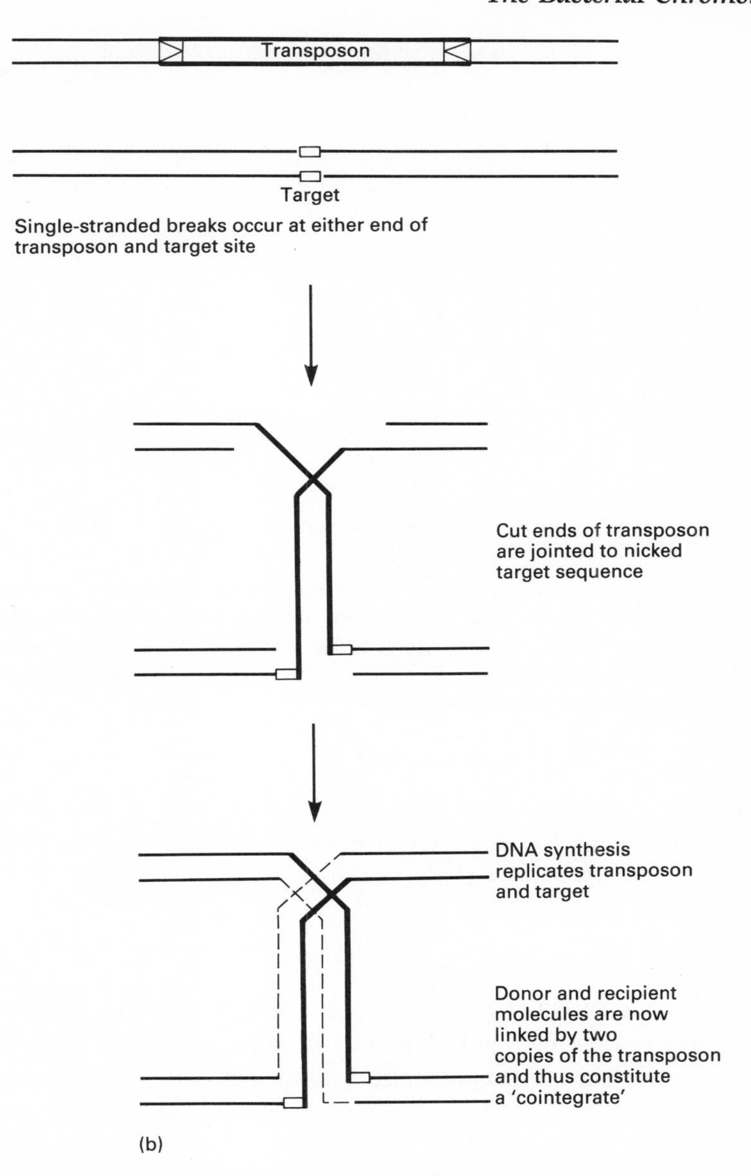

Fig. 2.23 (*continued*) (b) Replicative pathway. Single-stranded breaks occur at either end of the transposon and staggered cuts are made at the target site. The cut ends of the transposon are joined to the nicked target site, DNA synthesis replicates the transposon and the target. The result is a cointegrate molecule in which donor and recipient are linked by two copies of the transposon.

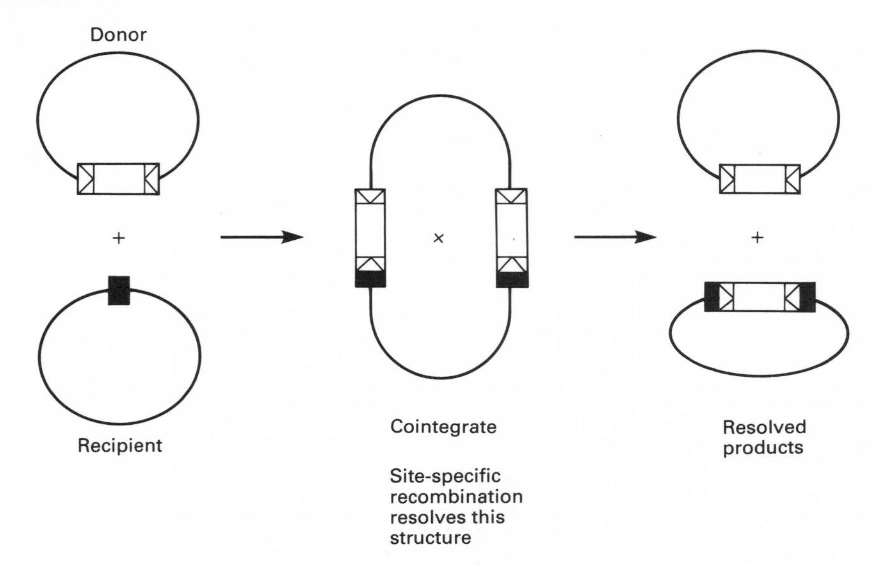

Fig. 2.24 Cointegrate formation and resolution. Replicative transposition generates a cointegrate structure in which the donor and recipient molecules (represented here by plasmids) are linked together by two copies of the transposon (centre). A site-specific recombination event involving resolution sites within the transposons yields the resolved products shown on the right.

breakage and reunion reactions without DNA loss or synthesis (Campbell, 1981; Craig, 1988; Craig & Kleckner, 1987; Plasterk, 1992; Sadowsky, 1986) (Fig. 2.25).

Conclusion

This chapter has sought to describe the bacterial chromosome as an environment capable of exerting multiple influences on the expression of the resident genetic material. These influences arise from the structure of the chromosome, its organization within the cell and its association with higher-order structures (i.e. chromatin). They are also the product of the biological functioning of the chromosome as a self-replicating genetic entity. Like other replicons, the chromosome is prone to rearrangements arising from transposition, general recombination and site-specific recombination. All these features can conspire

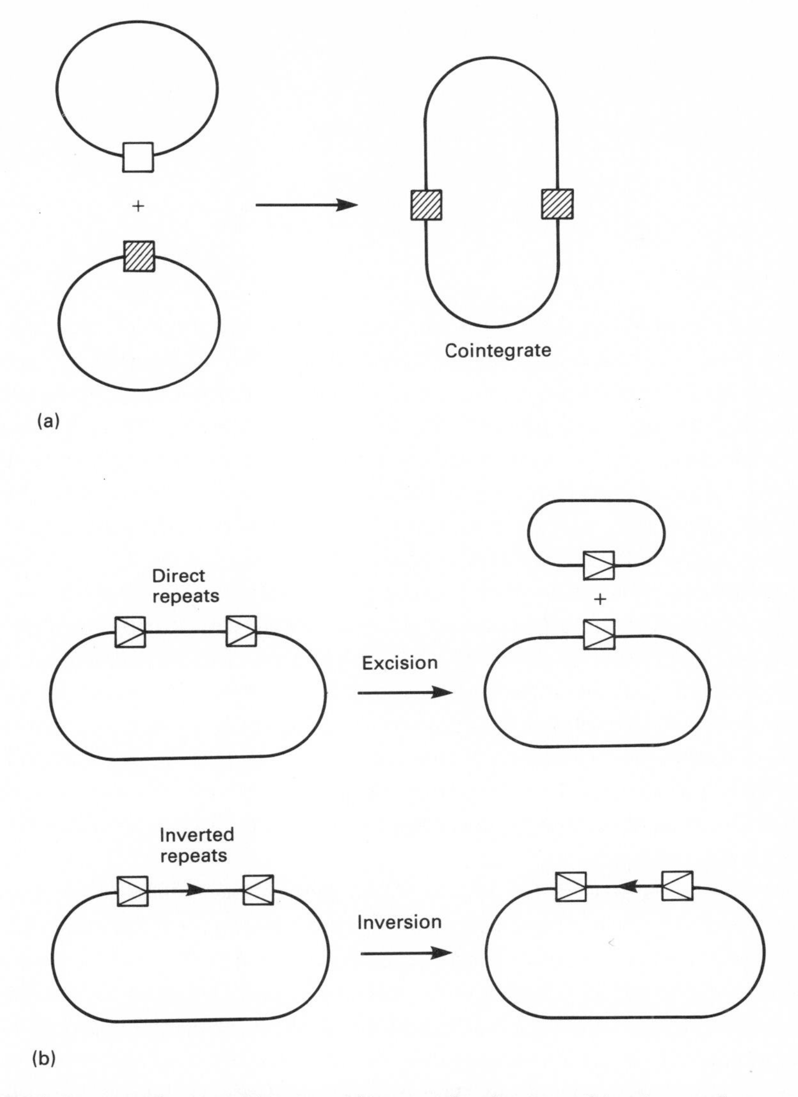

Fig. 2.25 Site-specific recombination. (a) Intermolecular recombination produces a cointegrate structure. This can be resolved by intramolecular recombination. (b) Intramolecular recombination. Here, recombination between directly repeated sequences resolves a cointegrate structure, whereas recombination between inverted repeats results in inversion of the intervening sequences.

to modulate gene expression. In the next chapter, the influence of plasmid-linkage on gene expression will be considered while bacteriophage- and transposon-linkage will be dealt with in Chapter 4. Once the reader is familiar with the general concepts discussed in these chapters, the later material dealing specifically with virulence gene regulation can be put much more satisfactorily into context.

3 The Bacterial Genome: Plasmids

Introduction

In the last chapter, the chromosome, the major component of the bacterial genome, was described as an environment for the expression of genetic information. In this chapter, plasmids, a minor genome component will be discussed. Although more modest than the chromosome in genetic complexity, the contributions of the extrachromosomal, autonomously replicating DNA molecules, known as plasmids, to bacterial virulence are frequently highly significant. They may carry the majority of the dedicated virulence genes of a pathogen, such that the organism is rendered non-pathogenic if the plasmid is lost. In addition, plasmids may make more subtle contributions to virulence. For example, some of them offer resident genes a vehicle with a copy number that may be higher (sometimes dramatically so) than that of the chromosome. Most of the plasmids that have been discovered in bacteria are covalently closed circular molecules which are maintained in a negatively supercoiled state (just like the chromosome). However, it must be pointed out that some plasmids have been discovered that are linear, having telomeric structures analogous to those found at the ends of eukaryotic chromosomes. Several of these plasmids encode virulence determinants.

Some plasmids possess the means to transfer themselves to other bacteria and so provide the genes they harbour with a mechanism for horizontal transfer. This transfer can be to cells of the same species or to cells of other bacterial species. In some cases, transfer even across kingdom boundaries is possible. Some plasmids are largely self-sufficient for replication and/or transfer while others depend on many chromosomally encoded factors to replicate and/or to transfer themselves horizontally. These different degrees of host dependency indicate variations in the evolution of autonomy among plasmids.

A number of very specialized plasmids in bacterial pathogens possess the means to transfer their DNA to the genome of a host, altering host gene expression in a manner that benefits the infecting bacterium. This behaviour is strongly reminiscent of that displayed by viruses. The distinction between plasmids and viruses can be very blurred, even to the extent that some plasmids are capable of becoming viruses. These combine features of both life-styles in a single molecule. Bacteriophage P1 is one of these. It functions as a general transducing phage but has the ability to form a lysogen with *Escherichia coli* by becoming a plasmid. Some plasmids may give up their autonomous life-style from time to time by recombining with the host chromosome to form cointegrates. Again, this is a type of behaviour shown by bacteriophage, for example, λ phage integrates with the *E. coli* chromosome in a reversible manner through a site-specific recombination mechanism. Plasmid integration can be reminiscent of this lysogenic phase in the life cycle of the temperate bacteriophages. P1, already cited as an example of a phage that can also be a plasmid, can, on rare occasions, integrate with the *E. coli* chromosome. Thus, classification of plasmids and phage according to their relationships with the host is not always straightforward.

From the point of view of gene regulation, plasmids may afford an environment in which constraints imposed upon chromosomal genes may not apply or in which new regulatory factors may be imposed which may not be appropriate to or as efficient when exerted on chromosomal genes. In pathogens, they provide a vehicle upon which virulence genes may be grouped for coordinated control and dissemination. Genes coding for factors required for several distinct phases of pathogenicity have been found on plasmids, including adhesins, invasins, secreted products that alter host cell signalling systems, toxins, iron uptake systems, etc. As will be seen later in this book, these plasmid-linked genes are often controlled by a specific plasmid-encoded regulator and by a chromosomal regulator with much wider effects on gene expression. In addition to conventional forms of regulation by proteins that affect transcription initiation, plasmid-encoded genes are also subject to control by mechanisms that involve DNA rearrangements. These can be site-specific or driven by homologous recombination.

A significant feature of plasmids which has kept them at the forefront of microbiological research for many years is their ability to carry and disseminate genes specifying resistance to antimicrobial agents. Plasmids of this type are widespread among the bacteria and pose very serious problems in clinical practice. Frequently, the genes encoding resistance functions are integral components of tranposable elements. This makes matters all the more serious because these genes can display mobility at two levels. They can exploit any transmissibility that the plasmid possesses to move themselves from strain to strain and can use transposition to move between replicons within any given strain. The subject of plasmid-encoded resistance to antimicrobial agents will not be treated in depth in this volume. Interested readers are referred to the comprehensive review by Foster (1983).

In this chapter, the general features of plasmids will be described using well-studied examples to illustrate important priciples. This will lead to a discussion of specific plasmids which contribute to bacterial virulence.

Plasmid maintenance

All plasmids require strategies to ensure faithful segregation of newly replicated copies to both daughter cells at cell division. Naturally occurring plasmids are lost at frequencies several orders of magnitude below those predicted from theoretical rates of loss that assume random segregation (reviewed in Williams & Thomas, 1992). In general, two types of mechanism are used. One relies on random segregation while ensuring that this occurs at maximum efficiency and the other exploits replication control to increase the chance of inheritance of a plasmid copy by each daughter cell.

Strategies based on efficient segregation

Plasmid multimer resolution systems maximize the number of independently segregating plasmids present at cell division. These systems usually operate via a site-specific recombination mechanism. Plasmids replicating via theta (Θ) intermediates exploit resolution mechanisms (as does the *E. coli* chromosome) while those replicating through the rolling circle mechanism have not been found to use this mechanism (Figs 3.1 & 3.2). Specific examples of plasmids with resolution systems include the F plasmid which has a multimer resolution system in which the *D* gene product resolves multimers across the *rsfF* site (Lane *et al.*, 1986). The plasmid prophage P1 possesses the Cre recombinase and its substrate, the *loxP* site (Austin *et al.*, 1981) and the ColE1 plasmid has its *cer* site at which the XerC recombinase (encoded by the *E. coli* gene *xerC*) acts (Stirling *et*

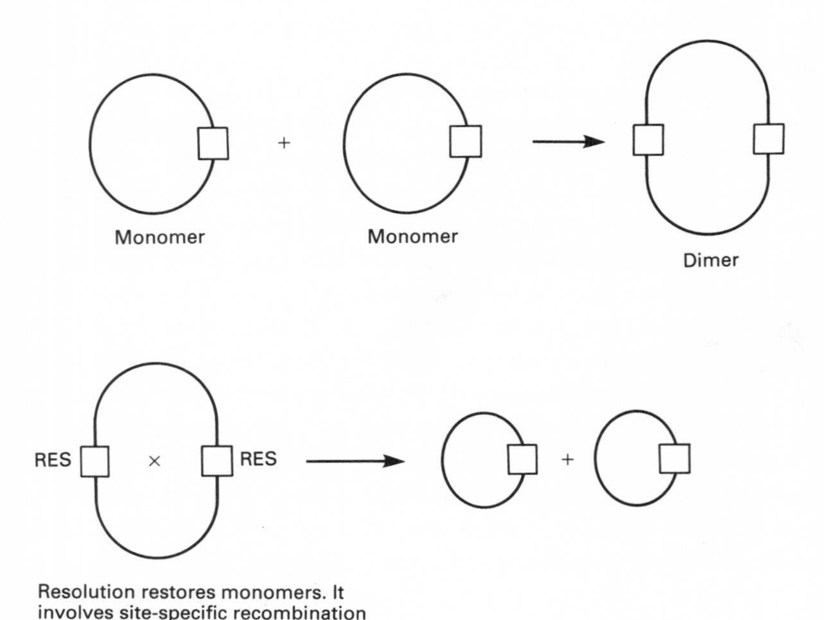

Fig. 3.1 Plasmid multimer resolution. Homologous recombination between plasmids of identical DNA sequence produces multimers. Some plasmids possess site-specific multimer resolution mechanisms in which the multimeric forms are resolved to monomeric forms.

al., 1988a). All of these site-specific recombinases belong to a group known as the integrase family (see Chapter 4).

The ColE1 XerC/*cer* system has been studied in great detail. This plasmid is a 6.6 kb element of narrow host range from *E. coli* which has a copy number of about 20 (Chan *et al.*, 1985). This system is required because multimerization of ColE1 reduces the number of segregating units at cell division leading to plasmid loss from the cell population (Summers & Sherratt, 1984). ColE1 dimers are resolved by a site-specific recombination event which involves a 280 bp *cis*-acting resolution site, *cer* (also see Chapter 4). Two directly repeated *cer* sites within a single plasmid molecule are the only ColE1 sequences needed for site-specific recombination (Summers & Sherratt, 1984). Recombination across these direct repeats separates the copies of ColE1. The recombinase is a member of the integrase family of site-specific recombinases and is encoded by the chromosomal *xerC* gene (Colloms *et al.*, 1990; Stirling *et al.*, 1988a). In addition to *xerC*, two other chromosomal genes are absolutely required for *cer* site-specific recombination. These are *argR* and *pepA* and their gene products are believed to play an accessory role in recombination at the *cer* site (Stirling *et al.*, 1988b; 1989a). In this way, ColE1 maintenance through multimer resolution is heavily dependent on host-encoded functions. As will be seen in the next section, ColE1 also possesses a copy number control system acting through DNA replication, as well as this very efficient multimer resolution system.

Plasmids like F harbour copies of transposons like gamma-delta which possess resolution systems of their own (see later). These may also contribute to plasmid multimer resolution. Plasmid pSC101 possesses a *cis*-acting *par* sequence which ensures stable inheritance of this relatively low-copy number plasmid. This sequence enhances plasmid replication and modulates plasmid copy number and superhelical density (Manen *et al.*, 1990). Plasmids deficient in *par* are unstable but can be stabilized by *topA* mutations which increase DNA supercoiling. Conversely, *par*-deficient pSC101 plasmids become further destabilized in *gyr* mutants or in strains treated with DNA gyrase-inhibiting antibiotics (Miller *et al.*, 1990a). This relationship demonstrates the importance of DNA topology to plasmid maintenance systems.

Some plasmids make sure that cells carrying copies of themselves are at an advantage by killing plasmid-free cells as these arise in the population. In general, this is achieved by a toxic product which is translated from a highly stable RNA molecule and then neutralized by a plasmid-encoded 'antidote'. If this stable RNA is inherited in the absence of the plasmid that coded for it (and the 'antidote'), the toxic product will kill that daughter cell (Fig. 3.3). Plasmid R1 encodes the *hok* mRNA whose translation to yield a toxic 52 amino acid product is prevented by the unstable antisense RNA *sok*, which acts indirectly on *hok* via the cotranscribed *mok* gene (Gerdes *et al.*, 1990; Thisted & Gerdes, 1992).

Some plasmids use active partition systems to solve the faithful segregation problem (Fig. 3.4). While the killer systems and the resolution systems reduce

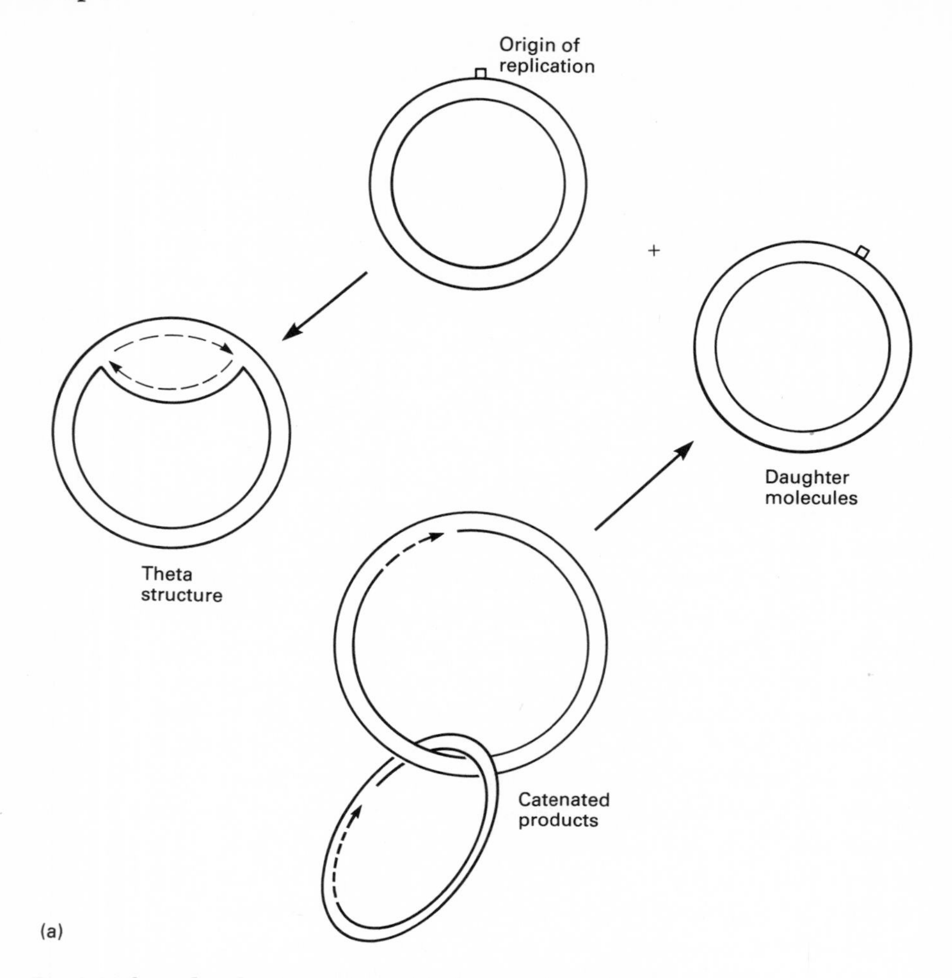

Fig. 3.2 Plasmid replication systems. (a) Theta mode replication.

the probability that a plasmid-free cell will arise, the active partition systems operate to ensure that faithful inheritance of plasmids occurs at cell division. The most extensively studied partition systems are those of the F plasmid (*sop*) and of the P1 prophage (*par*). These display a similar organization and consist of two *trans*-acting proteins A and B and a *cis*-acting centromere-like site. The B protein binds to the centromere; the A protein is a repressor of A and B gene expression and has an ATPase activity (reviewed in Williams & Thomas, 1992). Similar systems have been found on other plasmids, including those from *Agrobacterium tumefaciens* (Nishiguchi *et al.*, 1987; Tabata *et al.*, 1989). Some evidence suggests that the B proteins may link the plasmids to the cell membrane (Watanabe *et al.*, 1989). Histone-like protein IHF is required to fold the P1 centromere-like sequence into its active form (Funnel, 1988) and protein HU is necessary to establish mini-F and mini-P1 plasmids, demonstrating a role

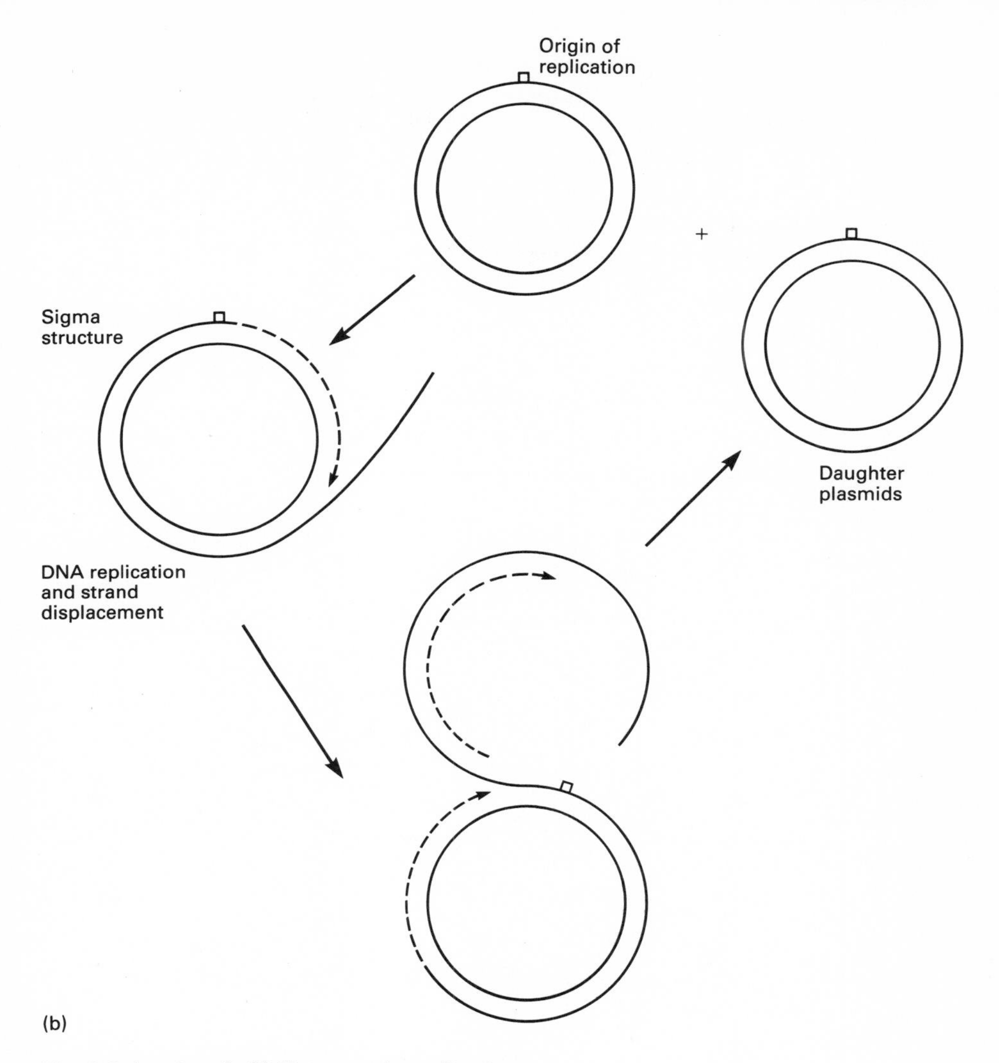

Fig. 3.2 (*continued*) (b) Sigma mode replication.

for host-encoded accessory proteins in plasmid partitioning systems. Furthermore, the efficiency of mini-F plasmid partitioning is sensitive to the degree of supercoiling of the plasmid DNA (Ogura *et al.*, 1990a, b). These data point once again to the importance of DNA topology in plasmid maintenance systems.

The means by which the plasmids are actively partitioned is believed to involve attachment to cellular structures which then segregate to either side of the cell division plane, dragging the plasmid copies towards the poles of the dividing cell. This is reminiscent of the mechanism by which chromosome partitioning is thought to be achieved. The membrane was once seen as an attractive candidate, as proposed in the original replicon hypothesis (Jacob *et al.*, 1963), although the rate of membrane elongation is now known to be too slow

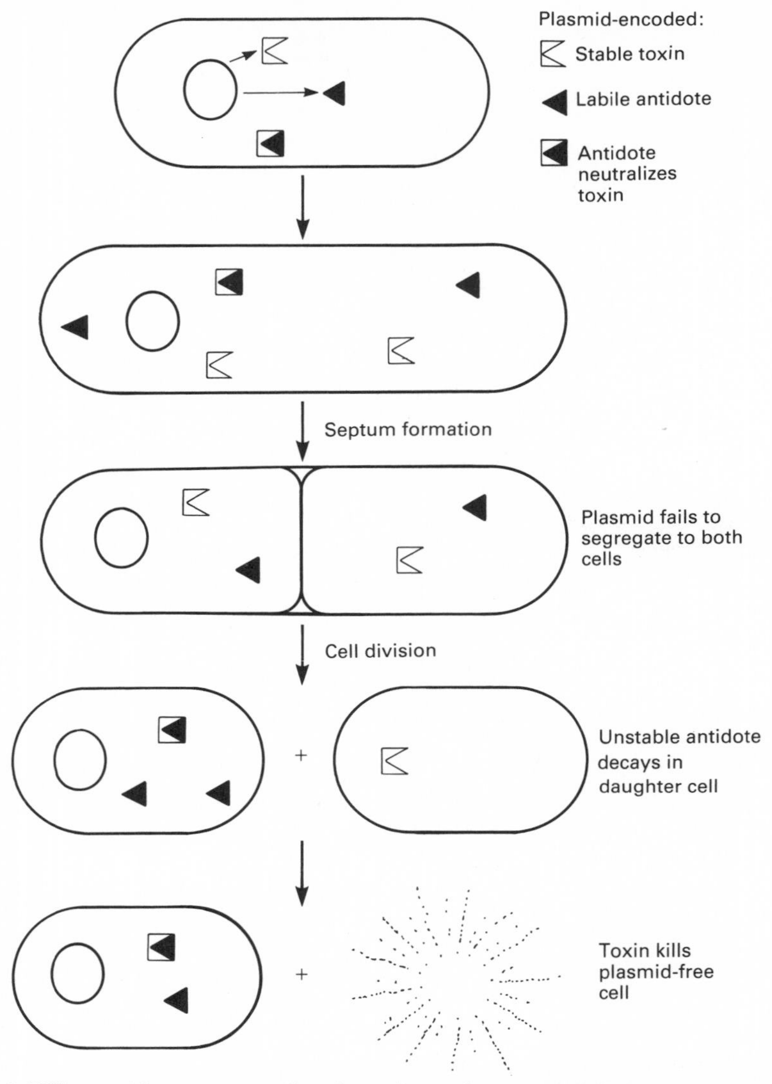

Fig. 3.3 Killer–antidote systems. The plasmid encodes a stable killer molecule and an unstable antidote molecule. If the plasmid is not segregated to both daughter cells at division, the cell without the plasmid soon loses the unstable antidote, allowing it to be killed by the stable killer function.

to provide the driving force for the rapid process of nucleoid segregation. Presumably, it is also too slow for plasmids. However, there is evidence for some cell envelope involvement in chromosome segregation. The *mukA* mutation affects chromosome partitioning in *E. coli* but not replication or cell division (Hiraga *et al.*, 1989) and *mukA* is an allele of *tolC*, a gene coding for an outer membrane protein. Unfortunately, while *mukA* mutations dramatically affects chromosome partitioning, they do not affect segregation of F (Hiraga *et al.*,

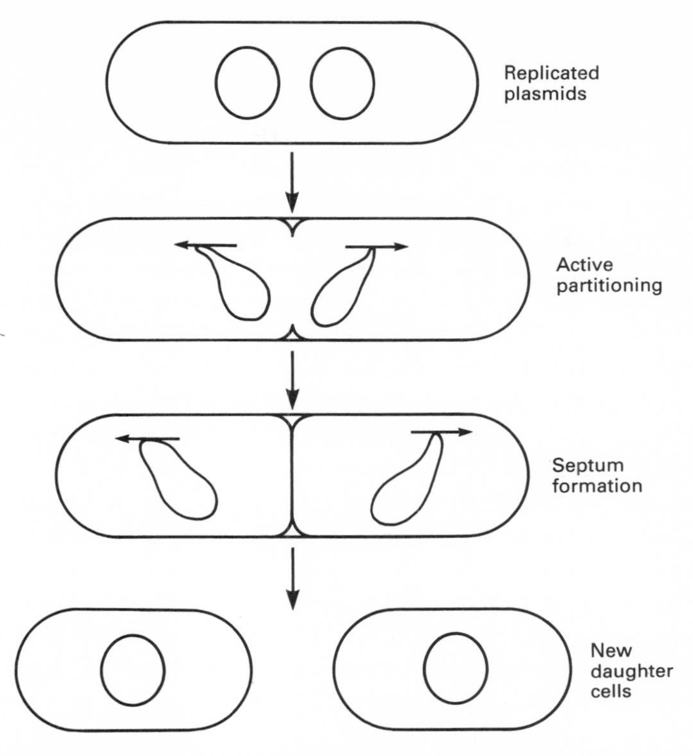

Fig. 3.4 Plasmid segregation by active partitioning. Here daughter plasmids are segregated actively before septum formation by a mechanism analogous to that responsible for chromosome segregation.

1989). Perhaps some plasmids supply their own membrane protein; for example, the SopB protein of F is believed to be membrane-associated (Watanabe *et al.*, 1989). However, it must be repeated that recent data from work on chromosome partitioning tend to rule out the membrane movement as a sensible candidate for the provision of motive force to the replicon during segregation.

Important clues concerning the molecular nature of the cellular apparatus used by segregating replicons have come from analyses of the *mukB* gene (Niki *et al.*, 1991). This encodes a product with homology to eukaryotic cytoskeletal proteins such as dynamin, a microtubule-associated mechanochemical enzyme (Obar *et al.*, 1990). MukB has a globular aminoterminal domain, a central coiled-coil domain and a potential DNA binding, 'zinc-finger'-like carboxyterminal domain (Hiraga, 1992; discussed at length in Chapter 2). Such proteins may drive nucleoid movement along filaments associated with the cell membrane. In eukaryotes, such motors move unidirectionally along polar filaments (actin or microtubules) (Vale & Goldstein, 1990). Attachment of replicated chromosomes or plasmids to pairs of antiparallel filaments would allow daughter molecules to move to opposite cellular poles (Fig. 2.19). Whether plasmids exploit this system for active partitioning is presently unknown.

Control of replication initiation

Another way of controlling plasmid maintenance is to regulate replication. Controls governing plasmid copy number are usually plasmid-inherited and concern the regulation of initiation of plasmid replication. These control mechanisms contribute to plasmid incompatibility because they frequently involve diffusible, *trans*-acting repressors of replication initiation which can suppress maintenance of plasmids with closely related replication origins. For this reason, it is difficult to establish two plasmids with closely related replication origins in the same cell; such plasmids are referred to as being 'incompatible'. This feature of plasmid biology has been used in plasmid classification. There are other factors that can lead to incompatibility, such as a need for a related segregation system. However, replication-based incompatibility is usually more stringent than that based on segregation (Kües & Stahl, 1989).

The replication systems of several plasmids has been studied in detail. Plasmid ColE1 is one of these (Fig. 3.5). Its replication is unidirectional and proceeds via a theta structure from the origin, *ori*. DNA gyrase (and presumably negative supercoiling) is required to open the DNA duplex at *ori* and may assist in the movement of the replication fork. DNA topoisomerase I is also required for replication, confirming the importance of controlling DNA supercoiling for efficient replication. RNA polymerase synthesizes the replication primer, RNA II, which is subject to processing by RNase H. RNA II forms an RNA–DNA hybrid with *ori*. Cleavage of RNA II at *ori* by RNase H permits priming of DNA synthesis by DNA polymerase I, resulting in leading strand synthesis. The initiation process is negatively regulated by an antisense RNA molecule called RNA I whose interactions with the RNA II primer are promoted by a plasmid-encoded protein called Rom (RNA one inhibition modulator) (Cesareni *et al.*, 1991) (Fig. 3.5). As with the multimer resolution system, the replication control mechanism is dependent on host-encoded functions (RNA and DNA polymerases).

ColE1 is a prototypic member of a class of plasmids that regulate their copy number through a similar mechanism. Many readers will have encountered these plasmids as DNA cloning vehicles. Other members of this class include P15A (the basis of the cloning vector pACYC184; Chang & Cohen, 1978) and pMB1 (the basis of cloning vector pBR322 and its derivatives; Bolivar *et al.*, 1977a, b; Peden, 1983; Sutcliffe, 1979) (reviewed in Kües & Stahl, 1989).

The 9.26 kb plasmid pSC101 exists in *E. coli* at about five copies per chromosome equivalent (Hasunuma & Sekiguchi, 1979). Like ColE1, it has a narrow host range, but it differs from ColE1 in several important respects in terms of its mode of replication initiation and shares several features with other non-ColE1 plasmids from all incompatibility groups. These common features include the possession of a plasmid-encoded Rep protein (an essential polypeptide for plasmid replication initiation), binding sites for the host-encoded DnaA protein, regions of A–T rich DNA and clusters of directly repeated sequences.

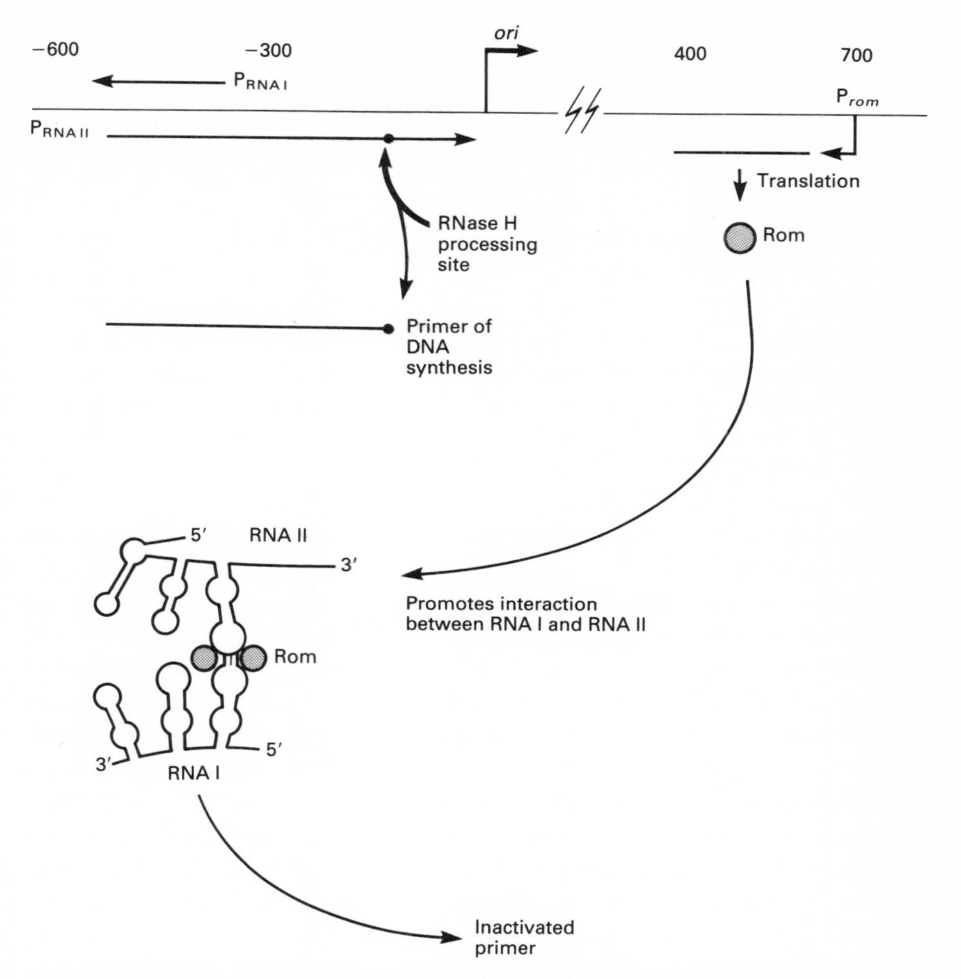

Fig. 3.5 Regulation of plasmid ColE1 replication. The minimal replication origin of ColE1 is shown. RNase H processed RNA II becomes a primer for DNA synthesis. This function is inactivated when RNA I interacts with RNA II, an event that is assisted by the Rom protein.

Rep-encoding plasmids like pSC101 do not require DNA polymerase I to prime their replication cycles, unlike the ColE1 group. This appears to give them a degree of independence from host-encoded functions. However, as will be seen later, this plasmid requires many replication functions (such as DnaA) which are also used by the host chromosome.

pSC101 replication is unidirectional and proceeds through a theta structure. The replication initiation site, *ori*, consists of a DnaA-binding site, an A–T rich DNA domain with an adjacent binding site for the host-encoded IHF protein, three direct repeats of a sequence to which the RepA protein binds and the gene *repA*, which codes for RepA and whose promoter region contains further RepA-

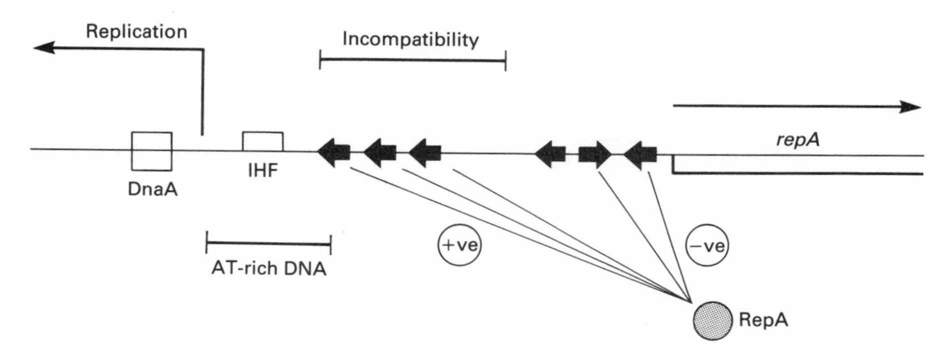

Fig. 3.6 The key regions of the replication origin of pSC101 are illustrated. These include the *repA* gene whose product binds to repeated sequences (arrows) within the origin, exerting negative effects on its own transcription and positive effects on DNA replication. Binding sites for DnaA and IHF are indicated.

binding sites, two of which are arranged as inverted repeats (Fig. 3.6) (reviewed in Kües & Stahl, 1989).

RepA positively controls plasmid replication by binding to its cognate sites within the origin where it is thought to initiate the formation of a replisome (by analogy with events at *oriC*, the origin of *E. coli* chromosome replication). Although DnaA is known to be essential for pSC101 replication, its role is not fully understood. IHF may serve to promote good contacts between pSC101-bound RepA and DnaA by bending the intervening DNA. In this way, a replisome could be formed composed of RepA, DnaA and the DnaB–DnaC complex. Two 13 bp repeats in the A–T rich DNA tract adjacent to the DnaA-binding site may be opened by the DnaB helicase prior to primer formation. Priming may then occur by the DnaG primer, an essential protein in pSC101 replication. An increase in the cellular concentration of RepA can negatively regulate pSC101 replication; RepA represses *repA* transcription (Fig. 3.6). The RepA protein can also determine incompatibility between pSC101 and related plasmids by *trans*-acting repression of *repA* transcription on the other plasmid.

Plasmid RK2 is an example of a high-molecular-weight plasmid whose maintenance strategies have been studied in some detail. It is a 60 kb IncP plasmid with a copy number of 4–7 per chromosome equivalent and it uses a theta mode of unidirectional replication. Its replication origin, *oriV*, is organized in a manner which is similar to that of pSC101, but several outlying genetic regions contribute to the control of events there (Kües & Stahl, 1989; Thomas & Smith, 1987) (Fig. 3.7). *oriV* includes two genes coding for *trans*-acting factors concerned with copy number control and incompatibility, *copA/incA* and *copB/incB*. The replication initiation proteins, A1 and A2 are encoded by genes outside *oriV*. These may interact with host-encoded proteins at *oriV* to form a replisome. RK2 has a very broad host range, a common feature of IncP group plasmids. When replicating in *E. coli*, DnaA, DnaG, DNA gyrase and DNA

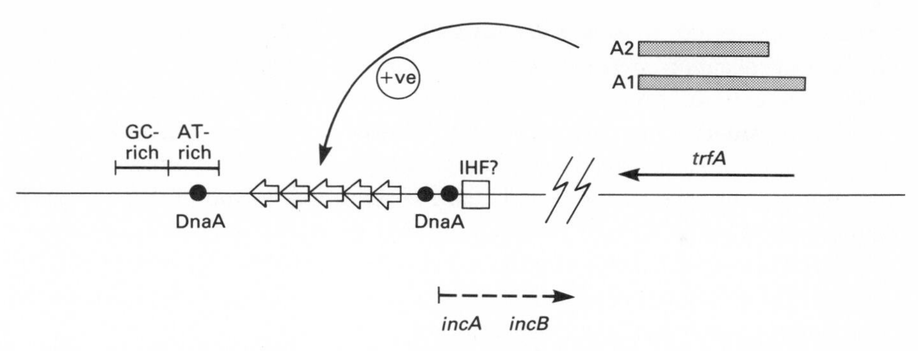

Fig. 3.7 The major components of the RK2 origin of replication are shown. Gene *trfA* encodes the A1 and A2 proteins which serve as replication functions at the directly repeated sequences (arrows) within the origin. The locations of DnaA binding sites and a possible IHF binding site are shown.

polymerase III are all required; IHF may also be needed. The broad host range of RK2 is a function of its multicomponent replication system which includes the *kil–kor* genes. Kil functions are host lethal or are inhibitory to plasmid maintenance when they are not controlled by the Kor (Kil override) functions. They represent a plasmid maintenance system analogous to the *hok/sok* system of plasmid R1, already discussed.

This short section on plasmid maintenance illustrates once again the principle that organisms seem to evolve a small number of ways of carrying out a particular function and then develop many variations on the basic plan. This theme was first encountered in the discussion of basic virulence mechanisms in Chapter 1 and will be seen again and again throughout this book.

Plasmid replication as a determinant of host range

An ability to replicate in the host cell is a prerequisite for plasmid maintenance in any given bacterial species. This ability is determined by *trans*-acting host functions and *cis*- and *trans*-acting plasmid-encoded factors. Replication of some plasmids (e.g. the IncP group) is supported in almost all bacteria while others are restricted to a very few organisms. While ColE1 cannot replicate in cell-free *Pseudomonas* extracts, partial replication is observed when these are supplemented with purified *E. coli* DNA gyrase and DNA polymerase I (Kües & Stahl, 1989). It is possible that pSC101 cannot replicate in *Pseudomonas* because the DnaB analogue from that species cannot perform the function of the *E. coli* protein in pSC101 replication. There is a paucity of firm molecular detail concerning determinants of host range which act at the plasmid replication origin.

The non-conjugative, mobilizable IncQ group plasmid RSF1010 has an extremely broad host range (Morales *et al.*, 1990; Scholz *et al.*, 1989); the only Gram-negative bacterial genus in which it has been reported as failing to become established is *Bacteroides* (Guiney *et al.*, 1985; Smith *et al.*, 1985). RSF1010 replication depends on three plasmid genes: *repC* coding for a replication initiator protein which binds to the *cis*-acting *oriV* site for DNA replication initiation; *repA* which codes for a DNA helicase analogous to the DnaB protein of *Escherichia coli*; and *repB* which codes for a primase which is analogous in function to the *E. coli* DnaG protein. Expression of these genes is tightly controlled by further plasmid-encoded functions (Frey & Bagdasarian, 1989; Haring *et al.*, 1985; Haring & Scherzinger, 1989). Thus, RSF1010 is independent of host-encoded functions (Scholz *et al.*, 1984) and this is believed to confer upon it its broad host range (Morales *et al.*, 1990).

Conjugative plasmids

Bacterial plasmids with the capacity to transmit themselves from one bacterial cell to another typically require three genetic functions to carry out this process. These are a transfer locus, *tra*, a mobilization locus, *mob*, and a sequence required *in cis* for transfer to proceed, called *oriT*. DNA transfer depends upon cell-to-cell contact and this is provided by plasmid-encoded surface structures called pili (singular: pilus). The first plasmid to be discovered was the F factor (Fertility factor) of *E. coli*, a 100 kb covalently closed circular molecule which harbours four transposable elements (IS2, IS*3a*, IS*3b* and gamma-delta) and codes for no selectable genetic markers (Willetts & Skurray, 1987) (Fig. 3.8). It was detected due to its ability to transfer chromosomal markers from cell to cell (Cavalli-Sforza *et al.*, 1953; Hayes, 1953). About a third of the F molecule, 33 kb, is devoted to the *tra* region which codes for the functions required for cell-to-cell transfer. Cells carrying a copy of F typically deploy three F pili on their surfaces. These are up to 3 μm long and composed of helically stacked pilin protein subunits giving a 8 nm diameter pilus with a 2 nm pore down through the centre (Folkhard *et al.*, 1979). The recognition site for this pilus on the recipient cell includes the OmpA outer membrane protein. The cells come into close apposition and finally dock. How this is achieved is not fully understood, but the pilus must undergo some structural change (probably involving shortening) if it is not to present a barrier to successful mating. ssDNA is then passed from the donor to the recipient cell, although the precise route is not known (Fig. 3.9). In the case of the self-transmissible plasmid RP4, a plasmid DNA primase protein is transferred along with the DNA (Rees & Wilkins, 1990).

Self-transmissible plasmids are particularly significant in terms of their ability to move DNA horizontally between strains of bacteria of the same species, between different species and even between organisms from different kingdoms. The insertion sequences found within F enable the plasmid to undergo homologous recombination with copies of themselves in the bacterial chromosome, with the resulting cointegrate structure having the self-transmissible characteristics of the autonomous F. This is the basis of 'Hfr' strains, in which chromosome transfer between cells is mediated via F conjugation, with the

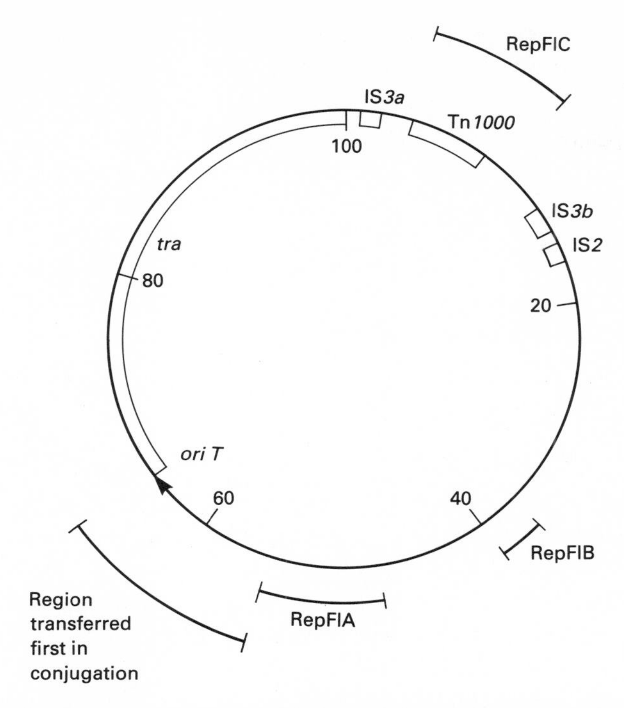

Fig. 3.8 Physical and genetic map of the F plasmid. The circular map of F is shown, indicating the locations of major genetic functions such as those concerned with replication, conjugal transfer of DNA (*tra* operon and the origin of transfer, *oriT*), and the position of transposable elements (transposon Tn*1000* and insertion sequences).

incoming DNA undergoing a High Frequency of Recombination with homologous sequences among the resident DNA. Excision of F from the chromosome can proceed via an imprecise event which generates a form of F that harbours flanking sequences of chromosomal DNA, some of which can be very extensive. These recombinant F elements, called F′, can also transfer bacterial DNA between cells and create merodiploid (partially diploid) strains in the process (Holloway & Low, 1987; Low, 1987).

The potential contribution of such processes to the evolution of the bacterial genome is enormous, although the relative stability of the genomes of bacteria such as *E. coli* and *S. typhimurium* suggests that this potential is held in check. Part of the barrier may be provided by the mismatch–repair apparatus (Rayssiguier *et al.*, 1989) and by restriction systems. However, some plasmids appear to have acquired the ability to defeat bacterial restriction barriers. For example, the IncN plasmid, pMK101, the IncI1 plasmid, ColIb-P9 and other IncN and IncI plasmids have been found to encode an antirestriction protein called Ard (alleviation of restriction of DNA). Similar antirestriction systems are used by bacteriophage, such as that encoded by the *ral* gene of phage λ (Zabeau *et al.*, 1980). In the case of ColIb-P9, the *ard* gene lies close to the origin of transfer and

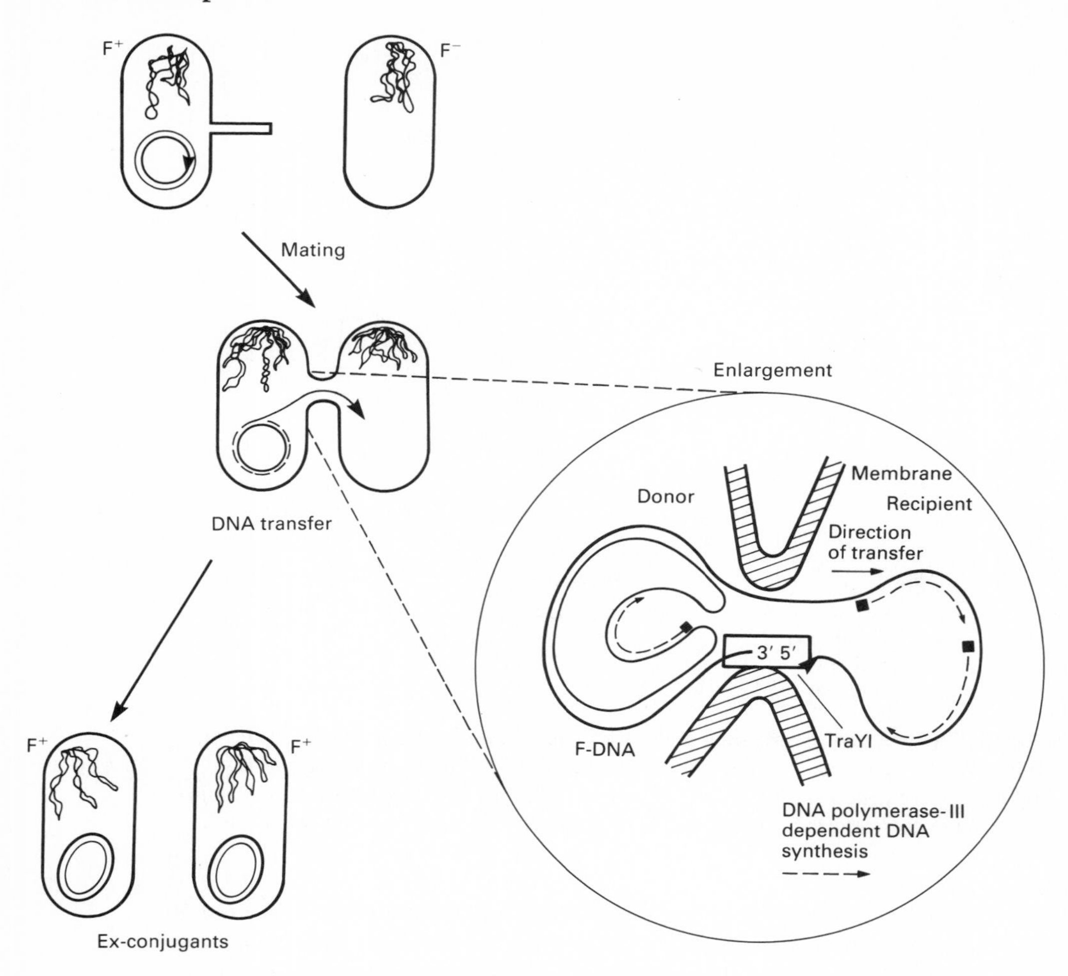

Fig. 3.9 The key events in F-mediated conjugation are shown. These include mating of an F^+ and an F^- cell, DNA transfer with concomitant replication of the displaced DNA strand and the separation of the exconjugants at the end of transfer. The conjugation bridge is shown in close up to illustrate the relative positions of the participating proteins and the pathway of the transferred DNA.

is part of the earliest region of plasmid DNA to be passed to the recipient in conjugation (Delver *et al.*, 1991).

Conjugation between Gram-positive bacteria is known to occur and will be discussed later (see section on conjugative transposition). Conjugation between Gram-positive and Gram-negative bacteria has also been reported (Mazodier *et al.*, 1989; Trieu-Cuot *et al.*, 1987, 1988). Recently, the ability of bacterial plasmids to transmit themselves to eukaryotes such as the yeasts *Saccharomyces cerevisiae* and *Schizosaccharomyces pombe* has been demonstrated (Amábile-Cuevas & Chicurel, 1992; Heinemann & Sprague, 1989; Heinemann, 1991; Sikorski *et al.*, 1990). This finding widens their scope for horizontal transfer of genetic information to species greatly separated from the bacteria in evolutionary terms. It has even been suggested that transfer between bacteria and animal cells may be possible (Heinemann, 1991). Specialized examples of interspecies plasmid DNA transfer form the basis of the interactions of *Agrobacterium tumefaciens* and *Rhizobium* spp. with their plant hosts (see later). Thus, conjugation may assist the spread of virulence determinants from pathogen to pathogen and provides the possibility that such factors may be placed in commensal organisms where, even if they do not create a novel pathogen, they may be stored before being passed to a dedicated pathogen.

The spread of genes coding for antibiotic resistance by self-transmissible plasmids continues to be a major cause for concern. These genes are frequently associated with transposable elements and can occur in different combinations on plasmids, affording to the host the ability to resist multiple antibiotics and other antimicrobial agents, such as heavy metals (Foster, 1983). The presence of these drug resistance determinants on plasmids of broad host range (such as members of the P incompatibility group) (Thomas & Smith, 1987) is particularly disquieting since they may be harboured by non-pathogenic bacteria living as commensals in the gut of healthy humans or animals and then be passed to any pathogens which subsequently infect the host (Wright, 1990).

Mobilization of non-conjugative plasmids

Plasmids like F, RK2 and R100 possess the means to transfer themselves from cell to cell via conjugation. Non-conjugative plasmids are usually much smaller than the self-transmissible types and lack the genetic machinery needed to code for cell-to-cell transfer. Nevertheless, some non-conjugative plasmids are equipped to exploit the conjugative apparatus of self-transmissible elements and are capable of being mobilized. For example, plasmid ColE1 possesses a *cis*-acting site called *nic* within the *bom* region at which a single-stranded nick is made by proteins encoded by the ColE1 *mob* genes (Finnegan & Sherratt, 1982) (Fig. 3.10). ColE1 can be mobilized by F and when this element has initiated mating, the nicked ColE1 strand forms a complex with a *mob*-encoded protein which leads the DNA into the recipient cell (Willetts & Wilkins, 1984).

Plasmids and bacterial virulence

In this section, plasmids with specific roles in bacterial virulence will be discussed. In many cases, the plasmids will be seen to harbour complex operons and regulons of virulence genes which are subject to sophisticated regulatory mechanisms. These genes and their regulation will be described in later chapters. Here, the nature of the plasmids themselves is discussed.

Plasmids in *Agrobacterium* species

Agrobacteria have been divided into three groups on the basis of their interactions with plant hosts. *Agrobacterium tumefaciens* produces crown gall tumours

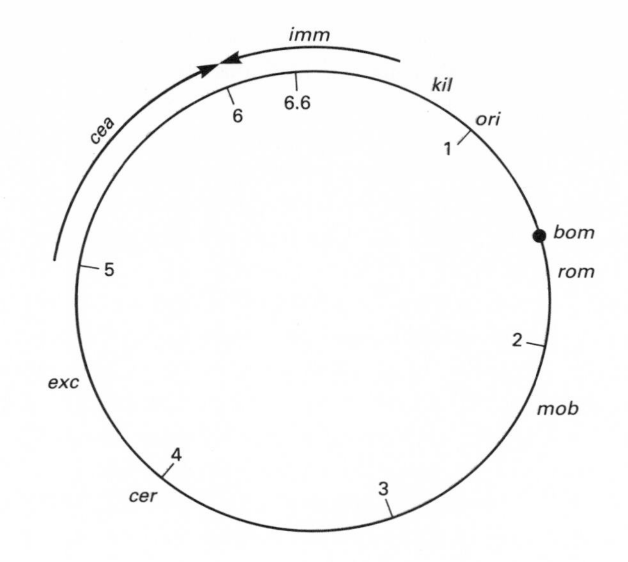

Gene	Function
bom	*cis*-acting mobilization site
cea	Colicin E1
cer	Cointegrate resolution site
exc	Incompatibility
imm	Colicin E1 immunity
kil	Kil function—colicin release
mob	Mobilisation. Acts at *nic*
nic	Nick site, cut by Mob protein
ori	Replication origin
rom	Copy number control

Fig. 3.10 Physical and genetic map of ColE1. A circular map of the 6.6 kb ColE1 plasmid is shown. The positions of the key genetic functions are shown on the map and their functions described above.

on plant stems and leaves, *A. rhizogenes* causes hairy root and *A. radiobacter* is avirulent (Cangelosi *et al.*, 1991; Hooykaas, 1989; Winans, 1992; Zambryski, 1988). Virulent Agrobacteria possess a single-copy 200 kb plasmid which encodes functions essential for infectivity. The plasmid in *A. tumefaciens* is called the Ti (tumour inducing) plasmid while that in *A. rhizogenes* is called the Ri (root inducing) plasmid. Not all the virulence genes are carried by the plasmids — chromosomal genes are also important. For example, chromosomal genes *chvA* and *chvB* are important for attachment of the bacterium to the plant (Douglas *et al.*, 1985).

A significant feature of agrobacterial infection is the transfer of bacterial DNA sequences to the plant host (Fig. 3.11). These sequences are carried on the Ti and Ri plasmids and are called T-DNA (tumour DNA) (Bevan & Chilton, 1982). The oncogenicity (*onc*) genes found in the T-DNA encode enzymes

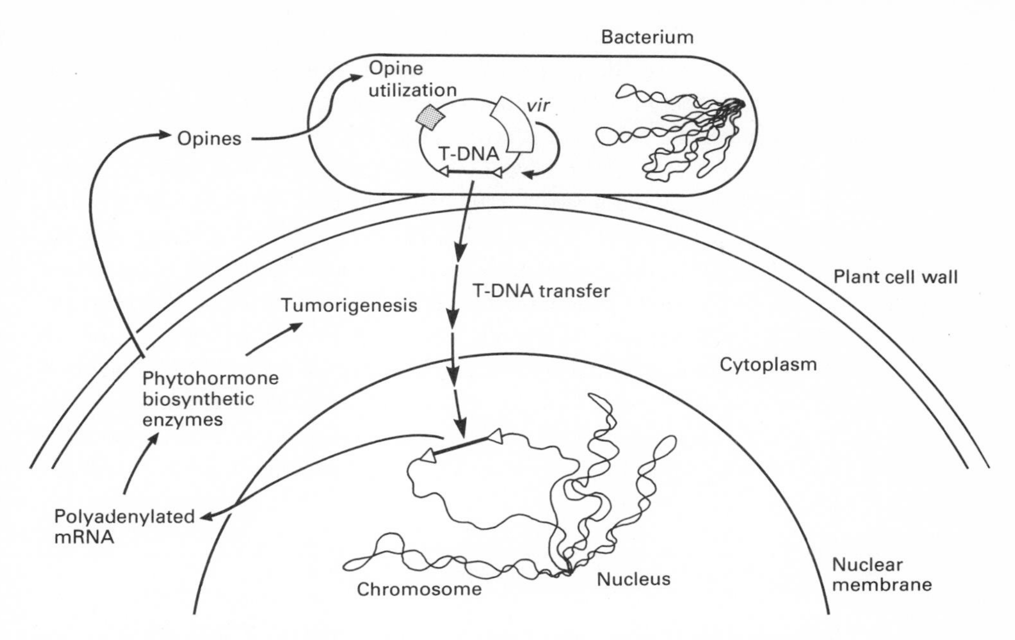

Fig. 3.11 Transfer of agrobacterial T-DNA to the plant nucleus. The interaction between *Agrobacterium* and its plant host is illustrated. Products of the *vir* region are required to activate T-DNA transfer from the bacterium to the plant nucleus. Expression of the prokaryotic DNA in the plant results in tumorigenesis and the secretion of opines which the bacterium uses as nutrients.

required for the biosynthesis of plant hormones such as the auxin, indole acetic acid and isopentenyl-AMP, a cytokinin. Production of these compounds within plant cells leads to unlimited proliferation or tumorigenesis. In addition to the *onc* genes, the T-DNA also carries genes for opine production. Expression of the bacterial genes in the plant is possible because these possess eukaryotic transcription initiation signals (TATA boxes, CAAT boxes and enhancer-like sequences at the 5′ end together with GT-rich transcription termination sites and AATAAA polyadenylation signals in the 3′ sequences) (Barker *et al.*, 1983; Memelink *et al.*, 1987).

Transmission of the T-DNA from the prokaryote to the plant is achieved via a mechanism that has some features in common with plasmid-mediated conjugation, an observation which may point to its evolutionary origin. No T-DNA sequences are involved in the control of DNA transmission (Leemans *et al.*, 1982). Instead, the genes required for DNA transfer are located in a distinct 40 kb region of the Ti plasmid called Vir. Vir contains a large number of virulence genes arranged in seven operons, some of which code for regulatory proteins (viz. *virA* and *virG*, see Chapter 7). The structural genes are not expressed during vegetative growth but are induced by plant wound exudates which include phenolic compounds such as acetosyringone and α-hydroxy-acetosyringone. T-DNA transfer requires flanking directly repeated sequences of

24 bp (Barker *et al.*, 1983). These are termed the border repeats and although they are interchangeable, they are differentiated from one another by their orientation with respect to the T region. Transfer is an orientated process, with the right border repeat being transferred first and the left border repeat last (Wang *et al.*, 1984). The right border repeat is close to a sequence called 'overdrive' that strongly enhances T-DNA transfer; mutants deficient in 'overdrive' are attenuated in plant virulence (Peralta *et al.*, 1986; Van Haaren *et al.*, 1987). T-DNA to be transferred is in a single-stranded form and is the lower strand of the T-DNA region (Stachel *et al.*, 1986). The VirD1 and VirD2 proteins, encoded by the *virD* operon in the Vir region, are a topoisomerase and an endonuclease respectively (Ghai & Das, 1989; Yanofsky *et al.*, 1986). Together, these proteins cut the T-strand DNA at a specific site at each border repeat (Fig. 3.12) (Wang *et al.*, 1987). 5′–3′ DNA synthesis displaces the T-strand DNA and this is then transferred to the plant. The VirD2 protein attaches covalently

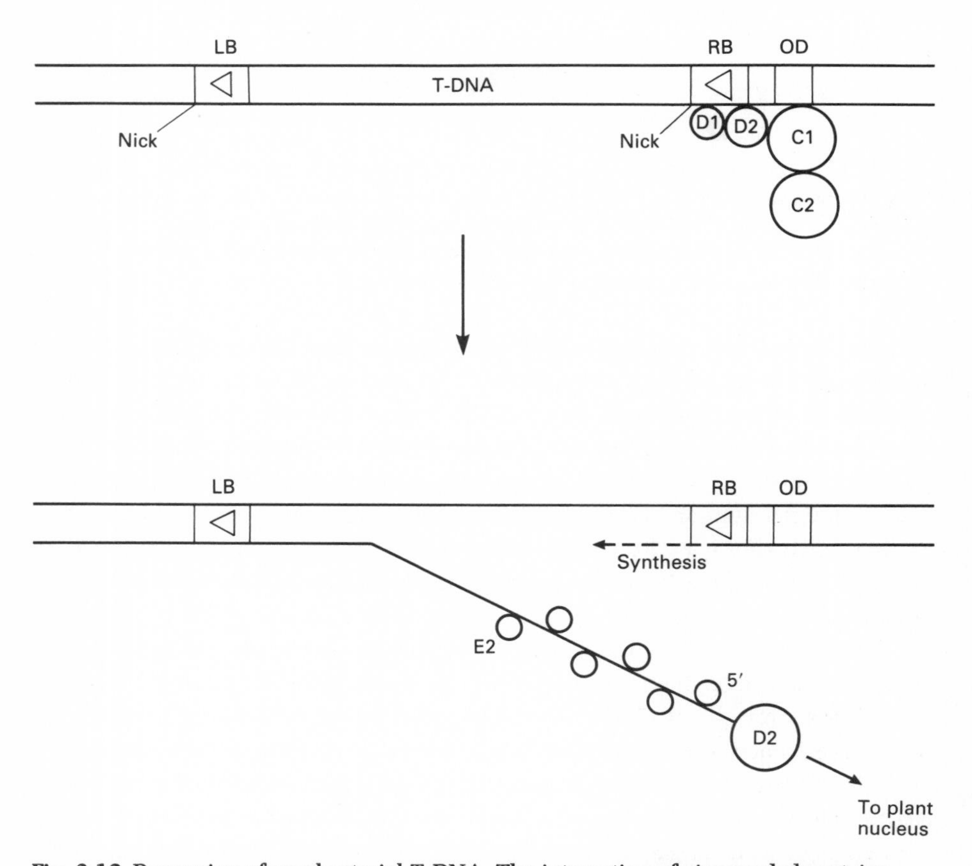

Fig. 3.12 Processing of agrobacterial T-DNA. The interaction of *vir*-encoded proteins VirD1, VirD2, VirC1, VirC2 and VirE2 with T-DNA is illustrated. The DNA is nicked at the left (LB) and right (RB) border repeats. VirD2 pilots the T-strand to the plant nucleus and VirE2 coats the ssDNA to protect it from degradation. The displaced strand is replaced by DNA synthesis. (OD, 'overdrive' enhancer sequence.)

to the 5′ end of the border sequences and may pilot the T-DNA into the plant cell (Howard *et al.*, 1992). The VirC1 and VirC2 proteins, encoded by the *virC* operon of the nopaline Ti plasmids, are required to transfer T-DNA to maize (Grimsley *et al.*, 1989). The VirC1 protein binds to the 'overdrive' sequence which is close to the 24 bp right border repeat (Toro *et al.*, 1989). VirE is a single-stranded DNA-binding protein which is thought to protect the T-DNA strand from degradation during its journey to the plant cell nucleus (Winans, 1992). Presumably, it coats the T-DNA strand, protecting it from nucleases. In addition to control by the Ti plasmid-encoded VirA/VirG system, the expression of the *virC* and *virD* operons is also regulated negatively by the chromosomal *ros* gene (Cooley *et al.*, 1991).

The relationship of this process to conjugal DNA transfer is strengthened by the observation that the 24 bp right border repeat can be replaced functionally by the *oriT* site of an IncQ plasmid (Buchanan-Wollaston *et al.*, 1987). The T-DNA appears to be inserted into the plant genome at random. Expression of the T-DNA genes in the plant is strongly context-dependent, however. This means that the genes do not work equally well at all sites. Furthermore, T-DNA gene expression can be silenced by DNA methylation (Hepburn *et al.*, 1983). This means that simply getting the T-DNA established in the plant genome does not guarantee that the genes will be expressed.

Although most work has dealt with the prokaryote–eukaryote transfer of T-DNA, it should be noted that Ti plasmid DNA can also be transferred between bacterial cells through a conjugal process that appears to be enhanced by a diffusible conjugation factor (CF) (Zhang & Kerr, 1991). Recently, this compound has been identified as an autoinducer molecule of the homoserine lactone family (Zhang *et al.*, 1993). Furthermore, TraR, the transcription activator required for the activation of *tra* gene expression, is a homologue of LuxR, the protein from the marine bacterium *Vibrio fischeri* which activates the bioluminescent *lux* operon in response to accumulation of homoserine lactone (Piper *et al.*, 1993). Regulation by homoserine lactones is discussed in Chapter 7. Agrobacterial Ti plasmid-mediated conjugation also requires the Vir region, suggesting a close relationship between the prokaryote–prokaryote and the prokaryote–eukaryote transfer mechanisms (Gelvin & Habeck, 1990; Steck & Kado, 1990).

The ColV plasmids of *Escherichia coli*

ColV plasmids are members of the IncFI group and code for the production of colicin V, a low-molecular-weight polypeptide which disrupts the membrane potential of the cytoplasmic membrane of target cells (Yang & Konisky, 1984). ColV plasmids also encode a low-molecular-weight immunity protein which protects producer cells from the effects of the colicin. This colicin is unusual in not being SOS-inducible and in having its own plasmid-encoded export system (Gilson *et al.*, 1987). Expression of colicin V is induced by iron limitation through the Fur iron-regulatory protein and mutants deficient in *cir*, *exbB* or *tonB* (genes required for iron-chelator uptake) are colicin V resistant (Chehade & Braun, 1988). Thus colicin V entry to cells and regulation of colicin V gene expression are intimately connected with bacterial iron metabolism. Why this should be is not clear but colicin V may assist the pathogen by enabling it to compete successfully with other bacteria while colonizing the mammalian gut (Smith & Huggins, 1978). Transposon mutagenesis has shown that bacteria harbouring mutant ColV plasmids unable to code for colicin V are just as pathogenic for mice as those able to specify the colicin (Quackenbush & Falkow, 1979). Thus, what contribution colicin V makes to virulence is uncertain.

The ColV plasmids display considerable diversity, range in size from 80 to 180 kb and are found predominantly among virulent enteric bacteria (Fig. 3.13). In addition to colicin V, these plasmids encode several virulence factors including the aerobactin iron uptake system, properties associated with serum resistance, and factors for resistance to phagocytosis and for intestinal cell adhesion (reviewed in Waters & Crosa, 1991; Williams & Roberts, 1989).

The aerobactin iron uptake system encoded by ColV plasmids is a major virulence determinant (Williams, 1979). This system is not exclusively plasmid-associated; aerobactin-encoding genes have been detected on the chromosomes of enteroinvasive *E. coli*, *E. coli* K1 and *Shigella flexneri* (Lawlor & Payne, 1984; Marolda *et al.*, 1987; Valvano & Crosa, 1984). Furthermore, aerobactin is encoded by a genetically distinct system in *Klebsiella pneumoniae* K1 and K2 strains and by yet another system in *Enterobacter cloacae* (Crosa *et al.*, 1988; Nassif and Sansonetti, 1986; Waters & Crosa, 1988). Aerobactin is superior to enterobactin (produced by most *E. coli* strains) in competition with serum transferrin for nutritional iron, making it a valuable virulence factor (Ford *et al.*, 1988).

On ColV plasmids, the aerobactin system genes are flanked by inverted copies of insertion sequence IS*1* and the genes lie between two plasmid replication regions called Rep I and Rep II (Perez-Casal & Crosa, 1984). Placing the aerobactin genes in such an essential region may help to maintain them on the ColV plasmids. Similarly the IS sequences may assist in the dissemination of the genes to the bacterial chromosome. However, IS*1*-mediated transposition of the aerobactin system genes is not believed to occur; instead the IS sequences are thought simply to participate in RecA-dependent recombination (Waters & Crosa, 1991).

A specific ColV locus, *iss* (increased serum survival) has been discovered to be required for serum resistance (Binns *et al.*, 1979). This sequence is homologous to *bor*, a bacteriophage λ gene which encodes a membrane protein contributing to serum resistance in lysogenic strains (Barondess & Beckwith, 1990). The transfer genes of the ColV plasmids have been implicated in resistance to serum killing. One of these codes for the TraT protein. TraT proteins are highly homologous membrane lipoproteins encoded by plasmids of incompatibility groups IncFI, IncFII and IncFIV. Together with TraS, the TraT protein mediates surface exclusion (Willetts & Skurray, 1987). This means that a second cell

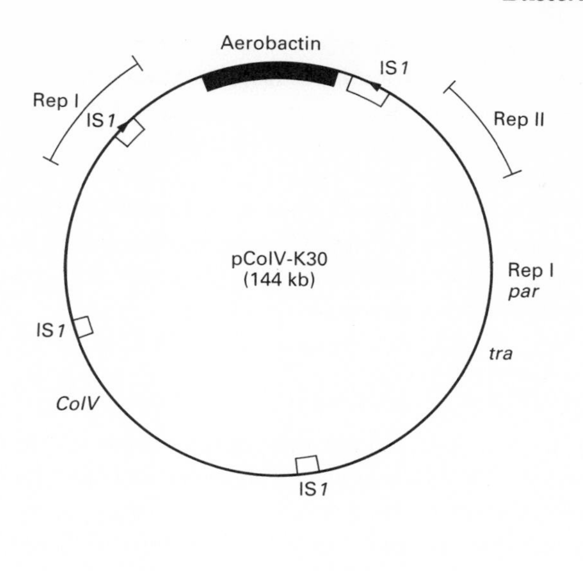

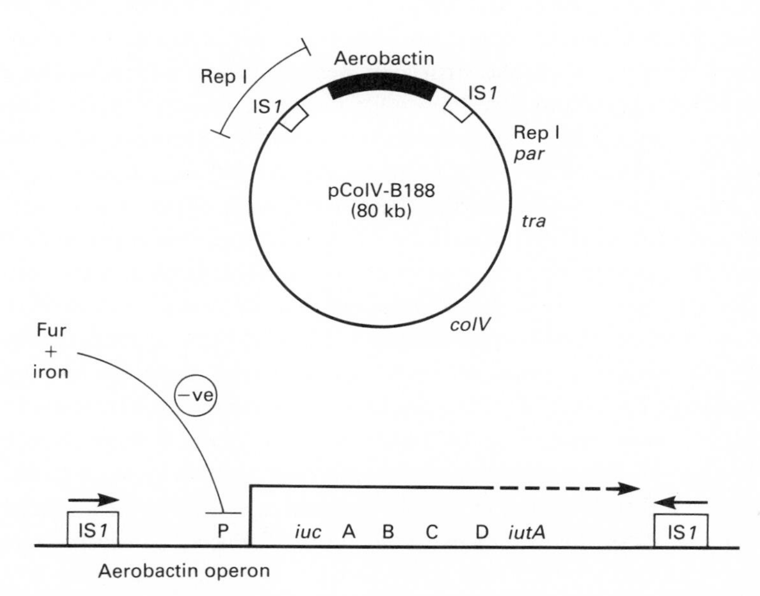

Fig. 3.13 Physical and genetic maps of ColV plasmids. Circular maps of the pColV-K30 (144 kb) and the pColV-B188 (80 kb) plasmids are shown. The positions of key genetic functions such as replication regions, aerobactin genes, transfer (*tra*) and partition (*par*) regions, colicin V genes are illustrated. The aerobactin genes are flanked by copies of the insertion sequence IS*1*. Aerobactin gene expression is regulated negatively by iron via the Fur protein (shown at the base of the figure).

carrying the same plasmid is unable to perform conjugative DNA transfer with the cell expressing the surface exclusion factors TraS and TraT. The *iss* gene product is also involved in surface exclusion (Chuba *et al.*, 1989). Plasmid-encoded TraT lipoprotein from *Salmonella typhimurium* increases serum resistance in both *S. typhimurium* and *E. coli* (Rhen & Sukupolvi, 1988), and may also contribute to antiphagocytic activity (Aguero *et al.*, 1984). Thus, these plasmid-encoded surface proteins perform tasks concerned with plasmid biology and with host cell virulence.

A plasmid-encoded iron uptake system analogous to aerobactin is found in the fish pathogen *Vibrio anguillarum*. This system is carried by plasmid pJM1 and encodes an iron siderophore called anguibactin (Crosa, 1989). Iron siderophores as virulence factors were discussed in Chapter 1.

The EAF plasmid of enteropathogenic *Escherichia coli*

Enteropathogenic *E. coli* (EPEC) are an important cause of childhood diarrhoea in impoverished communities in less developed countries. The bacteria invade epithelia of the lining of the small bowel. An early phase in invasion is attachment to the mucous membrane and the formation of microcolonies. These microcolonies consist of bacterial cells bound together by bacterially encoded rope-like bundles. The colonies are described as participating in localized adhesion (LA). These rope-like structures are composed of bundle-forming pili (Bfp) and their expression correlates with the presence in EPEC of a large (60 MDa) plasmid called EAF (for enteroadherent factor) (Giron *et al.*, 1991). Bfp from different EPEC isolates are antigenically related and show a structural relationship to the so-called toxin coregulated pilus (Tcp) of *Vibrio cholerae*. Recently, Sohel *et al.* (1993) cloned the *bfp* gene from the EAF plasmid of EPEC and determined the nucleotide sequence of the gene. Using the cloned gene as a probe, these investigators were able to show that the presence of *bfp* correlated with the presence of the EAF plasmid in all strains of EPEC capable of displaying the LA phenotype. In contrast, the gene was not detected in strains of enterotoxigenic *E. coli* (ETEC), enterohaemorrhagic *E. coli* (EHEC), uropathogenic *E. coli* (UPEC) or neonatal meningitis-causing *E. coli* (NMEC). Some hybridization was detected at low stringency with enteroaggregative (EAEC) strains and enteroinvasive (EIEC) strains of *E. coli*. Interestingly, the *bfp* probe detected a homologue in 13 strains of *Salmonella*, including *S. dublin*, *S. choleraesuis* and *S. typhimurium*. This is particularly significant given the data of Finlay & Falkow (1989, 1990) showing an LA-like phenotype for *S. choleraesuis* and *S. typhimurium* when growing on the surface of tissue culture cells. At the time of writing, it is not known if the *Salmonella* homologue is on the large virulence plasmid (see next section).

Salmonella virulence plasmids

While there is no plasmid involvement in human typhoid as caused by *Salmonella typhi*, low-copy-number, high-molecular-weight (80–100 kb) plasmids are important to the pathogenicity of many other serotypes of this genus

(Fig. 3.14). These include the mouse-virulent serovars *S. choleraesuis, S. dublin, S. enteritidis* and *S. typhimurium* (Baird *et al.*, 1985; Gulig & Curtiss, 1987, 1988; Jones *et al.*, 1982; Kawahara *et al.*, 1988; Lax *et al.*, 1990; Nakamura *et al.*, 1985; Williamson *et al.*, 1988) and the fowl-virulent serovars *S. gallinarum* and *S. pullorum* (Barrow *et al.*, 1987; Barrow & Lovell, 1988). Curing of the plasmids renders the bacteria less pathogenic while reintroduction of the plasmids restores full virulence. The plasmids appear to be required for successful infection of the reticuloendothelial system and seem to increase the growth rate of *Salmonella* in mice, particularly in intracellular niches (Gulig & Doyle, 1993).

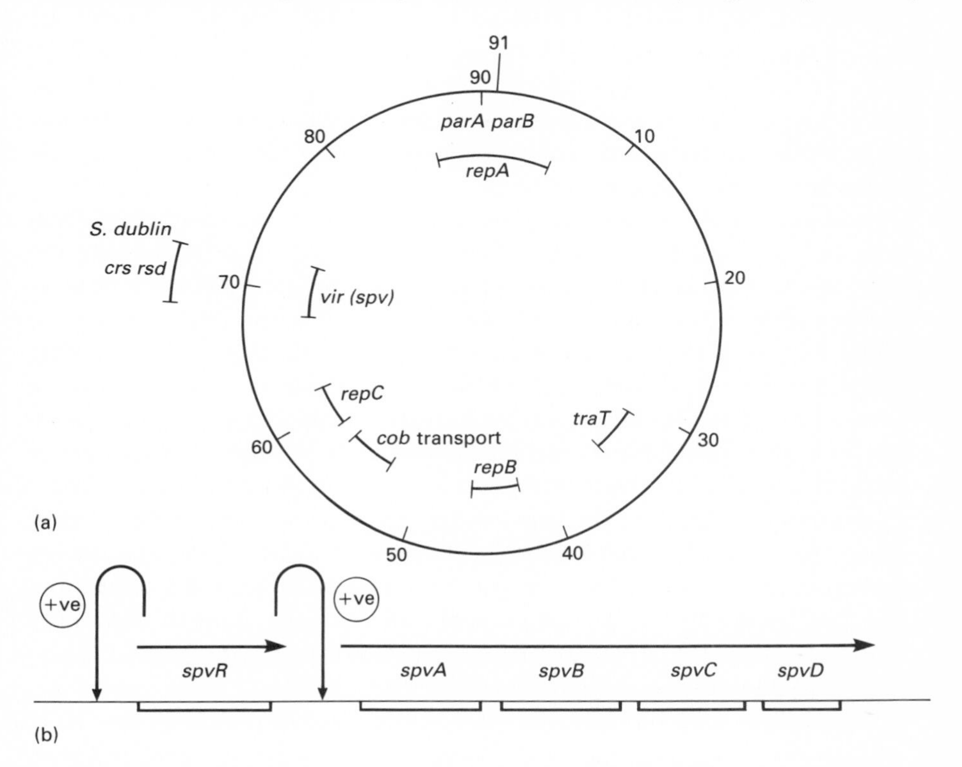

TACAGTATATCCTTTGCATAATACAGCTTATTGTTTACCAGTTACAGCTTATGATTAACATG
CTACAGTTTAGGTGATCA

(c)

Fig. 3.14 The virulence plasmid of *Salmonella.* (a) Physical and genetic map of the virulence plasmid. A circular map of the 91 kb element from *Salmonella typhimurium* is shown, together with the approximate locations of key genetic elements such as the *spv* (salmonella plasmid virulence) operon, replication functions (*repA, repB, repC*), cobalamin transport (*cob*) and various virulence-related sequences (*crs, rsd, traT*). (b) Structure of the *spv* operon. The operon consists of four structural genes (*spvABCD*) and a regulatory gene (*spvR*) with an autoregulatory function. (c) Nucleotide sequence of the *rsk* region of the plasmid. This includes a directly repeated element which contributes to the mouse virulence.

Plasmids from *S. dublin* and *S. typhimurium* have been studied in most detail at the molecular level. Large deletions of plasmid DNA have shown that only small portions are required for virulence. The precise number of plasmid regions and associated genes required for full virulence is unclear at the time of writing. Krause *et al.* (1991) have identified a 8.2 kb DNA sequence on the *S. dublin* Lane plasmid pSDL2 which is required for lethal salmonellosis in mice. Nucleotide sequence analysis has revealed six open reading frames, designated *vsdABCDEF*. In *S. typhimurium*, a plasmid gene called *vagA* or *mkaC* has been identified which is homologous to *vsdA* (Pullinger *et al.*, 1989; Taira & Rhen, 1989); the products inferred from the nucleotide sequences of these genes show homology to the DNA-binding proteins MetR and LysR (see Chapter 7 for a discussion of the LysR family of transcription regulators). This has led to the proposal that the *vsdA/vagA/mkaC* gene is a regulator of virulence gene expression and genetic evidence suggests that this is the case (Caldwell & Gulig, 1991; Fang *et al.*, 1991; Krause *et al.*, 1992; Taira *et al.*, 1991).

Difficulties in interpreting the molecular biology of the *Salmonella* plasmid virulence genes are compounded by the nomenclature used to describe the genes; the *S. typhimurium vagA/mkaC* gene is equivalent to ORF1 of Norel *et al.* (1989c). *vsdB* is equivalent to *mkaB* (Taira & Rhen, 1989) and ORF2 (Norel *et al.*, 1989c), *vsdC* to *mkaA* (Taira & Rhen, 1989) and *mkfB* (Norel *et al.*, 1989b) while *vsdD* is similar to *mkfA* (Norel *et al.*, 1989a) and *virA* (Gulig & Curtiss, 1988; Gulig & Chiodo, 1990). An homologous sequence to the virulence region from *S. dublin* Lane is found on plasmids in virulent *S. choleraesuis*, *S. enteritidis*, *S. gallinarum*, *S. naetved* as well as *S. typhimurium* (Roudier *et al.*, 1990). A unified nomenclature has now been adopted for the genes in this virulence system. These are now named *spv* genes, for *Salmonella* plasmid virulence genes. Under the unified nomenclature, the regulatory gene *spvR* is equivalent to *vsdA* of *S. dublin* and *mkaC*, *vagA* and ORF1 of *S. typhimurium*. It is also equivalent to *mba1* of *S. choleraesuis* (Krause *et al.*, 1992). The relationships between the original names and the *spv* designations are summarized in Table 3.1.

Downstream from the *spv* genes of *S. dublin* plasmid pDSL2, Krause *et al.* (1991) found a locus concerned with plasmid maintenance (Fig. 3.14). Deletions of this sequence lead to plasmid multimerization and instability. Plasmid pDSL2

Table 3.1 *Salmonella* plasmid *spv* operon nomenclature

	Alternative designation			
Gene	*S. typhimurium*		*S. dublin*	*S. choleraesius*
spvR	*mkaC* ORF1	*vagA*	*vsdA*	*mba1*
spvA	*mkaB*	ORF2	*vsdB*	*mba2*
spvB	*mkaA*	*mkfB*	*vsdC*	*mba3*
spvC	*mkaD*	*mkfA*	*vsdD*	*mba4*
spvD		*vsdE*		

carries a *trans*-acting resolvase gene, *rsd*, and a *cis*-acting resolution site, *crs*. The carboxy terminus of the *rsd* gene product shows homology to the carboxy-termini of members of the integrase family of site-specific recombinases (Table 4.1) and homologous sequences are found on the high-molecular-weight virulence plasmids of other *Salmonella* serovars (Krause & Guiney, 1991). The virulence plasmid of *S. typhimurium* possesses at least three other regions required for stable maintenance. Two, *repB* and *repC*, are functional replicons with homology to sequences on IncFI and IncFII plasmids. A third region, *par* (or *repA*, Michiels *et al.*, 1987) exhibits plasmid incompatibility and is a partitioning locus (Tinge & Curtiss, 1990).

The virulence plasmid of *S. typhimurium* carries the *rck* gene which codes for an outer membrane protein conferring high-level complement resistance when expressed in *E. coli* K-12 or in plasmid-cured rough *S. typhimurium* (Hackett *et al.*, 1987). Heffernan *et al.* (1992) have shown that the gene product (Rck) shares amino acid sequence homologies with PagC (an *S. typhimurium* chromosomally encoded protein required for survival in macrophages *in vitro* and for mouse virulence; Miller *et al.*, 1989a; Pulkkinen & Miller, 1991), Ail (a chromosomally encoded outer membrane protein from *Yersinia enterocolitica* involved in adherence to and invasion of epithelial cell lines; Miller *et al.*, 1990b), Lom (a bacteriophage λ-encoded outer membrane protein expressed in *E. coli*; Barondess & Beckwith, 1990) and OmpX (a chromosomally encoded outer membrane protein of *Enterobacter cloacae*; Stoorvogel *et al.*, 1991a,b). Of these, the two *S. typhimurium* proteins, PagC and Rck, are the most closely related (Heffernan *et al.*, 1992).

The *Salmonella typhimurium* virulence plasmid can integrate into the chromosome with significant consequences for the expression of the virulence phenotype. Strains with the plasmid in the integrated mode display enhanced susceptibility to serum killing. Plasmid sequences that reverse this phenotype have been identified by introducing the cloned DNA into strains harbouring integrated virulence plasmids and screening for restoration of normal levels of serum resistance. One of these, *rsk* (reduced serum killing) appears to be a 66 bp DNA sequence which includes two 10-mer direct repeats with 21 bp periodicity (Fig. 3.14). It has been suggested that this sequence may reduce serum killing in *S. typhimurium* strains with integrated plasmids by titrating a *trans*-acting regulatory factor (Vandenbosch *et al.*, 1989b). In addition to restoring serum resistance, *rsk* can also restore adhesion-invasion of HeLa cells and mouse virulence, both of which are also reduced by virulence plasmid integration (Vandenbosch *et al.*, 1989a).

Salmonella virulence plasmids are non-conjugative but harbour a gene coding for a TraT protein. This lipoprotein plays a role in *S. typhimurium* serum resistance and is closely related in sequence to the TraT proteins of incompatibility group F plasmids (Rhen & Sukupolvi, 1988; Sukupolvi *et al.*, 1990). Indirect evidence from early work suggested that the *S. typhimurium* virulence plasmid encoded an adherence system. This came from the observation that plasmid sequences harbour genes capable of complementing a defect in cobalamin uptake in *S. typhimurium* or *E. coli btuB* mutants. Nucleotide sequence analysis revealed that the complementing DNA includes a gene coding for a polypeptide with extensive homologies to the PapC and FaeD outer membrane proteins involved in the export and assembly of Pap pilus subunits and K88ab fimbrial subunits respectively (Rioux *et al.*, 1990). A thorough analysis of this region, which lies close to the *rck* gene, revealed a locus coding for all the functions required for expression of fimbriae. This system shows homology throughout its sequence to those coding for Pap and K88 fimbriae from *E. coli*. Introduction of recombinant plasmids harbouring cloned copies of these genes to *E. coli* results in expression of fimbriae on the bacterial surface, confirming that this is an active rather than a vestigial fimbrial system (Friedrich *et al.*, 1993).

Shigella virulence plasmids

Large plasmids play an essential role in the virulence of *Shigella flexneri* and *S. sonnei* (Sansonetti *et al.*, 1981, 1982). High-molecular-weight plasmids which are essentially homologous to those of *Shigella* are also required for virulence by enteroinvasive *Escherichia coli* (EIEC) (Hale *et al.*, 1983; Sansonetti *et al.*, 1983). The plasmids code for the ability to invade tissue culture cells (Hale, 1991; Maurelli & Sansonetti, 1988a; Watanabe, 1988; Yoshikawa *et al.*, 1988).

The virulence plasmids possess a *rep* locus which is essential for replication and for plasmid maintenance (Fig. 3.15). The Rep region of the 230 kb plasmid from *S. flexneri* 2a is homologous with the RepFIIA replicon family and exerts incompatibility with IncF and IncH1 plasmids (Makino *et al.*, 1988). The RepFIIA replicon is also found in the virulence plasmids of enteroinvasive and enterotoxigenic *E. coli* (Saadi *et al.*, 1987; Silva *et al.*, 1988). Clearly, these plasmids are likely to share a common origin in terms of their maintenance functions. The *Shigella* virulence plasmid carries a complex invasion gene regulon which is regulated in response to environmental stimuli by plasmid and chromosomally encoded gene products. This 230 kb plasmid also harbours three IS*1*-like elements (Hale, 1991).

Like the virulence plasmids of mouse-virulent *Salmonella* serovars, the 230 kb plasmid of *Shigella flexneri* can integrate into the bacterial chromosome. When this happens, a reduction in plasmid-encoded virulence gene expression is observed. A similar phenomenon is observed with the virulence plasmid of EIEC. Moreover, both the *S. flexneri* and the EIEC plasmids integrate into their respective chromosomes at identical genetic loci. Integration inactivates the *metB* gene and renders the cell auxotrophic for methionine. Precise excision of the plasmid from the chromosome restores virulence gene expression and methionine prototrophy (Zagaglia *et al.*, 1991). The integration and excision process is believed to involve homologous plasmid and chromosomal sequences since these are RecA-dependent events.

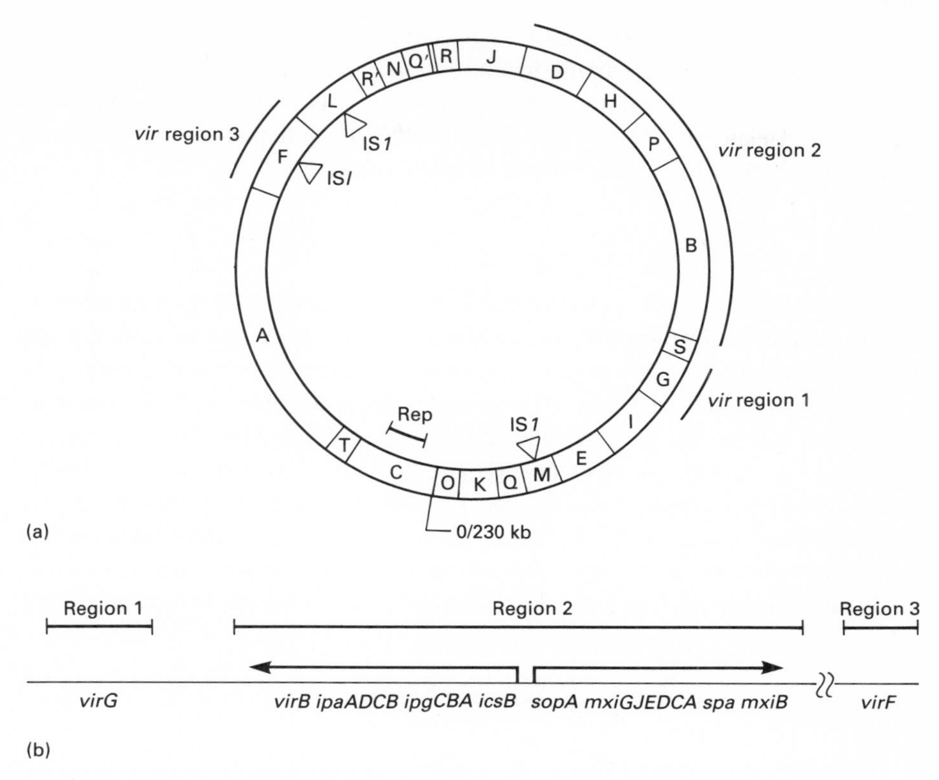

Fig. 3.15 The virulence plasmid of *Shigella flexneri.* (a) Map of the complete plasmid, PMYSH6000. The letters designate individual SalI DNA fragments. The positions of the three virulence *(vir)* regions are shown together with the location of the replication region and IS*1* elements. (b) Genetic map of the virulence regions. Region 1: *virG (icsA)*, concerned with intra- and intercellular spread of *Shigella.* Region 2: many genes, concerned with host cell invasion. Region 3: *virF*, coding for a transcription activator.

Plasmids in enterotoxigenic *Escherichia coli*

Enterotoxigenic strains of *E. coli* (ETEC) harbour plasmids encoding heat-labile enterotoxin (LT) or heat-stable enterotoxin (ST) I or both (Betley *et al.*, 1986). In general, the genes coding for enterotoxins and the transcription activator for expression of colonization factor antigens I and II (CFA-I and CFA-II) are located on the same plasmid (Caron *et al.*, 1989; Evans *et al.*, 1975; Penaranda *et al.*, 1983). Interestingly, the transcription activators on the ETEC virulence plasmids are closely related to others involved in the activation of plasmid-encoded virulence gene expression in *Shigella* and *Yersinia* (see Chapter 8). Plasmid pCG86 has been investigated in some detail and found to be conjugative. It carries genes for LT and ST and can be transferred from enterotoxigenic to non-enterotoxigenic *E. coli* in the gut of newly weaned piglets (Gyles *et al.*, 1977, 1978). This plasmid possesses two replicons; so does the related plasmid P307 (also encoding LT and ST). They have probably arisen as a result of plasmid fusion (Pickens *et al.*, 1984). The LT/ST plasmids from different ETEC isolates are highly related and display similarities to F. As discussed elsewhere (Chapter 5), the LT and ST genes have been found to be associated with actual or potential mobile DNA elements. Thus, these virulence factors may possess a high degree of mobility both within and between bacterial strains.

Plasmids in *Yersinia* species

Three species of *Yersinia*, called *Y. enterocolitica*, *Y. pestis* and *Y. pseudotuberculosis*, are facultative intracellular human pathogens. All three species release large amounts of proteins known as Yops (*Yersinia* outer membrane proteins) in response to growth restriction in the absence of calcium ions at 37°C. The Yop genes are located on a 70 kb plasmid and loss of the Yop release property correlates with loss of virulence (Cornelis, 1992; Cornelis *et al.*, 1989a) (Fig. 3.16). The plasmids are designated pYV and isolates from the three *Yersinia* species possess many highly conserved regions. However, the distribution of the *yop* genes varies for each of the plasmids, as does the location of the replication and partition regions. The pYV plasmids belong to incompatibility group FI and

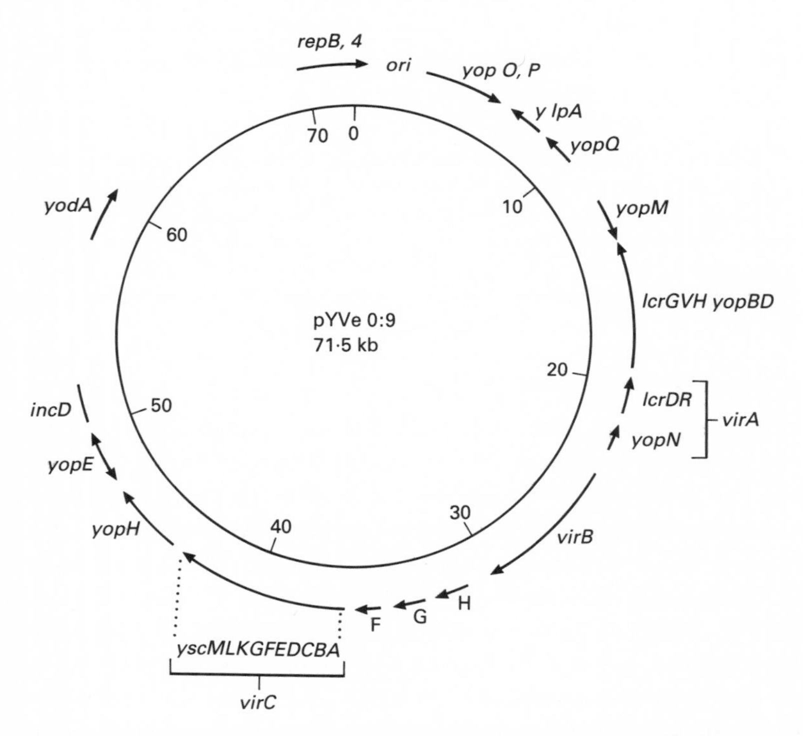

Fig. 3.16 Physical and genetic map of the pYV plasmid of *Yersinia enterocolitica* serotype O:9. The 71.5 kb plasmid is shown as a circular element on which the positions of the replication (*ori, rep*), incompatibility (*inc*) and virulence (*lcr, vir, yad, ylp, yop, ysc*) functions are indicated.

possess an *incD* incompatibility determinant. This makes them incompatible, for example, with the *E. coli* F factor (Biot & Cornelis, 1988; Portnoy & Falkow, 1981).

The minimum replication region of pYV is 68% homologous with the RepFIIA replicon of plasmid R100. Thus, the pYV plasmids share this feature with the virulence plasmids of *Salmonella typhimurium, Shigella flexneri*, enteroinvasive and enterotoxigenic *E. coli* and with the ColV plasmids. This strengthens further the impression that the maintenance functions of all these plasmids have arisen from a common source. The pYV plasmid encodes two replication proteins, RepA and RepB which are homologous to RepA1 and RepA2, respectively, of multiple antibiotic resistance plasmid R100. pYV also possesses sequences similar to those encoding copy number control RNAs in R100 (Vanooteghem & Cornelis, 1990).

In *Y. pestis*, a 9.5 kb plasmid called pPCP1, encodes a coagulase and a plasminogen activator (Ferber & Brubaker, 1981; Sodeinde & Goguen, 1988). The *pla* (plasminogen activator) gene encodes two proteins, derived from a common precursor polypeptide. The gene products are associated with the *Y. pestis* outer membrane and are responsible for *in vitro* degradation of Yops encoded by the *Y. pestis* 70 kb pYV plasmid (Sodeinde *et al.*, 1988). The *pla* gene product displays a high degree of homology with the *E. coli* OmpT protein and the *S. typhimurium* E protein. Like the *pla* gene product, both OmpT and the E protein are plasminogen activators, cleave specific outer membrane proteins in the strains which encode them and are post-translationally processed following secretion across the cytoplasmic membrane (Sodeine & Goguen, 1989). Inactivation of *pla* results in a dramatic reduction in the mouse virulence of *Y. pestis*. This observation has led to the proposal that the acquisition of the pPCP1 plasmid (presumably from the *Salmonella* lineage) marked an important step in the divergence of *Y. pestis* from the less virulent but, nonetheless closely related, species *Y. pseudotuberculosis* (Sodeinde *et al.*, 1992). This represents an important example of the ability of extrachromosomal genetic elements to influence the evolution of bacterial virulence.

Transferable plasmids of *Enterococcus (Streptococcus) faecalis*

Enterococcus (formerly *Streptococcus*) *faecalis* (Lancefield Group D) is a commensal organism of the human gut which has been associated with cases of endocarditis and with urinary tract infections (Caperon & Scott, 1991). It possesses a pheromone-inducible plasmid transfer system which is quite distinct from the conjugal transfer systems of Gram-negative bacteria. The pheromones are low-molecular-weight hydrophobic peptides and they are released by the recipient cell and induce the mating response in the donor (Figs 3.17 & 3.18). Most *E. faecalis* strains produce multiple pheromones and each pheromone induces a response only in cells harbouring a particular plasmid (Dunny, 1990). When the cell has acquired a plasmid associated with a particular pheromone,

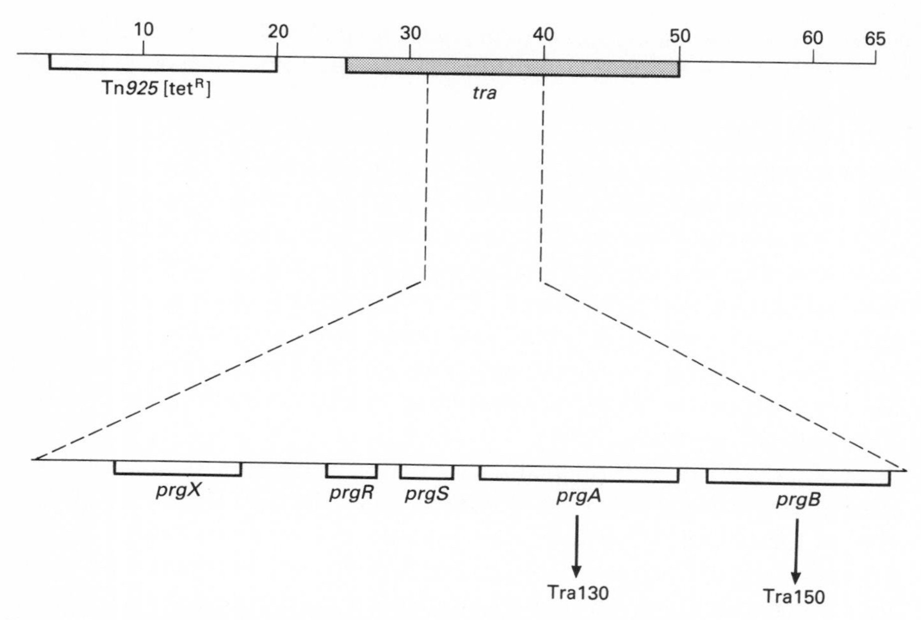

Fig. 3.17 Map of plasmid pCF10 of *Enterococcus faecalis*. A linear map is shown, calibrated in kb. The positions of the transposon Tn*925* (specifying tetracycline resistance, *tet*^{R}) and the transfer operon (*tra*) are shown. An expanded view of the *tra* region is shown at the base of the figure.

production of that pheromone ceases, although the cell continues to secrete unrelated pheromones (Dunny *et al.*, 1979). Pheromone-inducible genes on the plasmids encode surface antigens concerned with the distinct processes of surface exclusion and cell–cell aggregation. The latter is promoted by an 'aggregation substance' and is essential for close contact between mating cells. The aggregation substance encoded by plasmid pAD1 is 137 kDa and it appears as a dense layer of hair-like structures on the cell surface (Galli *et al.*, 1989; Wanner *et al.*, 1989). As with 'conventional' conjugative plasmids, these elements code for surface exclusion factors which prevent mating between cells harbouring the same plasmid. The surface exclusion protein of the 65 kb plasmid pCF10 is called Tra130 and is encoded by the *prgA* gene. This 130 kDa protein is associated with the cell surface and has regions of homology with other streptococcal surface proteins, including M protein (reviewed in Dunny, 1990). Plasmids studied to date are generally in the size range 54–71 kb (with the exception of the 9 kb plasmid pAMα1) and some encode single or multiple antibiotic resistance determinants while others encode a bacteriocin determinant or a haemolysin–bacteriocin determinant (Clewell & Weaver, 1989; Galli & Wirth, 1991).

The region of plasmid pCF10 coding for the sex pheromone binding function has now been characterized in detail. Three pheromone responsive genes (*prg*) have been identified. One, *prgW*, codes for a product showing homology to DNA

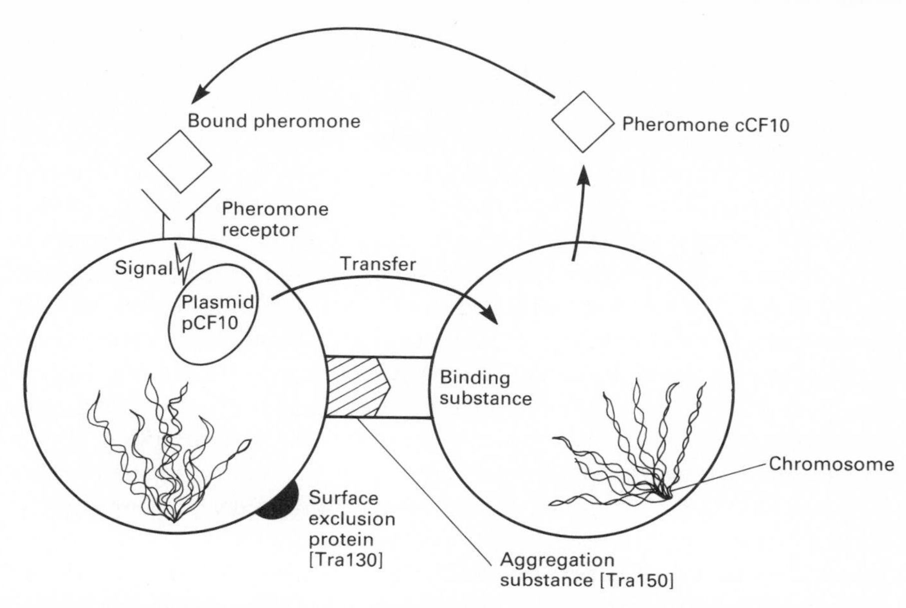

Fig. 3.18 Pheromone-induced conjugal transfer of plasmid pCF10 in *Enterococcus faecalis*. Two *Enterococcus faecalis* cells undergoing pCF10-mediated conjugation are illustrated. The recipient expresses the binding substance on its surface and secretes pheromone cCF10 into the medium. The donor cell expresses aggregation substance on its surface which binds the cells together and also expresses a receptor for the pheromone. Activated receptor/plasmid signalling results in plasmid DNA transfer to the recipient. The donor also specifies a surface exclusion function which prevents donor–donor mating.

binding proteins; another, *prgY*, is related to the *traB* gene of the pheromone-inducible plasmid pAD1; the third, *prgZ*, codes for a product showing homology to oligopeptide binding proteins found in Gram-positive and Gram-negative bacteria (Ruhfel *et al.*, 1993). The *traC* gene of the pheromone-inducible plasmid pAD1 also shows homology to these oligopeptide-binding proteins. The TraC protein is located on the cell surface and appears to bind the cAD1 sex pheromone (Tanimoto *et al.*, 1993).

Plasmids in *Staphylococcus aureus*

Plasmids in the Gram-positive pathogen *Staphylococcus aureus* are classified according to size, copy number and mode of replication. Class I plasmids are 1–5 kb in size and exist in 15–60 copies per chromosome equivalent (Novick, 1989, 1991). These plasmids replicate by an asymmetric rolling circle mechanism similar to that used by filamentous single-stranded coliphages such as M13 (Fig. 3.2). This mechanism is used by all known small plasmids from Gram-positive bacteria. All these plasmids encode a Rep protein which initiates DNA replication by introducing a site-specific nick in the leading strand replication origin. In one of the best-studied examples, pT181, the Rep protein is called RepC. Here, a RepC dimer binds via its carboxyterminus to its recognition site at the origin and promotes a structural transition in the DNA which causes a cruciform to extrude (Fig. 3.19). This cruciform includes the nick site and this site is brought into contact with the aminoterminus of Rep which nicks it and then forms a phosphotyrosine bond at the 5′ end of the DNA (Fig. 3.19) (Thomas *et al.*, 1988; Wong *et al.*, 1992). It then initiates replication by a 3′ extension from this point. Lagging strand replication is postponed until the *palA* sequence (Fig. 3.19) has become single-stranded and forms a hairpin which in turn initiates lagging strand replication (Gruss *et al.*, 1987). Plasmids from which *palA* has been deleted undergo normal leading strand replication but the lagging strand remains as a single-stranded molecule. Copy number control is achieved by antisense RNAs which cause attenuation of the Rep protein mRNA. If two plasmids share the same copy number control inhibitor and/or the same origin of leading strand DNA replication, they exert segregational incompatibility (Projan & Novick, 1986; Novick, 1989).

Class II plasmids are larger (15–30 kb) and have lower copy numbers (4–6 per chromosome equivalent) than Class I plasmids. They encode replication initiator proteins but replicate via theta (Θ) structures rather than rolling circles (Novick, 1991). Class III elements are even larger (30–60 kb) and carry a conjugal transfer (*tra*) region. Plasmids that do not obviously belong to these classes have been placed into a heterologous Class IV (Novick, 1989, 1991). Class I plasmids may be cryptic (e.g. pSN2 and pTCS1) but most carry a single marker for resistance to an antimicrobial agent. Class II and III plasmids harbour multiple resistance determinants.

Conjugal plasmids in *S. aureus* belong to Class III and encode resistance to gentamicin together with additional resistance determinants. In Class III plasmids pG01 and pCRG1600, the *tra* region has been found to consist of a 14 kb region flanked by copies of the IS*431* insertion sequence (also called IS*257*) (Fig. 3.20). Neither pili nor conjugative pheromones are involved in intraspecies conjugation in *S. aureus*. These conjugative plasmids can also mobilize the transfer of Class I plasmids. For example, the Class I plasmid pC221 can be mobilized by Class III conjugal plasmid pG01 via two pC221-encoded proteins MobA and MobB and a *cis*-acting pC221 sequence *oriT* which contains the nick site (Projan & Archer, 1989). MobA and *oriT* (but not MobB) are required for relaxation; all three are required for mobilization. Conjugative mobilization involves a relaxation complex composed of supercoiled plasmid DNA and protein similar to that seen with ColE1-like plasmids (Novick, 1976). As with mobilization of ColE1 by F, both participating plasmids play an active part. Despite the mechanistic similarities, there are no sequence homologies between the pC221 and the ColE1 systems. Mobilization of the Class I plasmid occurs 100 times more efficiently than conjugal transfer of the Class III plasmid itself (Novick, 1991). It should be noted that small *S. aureus* plasmids may be transferred via a phage-mediated conjugation mechanism which requires cell–

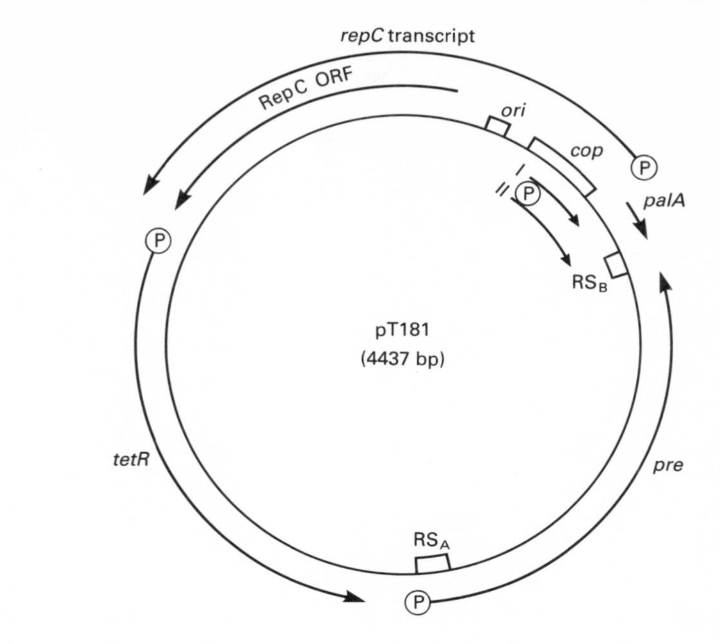

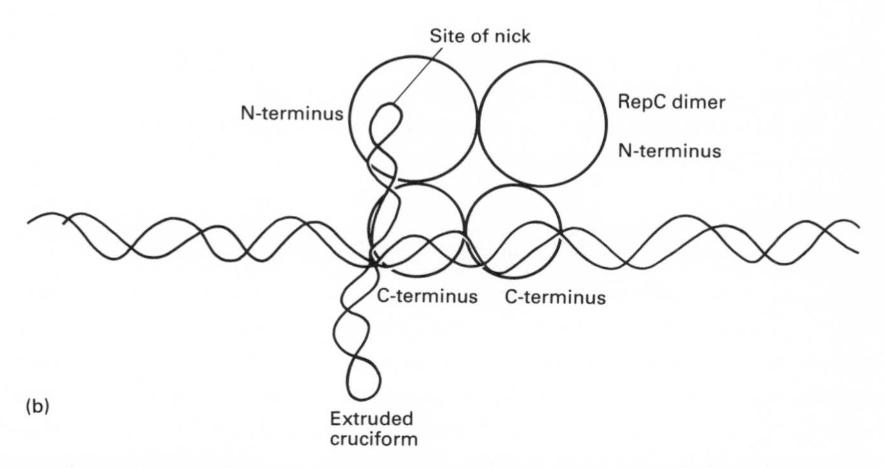

Fig. 3.19 Plasmid pT181 from *Staphylococcus aureus.* (a) Physical and genetic map of pT181 (4437 bp) showing its key features: replication functions (*ori, repC*), copy number control (*cop*), palindrome A (*palA*), site-specific recombination sites (RS_A and RS_B), site-specific recombinase gene (*pre*), tetracycline resistance gene (*tetR*). ORF, open reading frame; P, promoter. (b) Nicking of pT181 origin DNA by RepC. A RepC dimer is shown bound via its carboxyterminal domains to the plasmid origin. The extruded DNA cruciform is nicked by the aminoterminal domain of RepC.

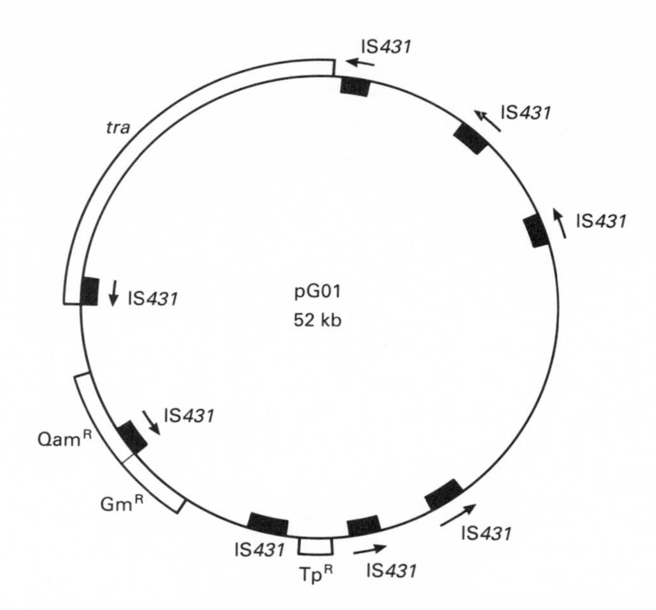

Fig. 3.20 Structure of plasmid pG01 from *Staphylococcus aureus*. The 52 kb plasmid pG01 harbours multiple copies of the insertion sequence IS*431*. It possesses genes for conjugal transfer (*tra*), resistance to quaternary ammonium compounds (Qam^R), resistance to gentamicin–kanamycin–tobramycin (Gm^R), resistance to trimethoprim (Tp^R).

cell contact and the presence of a transducing phage (Lacey, 1980). While pheromones are not required for Class III plasmid transfer, *S. aureus* elaborates a sex pheromone (cAM373) which permits conjugal transfer of plasmids from *Enterococcus* (*Streptococcus*) to *Staphylococcus* (Clewell *et al.*, 1985).

Plasmids in *Bacillus anthracis*

The Gram-positive pathogen, *Bacillus anthracis*, is the causative agent of anthrax. This is a fatal disease of domestic livestock and is caused by a multicomponent lethal toxin. Pathogenesis also requires a capsule. The toxin is composed of three parts; called oedema factor (89 kDa), protective antigen (85 kDa) and lethal factor (83 kDa). Oedema factor has intracellular calmodulin-dependent adenylate cyclase activity but is non-toxic when alone. It requires the protective antigen to induce increased cAMP levels. Lethal factor also works in concert with protective antigen, causing pulmonary oedema which is rapidly fatal.

B. anthracis maintains its major virulence determinants on two high-molecular-weight plasmids. The three toxin genes, *cya* (coding for oedema factor), *lef* (coding for lethal factor) and *pag* (coding for protective antigen) are located on a 174 kb plasmid called pXO1. Genes concerned with capsule biosynthesis are carried on a 90 kb plasmid called pXO2 (Mikesell *et al.*, 1983; Uchida *et al.*, 1985). Full virulence requires that both plasmids be present.

Linear plasmids of *Borrelia* species

The spirochaete *Borrelia hermsii* is the causative agent of tick-borne relapsing fever in North America (Barbour, 1989, 1990a,b). Spirochaetes are not particularly closely related to either Gram-negative or Gram-positive bacteria. Borreliae are entirely host associated. They are found in arthropods or in the vertebrate hosts upon which these arthropods (ticks or lice) feed. Vertebrate hosts infected by tick bite develop the symptoms of relapsing fever in which an initial crisis of fever is followed by an interval of 4–7 days of well-being followed by a fever relapse. This pattern is repeated many times. Each bout of fever correlates with the climax in cell numbers of a particular serotype of borrelia in the blood. As spirochaete numbers fall, host fever is relieved. Thus the key to understanding relapsing fever lies in an understanding of the mechanism of serotype variation.

Serotype specificity is determined by outer membrane proteins known as variable major proteins or VMPs. The genes coding for these VMPs are plasmid-linked. The plasmids are unusual in being linear molecules. VMP variation is caused by a genetic rearrangement which involves the transfer of *vmp* structural genes from a 'silent' site on one linear plasmid to an expression site on another (Plasterk *et al.*, 1985; Chapter 5). Each strain carries multiple copies of the plasmids which vary in size from 23 to 50 kb. The related species, *B. burgdorferi*, the cause of Lyme disease, also possesses linear plasmids (49 and 16 kb) which carry the coding sequences for the major surface antigens. The type strain, B31, also harbours a covalently closed *circular* plasmid of 27 kb. The linear plasmids of *B. burgdorferi*, and presumably those in *B. hermsii*, have covalently closed ends similar to the telomeres of eukaryotic chromosomes (Barbour, 1989; Barbour & Garon, 1987).

The linear and circular plasmids in *B. burgdorferi* have low copy numbers, approximately one per chromosome, indicating highly efficient mechanisms for plasmid maintenance (Hinnebusch & Barbour, 1992). The telomeric structures of these linear plasmids possess homology to the iridovirus which causes African swine fever (Hinnebusch & Barbour, 1991). Since this eukaryotic virus and *B. burgdorferi* share the same tick vector, this may indicate horizontal transfer of genetic information across the eukaryotic and prokaryotic kingdom boundary (Hinnebusch & Barbour, 1991). This is a very interesting possibility. It should be noted that linear plasmids are not confined to spirochaete bacteria, they have also been detected in the bacteria *Streptomyces rochei* and *Thiobacillus versutus* (Hirochika *et al.*, 1984; Wlodarczyk & Nowicka, 1988).

4 Genome Rearrangements: Reiterated Sequences, Transposition and Site-Specific Recombination

Introduction

It is now widely appreciated that genomes are not static but prone to a variety of changes which may result in variation in the expression of genetic information. These changes are grouped together under 'genome rearrangements'. Genome rearrangements may assist bacteria in adapting to new environmental cirumstances by allowing them to express newly acquired genetic information or previously cryptic endogenous information. Rearrangements can also alter the level of expression of genes by varying their copy number through amplification or deletion. Genes may be moved to different sites in the genome, for example from storage at so-called 'silent' sites to expression sites where they become coupled to expression signals. Rearrangements may involve very large changes in genome structure or changes in just a single base pair. Small, gene-specific changes can have wide-ranging implications for the fitness of the cell for a particular niche since an alteration to an important epitope in a surface-expressed protein may allow the mutant variant to evade the host defences. Some rearrangements appear to occur stochastically with the pressure of the environment then selecting derivatives carrying the best permutation. Others *appear* to be programmed and some involve specific DNA recombination proteins acting in a highly specific way with cognate DNA sequences. The 'programmed' rearrangements may only be triggered in response to a particular environmental signal or combination of signals. In this respect they resemble the environmentally responsive promoters whose level of transcription is modulated in response to overlapping signal inputs.

The topics of general recombination, site-specific recombination and transposition were reviewed briefly in Chapter 2 in terms of their impact on genome structure. This chapter deals with specific examples of each type of rearrangement and considers their consequences for the expression of genetic information in bacteria.

Reiterated sequences in bacterial genomes as substrates for general recombination

Bacterial genomes contain many examples of repetitive DNA sequences and these have the potential to act as substrates for the homologous recombination machinery of the cell, leading to genome rearrangements. Some of the best-studied examples include the palindromic REP (repetitive extragenic palindrome) sequences and the palindromic ERIC (enterobacterial repetitive intergenic consensus) sequences found in enteric bacteria (reviewed in Lupski & Weinstock, 1992) (Fig. 4.1).

REP sequences have been detected in the genomes of *E. coli* and *Salmonella typhimurium* (Higgins *et al.*, 1982; Stern *et al.*, 1984b), as well as those of *Enterobacter, Erwinia, Klebsiella, Proteus, Serratia* and *Shigella* (Dimri *et al.*, 1992). There has been much speculation in the literature about the *in vivo* role of REPs, including the possibility that they might be involved in promoting recombination (Allgood & Silhavy, 1988; Higgins *et al.*, 1982, 1988b; Petes & Hill, 1988). REP sequences have been shown to serve as recombination substrates in both RecA-dependent recombination and in a poorly understood RecA-independent system (Shyamala *et al.*, 1990).

REP sequences have been shown to bind several proteins. Specifically, REPs bind DNA gyrase in the presence of the histone-like protein, HU, and this has led to the suggestion that they may play a role in the higher-order organization of chromosome structure (Yang & Ames, 1989, 1990). Interestingly, the organization of the chromosome has been proposed as a barrier to homologous recombination at certain sites (see Chapter 3) so REPs may contribute to the control of recombination by acting both as recombination substrates and as part of the genome organization apparatus. REP sequences also bind DNA

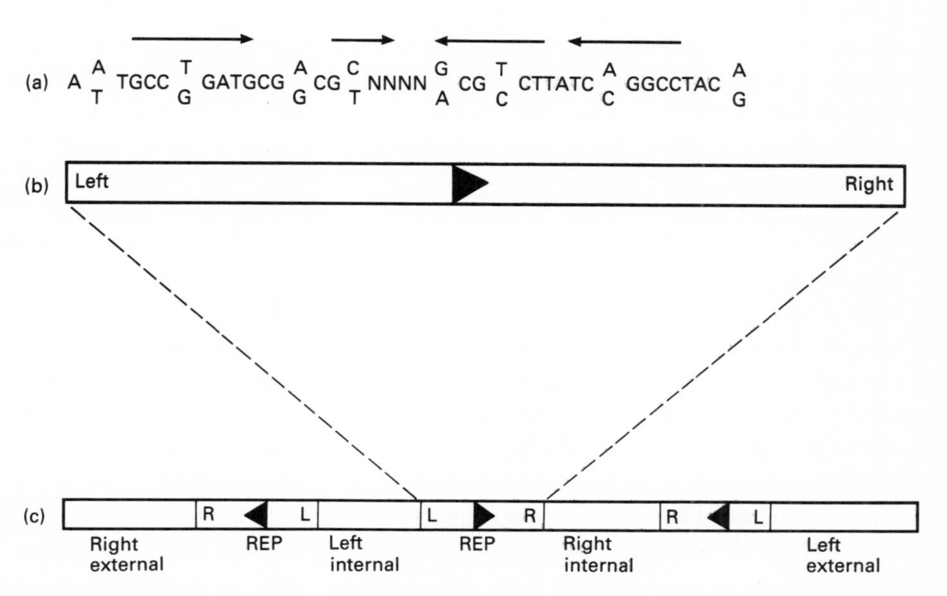

Fig. 4.1 Repetitive extragenic palidrome (REP) sequence. (a) Consensus sequence for the REP element. The linear sequence contains several inverted repeats. The repeat structure is preserved by sequence covariation. (b) Summary of the REP structure as a polarized bar with an arrow to indicate orientation and the left and right ends indicated. (c) BIME (bacterial interspersed mosaic element) structure. This is composed of REP sequences in association with other conserved sequences.

polymerase I (Pol I) and other (uncharacterized) protein complexes. This could reflect a specific interaction between transcribed REP and Pol I, since REP forms the 3′ end of about one quarter of the total mRNA in *E. coli* and an interaction with Pol I could be an initial step in the priming of DNA synthesis from the RNA template (Gilson *et al.*, 1990). Furthermore, REPs are found in association with other repetitive DNA sequences (including so-called L motifs) forming complex repeats called BIMEs, of which *E. coli* is thought to contain up to 500 copies (bacterial interspersed mosaic element) (Bachellier *et al.*, 1993; Gilson *et al.*, 1991) (Fig. 4.1).

ERIC sequences are found in *E. coli*, *S. typhimurium*, *Klebsiella pneumoniae*, *Vibrio cholerae* and *Yersinia pseudotuberculosis* (Hulton *et al.*, 1991) and are probably ubiquitous in Gram-negative bacteria (Versalovic *et al.*, 1991). Like REPs, they are associated with transcribed regions of the genome but no functions have been demonstrated for them experimentally. *Neisseria* species contain at least two types of repetitive sequence, a 1.1 kb species present in low amounts in *N. gonorrhoeae* and a 26 bp type which is abundant in *N. gonorrhoeae* and *N. meningitidis* (Correia et *al.*, 1986). Reiterated DNA sequences have been described in *Bordetella pertussis* (Glare *et al.*, 1990) and several families of repetitive DNA sequences have been detected in *Agrobacterium* and *Rhizobium* species, both on the chromosome and on plasmids (Flores *et al.*, 1987). Reiterated DNA sequences form an important component of the genomes of archaebacteria such as *Halobacterium halobium* and *H. volcanii* (Sapienza & Doolittle, 1982). The Gram-negative eubacterium *Haemophilus influenza* harbours approximately 600 copies of an 11 bp sequence (5′-AAGTGCGGTCA-3′) in its genome, although this motif appears to be concerned primarily with tagging DNA for uptake by natural transformation in this organism (Smith & Danner, 1981).

Mycoplasma pneumoniae has been shown to contain reiterated sequences of 300 bp called RepMP1 and of 400 bp called SDC1 (Colman *et al.*, 1990; Wenzel & Herrmann, 1988). Repeated sequences have also been detected in *Coxiella burnetti* (Hoover *et al.*, 1992), *Leptospira interrogans* (Zuerner & Bolin, 1988), *Mycobacterium leprae* (Clark-Curtiss & Docherty, 1989; Grosskinsky *et al.*, 1989), *Mycobacterium tuberculosis* (Ross *et al.*, 1992), *Mycoplasma pulmonis* (Bhugra & Dybvig, 1992), *Pneumocystis carinii* (Stringer et *al.*, 1991) and *Spiroplasma* species (Nur *et al.*, 1987). Repeated DNA sequences contribute to the genome instability typical of many species of the Gram-positive filamentous bacterium *Streptomyces* (for a review see Hopwood & Kieser, 1990).

In terms of size and DNA sequence conservation, the most significant repetitive sequences in *E. coli* are the rRNA operons (*rrn*) and the *rhs* genes. *E. coli* has seven *rrn* operons dispersed around the chromosome (Bachmann, 1990) and each is aligned with the orientation of DNA replication (Fig. 2.14). Deletion events involving recombination across *rrn* operons are more common than inversions by several orders of magnitude. This may be due to the effect of inverting these highly transcribed operons on the fitness of the cell (Brewer, 1988; Hill & Gray, 1988; see 'Colliding Polymerases' Chapter 2). These well-conserved sequences serve as recombination substrates leading to the formation of tandem duplications, deletions and inversions (Hill & Combriato, 1973; Hill *et al.*, 1977; Hill & Harnish, 1981; Petes & Hill, 1988). Similarly, recombinations between *rrn* operons in *Bacillus subtilis* have been shown to promote deletions of intervening chromosomal sequences (Loughney *et al.*, 1983; Widom *et al.*, 1988). The *rhs* (rearrangement hot spot) gene family of *E. coli* is composed of at least five conserved loci which contain an homologous 3.7 kb core sequence (Lin *et al.*, 1984; Sadowsky *et al.*, 1989). Recombinations between members of this family have been found to generate duplications of intervening chromosomal DNA sequences (Capage & Hill, 1979; Folk & Berg, 1971). At first glance, bacteria would appear to contain many reiterated DNA sequences capable of contributing to genome rearrangements and hence instability. However, most genomes appear to be quite stable. This indicates that there are forces at work which counteract the destabilizing features inherent in the genome.

Barriers to extensive reassortment of genomes

An analysis of the genomes of *E. coli* and *S. typhimurium* suggests that wholesale rearrangements of bacterial genomes are rare, at least in natural populations of these bacteria. These organisms are thought to have diverged from a common ancestor approximately 150 million years ago (Ochman & Wilson, 1987), yet a glance at their genetic maps reveals that the gene order of one is remarkably similar to that of the other. This indicates that there are constraints on the type and extent of genome rearrangements in these bacteria. This issue has been addressed experimentally in several laboratories. It appears that certain recombination events on the enteric chromosome are forbidden, although the sequences concerned can be made to recombine with each other when out of context. This observation led to the proposal that recombination can be prevented by structural features of the bacterial nucleoid in the area of the recombination substrates (Segall *et al.*, 1988). It may also be impeded by differential supercoiling of the domains in which the sequences to be recombined are located. However, as was pointed out in Chapter 2, data supporting the idea that chromosomal domains are supercoiled to different extents are not unambiguous. Alternatively, frequent passage of a replication fork through one of the participating sequences may make recombination undesirable because an inversion event may turn the fork around, sending it back to the origin and leaving the remainder of the DNA unreplicated (Mahan *et al.*, 1990). Other studies have identified the region in which chromosomal replication terminates as one in which recombinations may not proceed; once again, features concerned with the structure of the nucleoid are believed to constrain recombination in these zones (François *et al.*, 1990).

Genome rearrangements and gene dosage

In addition to gene dosage effects associated with distance from the origin of chromosomal replication, *oriC*, gene copy number can also be altered by amplifying or deleting copies of genes, either on the chromosome or other replicons. Such events are usually catalysed by recombination between directly repeated homologous sequences in the genome. In *E. coli*, repeat sequences as short as 15 bp (or less) have been implicated in the amplification or deletion of the chromosomal regions harbouring the *lac* genes (Albertini *et al.*, 1982; Farabaugh *et al.*, 1978; Tlsty *et al.*, 1984; Whoriskey *et al.*, 1987) and the *ampC* locus (Edlund & Normark, 1981). Duplications of chromosomal DNA sequences have also been demonstrated in *Salmonella typhimurium* by Anderson & Roth (1978a,b). Whether or not these assist in bacterial adaptation to new environments is not known. Nor is it known if these events are programmed or respond to specific environmental stimuli (apart from DNA damage or other inducers of the SOS response).

Site-specific recombination: transposition

Transposition involves the movement of discrete DNA sequences between sites either on the same replicon or on different replicons. In general, little target site sequence specificity is exhibited, although there are exceptions (e.g. Tn*7* in *E. coli*, see later). Two major transposition pathways have been described, the replicative pathway and the non-replicative pathway. The non-replicative pathway is also called the 'conservative' pathway, a nomenclature that leads to confusion with 'conservative site-specific recombination', which is quite a different process (see Chapter 2). In essence, the two transposition pathways differ in terms of the replication of the transposable element. In the replicative pathway, the element is duplicated and a copy remains at the donor site (Fig. 2.23). In the non-replicative pathway, no copy is left at the donor site since the element appears to excise itself from that site, leaving a gap that is probably lethal to the donor replicon unless repaired (Fig. 2.23). A further problem with the nomenclature is that a small amount of DNA replication does take place in the 'non-replicative' pathway, although this occurs only at the ends of the inserting element (Fig. 2.23).

Transposable elements

The contributions of transposable elements to the rearrangement of the genomes of bacteria, pathogenic or otherwise, have been well documented. The details of their structures and transposition mechanisms may vary but all are characterized by the possession of genes coding for transposase functions which catalyse their movement between replicons. Transposable elements can negatively affect the expression of genes by inactivating them by insertion, either into the coding sequence, or into nearby non-coding regulatory sequences where they disrupt regulatory protein binding sites or alter the topology of the DNA in a manner unfavourable for gene promoter function. They may transpose into corresponding regions of positive regulatory genes, inactivating them and their dependent structural genes. They can activate non-expressed genes by insertional inactivation of repressor genes and by inserting into upstream regulatory regions from which they can activate transcription either by changing DNA topology or by supplying a promoter.

Composite transposons may harbour genes coding for functions that confer desirable phenotypes on the cell, such as resistance to antimicrobial agents. Some may even harbour virulence factors, such as toxins (for example, see the section on amplification of *Vibrio chloerae* toxin genes in Chapter 5). Transposable elements can serve as regions of homology for genome rearrangement via general recombination (Louarn *et al.*, 1985; Savic *et al.*, 1983; Timmons *et al.*, 1986). This can involve the same permutations as are possible when other repetitive sequences serve as recombination substrates, i.e. inversions, tandem amplifications, deletions and intramolecular recombinations leading to cointegrate formation. The nature of the recombination products is determined by the location and relative orientation of the homologous sequences (i.e. intermolecuar recombination leading to cointegrate formation, intramolecular recombination leading to cointegrate resolution/deletion or inversion of intervening sequences).

The insertion sequences

In terms of genetic complexity, the simplest transposable elements are those known as the insertion sequences (abbreviated to IS). These consist of just sufficient genetic material to code for their own transposition. The enzyme which catalyses transposition of any transposable element is called its 'transposase'. Insertion sequences will typically code for this activity and also carry *cis*-acting sites at their termini where the transposase acts. These terminal sequences are usually organized as inverted repeats. Recent reviews have listed over 40 IS elements in Gram-negative bacteria (Galas & Chandler, 1989), 12 in Gram-positive bacteria (Murphy, 1989) and 13 in the *Halobacteria* species of archebacteria (Charlebois & Doolittle, 1989). In Gram-negative bacteria a 768 bp IS element called IS*1* has been studied in great detail. It occurs at between 2 and 40 copies in the genome of some *Shigella* species while other species may harbour up to 200 copies (Nyman *et al.*, 1981). There are up to 17 copies in some strains of *E. coli* (Hu & Deonier, 1981) and it is also found in *Serratia marcescens* (two copies, Nyman *et al.*, 1981) and *Klebsiella aerogenes* (one copy, Nyman *et al.*, 1981).

Despite its short length, IS*1* has eight potential open reading frames within it and two of these, *insA* and *insB* code for the transposition functions (Machida *et al.*, 1984) (Fig. 4.2). The active transposase of IS*1* consists of a fusion protein

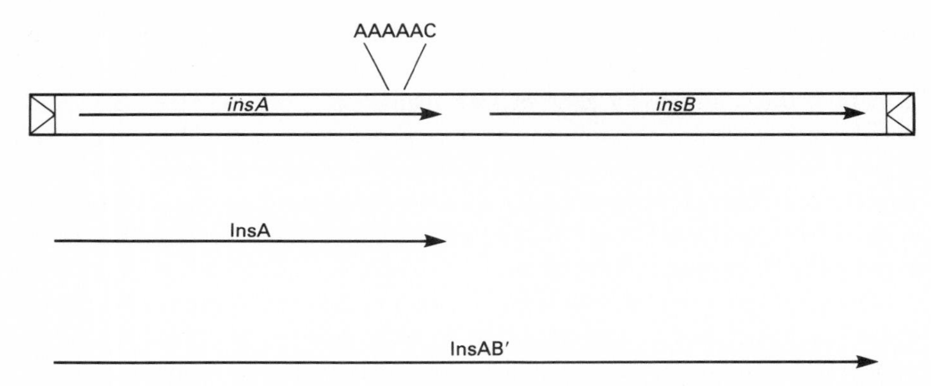

Fig. 4.2 Insertion sequence IS*1* is a 768 bp insertion sequence with 20–23 bp terminal inverted repeats. InsA, transposition inhibitor; InsAB′, active transposase; AAAAAC, translationally slippery site involved in frame-shift event required for InsAB′ formation.

made up from InsA and the product of the *insB* gene. Interestingly, the InsB protein does not have an independent existence. This is because the fusion protein is made by a translational slippage during the translation of the *insA* mRNA that introduces a −1 frameshift into the reading frame, fusing *insA* to *insB* (Fig. 4.2) (Sekine & Ohtsubo, 1989). Besides generating an active transposase, this fusion event has important regulatory side benefits. Since the DNA-binding domain of the fused InsAB′ protein lies in the InsA component, both InsAB′ and InsA itself are capable of binding to the same DNA sequence. Consequently, InsA and the fusion protein InsAB′ compete for a binding site at the inverted repeats of IS*1*. If InsAB′ wins, the result is transposition, if InsA binds, InsAB′ is excluded and transposition is prevented (Plasterk, 1991). Translational frameshifting has been found to regulate transposition in several other bacterial insertion sequences (reviewed in Chandler & Fayet, 1993). It is also used in eukaryotic retroviruses such as HIV.

As with all transposable elements, insertion of IS*1* into a DNA sequence causes a target site duplication. This is simply an outcome of the mechanism by which transposition into the target site occurs (Fig. 2.23). For IS*1*, the duplicated region is typically 9 bp but can vary in length from 7 to 14 bp (Galas & Chandler, 1989). In addition to its own transposase function, IS*1* requires certain host factors for transposition. The histone-like, DNA-bending protein, integration host factor (IHF), binds specifically to the termini of IS*1* where it may function cooperatively with InsA, although the role of IHF in IS*1* transposition *in vivo* remains obscure (Gamas *et al.*, 1985, 1987). IS*1* elements exhibit little sequence preference in their target DNAs although some insertion hot-spots have been detected. These sites appear to include regions of bent DNA suggesting that DNA structural features may influence the choice of insertion site (Galas & Chandler, 1989). IS*1* transposes via the so-called replicative transposition pathway (Fig. 2.23) and can form co-integrates between independent replicons (Biel & Berg, 1984; Galas & Chandler, 1982).

The role of IS sequences in promoting cointegrate formation via homologous recombination has been quite important to the development of modern molecular genetics. This is because the F plasmid of *E. coli* harbours copies of IS2, IS*3a*, IS*3b* and a copy of the transposon Tn*1000*, also known as gamma-delta (Willets & Skurray, 1987). These permit F to form cointegrates with the chromosome via homologous recombination involving copies of these IS elements. This results in the formation of the Hfr strains and F′ elements which have been so central to bacterial genetic analysis (Chapter 4).

Composite transposons

Composite transposons arise when IS elements form repeats flanking unique sequences of DNA. These unique sequences frequently code for important functions which alter significantly the phenotypes expressed by the bacterium that harbours them. These functions may specify resistance to antimicrobial agents such as antibiotics or heavy metals or they may be protein toxins which improve the virulence of a pathogenic bacterium which acquires a copy of the transposon. Some transposons code for just one such function while others may carry complex operons coding for multiple functions. A great deal of research has been carried out into the nature and evolution of composite transposons and a brief review of some of the best-characterized examples is given next.

Transposon Tn*9* and the *r-det* of plasmid R100

Transposon Tn*9* is a 2.6 kb sequence composed of directly repeated IS*1* elements flanking a region of DNA that includes a gene coding for chloramphenicol acetyltransferase (Alton & Vapnek, 1979; Berg & Berg, 1987). This makes it one of the smaller composite transposons (Fig. 4.3). Two directly repeated copies of IS*1* flank a region of DNA in the multiple drug resistance plasmid R100 which includes genes coding for resistance to chloramphenicol, mercury, streptomycin and sulphonamides (Fig. 4.4). This region is referred to as *r-det* (for resistance determinant) and it is amplifiable by a tandem duplication mechanism involving homologous recombination across the IS*1* direct repeats (Chandler *et al.*, 1977; Foster, 1983; Hu *et al.*, 1975). The amplified *r-det* is unstable in the absence of selection and deletions restore its copy number to one. Although it is maintained stably in *E. coli*, if R100 is moved to *Salmonella typhimurium*, the *r-det* can be lost by a RecA-promoted deletion across the

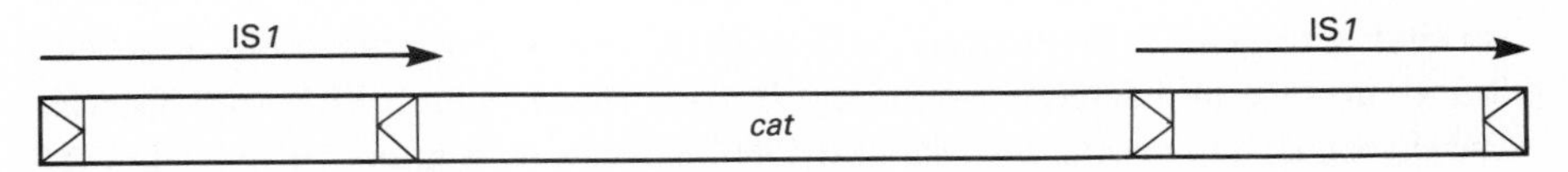

Fig. 4.3 Transposon Tn*9* is a composite transposon consisting of directly repeated copies of insertion sequence IS*1* flanking the gene coding for type 1 chloramphenicol acetyltransferase (*cat*).

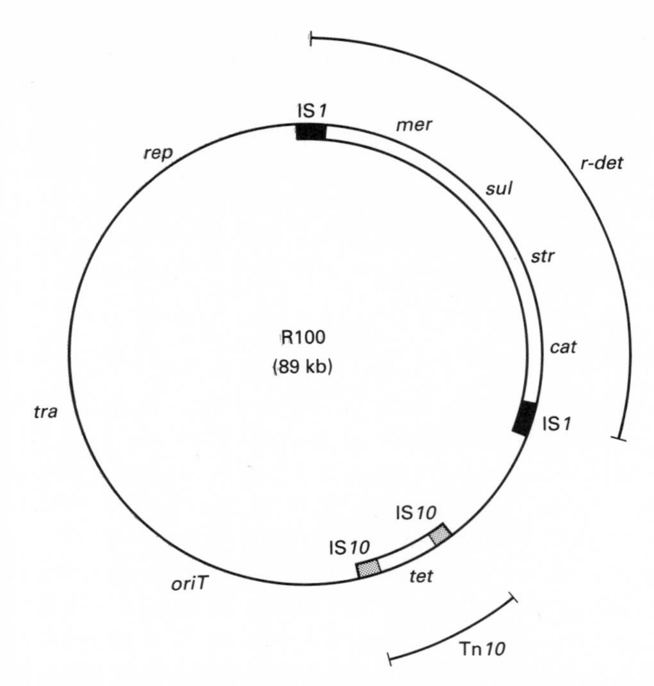

Fig. 4.4 *r-det* element of plasmid R100. R100 is a 89 kb multiple drug resistance plasmid carrying genes for resistance to chloramphenicol (*cat*), mercuric ions (*mer*), streptomycin (*str*), sulphonamides (*sul*) and tetracycline (*tet*). The *tet* gene is a component of transposon Tn*10* while the remaining antimicrobial agent resistance genes are components of the *r-det* element, a large composite transposon flanked by copies of insertion sequence IS*1*. The other features of R100 are: *oriT*, the origin of conjugal transfer; *rep*, the replication region; *tra*, the conjugal transfer operon.

directly repeated IS*1* elements (Galas & Chandler, 1989). Tn*9* and the *r-det* are closely related and it is thought that Tn*9* evolved from *r-det* by a deletion that removed the other *r-det* antimicrobial resistance genes (Iida & Arber, 1977). This is an example of a refinement process taking place within a composite transposon leading to the emergence of a distinct transposable element.

Transposon Tn*10*

The plasmid R100 also harbours the composite transposon Tn*10* (Foster, 1983) (Figs 4.4 & 4.5). This is a 9.3 kb element in which inverted copies of the insertion sequence IS*10* flank a unique region of DNA harbouring genes involved in specifying resistance to tetracycline (Kleckner, 1989). This transposon has been found in several bacteria, including *E. coli*, *Haemophilus*, *Klebsiella*, *Proteus*, *Pseudomonas*, *Salmonella*, *Shigella* and *Vibrio* species (Kleckner, 1989). The IS*10* elements in Tn*10* are not equivalent; IS*10*-Left is defective for transposition and so the composite transposon depends on IS*10*-Right to provide active transposase (Fig. 4.5). Tn*10* insertions appear to be near-random, although hot-spots have

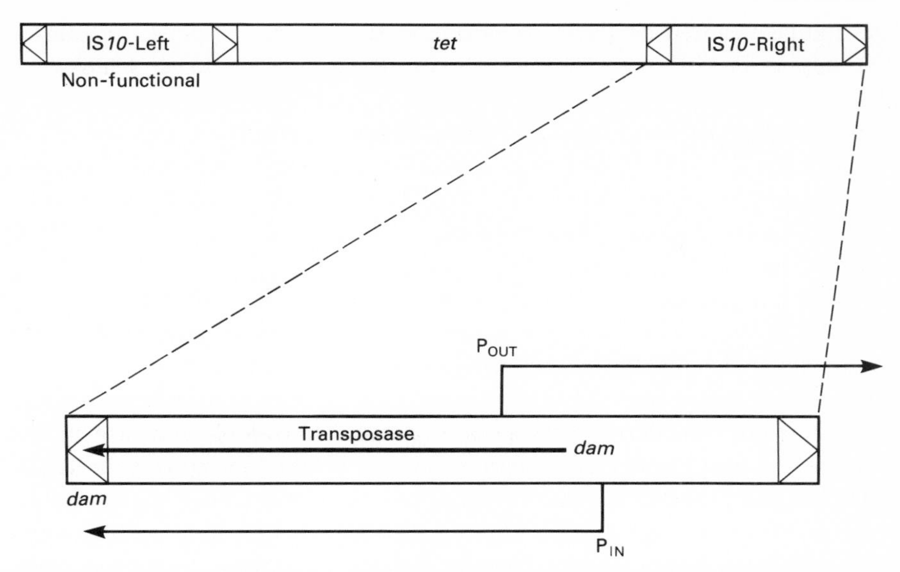

Fig. 4.5 Transposon Tn*10* is a composite transposon consisting of inverted copies of insertion sequence IS*10* flanking unique DNA which includes genes specifying tetracycline resistance (*tet*). IS*10*-Left is transpositionally non-functional. An expanded view of IS*10*-Right is shown, indicating the transposase gene, the pIN and pOUT promoters and the sites for Dam methylation (*dam*).

been reported (Kleckner *et al.*, 1979). Tn*10* insertions can inactivate gene expression by disrupting the coding sequence or by polarity. However, Tn*10* insertions can also cause gene activation from the outward reading P_{OUT} transposon promoters (Ciampi *et al.*, 1982; Simons *et al.*, 1983), especially when the P_{OUT} promoter of IS*10*-Right is used (Kleckner, 1989) (Fig. 4.5).

Tn*10*, like other composite transposons, has the ability to promote so-called 'inside out' transposition events. Here, the IS sequences serve as substrates for transposase as usual but instead of the region of the transposon harbouring the antibiotic resistance genes, the flanking replicon sequences are transposed to a new site. Tn*10* can also promote intramolecular deletion and inversion events which rearrange the host genome. IS*10* insertion sequences alone can promote a rearrangement known as adjacent deletion and Tn*10* can also cause this type of deletion (Kleckner, 1989). Thus, these transposable elements have considerable potential to alter the structure of the genome in ways that are much more radical than by simple insertion mutations.

The best evidence available indicates that Tn*10* transposes via a non-replicative mechanism; consequently, Tn*10* and IS*10* do not promote cointegrate formation (Bender & Kleckner, 1986). Tn*10* insertion produces a 9 bp duplication at the target site (Kleckner, 1979) and the non-replicative transposition mode leaves a double-stranded break in the donor molecule (Kleckner, 1989). This break in the donor molecule is not repaired by transposon-encoded functions and unless it is sealed by host-encoded functions, the molecule will be lost. Damage caused by Tn*10* excision mobilizes the SOS response, possibly a first step in dealing with the DNA break (Roberts & Kleckner, 1988). The histone-like proteins HU and IHF contribute to Tn*10* transposition *in vivo* and a binding site for IHF is found next to the inverted repeat at the outside end of IS*10* (Kleckner, 1989). Thus, like IS*1*, translocation of this transposable element is influenced by these modulators of DNA topology. DNA gyrase may also be required for Tn*10* transposition, suggesting that the process is sensitive to DNA supercoiling.

Tn*10* transposition is also under the control of *dam* methylation. One 5′-GATC-3′ site for DNA adenine methylase (encoded by *dam*) lies at the −10 box of the transposase promoter (pIN) in IS*10* and another is located at the transposase binding site (Fig. 4.5). Failure to methylate these sites enhances expression and function of transposase, causing a 100-fold increase in the frequency of Tn*10* and IS*10* transposition and a 500-fold increase in Tn*10*-promoted deletions and inversions (Roberts *et al.*, 1985). Under normal conditions, in a cell wild-type for *dam*, the two 5′-GATC-3′ sites become transiently hemimethylated during DNA replication. This enhances their biological activity, coupling transposase gene expression to activation of the transposase binding site and coupling the whole transposition system to the DNA replication cycle of the host cell. Differential activation of Dam sites means that only one of the daughter Tn*10* elements created by the passage of the replication fork is proficient to transpose; an important caveat given a transposition mechanism which produces a double-stranded gap in the donor molecule! The timing of replication also means that transposase will preferentially bind to its own IS*10* element because this will be hemimethylated at the time of pIN activation. Action of transposase *in cis* is also a function of transposase protein which does not diffuse freely through the cell but remains at the IS*10* element that encodes it (Morisato *et al.*, 1983).

Tn*10* transposition is negatively regulated by antisense RNA transcribed from the pOUT promoter which has sequence complementarity with the 5′ end of the transposase transcript from the pIN promoter (Fig. 4.5). By base pairing with the pIN transcript, pOUT RNA sequesters the ribosome-binding site needed for transposase to be translated (Fig. 4.5). This *trans* effect increases as the number of copies of IS*10* in the cell increases and is known as 'multicopy inhibition' (Simons & Kleckner, 1983).

Transposon Tn*5*

The composite transposon Tn5 and its derivatives have been used widely in genetic analyses of many bacteria, including several major pathogens (Berg, 1989; Berg & Berg, 1983; Berg *et al.*, 1989). Tn5 is a 5.8 kb element composed of 1533 bp IS*50* inverted repeats flanking a region of unique DNA that includes genes specifying resistance to kanamycin, bleomycin and streptomycin, although the last of these is not expressed in *E. coli* (deVos *et al.*, 1984; Selvaraj & Iyer, 1984). The antibiotic resistance genes are all transcribed from a promoter

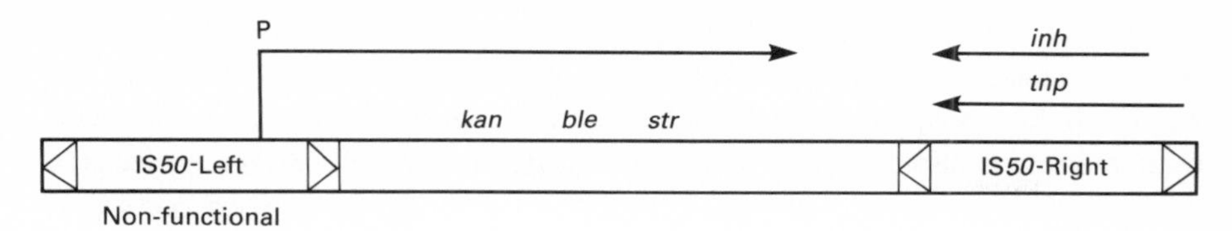

Fig. 4.6 Transposon Tn5 is a composite transposon consisting of inverted copies of insertion sequence IS*50* flanking unique DNA coding for resistance to kanamycin (*kan*), bleomycin (*ble*) and streptomycin (*str*). The IS*50*-Left element is transpositionally non-functional but harbours the promoter for expression of the drug resistance genes. IS*50*-Right encodes transposase (*tnp*) and transposase inhibitor (*inh*).

in the left-hand IS*50* element, IS*50*-Left (Fig. 4.6). As with the IS elements bracketing Tn*10*, IS*50*-Left and IS*50*-Right are non-equivalent. Only IS*50*-Right encodes active transposase and this IS is responsible for the transposition of the whole transposon. In IS*50*-Left, a single base substitution has generated a UAA translation stop codon in the transposase open reading frame and simultaneously created the promoter for expression of the drug resistance genes in the unique sequence (Rothstein & Reznikoff, 1981) (Fig. 4.6). This is a fascinating example of a single base pair change in a DNA sequence simultaneously creating a novel operon and a novel composite transposable element.

Tn5 transposes via a non-replicative mechanism (Fig. 2.23) and generates 9 bp repeats at the site of insertion in the target DNA (Berg, 1989). IS*50*R encodes both the transposase and the transposition inhibitor protein from the same open reading frame from staggered translation start sites (Isberg *et al.*, 1982). As with the transposase in Tn*10*, Tn5 transposase works best in *cis* (Isberg *et al.*, 1982; Johnson *et al.*, 1982). In contrast, the inhibitor is *trans*-acting, but the mechanism of inhibition is unknown (Berg, 1989). Tn5 shares with Tn*10* a sensitivity to Dam methylation; 5′-GATC-3′ sequences overlap the transposase promoter and are found at the inside (I) end of IS*50* but not at the functionally distinct outside (O) end. Methylation of the Dam site in the transposase promoter is believed to reduce expression of the transposase gene and methylation of the I-end Dam site is thought to prevent its use as a substrate for transposition (Berg, 1989; Yin *et al.*, 1988).

Host factors implicated in the control of Tn5 transposition include integration host factor (Makris *et al.*, 1988), DNA gyrase (Isberg & Syvanen, 1982), DnaA (Yin & Reznikoff, 1987), FIS (Weinreich & Reznikoff, 1992) and Lon (Sasakawa *et al.*, 1987). Some of these factors link Tn5 transposition to the degree of DNA supercoiling in the cell and to the cell cycle. Thus, the transposon is not a completely independent agent but is keyed into major processes affecting DNA metabolism in the cell.

The Tn*3* family of transposable elements

Transposon Tn*3* represents a distinct family of bacterial transposable elements which contains more than 20 individually numbered members (listed in

Sherratt, 1989). Tn*3* is a 4957 bp element with 38 bp terminal repeats and it specifies resistance to the β-lactam antibiotic, ampicillin. It is related to the gamma-delta transposon (alias Tn*1000*) already mentioned in conjunction with the F plasmid (Chapter 3), although unlike Tn*3*, gamma-delta does not possess a drug resistance marker. Transposition of this family of transposons occurs via cointegrates, i.e. by a replicative mechanism (Figs 2.23 & 2.24), distinguishing its members from other transposons such as Tn*5* or Tn*10*. The Tn*3*-like transposons possess the means to resolve the cointegrate intermediates themselves, using a protein called resolvase to do it. Tn*3* is organized such that the transposase gene and the resolvase gene are divergently transcribed from promoters within a specialized region called *res* (Fig. 4.7). This sequence is bifunctional, serving as a regulatory site for control of transcription of the transposase and resolvase genes (*tnpA* and *tnpR* respectively) and as the site required for the recombination event that resolves the cointegrates (Fig. 4.7). Not all Tn*3* family members are organized according to this pattern. A sub-group known as the Tn*501* group (named after its prototypic member) has the resolvase and transposase genes organized as an operon with *res* in the regulatory region of the resolvase gene (Fig. 4.7).

Several theories have been advanced to describe the precise topological detail of Tn*3* cointegrate resolution and expositions of the currently favoured 'two-step synapsis' as well as the now out-of-favour 'tracking' and 'slithering' models may be found in Benjamin *et al.* (1985), Gellert & Nash (1987), Hatfull *et al.*

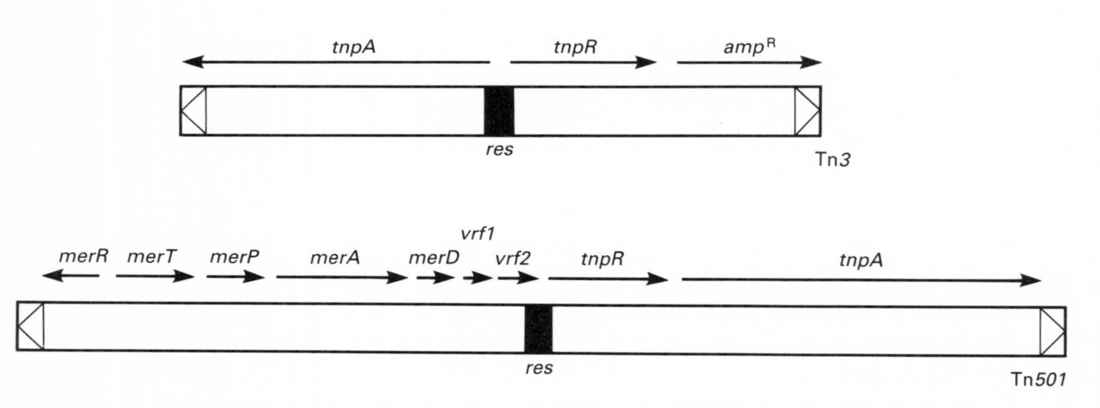

Fig. 4.7 The Tn*3* family of transposons. Two well-characterized members of the Tn*3* family are illustrated. Tn*3* itself carries divergently transcribed operons coding for transposase (*tnpA*, the leftward-reading gene) and resolvase (*tnpR*) and ampicillin resistance (β-lactamase) (*amp*^{R}, the rightward reading operon). The operons are transcribed from a central region which includes the resolution (*res*) site, which is essential for the site-specific recombination event by which cointegrate structures are resolved. Tn*501* possesses genes for transposase (*tnpA*) and resolvase (*tnpR*) which are transcribed in the same direction. It also harbours a mercuric ion resistance (*mer*) operon. The *res* site is located 5′ to the *tnpR* gene. *vrf*: an open reading frame of unknown function.

(1988), Krasnow & Cozzarelli, (1983), Stark *et al.* (1989a,b). The resolution step involves a site-specific recombination event of the 'conservative' rather than the transpositional category. Resolvase, the recombinase responsible for this process, is a member of a family of site-specific recombinases known as the 'invertases'. These are so-called because they catalyse principally DNA inversion reactions. This group is described later.

Tn*3*-like transposons have been detected in the Gram-positive cocci, *Staphylococcus aureus* and *Streptococcus* (now *Enterococcus*) *faecalis*. These elements, Tn*551* and Tn*917*, respectively, specify resistance to MLS (macrolide-lincosamide-streptogramin B) antibiotics, transpose via a replicative mechanism and generate 5 bp repeats in the target DNA (Murphy, 1989). A Tn*3*-like cryptic transposon in *Bacillus thuringiensis*, Tn*4430*, has a transposase protein similar to other members of the family but a resolvase function that is related to the integrase protein of bacteriophage λ (Murphy, 1989; see later for a discussion of the integrase family of site-specific recombinases). Tn*3*-related elements have also been found in *Streptomyces fradiae* (Olson & Chung, 1988) and *Clostridium perfringens* (Abraham & Rood, 1987).

Transposons with insertion site specificity

Some transposons resemble phage in their insertion site specificity. In Gram-negative bacteria, Tn*7* (which codes for resistance to the antibiotics trimethoprim, spectinomycin and streptomycin) inserts at specific sites in the genomes of *Agrobacterium tumefaciens, Caulobacter crescentus, Escherichia coli, Klebsiella pneumoniae, Pseudomonas aeruginosa, Ps. fluorescens, Ps. solanacearum, Rhizobium meliloti, Rhodopseudomonas capsulata, Salmonella typhimurium, Serratia marcescens, Vibrio* species and *Xanthomonas campestris*. (For references, see the review by Craig, 1989.) In fact, so similar is Tn*7* to a bacteriophage in terms of its relationship with the host genome, that its insertion site in *E. coli* is referred to as *att*Tn*7* (Lichenstein & Brenner, 1982). The mechanism through which Tn*7* transposes remains obscure, although cointegrate formation is thought to be unlikely (Craig, 1989).

A site-specific transposon called Tn*554* has been found in the Gram-positive organism *Staphylococcus aureus*. This element specifies resistance to MLS antibiotics and spectinomycin and inserts at a unique site called *att*Tn*554* in one orientation (Krolewski *et al.*, 1981; Philips & Novick, 1979). Tn*554* differs from bacteriophages in that it does not seem to excise from the donor during transposition (Murphy, 1989).

Bacteriophage Mu

In contrast to transposons Tn*7* and Tn*554* which have some superficial similarities to bacteriophages, Mu *is* a true temperate bacteriophage which can also function as a transposon. The viral DNA is 37 kb long but the virion includes 50–150 bp of host DNA covalently linked to the 'c' (left) end and from 500 bp to 3 kb attached to the S (right) end. During lytic growth, Mu uses the replicative mechanism of transposition via cointegrates (Fig. 4.8). To integrate with the host genome following infection, Mu uses the non-replicative transposition pathway (Craigie *et al.*, 1988; Pato, 1989; Symonds *et al.*, 1987; Toussaint & Résibois, 1983). It inserts into target DNA in a near-random manner, making it an excellent mutagen (hence its name, from mutator) and it creates 5 bp direct repeats at the site of insertion (Allet, 1979).

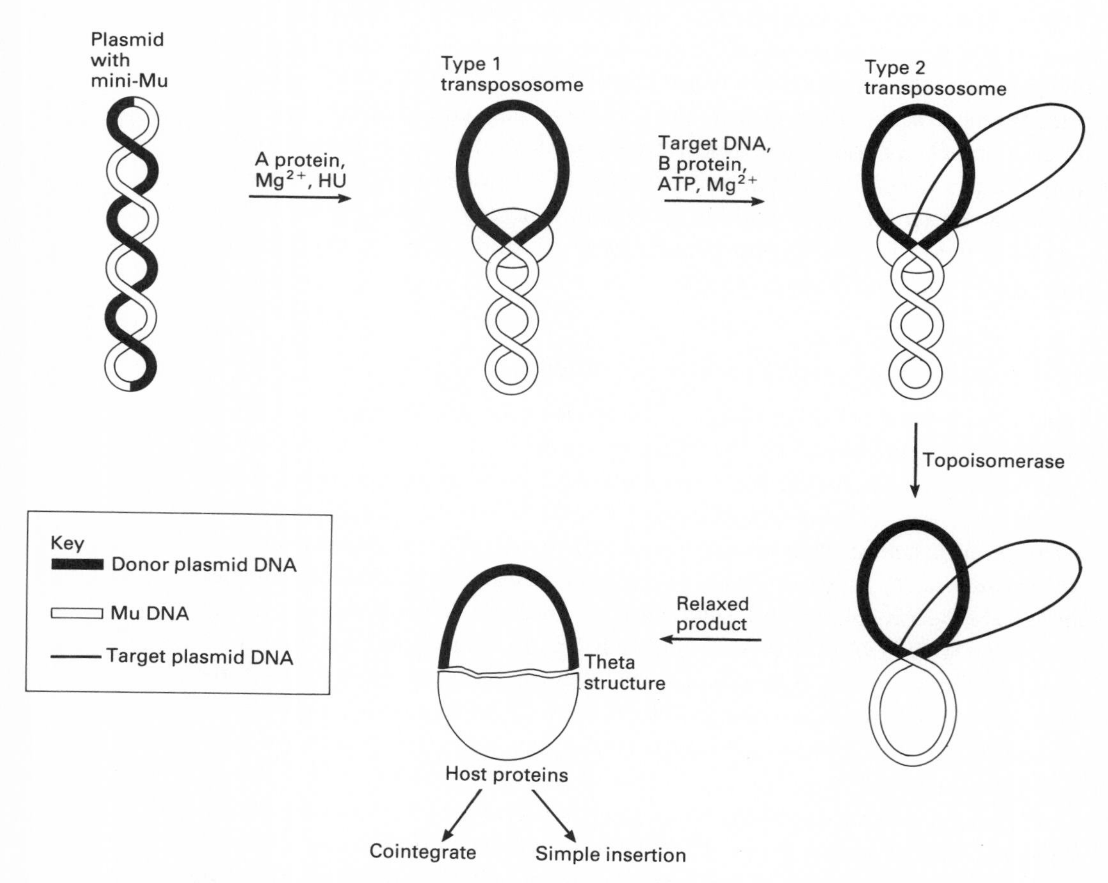

Fig. 4.8 Mu transposition pathway. In the presence of Mu-encoded A protein, histone-like protein HU and Mg^{2+} ions, a plasmid harbouring a bacteriophage Mu insertion is converted to the type 1 transpososome. Here, the plasmid DNA is shown as a relaxed circle with the Mu DNA as plectonemically wrapped supercoils. This structure is stabilized by the bound A protein. Adding target DNA, ATP, Mu-encoded B protein and Mg^{2+} promotes the formation of the type 2 transpososome in which a synaptic complex is formed between the Mu-donor molecule and the target DNA. The supercoiled Mu DNA is relaxed by topoisomerase activity, giving rise to a theta structure which can be resolved by host-encoded functions to yield either a simple insertion in the recipient molecule or a Mu-linked donor–recipient cointegrate. Based on the data of Surete *et al.*, (1987).

The key components of Mu for transposition are the ends and the A and B proteins. A is the transposase and B is a non-specific DNA-binding protein with ATPase activity. Mu transposition proceeds via two intermediate 'transpososome' stages. The first stage requires supercoiled phage DNA, histone-like protein HU, Mg^{2+}; the second requires the product of the first stage (transpososome one), target DNA, B protein ATP and Mg^{2+} (Fig. 4.8). The transposition pathway and the topology of the participating DNA molecules have been described in detail by Chaconas & Surette (1988); Pato (1989) and Surette *et al.* (1987).

Replicons already harbouring a copy of Mu are unlikely to accept a second copy, a phenomenon known as transposition immunity. A mechanism to explain Mu transposition immunity has been proposed in which the B protein fails to associate stably with immune molecules due to the presence of A protein and ATP. Thus, immune molecules are blocked at stage one of the transposition process (Adzuma & Mizuuchi, 1988; Symonds, 1988). In addition to HU, the histone-like protein IHF is required for Mu transposition. This is because IHF is needed for transcription of Mu genes (Krause & Higgins, 1986). Bacteriophage Mu has been heavily exploited as a tool in genetic research and Mu-like phages have been detected in several organisms, including *Pseudomonas* (Krylov *et al.*, 1979) and *Vibrio cholerae* (Gerdes & Romig, 1975; Mekalanos *et al.*, 1982).

Conjugative transposons

Some transposable elements have the genetic capacity to promote their own transfer from cell to cell via a conjugative mechanism. Examples have been found in both Gram-positive bacteria (reviewed in Clewell & Gawron-Burke, 1986; Clewell *et al.*, 1988; Murphy, 1989; Scott, 1992) and Gram-negative bacteria (Hecht & Malamy, 1989). In the Gram-positive bacteria, these elements are large (16–60 kb) and always encode resistance to tetracycline and, in some cases, resistance to additional antibiotics. The Gram-negative element Tn*4399* has a tetracycline resistance determinant and, unlike the Gram-positive transposons, can mobilize non-conjugative plasmids in *cis* (Hecht & Malamy, 1989). Among the Gram-positive organisms, Tn*916*, Tn*918*, Tn*919*, Tn*925*, Tn*1545* and Tn*3951* have been detected in species of *Streptococcus* but most exhibit a broad host range (Christie *et al.*, 1987; Clewell *et al.*, 1985; Courvalin & Carlier, 1986, 1987; Fitzgerald & Clewell, 1985; Vijayakumar *et al.*, 1986).

Tn*916* is the Gram-positive element that has been studied in greatest detail. It is 16.4 kb long with over half of the DNA being concerned with conjugative transfer. It is self-transmissible during mating and can transpose to both conjugative and non-conjugative plasmids or to the chromosome in the recipient cell. Donor–recipient cell contact may provide the signal for the onset of transposition. Tn*916* forms covalently closed supercoiled circular intermediates during transposition (Fig. 4.9), does not generate target site sequence duplications and excises via a mechanism that generates a heteroduplex in the circular intermediate (Scott *et al.*, 1988; Caparon & Scott, 1989). It is possible that

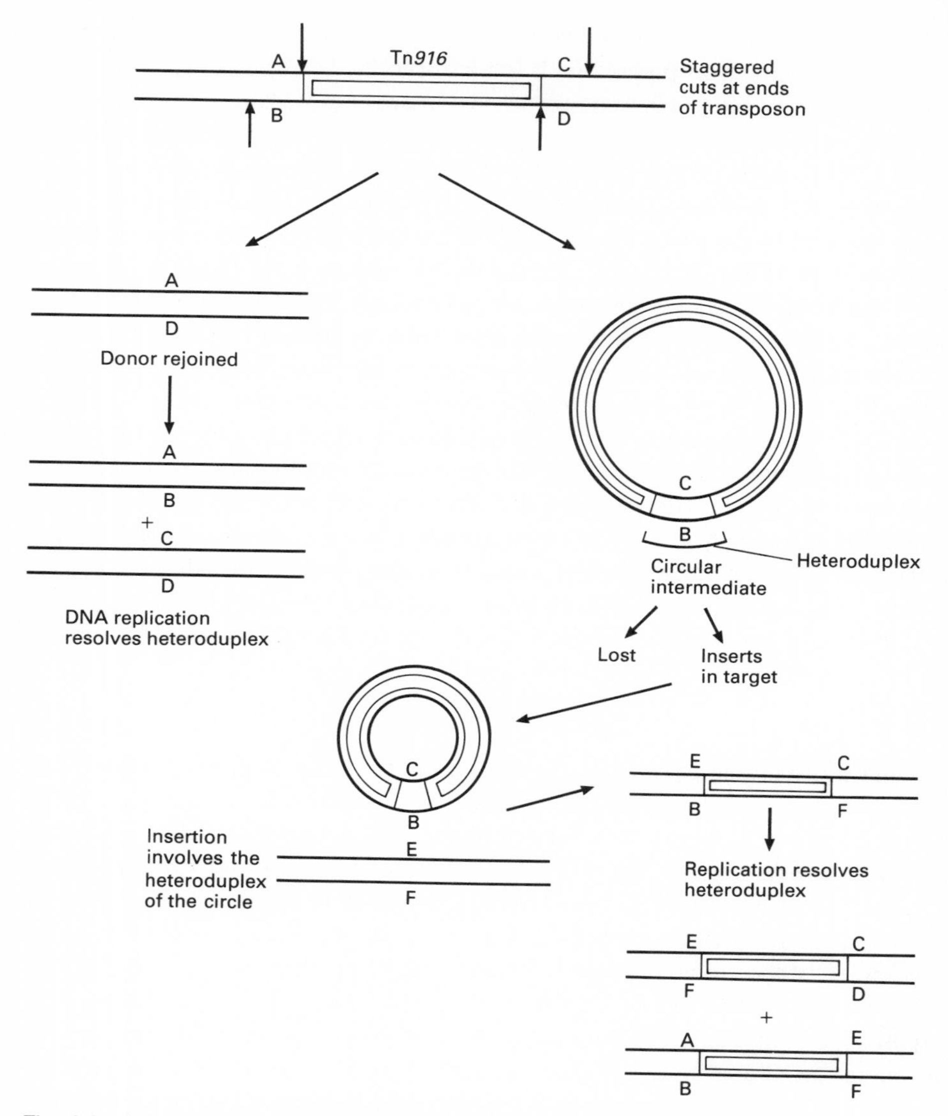

Fig. 4.9 Model of Tn*916* transposition. Transposon Tn*916* is excised from the donor molecule by staggered cuts at its ends. The donor is rejoined to form a heteroduplex, which is resolved by subsequent rounds of DNA replication. The excised transposon forms a circular transposition intermediate, closed at a heteroduplex. This molecule can be lost or it can insert into a target molecule. The heteroduplex is required for the insertion event. This results in the formation of further heteroduplexes between the inserted transposon and the recipient molecule. These are resolved by DNA replication. Based on the data of Caparon & Scott (1989).

host-encoded functions are required for transposition, for example, once conjugative transposons become established in *Lactococcus lactis*, they cannot be donated to further recipients. This could be explained if this bacterium lacks the requisite accessory factors for conjugative transposition (Bringel *et al.*, 1991).

Site-specific recombination: conservative recombination

Conservative site-specific recombination involves DNA breakage and reunion reactions of specific base pairs within specific sequences and is catalysed by specialized enzymes (site-specific recombinases) which bind to and act at these sequences. In these reactions, there is no net loss or gain of DNA and no DNA synthesis. The products of the reaction depend on the locations and the orientations of the participating DNA sequences. Site-specific recombination can be intermolecular or intramolecular and, for intramolecular events, can involve sequences arranged in directly repeated or inverted orientation. Intermolecular recombination leads to replicon fusion through cointegrate formation (Fig. 2.25). Intramolecular recombination across directly repeated sequences produces a deletion of the intervening sequences, as happens during cointegrate resolution (Fig. 2.24). Such resolutions restore independence to replicons which had previously fused to form a cointegrate structure. Recombination across inverted repeats causes the intervening sequences to invert (reviewed in Stark *et al.*, 1992) (Fig. 2.25). This effect is exploited in several 'genetic switch' systems such as those described later in this chapter (reviewed in Plasterk, 1992).

Integrase systems

This group of site-specific recombination systems is operated by members of a family of recombinases showing homology to the Int (integrase) proteins of

Table 4.1 Alignment of homologous carboxyterminal regions of members of the integrase family of site-specific recombinases. Data come from Argos *et al.* (1986); Colloms *et al.* (1990); Dorman & Higgins (1987); Murphy (1989); Krause & Guiney (1991)

E. coli	FimB	HMLRHSCGFALANMG–IDTRLIQDYLGHRN–IRHT–VWYTA
E. coli	FimE	HMLRHACGYELAERG–ADTRLIQDYLGHRN–IRHT–VRYTA
E. Coli	XerC	HKLRHSFATHMLESS–GDLRGVQELLGHAN–LSTT–QIYTH
Phage	λ	HELRSLSA–RLYEKQ–ISDKFAQHLLGHKS–DTMA–SQYR–
Phage	P1	HSARVGAARDMARAG–VSIPEIMQAGGWTN–VNIV–MNYIR
Phage	P2	HALRHSFATHFMING–GSIITLQRILGHTR–IEQT–MVYAH
Phage	P4	HGFRTMARGALGESGLWSDDAIERQLSHSERNNVR–AAYIH
Phage	P22	HDLRHTWASWLVQAG–VPISVLQEMGGWES–IEMV–RRYAH
Phage	Phi80	HDMRRTIATNLSELG–CPPHVIEKLLGHQM–VGVM–AHYNL
Phage	186	HVLRHTFASHFMMNG–GNILVLQRVLGHTD–IKMT–MRYAH
FD-Protein		HTFRHSYAMHMLYAG–IPLKVLQSLMGHKS–ISST–EVYTK
R46	Orf3	HTLRHSFATALLRSG–YDIRTVQDLLGHSD–VSTT–MIYTH
pSDL2	Rsd	HTFRHSYAMHMLYAG–IPLKVLQALMGHKS–VSST–EVYTK
Yeast	Flp	HIGRHLMTSFLSMKGLTELTNVVGNWSDKRASAVARTTYTH
Tn*554*	TnpA	HMLRHTHATQLIREG–WDVAFVQKRLGHAH–VQTTLNTYVH
Tn*554*	TnpB	HAFRHTVGTRMINNG–MPQHIVQKFLGHES–PEMT–SRYAH
Tn*2603*		HTLRHSFATALLRSG–YDIRTVQDLLGHSD–VSTT–MIYTH
Tn*4430*	TnpI	HQLRHFFCTNAIEKG–FSIHEVANQAGHSN–IHTT–LLYTN
Consensus		H LRH G GH Y

bacteriophage λ and related bacterial viruses (Table 4.1). These proteins catalyse both intermolecular and intramolecular recombination events and can recombine pairs of repeat sequences which are directly or inversely oriented on the same DNA molecule.

Bacteriophage λ integration and excision

This section begins with a discussion of the λ site-specific recombination system because it is understood in detail at the molecular level and many of its features are recapitulated in other, less well-understood systems, including many involved directly in bacterial virulence. Studies of λ recombination have identified important cofactors encoded by the bacterial host which play key roles in organizing the local architecture of DNA. These cofactors, such as the integration host factor, are now recognized as playing vital roles in many other processes, including the regulation of transcription.

Before describing the site-specific recombination system of λ, it is necessary to place this system in the context of the bacteriophage life cycle. λ is a temperate phage which parasitizes *E. coli* K-12. The λ virion binds to a specific receptor (the LamB protein) on the surface of the host and injects linear DNA into the cytoplasm. This DNA is circularized by DNA ligase, supercoiled by DNA gyrase and can then follow one of two pathways, the lytic or the lysogenic. The lytic pathway involves replication of the λ DNA by a rolling circle mechanism, packaging of the viral DNA into new phage heads and then lysis of the host with release of the virus particles to the external medium. In the lysogenic pathway the viral DNA becomes part of the host genome by a site-specific recombination reaction and thereafter is replicated along with the rest of the bacterial chromosome. In this integrated state the λ phage DNA is referred to as 'prophage' and it is maintained in this state by high concentrations of a λ-encoded protein called the cI repressor. Excision of the prophage can be triggered environmentally, e.g. by DNA damage leading to activation of the SOS response, with subsequent cleavage of the λ cI repressor in the presence of RecA. The excised phage DNA then enters the lytic cycle.

λ integrates at a specific site on the *E. coli* chromosome called *att*λ via a recombination mechanism involving a viral sequence called *attP* and a chromosomal sequence called *attB* (for reviews see Hendrix *et al.*, 1983; Landy, 1989; Thompson & Landy 1989) (Fig. 4.10). The core of each site consists of 15 bp of identical, non-palindromic sequence which confers polarity as well as site-specificity on the recombination event. Integration is catalysed by a phage-encoded recombinase called Int (the 'integrase') which binds to specific sequences in *attB* and *attP*. Recombination is stimulated by supercoiling of the phage DNA; the bacterial DNA need not be supercoiled. Int requires the host-encoded integration host factor (IHF) to catalyse recombination at normal efficiency. This site-specific DNA-binding protein has binding sites among the Int-binding sites in *attP* (Fig. 4.10). The function of IHF appears to be to fold up the *attP* DNA into a higher-order structure (called the 'intasome') which

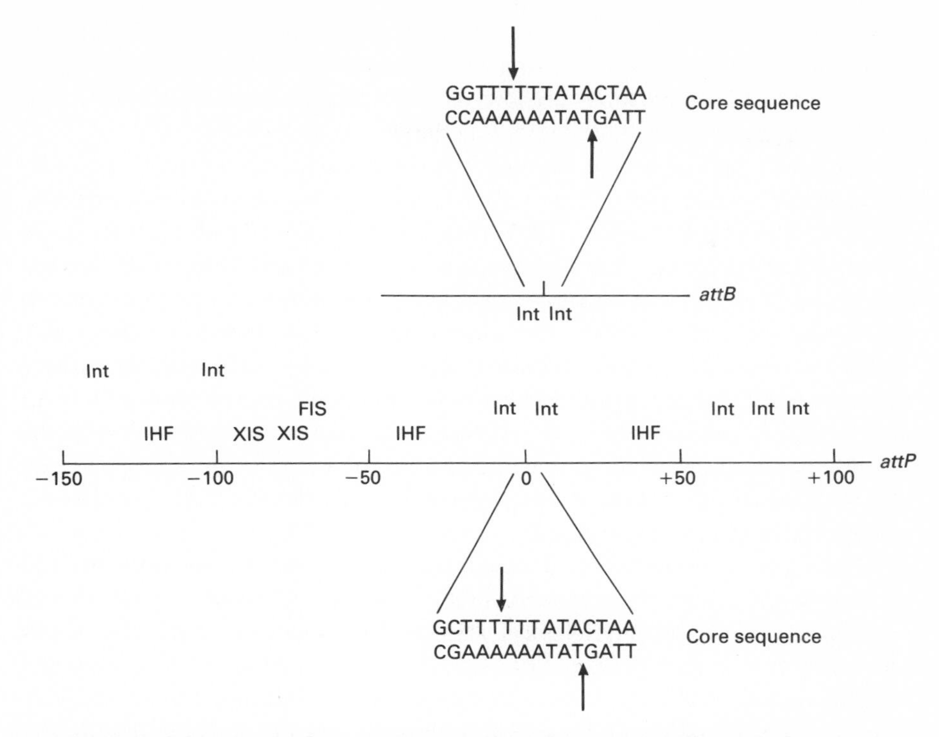

Fig. 4.10 Lambda site-specific recombination site. The *attP* site in bacteriophage lambda includes a core sequence which has an homologous counterpart on the *Escherichia coli* chromosome at the *attB* site. The *attB* site binds lambda-encoded integrase protein which catalyses the formation of staggered cuts within the core sequence. Corresponding cuts are made within the core sequence within *attP* on the phage. In addition to the core sequence, *attP* possesses multiple binding sites for structural and catalytic proteins required to promote site-specific recombination events involved in lambda integration and excision. Int, integrase; IHF, integration host factor; FIS, factor for inversion stimulation; XIS, excisionase. Distances are calibrated in bp. Arrows show cleavage sites in core sequence.

provides a cleft in which *attB* may be captured by the Int protein bound to *attP* (Fig. 4.10). It has been suggested that λ integration may be linked to the physiology of the host cell by IHF availability since this protein appears to be most abundant at the onset of the stationary phase of the growth cycle (Bushman *et al.*, 1985; Chapter 2).

The reverse reaction (excision) requires Int, IHF, Xis (the excisionase, another phage-encoded protein) and another host factor called FIS (factor for inversion stimulation, described in Chapter 3). The Xis and FIS proteins have specific binding sites on the DNA of the integrated phage (Fig. 4.10). There are two adjacent Xis-binding sites to which the protein binds cooperatively. There is a single FIS site and this overlaps one of the Xis sites (Fig. 4.10). Xis binding introduces a bend of up to 140° into the DNA and this bending is believed to be necessary for the DNA sequences *attL* and *attR* to find each other so that recombination can proceed (Landy, 1989; Thompson & Landy, 1988). FIS also alters the pathway of the DNA duplex, bending it by up to 95° (Thompson & Landy, 1989). When Xis levels are low, as in the case of a weak induction of the SOS response, FIS can stimulate Xis function by up to 20-fold.

FIS levels appear to be regulated metabolically; the protein is almost undetectable in stationary phase cells whereas it is abundant in exponentially growing cells (Thompson *et al.*, 1987; Chapter 2). Thus FIS levels and IHF levels are inversely regulated. This is significant for λ since high levels of IHF inhibit excision, a situation that occurs *in vivo* under conditions where FIS is limiting (Bushman *et al.*, 1984, 1985). The physiological link between FIS, IHF and λ excision conspire to prevent excision at a point in the growth cycle (stationary phase) at which the cell is probably least able to support the lytic pathway. In an emergency, λ can override this physiological control system by exploiting conditions that produce an abundance of Xis. For example, following a strong SOS response in which most of the cI repressor is cleaved, Xis levels are elevated and the excision of λ can proceed.

λ site-specific recombination illustrates the use of both overlapping and countervailing regulatory motifs to link a discrete molecular event (*att* site recombination) to the physiological state of the host cell. Thus, IHF levels participate in the lytic/lysogenic decision: high IHF level favour Int-promoted integration and these occur in non-growing cells. Non-growing cells are unlikely to support a successful lytic burst. Put anthropocentrically, it is prudent for the phage to tag itself to the chromosome until better times arrive. Once in the chromosome, the phage is likely to stay there unless the host is damaged. A low-level induction of the SOS response permits a modest build-up in Xis levels. If this happens in growing cells, the availability of FIS and the absence of the inhibitory IHF will favour excision. If a mild SOS induction occurs in non-growing cells, the high IHF and low FIS levels will restrain excision. However, if the cell is severely damaged, λ has the ability to override the IHF–FIS balanced control of excision. The high levels of Xis produced when all of the cI molecules are cleaved by RecA will promote λ excision regardless of the condition of the cell since high levels of Xis permit cooperative binding of that protein to the two adjacent Xis-binding sites without the need for FIS enhancement (Fig. 4.10). This may be regarded as an emergency evacuation procedure used by λ in the event of massive damage to cellular DNA.

Other integrase systems

A family of site-specific recombination systems which contain 'integrase-like' recombinases has been described. Most of the genetic elements which exploit these recombinases are lambdoid bacteriophage Int proteins (such as those of ϕ80 and P22) but the group also includes the Cre system of phage P1 which is concerned with the resolution of P1 prophage plasmid dimers and which reacts with specific sites called *lox* within the phage genome (Hoess & Abremski, 1984; Chapter 3), the Fim recombinases which regulate Type 1 fimbrial gene expression through inversion of a promoter-carrying DNA segment in *E. coli* (Dorman & Higgins, 1987; Eisenstein *et al.*, 1987; Chapter 5) and the Flp protein which regulates site-specific recombination on the 2 μm plasmid of the yeast *Saccharomyces cerevisiae* (Argos *et al.*, 1986). Family members possess a small number of conserved amino acids in the carboxy-termini of the recombinases, which may be the location of the catalytic centre (Table 4.1). Int, Flp and Cre are also known to interact with DNA in a similar way (reviewed in Craig, 1988). Although the Fim recombinases and the Int proteins encoded by bacteriophages λ, ϕ80 and P22 all require IHF (Friedman, 1988a), host-encoded cofactor requirements are not constant among the group.

Integrons

Genes coding for antibiotic resistance in transposons and plasmids of Gram-negative bacteria are frequently found at a unique site within a conserved DNA sequence. The conserved regions extend for 1.36 kb 5′ to the drug resistance genes and for 2 kb on the 3′ side. These flanking sequences, together with their intervening segment of variable DNA constitute an 'integron', a potentially mobile DNA element (Stokes & Hall, 1989). The genes of the variable central region are thought to be cotranscribed from a common promoter within the 5′ conserved segment. Each inserted drug resistance gene is associated at its 3′ end with a copy of an imperfectly repeated sequence called the '59-base element' (Fig. 4.11). The members of the 59-base element family serve as substrates for site-specific recombination events catalysed by the integron-encoded 'integrase'. Integron 'integrase' is encoded by the *int* gene within the 5′ conserved element and the protein shows homology to the integrase family of site-specific recombinases (Argos *et al.*, 1986; Stokes & Hall, 1989) (Table 4.1). Based on data from DNA sequence comparisons, it has been proposed that a related gene insertion system is associated with transposon Tn7 (Hall *et al.*, 1991).

Hall *et al.* (1991) have defined three conserved features shared by individual drug resistance gene inserts. These are the coding sequence of the gene up to the last 7 bp of the 59-base element associated with the 3′ end of the gene and the core recombination site located at the 5′ end. These features constitute a gene cassette. Their structure can be explained if the cassette forms a circle in which the 5′ core site and all but the last 7 bp of 59-base element are covalently associated. These circularized gene cassettes then integrate into integrons through a single site-specific recombination event across the core site of the circle and a core site within the integron (Collis & Hall, 1992b) (Fig. 4.11). Integrase-promoted deletion events can remove gene cassettes from the integron and these can then integrate elsewhere. Excised cassettes exist as covalently closed supercoiled circles (Collis & Hall, 1992a). Deletion and duplication events can be promoted by integrase in which cointegrate structures are formed and then resolved by integrase using different pairs of sites; the frequency of deletion is ten times higher than that of duplications (Collis & Hall, 1992b). Thus, individual gene cassettes can be mobilized by integrase-mediated excision, with

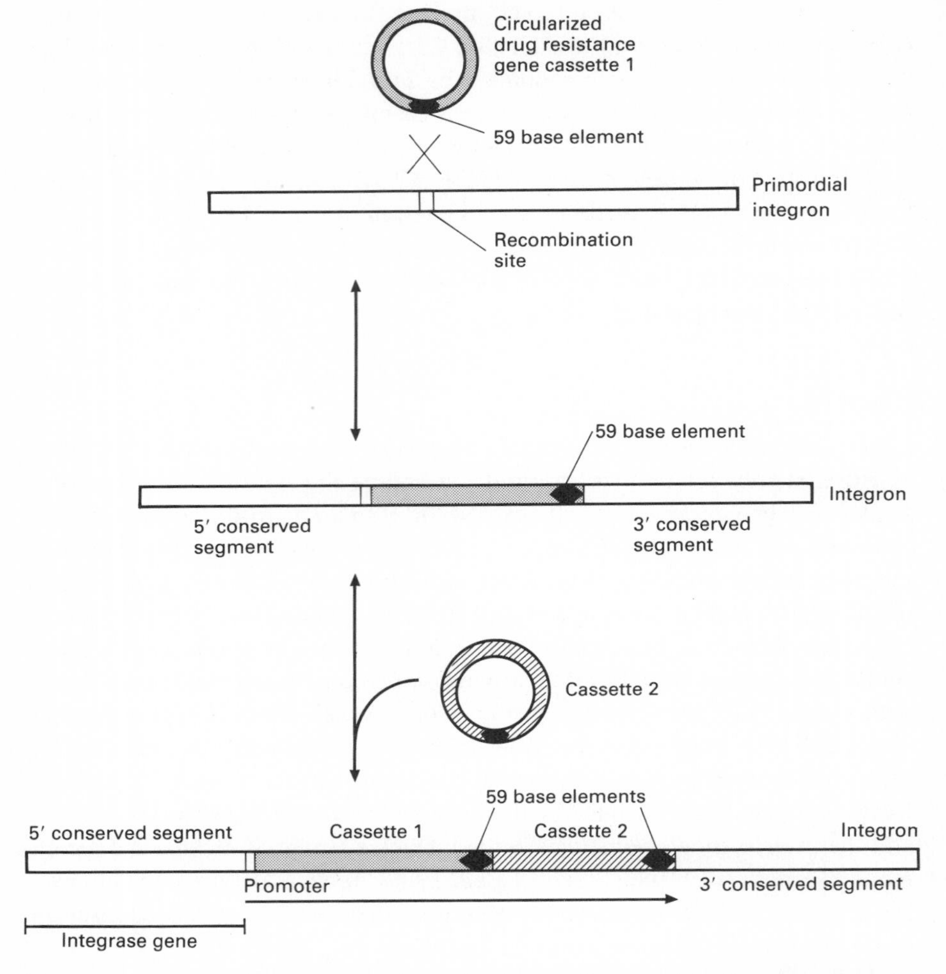

Fig. 4.11 Evolution of integron complexity by site-specific recombination. Circularized drug resistance gene cassette 1 inserts into a primordial integron at the recombination site via an event that involves the 59-base element. Further cassettes can enter the developing integron by such recombination events. They can also leave by a reversal of the process. Drug resistance genes are expressed as an operon from a common promoter within the 5′ conserved segment. This DNA sequence also encodes the integrase protein which catalyses the recombination events.

profound consequences for the dissemination of antibiotic resistance genes in the bacterial population.

Invertase systems

Site-specific recombinases of the invertase family catalyse DNA inversion reactions which alter gene expression (Craig, 1985, 1988; Glasgow *et al.*, 1989; Van de Putte & Goosen, 1992). These enzymes only work efficiently with sites that are inversely oriented on the same molecule. Unlike some transposases, the invertases work well *in trans*. The prototypic members of the family are Cin (bacteriophage P1 tail fibre expression), Hin (flagellar phase variation in *Salmonella typhimurium*), Gin (bacteriophage Mu tail fibre expression) and Pin (encoded by the e14 defective prophage in *E. coli*) (Hiestand-Nauer & Iida, 1983; Plasterk *et al*., 1983; Plasterk & van de Putte, 1985; Zeig & Simon, 1980). These proteins are up to 70% homologous in amino acid sequence and are functionally interchangeable (Iida *et al*., 1982; Kamp & Kahmann, 1981; Kutsukake & Iino, 1980). They also show significant homology to the resolvase proteins encoded by Tn*3*-like transposons (see earlier). Invertases recognize a conserved 26 bp DNA sequence made up of two 12 bp imperfect inverted repeats and a 2 bp core (Iida *et al*., 1984; Mertens *et al*., 1988; Plasterk & van de Putte, 1985). Efficient recombination also requires enhancer sequences in *cis* and these have been shown to be interchangeable for some systems (Huber *et al*., 1985; Johnson & Simon, 1985). The protein that binds to the Hin and Gin enhancers is FIS (Factor for Inversion Stimulation), which is also involved in modulating the excision step in λ site-specific recombination (Johnson *et al*., 1987; Koch & Kahman, 1986; Thompson *et al*., 1987). FIS binding to the enhancer is thought to organize the Hin recombination sites into a conformation required for a productive recombinational synapse (Johnson *et al*., 1987) (Fig. 4.12). The histone-like protein HU also stimulates recombination by the invertase protein Hin with the stimulatory effect depending on the distance between the enhancer and the recombination site (Johnson *et al*., 1986b, 1987) (Fig. 4.13). The relative positions of the recombination sites, enhancers and genes coding for the recombinases vary among the prototypic invertase systems (Fig. 4.12).

Complementation studies have permitted the identification of functioning invertase systems in *Shigella boydii* (*pinB*) and *S. dysenteriae* (*pinD*) and genetic transduction studies have detected a vestigial system in *S. flexneri* (*pinF*). The *S. boydii pinB* system includes a segment of invertible DNA (the B-loop) and shows considerable nucleotide sequence homology to the Mu *gin*-G-loop system. The defective *pinF* gene in *S. flexneri* appears to be closely related to the *E. coli pin*-P-loop system in cryptic phage e14, suggesting that this *Shigella* species is the more closely related to *E. coli* K-12 (Tominaga *et al*., 1991).

In the Gram-positive pathogen, *Staphylococcus aureus*, plasmids of classes II and III (see Chapter 3) undergo site-specific inversion of a 2.2 kb DNA segment, flanked by a 650 bp inverted repeat (IR) (Murphy & Novick, 1979). The recombinase, Bin, is encoded by the IR sequence and has homology to the invertase family members (Rowland & Dyke, 1988). Class II *S. aureus* plasmids with one copy of the IR undergo high-efficiency recombination at this site. A transposon, Tn552, with a copy of the IR, transposes preferentially to a site adjacent to the plasmid-linked IR such that the transposon IR copy and the plasmid-resident IR copy are in inverted orientation and can undergo the inversion event catalysed by Bin. Bin can also resolve transposon-generated plasmid cointegrate structures, using the resolution site within the IR (Novick, 1989).

DNA hybridization experiments using the *E. coli* e14 *pin* gene or the *S. typhiumurium hin* gene as probes have identified a homologue in *Bordetella*

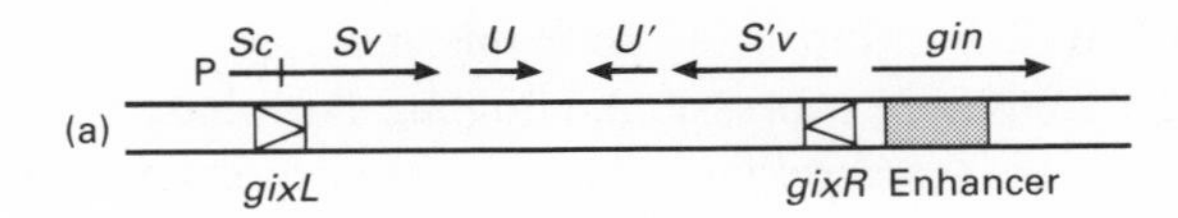

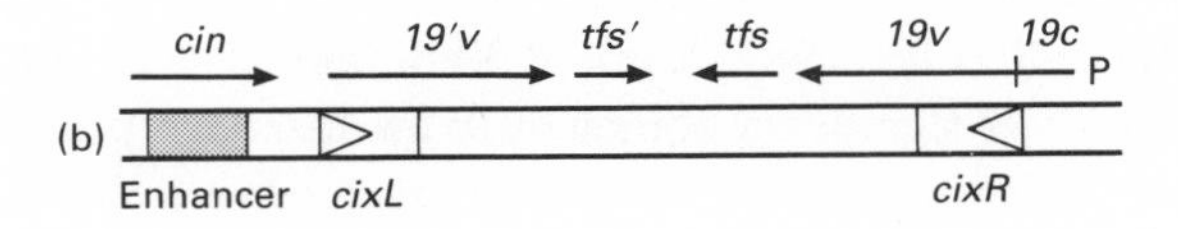

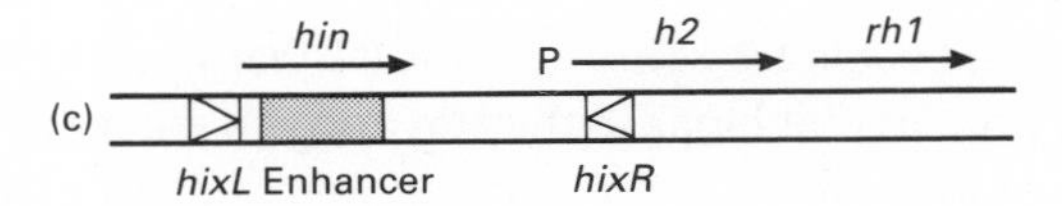

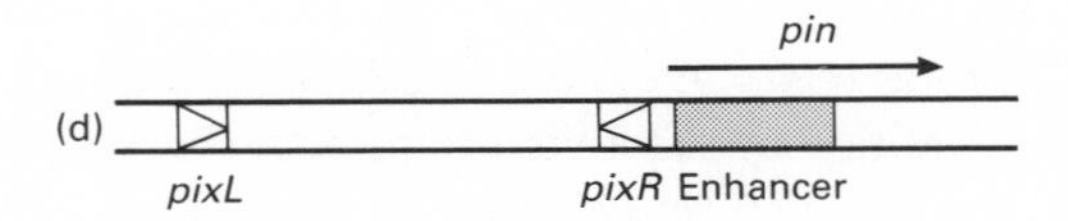

Fig. 4.12 Invertase systems. (a) The bacteriophage Mu Gin system. This is concerned with varying the expression of phage particle tail fibres, encoded by either the *S* and *U* genes or the *S′* and *U′* genes. The recombinase is encoded by the *gin* gene. (b) The bacteriophage P1 Cin system. This is also concerned with phage tail fibre variation. The tails fibres are expressed either by the *19* and *tfs* genes or the *19′* and *tfs′* genes. The recombinase is encoded by the *cin* gene. (c) The *Salmonella typhimurium* Hin system. Orientation of the promoter-carrying DNA segment determines whether or not the H2 flagellar antigen and the *h1* gene repressor are expressed. The recombinase is encoded by the *hin* gene. (d) The *Escherichia coli* Pin system. This system has no clear function. Its recombinase is encoded by the *pin* gene.

pertussis. However, the nature and the role of this putative *B. pertussis* invertase gene are unknown (Foxall *et al.*, 1990).

Concluding remarks

The purpose of this chapter is to introduce the reader to the different types of gene rearrangement events which occur in bacteria due to homologous or to site-specific recombination. In the case of the latter, some of the best-understood systems have been described here. While few of these systems are essential for bacterial virulence, many systems that do contribute to virulence use regulatory mechanisms that are related to these systems. In the cases of these virulence functions, the available molecular detail often lags behind that of the 'model' systems from organisms like *E. coli* K-12. To appreciate what is occurring in these

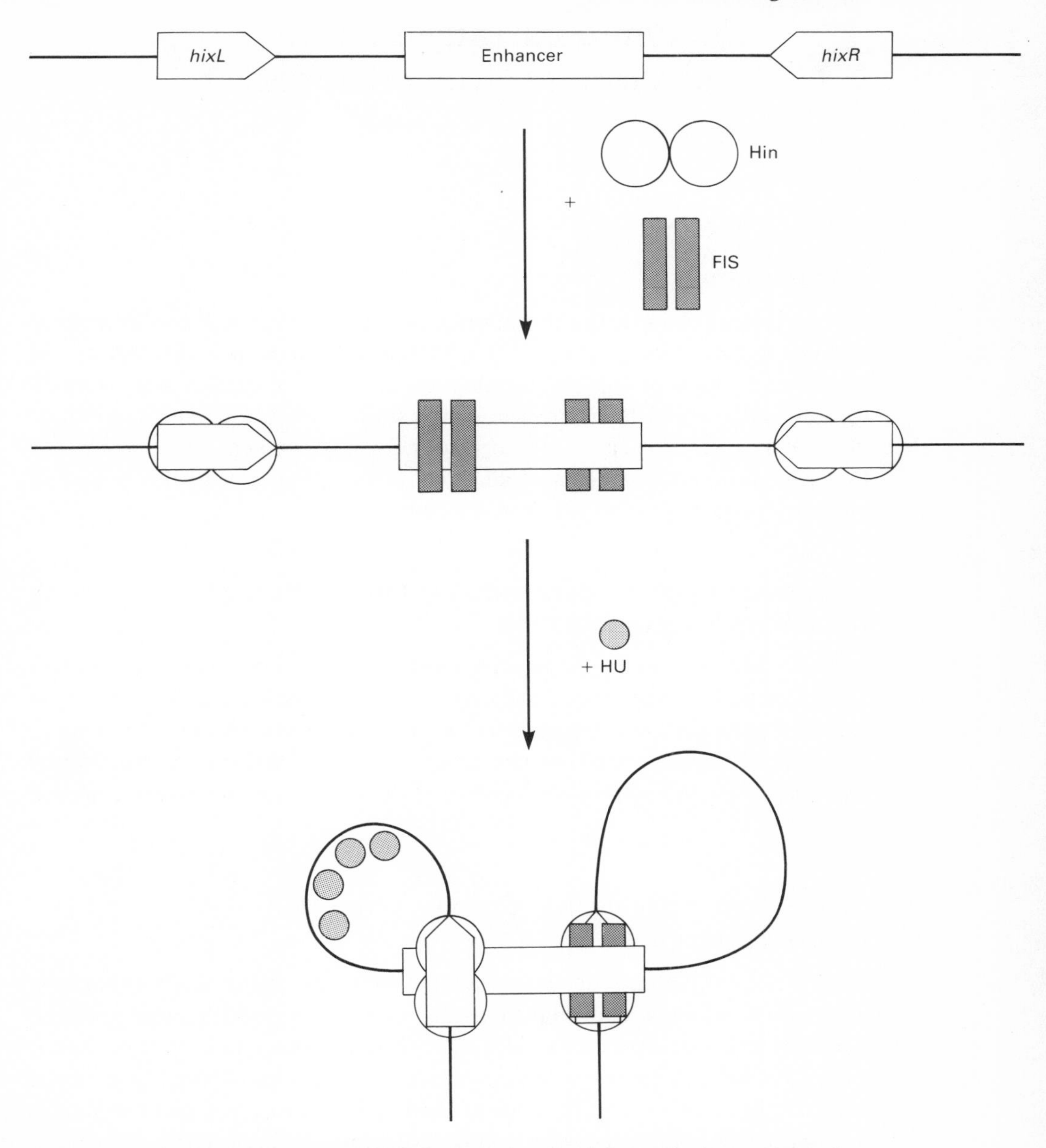

Fig. 4.13 The Hin invertasome. Simplified representation showing the interactions of the Hin invertase and the FIS and HU histone-like proteins with the Hin system *hix* sites and the recombination enhancer.

less well-characterized systems, it is necessary to be equipped with information about the model systems. Apart from giving clues about what may be happening within the genomes of the pathogens, a study of the model systems also allows some conclusions to be drawn about the evolution of these systems and their adaptation by different bacteria to solve different problems in gene regulation.

5 Genome Rearrangements and Virulence Gene Expression

Introduction

In the previous chapter, the concept of genome rearrangement was introduced and the mechanisms by which it is achieved in bacteria were discussed. The emphasis throughout was on so-called 'model' systems such as λ site-specific recombination where the systems are understood in minute detail but have little direct relevance to bacterial pathogenicity. In this chapter, the contributions that genome rearrangements of all kinds make specifically to the control of virulence gene expression will be discussed.

Variation in genetic expression achieved through homologous recombination

As was seen in Chapter 4, genome rearrangements can occur by different mechanisms. This first section will look at genetic rearrangements occurring in bacterial pathogens via homologous (or general) recombination. This type of recombination was covered in some detail in Chapter 2 and readers are referred there for information on the recombination mechanism and the factors required to catalyse it.

Genetic rearrangements in the pilin genes of *Neisseria gonorrhoeae*

The pili of *Neisseria gonorrhoeae* are essential for initial interactions with the host where they recognize and bind to specific receptor sites on human epithelia. These pili are composed of a 17 kDa pilin subunit protein and this is subject to a high degree of sequence variation. Pilin contains regions of variable sequence known as minicassettes which consist of both a hypervariable immunodominant region near the carboxyterminus and several semivariable regions in the central portion and near the aminoterminus of the protein (Fig. 5.1). Fluctuations in the sequences of these minicassettes produce antigenic variation and this may assist the pathogen in evading the host defences, in spreading from site to site and in attaching to different types of host cells (Meyer *et al.*, 1988, 1990). Sequences outside the minicassettes appear to be highly conserved at the protein and the DNA levels (Haas & Meyer, 1986) (Fig. 5.1).

Antigenic and phase variation in *N. gonorrhoeae* pilin expression occurs at the DNA level. Each cell in the population carries at least one structural gene for

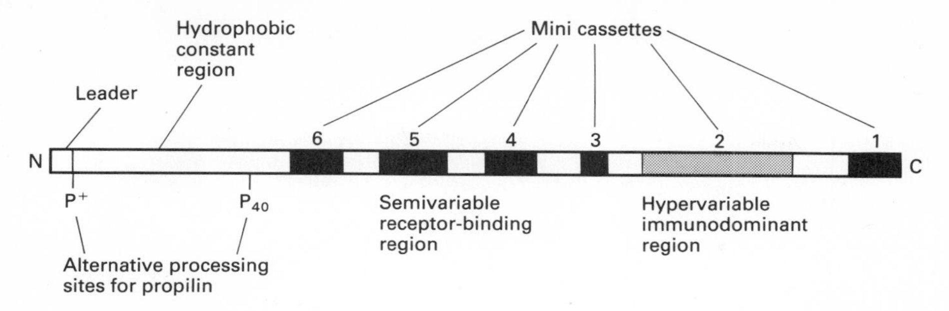

Fig. 5.1 Propilin from *Neisseria*. The propilin molecule possesses alternative processing sites (P^+ and P_{40}) and six minicassettes, five are semivariable and one (number 2) is hypervariable, defining the immunodominant region.

pilin expression. This gene, *pilE*, may be accompanied by a second expression gene in rare cases (Meyer *et al.*, 1984). Silent copies of the pilin gene, called *pilS1*, *2* etc., are located at other sites in the genome. These are partial genes which lack the conserved amino-terminal sequences but which contain minicassettes of variant sequence (Haas & Meyer, 1986). The silent genes provide a pool of genetic information for use in pilin gene variation. This pool may be exploited by the cell which contains it or it may become available to other cells should lysis occur and the DNA be taken up through transformation (see later).

Intracellular variation involves a transfer of minicassette DNA from a silent locus to *pilE* where an homologous recombination event replaces the existing *pilE* minicassette DNA by the material from the silent locus (Robertson & Meyer, 1992) (Fig. 5.2). The process is RecA-dependent and involves recombination at the intercassette conserved DNA sequences (Haas & Meyer, 1986; Koomey *et al.*, 1987). The discovery, in the vicinity of both gonococcal and meningococcal *pil* genes, of DNA sequences resembling those recognized by site-specific recombinases such as the Tn*3*-encoded resolvase has prompted the suggestion that additional factors may serve to regulate the recombination process (Meyer *et al.*, 1988; Saunders, 1989). This suggestion is supported by the observation that a set of *pil* silent locus repeat sequences called RS1 has homology to the Hin enhancer region involved in *Salmonella typhimurium* flagellar phase variation (Meyer, 1990). This *S. typhimurium* enhancer region is a binding site for the FIS (Factor for Inversion Stimulation) nucleoid-associated protein (see Chapters 2 and 4).

Pilin phase variation involves switching between piliated and non-piliated states. Most non-piliated variants of the intensively studied *N. gonorrhoeae* strain MS11 are so-called S-phase variants which arise due to non-standard processing of propilin. The propilin molecule has a seven amino acid leader sequence followed by a conserved hydrophobic region that normally forms part of the mature pilin protein. For this to happen, the molecule must be processed at amino acid position 1 (p^+) (Fig. 5.1). However, some minicassette arrangements result in processing at position 40, producing a protein that cannot be incorporated with the pilus (Fig. 5.1). Instead, this pilin, called S-pilin, is

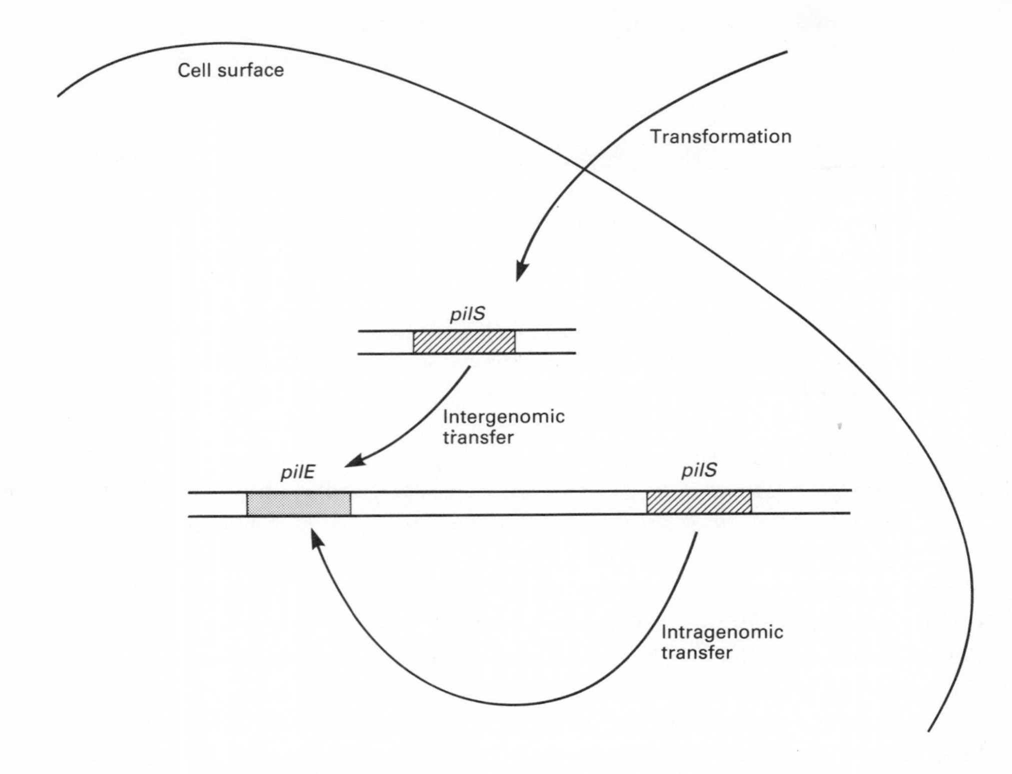

Fig. 5.2 Two routes to pilin variation in *Neisseria*. Pilin gene DNA can be moved from silent to expression sites via an intragenomic event or through an intergenomic event that requires uptake of DNA by transformation.

secreted into the medium as a soluble protein (Haas *et al.*, 1987). A second type of non-piliated variant is known as the L-phase variant. Here, unequal recombination between *pilS* and *pilE* can generate multiple, tandem, in-frame copies of the *pilS* sequence at the expression site. These express an elongated pilin, L-pilin, which is neither secreted nor assembled into pili (Meyer *et al.*, 1990).

There are two possible routes by which pilin variation might occur, one involving intracellular reciprocal exchange (homologous recombination) between silent and expression loci and the other involving intercellular exchange of DNA through autolysis of the donor cell followed by DNA uptake by the naturally transformable recipient (Fig. 5.2). The transformation-mediated mechanism would mimic intracellular gene conversion because the genetic exchange would appear to be non-reciprocal in terms of the sequences of the silent and expression loci within the recipient cell. For this reason, Meyer and colleagues (1990) believe that many of the earlier reports of gene conversion in *Neisseria* may be explained by the transformation phenomenon.

Experimental data indicate that intercellular DNA transfer leading to pilin phase variation occurs *in vitro* when the culture is in stationary phase, a condition that leads to the onset of autolysis (Gibbs *et al.*, 1989). When this system operates, lysing cells liberate copies of pilin gene DNA which are then taken up by their naturally transformable neighbours who then incorporate the genetic material with their genomes by homologous recombination. The role of extracellular DNA in pilin phase variation is confirmed by the observations that phase variation is reduced by DNase treatment of post-exponential phase cells and that *dud1* mutants (deficient in uptake of DNA) also show reduced phase variation in post-exponential growth (Gibbs *et al.*, 1989; Seifert *et al.*, 1988) (Fig. 5.2). How relevant intercellular DNA transfer is to pilin phase variation *in vivo* is uncertain. The issue of which pathway is the more relevant has been complicated by data from work by Zhang *et al.* (1992) using precisely engineered mutations in the pilin genes and growth conditions that prevent cell-to-cell transfer of DNA through the medium. They have produced results compatible with pilin phase variation occurring via intracellular non-reciprocal DNA rearrangements (i.e. gene conversion).

See also the section on neisserial PilC expression under 'Phase variation mechanisms involving oligonucleotide repeat sequences' later.

Antigenic variation and the streptococcal M protein

Streptococcus pyogenes has on its surface a fibrous layer composed of the antigenically variable M protein (Scott, 1990). M proteins are dimeric molecules in which each partner twists about the other by forming a coiled-coil structure with a seven amino acid periodicity in which the first and fourth residues are non-polar and the seventh is charged. The M protein is encoded by the *emm* gene and most group A strains possess just one copy of this gene, although it is possible that some strains may carry a second gene (Haanes *et al.*, 1992; Kehoe *et al.*, 1985, 1987; O'Toole *et al.*, 1992). Transcription of the *emm* gene is regulated positively by the Mry protein encoded by the *mry* gene located immediately upstream of and transcribed in the same direction as the *emm* gene (Perez-Casal *et al.*, 1991). Transcription of *emm* is environmentally modulated through Mry; specifically, *emm* is induced by growth in a carbon dioxide atmosphere (Caparon *et al.*, 1992; Chapter 8).

One serotype, M6, has been subjected to minute analysis at the molecular level and a mechanism of antigenic variation proposed. The DNA sequence of the *emm6* gene reveals extensively reiterated regions which would form ideal substrates for homologous recombination and could explain the variability in size of the different M proteins. By examining the DNA sequences of variants of a prototypic M6 serotype, it was discovered that these could indeed be explained by homologous recombination across the repeated sequences (Hollingshead *et al.*, 1987; Jones *et al.*, 1988).

Antigenic variation in *Borrelia vmp* genes

The phenomenon of spirochaete surface antigen variation has been studied in most detail in *Borrelia hermsii* (the cause of relapsing fever) (Barbour, 1990a, b;

Chapter 3). Periodic relapses of fever occurring at 4–7 day intervals correlate with serotype variation in the spirochaete. By varying its cell surface antigen composition, new serotypes of the pathogen are able to undergo rapid population expansion in the host because they look 'new' to the host defences. These antigens are the outer membrane-located variable major proteins, or Vmps, and they carry the serotype-specific epitopes.

Two serotypes, 7 and 21, have been studied in detail at the molecular level and the amino acid sequences of their Vmps, Vmp-7 and Vmp-21, respectively, have been used to design oligonucleotide probes for the detection of their genes. In Vmp-7 serotype cells, the probe detected two copies of the *vmp-7* gene; other serotypes had only one copy of *vmp-7*. Similarly, the Vmp-21 probe detected two copies of the *vmp-21* gene in Vmp-21 cells but only one copy in other serotypes. At the mRNA level, Northern blots detected *vmp-7* mRNA only in Vmp-7 serotype cells and *vmp-21* mRNA was detected only in Vmp-21 serotype cells (Meier *et al.*, 1985; Plasterk *et al.*, 1985). These data were consistent with the existence in Vmp-7 serotypes of a silent and an expressed copy of the *vmp-7* gene but only a silent copy in other serotypes. Similarly, Vmp-21 strains contained a silent and an expressed copy of *vmp-21* while other serotypes possessed just a silent copy.

The genes encoding the outer membrane-located variable major proteins are carried on linear plasmids (see Chapter 3). Whether or not a particular *vmp* gene is expressed is a function of the sequences 5′ to the coding region. A sequence of up to 7 kb, known as the 'central expression region', which is unique to the expression site, is required for *vmp* expression. This region contains a σ^{70} promoter sequence, a poly(dT.dA) sequence and several elements that resemble insertion sequences (Barbour *et al.*, 1991a,b). These insertion sequence-like elements are not found elsewhere in the *Borrelia* genome (Barbour *et al.*, 1991a). Genes at both silent and expression sites share features known as 'downstream homology sequences' (DHS) and 'upstream homology sequences' (UHS), with the UHS containing the transcription start site for the *vmp* gene (Fig. 5.3).

The expression sites occupy subtelomeric locations on the linear plasmids, with the promoters being typically 1–2.5 kb from the telomere (Kitten & Barbour, 1990; Restrepo *et al.*, 1992; Fig. 5.3). The silent sites are more distantly located from the telomeres (Kitten & Barbour, 1990). The gene at the expression site can be replaced by a unidirectional transfer of a different *vmp* gene from another linear plasmid. The DHS and UHS regions of sequence similarity shared by the silent and expression sites seem to provide substrates for the recombination event in which the genetic switch occurs (Fig. 5.3) (Barbour, 1989, 1990a,b; Kitten & Barbour, 1990), although any proposed mechanism must take into account the unidirectional nature of the DNA transfer. Little is known about the enzymes required to catalyse the DNA exchange although it is possible that a transposase associated with the insertion sequence-like elements might be involved (Barbour *et al.*, 1991a).

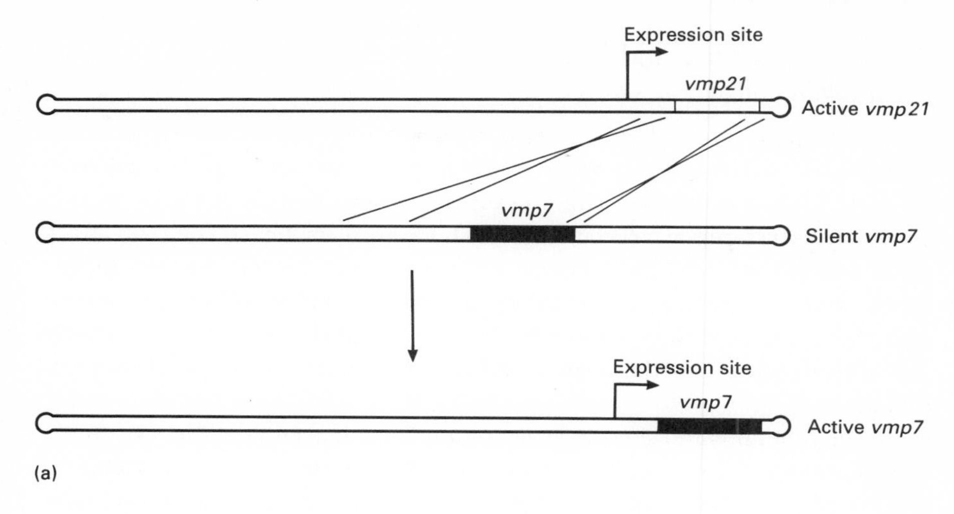

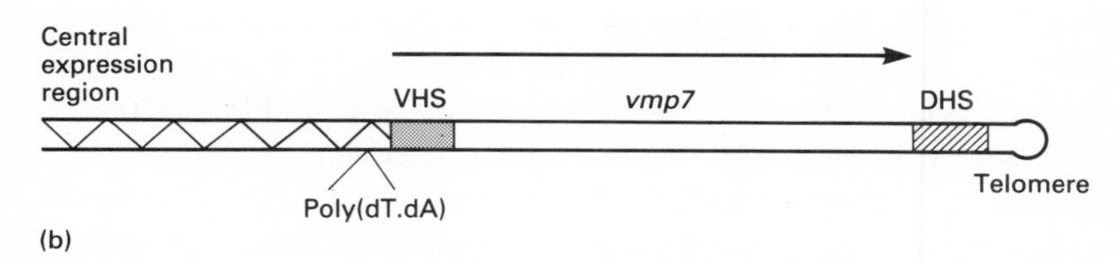

Fig. 5.3 Vmp variation in *Borrelia hermsii*. (a) Simplified model of Vmp21 to Vmp7 expression transition involving a recombination event between two linear plasmids. (b) Key features of *vmp* expression site close to the telomere of a *B. hermsii* linear plasmid.

Variable expression of outer surface proteins in *Borrelia burgdorferi*

Borrelia burgdorferi is the causative agent of Lyme disease and its ability to evade host defences and establish a chronic infection may involve changes in cell surface-expressed proteins, by analogy with the variable expression of outer membrane Vmp proteins in the relapsing fever organism, *B. hermsii* (see earlier). The expression of outer surface proteins (Osps) OspA (31 kDa) and OspB (34 kDa) of *B. burgdorferi* varies both clonally within an individual strain and among strains (Bundoc & Barbour, 1989; Schwan & Burgdorfer, 1987). The *osp* genes are located at a single genetic locus on a 49 kb double-stranded DNA linear plasmid where they are organized as an operon, *ospAB* (Barbour & Garon, 1987; Howe *et al.*, 1985, 1986). The amino acid sequences of OspA and OspB show a high degree of similarity and resemble prokaryotic lipoproteins (Bergström *et al.*, 1989). Antigenic variation among the Osp proteins can occur by homologous recombination either between *osp* genes on the same plasmid or genes on different copies of the 49 kb plasmid (Rosa *et al.*, 1992).

Genome instability and haemolysin expression in *Escherichia coli*

Haemolysins have a tradition of being regarded as virulence factors, although evidence that this is so varies greatly from system to system. They are produced by both Gram-positive and Gram-negative bacteria and most cause lysis of erythrocytes by forming a pore in the lipid biliayere (Chapter 1). They can attack other mammalian cells by a similar mechanism and for this reason are also called cytolysins. Haemolysin-producing strains of *E. coli* can infect humans and animals. In humans they frequently attack sites outside the intestine, notably the urinary tract, and they can cause cystitis, pyelonephritis or even septicaemia.

In the human pathogen, the genes coding for α-haemolysin, *hly*, are on the chromosome and are organized as an operon. The organization and regulation of this operon are discussed in Chapter 8. Genome rearrangement can have important consequences for this system because the *hly* genes on the chromosome of uropathogenic *E. coli* strains are subject to deletion. Studies on one strain (*E. coli* 536) showed two *hly* loci on the chromosome which were lost at frequencies of 10^{-3} to 10^{-4}. One *hly* insert was very large (75 kb in length) and was flanked by directly repeated sequences of 16 bp (5′-TTCGACTCCT-GTGATC-3′) (Knapp *et al.*, 1986). Recombination between these repeat sequences deleted the intervening *hly* DNA, leaving just a single copy of the 16 bp repeat on the chromosome. Loss of *hly* also correlated with a loss of an adhesin (S-type fimbriae) and a reduction in serum resistance. Perhaps this occurred because an essential selective pressure had been removed when the bacteria were cultured under laboratory conditions.

Plasmid-encoded haemolysin genes isolated from *E. coli* strains involved in animal intestinal infections are structurally identical to the chromosomal genes of human isolates but differ in being frequently flanked by copies of the insertion sequences IS2 and IS92. These IS sequences may contribute to the spread of the genes (Hess *et al.*, 1986; Knapp *et al.*, 1985).

Genome instability and urease expression in *Escherichia coli*

Urease catalyses the hydrolysis of urea to ammonia and carbon dioxide and is thought to be a virulence factor in bacteria infecting the urinary tract. In very rare cases, clinical isolates of uropathogenic *E. coli* strains specifying urease activity have been detected. These harbour the urease genes on the chromosome. Furthermore, urease-minus segregants arise at high frequency when these strains are cultured in the laboratory. These are found to have undergone a specific deletion which removes the urease genes from the chromosome (Collins & Falkow, 1988). How the urease genes get into the strains in the first place is not understood. It would appear that some selective pressure associated with growth in association with the host is required if the genes are to be maintained on the chromosome. In the absence of this pressure (such as during laboratory growth) the genes are lost.

The role of RecA and RecBC in *Salmonella* virulence

The general recombination system of *E. coli* was described in Chapter 2. Available evidence suggests that the equivalent system in *Salmonella typhimurium* is similar. Genetic studies with *S. typhimurium* mutants deficient in RecA or RecBC showed that these are essential functions for full virulence in mice. It has been suggested that the recombination system performs a protective function *in vivo* by repairing damage caused to DNA by oxidative bursts within phagocytes (Buchmeier *et al.*, 1993). It is also possible that particular genome rearrangements, which depend on an active general recombination system, are required as part of the process of adapting to host-associated life. In the mutants, these processes would not be possible and host adaptation would be compromised.

Variation in genetic expression achieved through site-specific recombination

In addition to genome rearrangements which depend upon the general recombination system, bacteria use system-specific mechanisms of site-specific recombination to vary the expression of genes required for virulence. Several of these have been studied in detail and much progress has been made in systems using components related to those previously studied as part of site-specific recombination mechanisms controlling bacterial and bacteriophage gene expression in *E. coli* K-12 or *S. typhimurum* LT-2. Several of these systems were reviewed in the last chapter.

Type 1 fimbrial phase variation in *Escherichia coli*

Type 1 fimbriae, produced by *E. coli* and other Enterobacteriaceae promote mannose-sensitive attachment to eukaryotic cells. How important these structures are to virulence is uncertain, although they appear to make a significant contribution to the ability of *S. typhimurium* to infect mice (Ernst *et al.*, 1990).

In *E. coli*, the *fimA* gene encodes the fimbrial subunit and the promoter for this gene is carried on a 314 bp invertible segment of DNA (Abraham *et al.*, 1985). When the promoter is directed towards *fimA* the gene is expressed and the cells are fimbriated (ON phase); when the promoter is directed away from *fimA*, the gene is not expressed and the cells are non-fimbriated (OFF phase) (Fig. 5.4). The inversion of the 314 bp promoter segment occurs by a site-specific recombination event involving 9 bp inverted repeat sequences flanking the segment. The recombinases are encoded by two closely linked genes, *fimB* and *fimE*, located 5′ to *fimA* (Fig. 5.4).

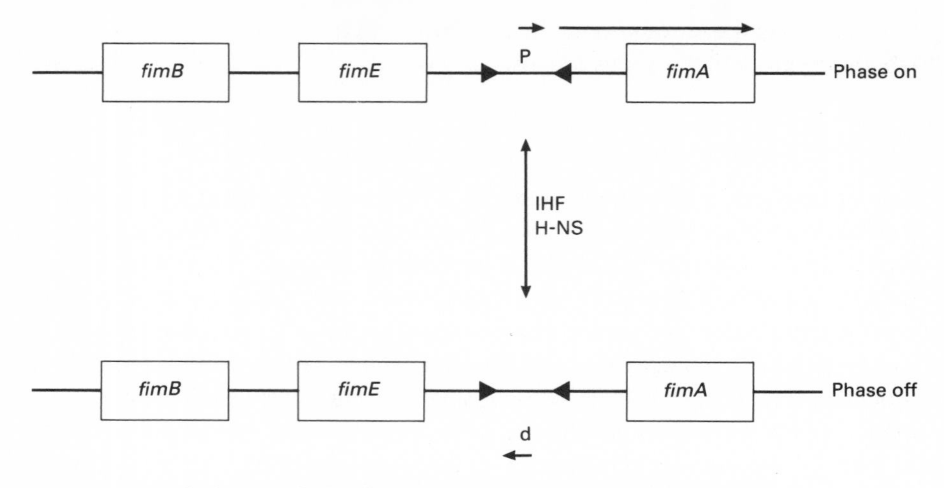

Fig. 5.4 Control of type 1 fimbrial subunit gene expression by site-specific recombination in *Escherichia coli.* Transcription of *fimA*, the gene coding for the fimbrial subunit, is dependent on the orientation of the promoter DNA segment. The site-specific recombination event is catalysed by the products of the *fimB* and *fimE* genes and is modulated by the IHF and H-NS proteins.

The proteins FimB and FimE are highly homologous (Klemm, 1986). FimB acts on the 314 bp segment in an orientation-independent manner, while FimE acts mainly to switch the system from the ON phase to the OFF phase (McClain *et al.*, 1991). Such directional site-specific recombination is highly reminiscent of λ integration and excision. Indeed, the FimB and FimE proteins possess significant amino acid homologies to the bacteriophage λ Int protein over a region that is conserved throughout the integrase family (Table 4.1). Furthermore, the *fim* site-specific recombination system shares with λ-encoded Int a dependency on the integration host factor (IHF). In the case of *fim*, the frequency of switching falls from one per thousand in wild-type cells to almost zero in IHF-deficient mutants. Two matches to the consensus IHF-binding site sequence are located at or near to the 314 bp invertible segment (Dorman & Higgins, 1987; Eisenstein *et al.*, 1987). In addition, the *fimA* promoter appears to require IHF for full transcription efficiency (Dorman & Higgins, 1987).

A second nucleoid-associated protein, H-NS, is involved in *fimA* site-specific recombination (Higgins *et al.*, 1988a; Kawula & Orndorff, 1991). Mutants lacking H-NS undergo *fimA* switching at a 100-fold accelerated rate, implying that the role of H-NS is to inhibit the inversion event. H-NS and IHF were described in Chapter 2.

Pilus phase variation in *Moraxella bovis*

Moraxella bovis causes infectious bovine keratoconjunctivitis and pili are thought to be required for attachment to the corneal epithelium. Individual strains can make more than one type of pilin and transitions between the two states do not

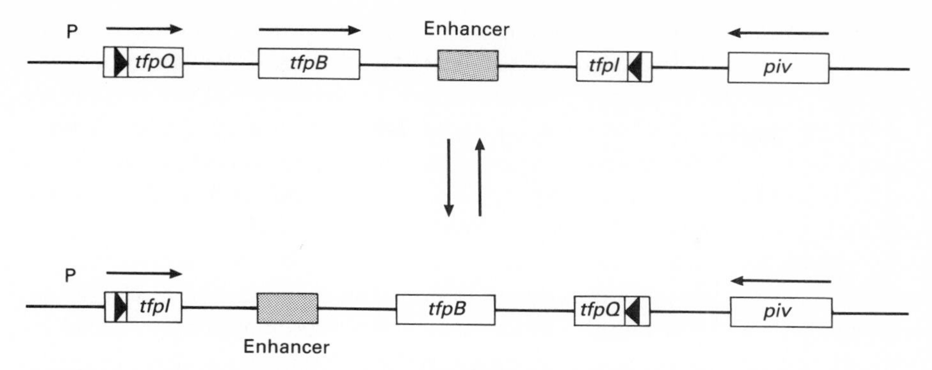

Fig. 5.5 The Q pilin–I pilin switch in *Moraxella*. Pilin variation in *Moraxella* species involves a site-specific recombination system. The filled arrow heads represent inverted repeats found within the pilin subunit genes: *tfpQ*, Q pilin; *tfpI*, I pilin. The *piv* gene encodes the putative recombinase.

proceed via a non-piliated intermediate. *M. bovis* strain Epp63 synthesizes two antigenically distinct pilin proteins, originally called alpha and beta pilin but now renamed I and Q pilin (Marrs *et al.*, 1988, 1990). Data from Southern hybridization experiments in which *M. bovis* genomic DNA was probed with a Q (beta) pilin gene show that the switch from the expression of Q to I pilin involves the inversion of a 2.1 kb region of DNA (Marrs *et al.*, 1988). The pilin genes are arranged in a convergent manner on the chromosome and only one has an active promoter. The recombination sites lie within the coding sequences for the aminoterminal domains of the proteins. The switch moves the pilin genes alternatively into a position from which they can be transcribed from the single promoter (Fig. 5.5).

The site-specific recombination event occurs within a 26 bp sequence which displays sequence similarity to the left inverted repeat (*hixL*) involved in *Salmonella typhimurium* flagellar phase variation; a further 60 bp of *M. bovis* DNA 3′ to the I pilin gene shows 50% homology to the *hin* enhancer of the same *S. typhimurium* inversion system (Fulks *et al.*, 1990; Chapter 4). The invertase gene, *piv*, is outside of, and is transcribed towards, the invertible region (Fig. 5.5). Piv has no significant homology with the amino acid sequences of the invertase proteins, despite the Hin-like nature of the recombination site and putative recombinational enhancer (Marrs *et al.*, 1990). A similar pilin expression control system exists in *M. lacunata*, which is a human pathogen (Marrs *et al.*, 1990; Rozsa & Marrs, 1991).

Genome rearrangements involving actual or putative mobile genetic elements

Mobile genetic elements (insertion sequences or transposons) are found frequently in association with bacterial virulence genes. In some cases, evidence has been obtained that these elements contribute to the mobilization of these genes, allowing them to move within or between replicons. In other cases, the transposon has only been detected as part of an analysis of the structure of the region in which the virulence gene lies. It may be detected by DNA sequencing or by hybridization followed by sequencing. Frequently, homology to a known mobile element represents the only indication that the newly discovered element is a transposon. In some cases, more detailed analysis reveals that the transposon-like sequence is really a vestigial mobile element. For example, see the description in this section of insertion sequence IS*1016* which is found in association with the capsule genes of *Haemophilus influenzae*. In this case, the mobile phase of the element's natural history expired some time ago, although not before it contributed to the movement of the accompanying virulence genes.

Thus, studies of transposon association with virulence genes are characterized by a graded data set. The best evidence shows true transposition of virulence genes. Less satisfactory evidence shows known transposable elements with no mutations in association with the virulence genes in a configuration that is suggestive of a composite transposon. Even less satisfactory is evidence showing sequences that display a relationship to a known transposable element but have no known history of transposition themselves. Examples of all of these can be found in the literature and some are described in this section.

Also described in this section are associations between virulence genes and bacteriophage. In some instances, the gene forms a component of the virus. When a bacterium is infected by the virus, it acquires the virulence factor encoded by it. Some toxin genes have such viral associations in Gram-positive and in Gram-negative bacteria. In other cases, virulence gene expression is affected profoundly by the virus, as when the gene includes the virus attachment site. Here, formation of a lysogen inactivates the gene. If that virus carries a second virulence gene coding for a distinct virulence factor, the bacterium will be seen to have undergone a genetic switch between expression of alternative virulence genes. This is called 'phage conversion' and examples of it are given below.

Amplification of *Vibrio cholerae* toxin genes

Vibrio cholerae is a Gram-negative comma-shaped bacterium and is the aetiological agent of cholera. The organisms can be water-borne or food-borne. Ingested bacteria pass through the stomach and colonize the upper part of the small intestine. Here, a heat-labile enterotoxin is produced, which enters mammalian cells where it increases the activity of adenylate cyclase, elevating intracellular levels of cyclic adenosine monophosphate (cAMP) leading to a loss of ions and water and hence the onset of watery diarrhoea (Betley *et al.*, 1986; Taylor, 1989). This toxin is composed of two subunits, A and B, assembled into an active toxin molecule in the ratio of one A molecule to five B molecules. The B subunits are required for toxin excretion from the bacterial cell and for binding to the ganglioside GM_I host cell surface receptor, gaining entry for the toxin to the host cell (Hirst *et al.*, 1984). The A subunit must be proteolytically cleaved in order to become activated and the product of this cleavage, an aminoterminal peptide called A_1, deregulates mammalian adenylate cyclase (Mekalanos *et al.*, 1979). It does this by catalysing the transfer of an ADP-ribose moiety from nicotine adenine dinucleotide (NAD) to the G_s control component of adenylate cyclase (Cassel & Pfeuffer, 1978; Gill & Meren, 1978).

There are two biotypes of *V. cholerae* that cause disease, El Tor and Classical. The genes coding for cholera toxin, *ctxA* and *ctxB*, are organized as an operon, *ctxAB*, and most El Tor strains possess one copy of this on their chromosome (Mekalanos *et al.*, 1983). The El Tor strain RV79 has two copies, tandemly duplicated on the chromosome and linked to the *nal* locus (Sporecke *et al.*, 1984). Classical strains have two copies of *ctxAB*, one in the same position as in El Tor, and another at a distant site on the chromosome (Mekalanos, 1983). Both El Tor and Classical strains contain 6 kb of conserved DNA upstream of *ctxAB*, indicating that this operon forms part of a larger conserved genetic element (Mekalanos, 1983). Moreover, non-toxigenic isolates of *V. cholerae* have been found to lack both *ctxAB* and this 6 kb of conserved sequence (Miller & Mekalanos, 1984). This 6 kb of conserved DNA includes a repetitive sequence called RS1. RS1 elements are 2.7 kb long and RS1 direct repeats flank a 4.3 kb core region of unique DNA which includes the *ctx* operon (Mekalanos, 1983). These features give the *ctx* element the characteristics of a composite transposon. RS1 has been shown to transpose and also to participate in a *recA*-dependent recombination reaction which amplifies *ctx* as tandem repeats (Goldberg & Mekalanos, 1986b). This amplification process is reversible and can lead to a reduction in the number of *ctx* copies on the chromosome. The trigger for *ctx* amplification is associated with growth in the intestine (Mekalanos, 1983). Thus, this pathogen can use unequal cross-over events of RS1 elements, catalysed by RecA-promoted homologous recombination, to acquire additional copies of *ctx* (Fig. 5.6). This might happen in response to environmental signals (temperature, osmotic stress, anaerobic conditions, nutrient availablility, pH, etc.) encountered at the site of infection and then be selected for because toxin gene amplification gives the bacterium an advantage if stimulation of water and electrolyte loss from the host provides a nutrient that is otherwise limiting. Genetic studies have permitted the identification and mutagenesis of the *V. cholerae recA* gene (Goldberg & Mekalanos, 1986a; Hamood *et al.*, 1986; Paul *et al.*, 1986) and *in vivo* studies with a *recA* mutant of *V. cholerae* have shown that this strain is reduced in its ability to colonize the host and also in its immunogenicity (Ketley *et al.*, 1990).

Insertion sequences associated with *Escherichia coli* and *Salmonella* aerobactin genes

Genes coding for the iron siderophore aerobactin have been identified on ColV plasmids in *E. coli* and on F1*me* plasmids in isolates of epidemic *Salmonella*

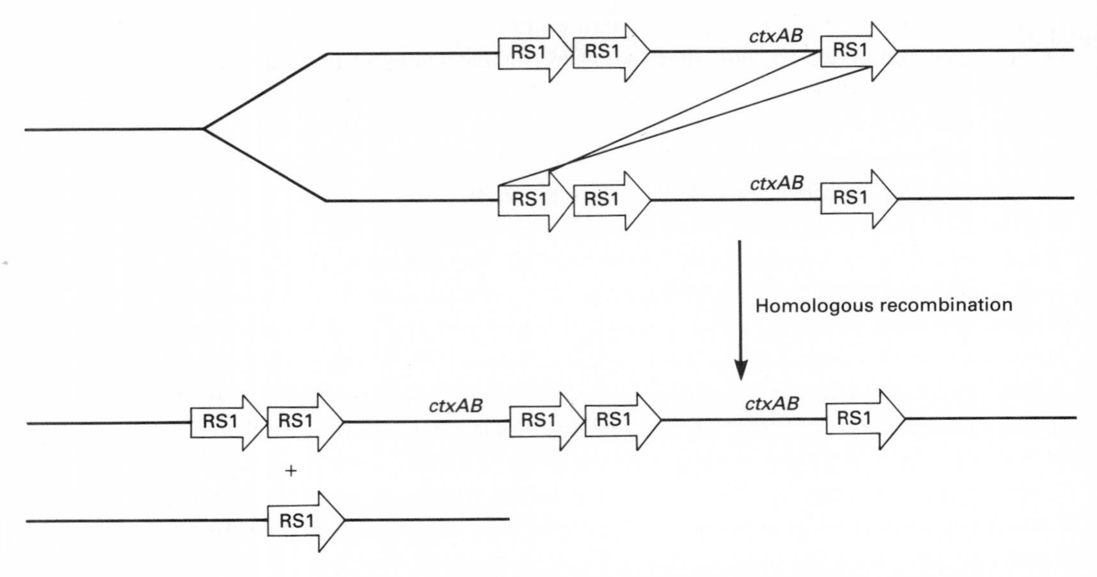

Fig. 5.6 Mechanism of cholera toxin gene amplification. The *ctxAB* operon of *Vibrio cholerae* is flanked by directly repeated copies of the RS1 sequence. Homologous recombination between these following DNA replication can result in tandem amplification of *ctxAB*. Based on the data of Mekalanos (1983).

strains (Chapter 3). In the aerobactin system of *S. wien*, the structural genes are flanked by inverted copies of the insertion sequence IS*1*, located immediately upstream and downstream of the coding region (Colonna *et al.*, 1985). These genes appear to be identical to the *E. coli* aerobactin genes from plasmid ColV (discussed in Chapter 3; illustrated in Fig. 3.13), which is indicative of a common origin. The orientation of the aerobactin genes on the F1*me* plasmids varies from isolate to isolate and in some strains is not plasmid-associated, presumably being located on the chromosome (McDougall & Neilands, 1984). Similarly, aerobactin genes in *E. coli* (and in *Shigella flexneri*) have been found on the chromosome (Lawlor & Payne, 1984; Marolda *et al.*, 1987; Valvano and Crosa, 1984). Although the structure of the aerobactin-IS*1* system is suggestive of a composite transposon, transposition has not been demonstrated (Payne and Lawlor, 1990). However, the presence of IS*1*, which is widely distributed among Gram-negative bacteria, could afford opportunities for homologous recombination. This could allow the aerobactin system to move between plasmids and chromosome without the need for transposition.

Escherichia coli enterotoxin genes are associated with mobile genetic elements

Enterotoxigenic *E. coli*, which cause diarrhoeal diseases in humans and animals, elaborate two types of toxin. One is a heat-labile toxin, LT, which functions in a similar manner to cholera toxin. The other is a heat-stable toxin, ST, which has a distinct mode of action. The genes coding for these toxins are plasmid-linked and both types may be encoded by the same plasmid. The ST toxins occur in two forms, the methanol-soluble STI and the methanol-insoluble STII (or STa and STb). The gene coding for STI is flanked by inverted repeats of IS*1*, forming a compound transposon called Tn*1681* (So *et al.*, 1979). This transposon is frequently found in association with the gene cluster coding for the 987P adhesin in enterotoxigenic *E. coli* strains (Klaasen *et al.*, 1990).

The gene for STII is also flanked by inverted repeat sequences, forming a 9 kb active compound transposon. The nature of the sequences flanking the STII gene can vary from isolate to isolate, indicating that the gene is not always harboured within the same transposable element (Lee *et al.*, 1985). LT genes have also been found to be flanked by inverted DNA repeats (Yamamoto & Yokota, 1981). The LT genes, which are closely related to the cholera toxin genes, have been described as foreign to *E. coli* (Yamamoto *et al.*, 1987) and the inverted repeats may have played a role in their movement between and within species.

Bacteriophage involvement in diphtheria toxin expression

Diphtheria toxin is expressed in *Corynebacterium diphtheriae* during late log phase or stationary phase growth in low iron medium (Pappenheimer, 1977). The importance of iron in bacterial infections has already been discussed (Chapter 1) and its role as a regulator of gene expression will be described in Chapters 7 and 8. Since toxins damage host cells, induction of their expression by low iron has generally been rationalized as a mechanism for releasing the metal from host cells at periods when the bacterium is particularly starved for it.

The structural gene for the toxin, *tox*, is carried within the genome of a family of related bacteriophages, which includes the β corynephage (Groman, 1984). Diphtheria toxin is only expressed in *C. diphtheriae* strains infected or lysogenized by the *tox*-carrying bacteriophage, and transcription of the *tox* gene is regulated in response to iron by the chromosomally encoded DtxR transcription regulator (Boyd *et al.*, 1990; Schmitt & Holmes, 1991; Chapter 8).

Bacteriophage involvement in *Escherichia coli* enterohaemolysin production

Enteropathogenic *E. coli* strains of serogroups O26 and O111 produce a haemolysin in association with vero-toxin (also called shiga-like toxin) which is distinct from α-haemolysin (Beutin *et al.*, 1989). These haemolysin-producing *E. coli* serotypes cause enteric disease, in contrast to α-haemolysin-producing strains which are primarily causes of extraintestinal infections. For this reason, Beutin and colleagues (1990) have named this toxin enterohaemolysin. It is genetically, phenotypically and serologically distinct from α-haemolysin. These

E. coli strains are lysogenic for a temperate bacteriophage, ϕC3888, which either encodes the haemolysin or regulatory functions required for haemolysin expression (Beutin *et al.*, 1990). Such phage-association has not been described for α-haemolysin.

Bacteriophage involvement in shiga-like toxin expression

E. coli strains produce two related shiga-like toxins, called SLT-1 and SLT-2. Strains producing shiga-like toxins have been associated with diarrhoea, haemorrhagic colitis and haemolytic uremic syndrome (De Grandis *et al.*, 1987). The genes coding for shiga-like toxin 1 are carried by a lambdoid bacteriophage which is capable of lysogenizing *E. coli* K-12. The ability of the phage to mobilize the toxin genes from one strain to another is known as toxin conversion and is probably an important factor in toxin gene dissemination. The structures of the SLT-1 phages H-19B and 933J are very similar to that of bacteriophage λ (including the *int* and *xis* gene regions). It has been pointed out that because the toxin genes are not located near to the *att* site in these phage, it is unlikely that the phage acquired them from the chromosome during an aberrant excision event. Instead, they may have become phage-associated due to a transposition event or following a complex multistep event in which chromosomal sequences replaced some viral DNA in the prophage (Huang *et al.*, 1987; Newland *et al.*, 1985).

Bacteriophage involvement in *Staphylococcus aureus* toxin expression

Strains of *Staphylococcus aureus* are frequently multiply lysogenic for temperate bacteriophage and these phage can affect the expression of several secreted proteins with important roles in virulence. Expression of these extracellular toxins and enzymes can be affected positively or negatively by bacteriophage lysogenization. In negative phage conversion, lysogenization by phage abolishes expression of β-lysin (β-toxin, a phospholipase C which acts on sphingomyelin and lysolecithin) and lipase. Conversely, positive phage conversion results in the acquisition of the ability to express staphylokinase (a fibrin digesting activity) and enterotoxin A (see later).

Group F bacteriophage mediate negative conversion of β-lysin expression and positive conversion of staphylokinase activity. These are classed as double-converting phages. Positive conversion for staphylokinase activity is mediated by phage of serotype B (Kondo & Fujise, 1977). Negative conversion by serotype A phage leads to loss of β-lysin expression (Coleman *et al.*, 1986). Conversion of β-lysin (negative) and staphylokinase (positive) by serotype F double-converting phage ϕ13 and conversion of β-lysin (negative) by serotype A single-converting phage ϕ42E is caused by insertional inactivation of the chromosomal gene for β-lysin by phage DNA during prophage formation (Coleman *et al.*, 1986).

Inactivation of β-lysin expression involves site- and orientation-specific insertion of phage ϕ13 into the 5′ end of the *hlb* gene. Recombination occurs within a 14 bp core sequence of host DNA which is identical with the *attP* core sequence within the phage (Coleman, 1990; Coleman *et al.*, 1991). Phage ϕ13 carries the staphylokinase gene (*sak*) within its genome near to the phage attachment site (*attP*) (Coleman *et al.*, 1989). Lipase converting phage L54a insertionally inactivates the chromosomal lipase gene during prophage formation (Lee & Iandolo, 1986a,b).

Positive conversion for staphylokinase and enterotoxin A is due to the acquisition of temperate phage harbouring the genes for these virulence factors within the phage genome (Betley & Mekalanos, 1985; Sako *et al.*, 1983). Coleman *et al.* (1989) have described phage capable of triple conversion in which β-lysin gene expression is inactivated by phage genome insertion and the phage provides the host with the genes for staphylokinase (*sak*) and enterotoxin A (*entA*) expression. In these triple conversion phage, both the *entA* and the *sak* genes map close to the *attP* site. This arrangement contrasts with that obtaining in the shiga-like toxin phage of *E. coli* and suggests that the staphylococcal phage acquired the virulence genes from the bacterial chromosome via an aberrant excision event.

Att-site proximal locations have also been detected for the diphtheria toxin genes in corynephage β of *Corynebacterium diphtheriae* and for the erythrogenic toxin in *Streptococcus pyogenes* phage T12 (Johnson *et al.*, 1986a; Laird & Groman, 1976). These findings suggest that virulence gene acquisition by temperate phage is not unusual in pathogenic bacteria and affords the genes a degree of mobility they would otherwise lack.

Bacteriophage involvement in streptococcal toxin expression

The pyrogenic exotoxins of *Streptococcus pyogenes* are distantly related to the staphylococcal enterotoxins but have not been studied in as much detail. They are known by a variety of names (Dick toxin, scarlet fever toxin, erythrogenic toxin, streptococcal exotoxin, blastogen and mitogen; Yu & Ferretti, 1991) and are implicated strongly in the induction of heart damage (myocardial necrosis) during rheumatic fever and in the pathogenesis of scarlet fever (reviewed in Hewitt *et al.*, 1992). The *S. pyogenes* endotoxins A and C are each encoded by different bacteriophage (Goshorn & Schlievert, 1988; Johnson *et al.*, 1986a; Weeks & Ferretti, 1986). This association may explain the distribution of the toxins among the group A streptococci since this is likely to be influenced strongly by the host ranges of the bacteriophage (Yu & Ferretti, 1991).

Capsule gene cluster in *Haemophilus influenzae*

Capsular polysaccharide is a major determinant of virulence in *Haemophilus influenzae* infections. There are six capsular serotypes, a, b, c, d, e and f, of which type b strains are the most virulent, giving the highest levels of bacteraemia and causing more than 95% of invasive disease. Capsular polysaccharide protects the bacterium from complement-mediated host defences and enables the organism to resist phagocytosis (Moxon & Kroll, 1990). The *H. influenzae cap* locus is divided into three distinct regions. The central, serotype-specific Region 2 is flanked by Regions 1 and 3, which are found in all serotypes (Fig. 5.8) (Kroll *et al.*, 1989, 1991). Region 1 includes the *bex* gene cluster, which is involved in polysaccharide export (Kroll *et al.*, 1990). Musser *et al.* (1988) have used multilocus electrophoretic typing of Cap$^+$ *Haemophilus* strains to identify two major phylogenetic divisions. Divison 1 type b strains possess a duplication of the *cap* genes with two directly repeated copies of about 17 kb of DNA on either side of a 1.2 kb bridge (linker) region (Fig. 5.7) (Kroll *et al.*, 1988). Although the duplication is found in type b clinical isolates from all over the world (Allan *et al.*, 1988), it is not essential for virulence; a single copy of the 17 kb segment with

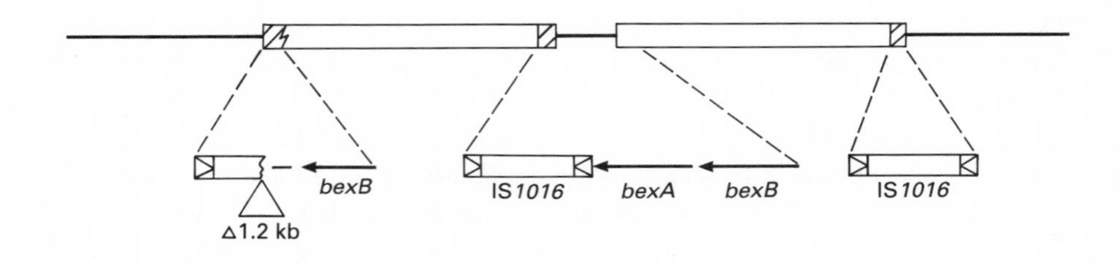

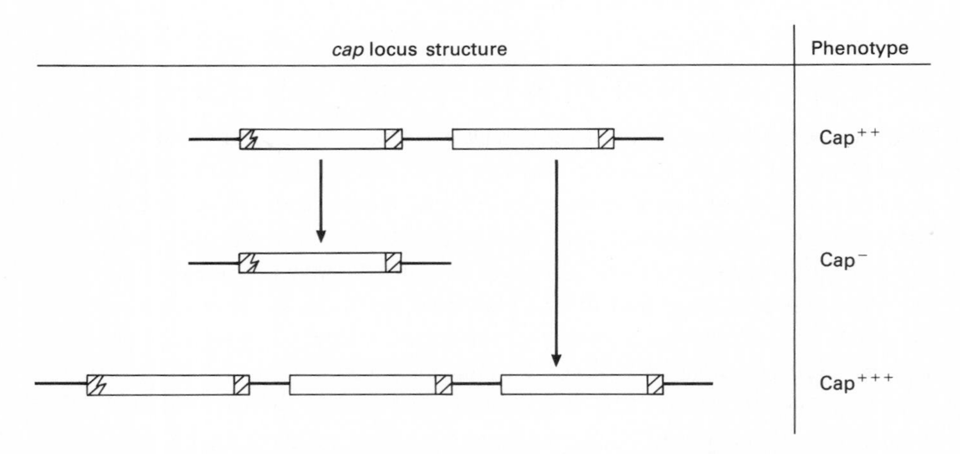

Fig. 5.7 Amplification and reduction of *cap* copy number by unequal recombination in *Haemophilus influenzae.* The minimal *cap* locus structure capable of specifying a Cap$^+$ phenotype is shown at the top of the figure. The left end has suffered a 1.2 kb deletion which has removed the *bexA* gene and part of the IS*1016* sequence. Transport function is provided by the intact *bex* operon in the centre of the locus. Variation in the copy number of the *cap* gene cluster correlates with variation in the strength of the Cap phenotype.

the 1.2 kb bridge region constitutes a reduced locus capable of conferring a virulent phenotype (Kroll & Moxon, 1988). The bridge region contains the most 3′ gene of the *bex* locus (*bexA*) whose product is essential for polysaccharide export. Recombination between the duplicated sequences flanking *bexA* reduces *cap* to a single copy with the concomitant loss of *bexA* rendering the strain Cap$^-$ (Fig. 5.7).

The *cap* locus in division 1 strains is flanked by copies of insertion sequence IS*1016*. This gives the *cap* locus the structure of a compound transposon. However, the IS*1016* element is not capable of transposition. It is composed of a 673 bp core sequence flanked by 19 bp inverted repeats. Nucleotide sequence analysis of six separate IS*1016* elements has shown that while each may once have encoded a transposase, point mutations in each element now prevent its expression. Thus, IS*1016* seems to be a vestigial transposable element (Kroll *et al.*, 1991). The amino acid sequence of the putative transposase has no homologies to those of other transposases but the 19 bp inverted repeat shows similarities to those flanking transposons of the Tn*3* family. Kroll *et al.* (1991) have suggested that the most likely sequence of events is one in which an active copy of IS*1016* transposed to either side of the *cap* gene cluster and then the resulting compound transposon moved to its current site on the *Haemophilus influenzae* chromosome. The transposase genes of the IS*1016* elements then degenerated. The acquisition of different serotype-specific genes occurred after the end of the transposition period in *cap* evolution in the division 1 strains. The defunct IS sequences continue to play a role because they provide directly repeated stretches of homologous DNA which can participate in RecA-dependent general recombination, leading to amplification and reduction of *cap* gene cluster copy number. Transient amplification of capsule production may assist survival in desiccating conditions while the bacterium is moving between hosts. Amplification of *cap* appears to correlate with enhanced virulence. Division 2 strains also possess the IS*1016* element, but not in association with *cap* (Kroll *et al.*, 1991).

Type b strains are forced to maintain *cap* in a duplicated state because of the need to maintain the unique copy of the *bexA* gene in the bridge region. Thus the ground state for *cap* amplification in type b strains is a duplicated *cap* locus. It has been proposed that this may explain the prevalence of type b strains in human infections (Kroll *et al.*, 1991).

The importance of natural competence in DNA uptake leading to genetic diversity in Neisseriae has already been described in this chapter. *Haemophilus influenzae* is also naturally competent but a restriction is imposed on DNA uptake by the requirement for a particular sequence to be present in DNA taken up from the environment. This sequence, 5′-AAGTGCGGTCA-3′ tags the DNA as 'self' since it occurs 600 times in the *H. influenzae* genome but is statistically likely to occur at most once in the genomes of other bacteria (Chater & Hopwood, 1989). The importance of natural competence to *cap* evolution *in vivo* is unknown.

Capsular genetic loci of *Escherichia coli*: similarities to the *cap* system of *Haemophilus influenzae*

The organization of the *cap* locus of *Haemophilus influenzae* is repeated in the *kps* loci of pathogenic *E. coli*; i.e. there is a central serotype-specific region flanked by regions of conserved sequence found in all serotypes (Boulnois *et al.*, 1987; Kroll *et al.*, 1988, 1989; Silver *et al.*, 1984). In *E. coli* K1, the three regions making up the *kps* locus are referred to as Region 3 (2.8 kb, perhaps involved in assembly and/or regulation), Region 2 (5.8 kb, serotype-specific and involved in synthesis, activation and polymerization of the extracellular polysaccharide) and Region 1 (6–8 kb, perhaps involved in assembly and/or translocation) (Fig. 5.8) (Vimr *et al.*, 1989; Chapter 1). Such a system would make an ideal candidate for cassette-type gene replacement, affording the cell the potential to switch from one capsule type to another. However, there is no evidence that such a system operates. A similar system coding for capsule polysaccharide, with similar possibilities for genetic exchange, exists in *Neisseria meningitidis* group B. Apart from the serotype-specific polymerization region, this system has counterparts in other meningococcal serotypes and in other *Neisseria* species (Frosch *et al.*, 1989; Meyer *et al.*, 1988).

Phase variation mechanisms involving oligonucleotide repeat sequences

The genome rearrangements discussed hitherto in this chapter have involved gross reordering of sequences through either general recombination or site-specific exchange. Mechanisms acting through changes involving as few as one base pair also occur and are much more subtle (and difficult to detect) than the others. They share with the recombination systems an element of unpredictability. One can measure the frequency at which they will occur under a given set of conditions but cannot say which cell in the population will experience the change in nucleotide sequence.

Mechanisms based on single base pair or short oligonucleotide variations within tracts of reiterated sequence are akin to point mutations or oligonucle-

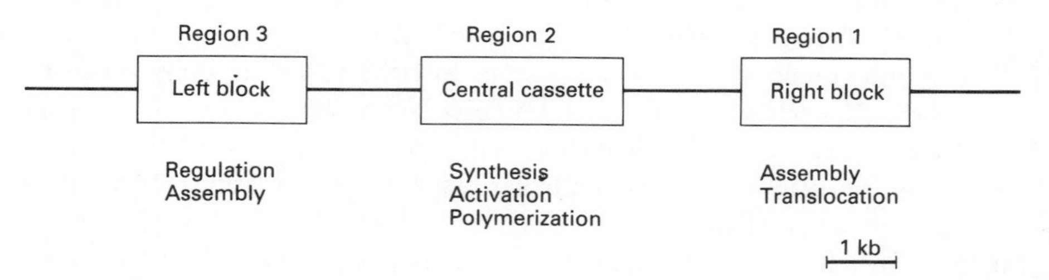

Fig. 5.8 Structure of the *kps* locus of *Escherichia coli* K-1. This capsule locus is organized in a similar manner to the *Haemophilus influenzae cap* locus. There are three genetically distinct regions whose functions are summarized in the figure.

otide deletions or amplifications. The forces that drive these depend on the special nature of the DNA sequence affected. For example, polypyrimidine–polypurine tracts have a tendency to loop out, which could make them susceptible to deletion or amplification during DNA replication. This tendency could be exacerbated by increases in DNA supercoiling, perhaps triggered by environmental stress (see Chapter 2). Such tracts are involved in many of the systems discussed in this section. In these systems, bacteria will be seen to exploit the unstable properties of specific DNA sequences in order to introduce an element of randomness into the expression of some of their virulence genes.

Pilin phase variation in *Neisseria gonorrhoeae*

The contribution of pili to virulence in *Neisseria gonorrhoeae* was described earlier in this chapter. In addition to the pilus genetic variation mechanisms driven by homologous recombination, a further mechanism of gonococcal pilus phase variation has been described which involves frameshift mutations in *pilC*, a gene required for pilus assembly (Jonsson *et al.*, 1991). Expression of the PilC outer membrane protein is controlled at the translational level by frameshifting within a run of 13 guanosine residues in the signal peptide-coding region. Eleven, 12 or 14 guanosine residues result in non-expression of PilC; 10 guanosines (that is, a number resulting from a three-residue, in-frame deletion within the poly(G) tract) permit expression (Fig. 5.9). In the absence of PilC, the cells are unpiliated.

Fimbrial phase variation in *Bordetella pertussis*

Bordetella pertussis, the whooping cough agent, produces two serologically distinct fimbriae. Transcription of the fimbrial subunit genes (*fim*) is under the positive control of the pleiotropic *bvg* locus (see Chapter 8). The transcription of the individual *fim* genes is also regulated by phase variation which involves switching between a high and a low state of transcription. This may involve a conserved region of DNA found within the regulatory regions of the *fim* genes which is thought to play a role both in Bvg-directed expression and phase

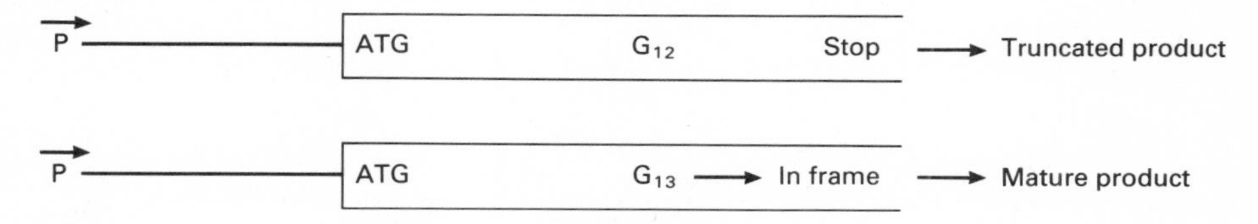

Fig. 5.9 Phase variation in *pilC* expression in *Neisseria gonorrhoeae*. The *pilC* gene alternates between expressed and non-expressed states due to variation in the length of a poly(G) tract within the open reading frame. For example, 12 G residues results in a truncated gene product whereas 13 G residues results in a wild-type, mature protein. P represents the transcriptional promoter and ATG the translation initiation codon.

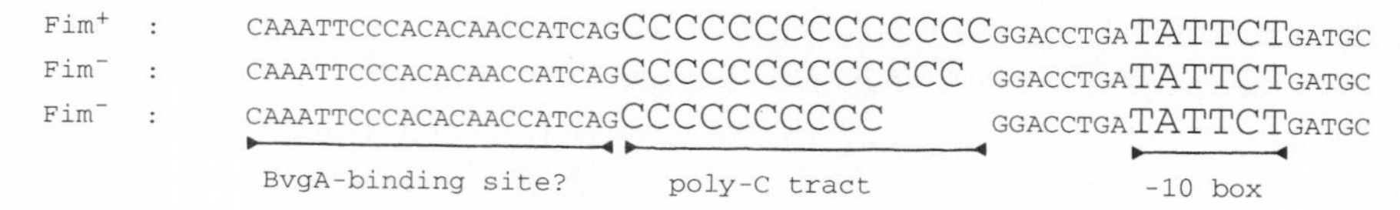

Fig. 5.10 Phase variation in *fim3* gene expression in *Bordetella pertussis.* The promoter region of the *fim3* gene includes a poly(C) tract. Variation in the length of this tract affects *fim3* promoter function severely, perhaps by altering the interaction of RNA polymerase with a bound upstream transcription factor (BvgA?).

variation. In each case, the promoter region contains a run of approximately 15 cytosine residues and it is proposed that phase variation occurs through the insertion or deletion of small numbers of cytosines (Fig. 5.10). The resulting change in the distance between the binding site for the BvgA transcription activator and that for RNA polymerase is believed to modulate *fim* transcription (Willems *et al.*, 1990).

Genetic variation and *opa* gene expression in *Neisseria gonorrhoeae*

The opacity proteins (also called Protein II or, in the case of meningococci, Class 5 proteins) are major components of the outer membrane and are involved in interactions with host phagocytic cells (Bhat *et al.*, 1991; Connell *et al.*, 1990; Virji & Heckels, 1986). These proteins are subject to antigenic variation but the mechanism is distinct from that underlying antigenic variation in the pilin proteins. At least 10 or 11 *opa* genes are located in the gonococcal genome, with some being linked to *pil* genes (Brooks *et al.*, 1991). Although the genes are highly conserved, major differences in amino acid sequence occur between them in two hypervariable regions, called HV1 and HV2 (Connell *et al.*, 1990; Fig. 5.11). Phylogenic analysis of the *opa* gene sequences indicates that this gene family has evolved through partial inter-*opa* gene recombination events, gene duplications and gene replacements (Bhat *et al.*, 1991). Each *opa* gene is complete and is transcribed from its own promoter (Stern *et al.*, 1984a, 1986; Stern & Meyer, 1987). Although transcription of the genes is constitutive, the mRNA is not always translated. In some cells at a given time none of the *opa*

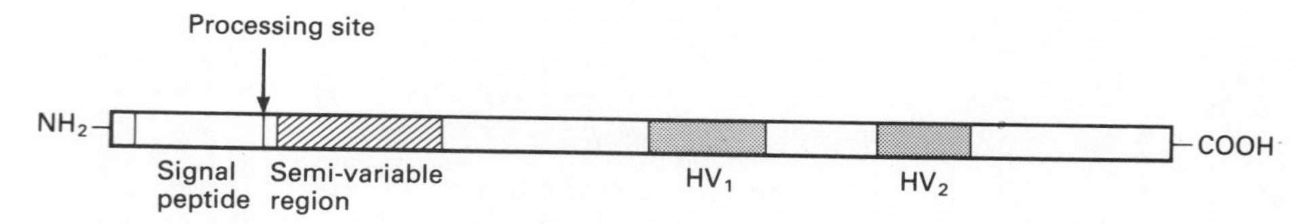

Fig. 5.11 Structure of the Opa protein of *Neisseria gonorrhoeae.* The mature protein possesses an aminoterminal semivariable region and two hypervariable regions (HV1 and HV2).

mRNA is translated. The explanation for this situation comes from an analysis of the structure of the 5′ portion of the gene.

The 5′ end of each *opa* gene contains a number of directly repeated pentameric pyrimidine units with the sequence 5′-CTCTT-3′ known as CR (coding repeat) units (Fig. 5.12). These repeats code for the hydrophobic core of the Opa leader peptide. The number of CR units located in the part of *opa* which codes for the opacity protein varies from seven to 28 and the number of CR units determines whether or not the *opa* message is translated. Each *opa* gene in the cell switches on or off independently. Numbers of CR units (e.g. nine) which place the leader peptide in frame with the rest of the coding region result in a translatable message, while numbers which place the leader out of frame with the remainder of the coding sequence result in a failure to express the protein (Fig. 5.12). Thus phase changes result from the addition or the deletion of CR units. These events are thought to arise as a result of CR DNA slippage during replication of the homopurine–homopyrimidine region (Stern *et al.*, 1986) and the process is RecA-independent (Belland, 1991; Murphy *et al.*, 1989; Robertson & Meyer, 1992). Slippage could occur if such DNA sequences loop out as they do in highly supercoiled DNA (Wells, 1988), a situation which can be triggered by environmental conditions (Dorman, 1991). Certainly, *in vitro* work shows that the CR repeats form triple-stranded or H-DNA (Htun & Dahlberg, 1989; Johnston, 1988; Mirkin *et al.*, 1987) in the presence of negative supercoiling under acid conditions (Belland, 1991). Since the process of CR deletion is independent of RecA and can proceed in *E. coli*, it is likely that specialized *Neisseria* proteins are not necessary (Meyer *et al.*, 1988). Further variation can arise due to non-reciprocal recombination between different *opa* genes, in a manner analogous to that of pilin antigenic variation (Connell *et al.*, 1988; see earlier).

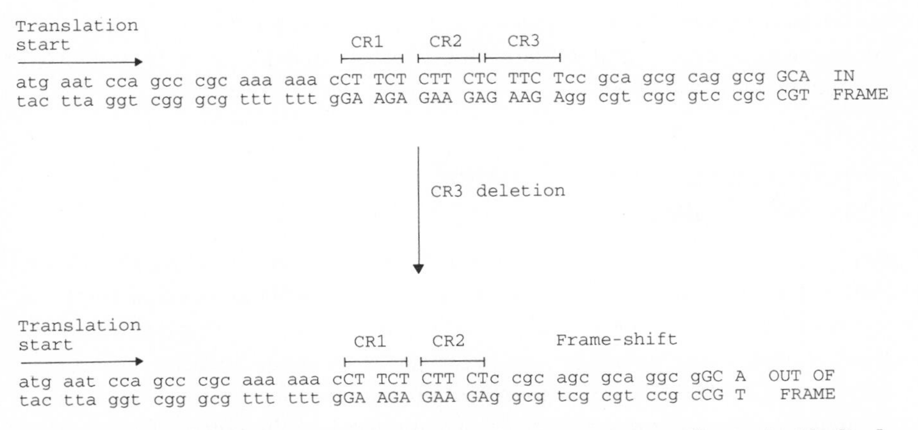

Fig. 5.12 Phase variation in *opa* gene expression in *Neisseria gonorrhoeae*. An idealized 5′ *opa* gene sequence is shown, with just three CR repeats. This sequence translates into a biologically active protein. Removing one CR repeat is equivalent to a frame-shift mutation.

Mycoplasma surface antigenic and phase variation

Mycoplasmas are related to Gram-positive eubacteria, possess small AT-rich genomes and are characteristically unable to synthesize a eubacterial cell wall. There are about 80 species and most are parasites of humans and animals. In addition, they are persistent and common contaminants of laboratory cell line cultures. *Mycoplasma pneumoniae* is a cause of atypical pneumoniae in man, *M. gallisepticum* causes chronic respiratory disease in chickens, *M. pulmonis* is a cause of chronic respiratory disease in rodents, *M. hypopneumoniae* is associated with pneumoniae in pigs. This is far from being an exhaustive list. Those interested in a detailed treatment of these strange organisms are referred to the review by Taylor-Robinson (1990).

The mycoplasma cell is surrounded by a single membrane containing lipoproteins that undergo antigenic variation and high-frequency phase variation. Molecular analysis of the variant lipoproteins (Vlps) of the swine parasite *Mycoplasma hyorhinis* has revealed the nature of surface antigenic variation and phase variation. Here, a cluster of *vlp* genes (*vlpA*, *vlpB* and *vlpC*) located at a single chromosomal site encode products with conserved amino-terminal domains for membrane insertion and lipoprotein processing and extracellular domains of variable size generated by the loss or gain of repetitive intragenic coding regions. The mechanism appears to involve precise insertion or deletion of repetitive coding regions (Yogev *et al.*, 1991).

No evidence has been found for chromosome rearrangements or frame-shift mutations within the *vlp* genes leading to variation between phase ON (expression of Vlps) and phase OFF (non-expression). Instead, the switch from ON to OFF phase appears to be to be due to variation in the length of a tract of adenine residues between the −10 and −35 region of the *vlp* promoter (Yogev *et al.*, 1991). With 17 adenines, the promoter is functional, with 18 or 20 adenines, no transcription occurs. Such changes in spacing between the −10 and −35 regions might be expected to influence the interaction of the promoter with RNA polymerase (see Chapter 6).

Lipopolysaccharide phase variation in *Haemophilus influenzae*

Haemophilus influenzae is commonly found as a commensal organism in the upper respiratory tract of humans. It can become pathogenic, and the form expressing the b serotype capsule is very pathogenic for young children, in whom it can cause menigitis, pneumoniae and other invasive infections. Lipopolysaccharide (LPS) is also an important virulence factor in this bacterium. This macromolecule forms an important component of the outer membranes of Gram-negative bacteria and is likely to be a target for the host defences (reviewed in Moxon & Maskell, 1992).

Structural variation in the LPS of *H. influenzae* involves high-frequency acquisition and loss of carbohydrate molecules. Two LPS epitopes are encoded by the *lic-1* genetic locus, which consists of the *lic1ABCD* operon (Weiser *et al.*, 1989). In this system, phase variation is a function of the number of copies (usually approximately 30) of the tetrameric sequence 5′-CAAT-3′ found in the 5′ end of the open reading frame of the first gene, *lic1A*. These 4 bp units generate a translational switch by moving the downstream reading frame in and out of phase (Fig. 5.13). Expression varies between three levels of expression: high, low and none. Zero expression correlates with 29 5′-CAAT-3′ repeats (i.e. $29 + / - 3n$); here, no translation initiation codons are placed in frame with the *lic1A* open reading frame. Low expression and high expression correlate with 30 and 31 repeats, respectively. In the former case, a single translation initiation codon is in frame with the *lic1A* open reading frame. In the case of high level expression, two translation initiation codons are in frame.

Further *lic* loci, *lic2* and *lic3*, have been identified by probing the genome with the oligonucleotide sequence $(5'\text{-CAAT-}3')_5$ (Weiser *et al.*, 1989). The *lic2* locus is composed of a single open reading frame while *lic3* consists of four (Maskell *et al.*, 1991). There are multiple copies of the 5′-CAAT-3′ repeat at the 5′ end of the *lic2* open reading frame and at the 5′ end of the first gene, *lic3A*, of the *lic3* locus. *lic3A* has two possible translational start codons located 1 and 15 bp upstream of the 5′-CAAT-3′ repeats. This gene is subject to three levels of control, high, low (the 'bull's eye' phenotype, so-called due to the appearance of colonies with a *lic3A–lacZ* fusion on indicator plates) and none (Moxon & Maskell, 1992; Szabo *et al.*, 1992). High-level expression involves the use of one start codon (that which is 1 bp upstream of the 5′-CAAT-3′ repeats). Low-level expression (bull's eye) correlates with the use of either ATG codon, while zero expression is due to use of any of the three possible reading frames. Thus, in addition to the contribution of the 5′-CAAT-3′ repeats, another regulatory

AATATAAAAATGAATACAAAAATGCTATG-$(CAAT)_N$-CAAATTGT
1 2 3

Number of repeats	Phenotype	In-frame start codon used
$(CAAT)_{29}$	–	None
$(CAAT)_{30}$	+	3
$(CAAT)_{31}$	++++	1 and 2

Fig. 5.13 Phase variation of lipopolysaccharide expression in *Haemophilus influenzae*. The sequence at the translational start site of the first gene at the *lic1* locus in *H. influenzae* is shown. There are three possible translation start signals, two of which (1 and 2) are in frame and there is a $(CAAT)_N$ sequence downstream of the ATG codons. Variation in the number of CAAT repeats varies the reading frame and so alters the expression of the locus over three levels: none, low and high, as summarized in the figure.

mechanism, possibly transcriptional, is involved in controlling the expression of *lic3A*.

The *yopA* gene of *Yersinia*

Yersinia pestis is a highly virulent human pathogen which, like its less virulent relative, *Y. pseudotuberculosis*, carries a copy of the *yopA* gene which encodes an outer membrane protein. Both organisms also harbour a copy of the *inv* gene, coding for the invasin protein (Bolin *et al.*, 1982; Isberg & Falkow, 1985). However, *Y. pestis* expresses Inv but not YopA, whereas *Y. pseudotuberculosis* expresses both Inv and YopA. The *Y. pestis yopA* gene contains a 1 bp deletion in a polyadenosine tract which results in a translational reading frame shift. Restoration of *Y. pestis* to $YopA^{+}$ reduces the virulence of the organism, while mutating the *yopA* and *inv* genes of *Y. pseudotuberculosis* greatly increases its virulence (Rosqvist *et al.*, 1988). Thus, *Yersinia* species have the potential to use alterations to the sequences of specific genes in order to switch between different virulence levels. The underlying mechanisms and selective pressures that govern this process are unknown.

6 Transcriptional Regulation

Introduction

Regulation at the level of transcription represents a highly efficient method of controlling gene expression. By preventing transcription, the later stages in gene expression are all automatically prevented too. Transcription control also offers a mechanism for rapid regulation in response to sudden changes in the environment. Moreover, it allows groups of genes to be regulated simultaneously. This is of particular importance when mounting a response to a complex series of stimuli, such as those encountered by an infecting bacterium as it adapts to a host-associated niche.

The important steps in transcription initiation were considered in Chapter 2. Here, mechanisms for controlling this process negatively through repression or positively through activation will be considered. Bacterial pathogens appear to use most of the recognized control modes for the regulation of virulence gene expression. In most cases, the biochemical nature of these control elements was first worked out in most detail for genes contributing to 'housekeeping' functions and their involvement in virulence gene regulation was discovered subsequently, often through the recognition of structural similarities or homologies between the housekeeping gene regulators and those controlling expression of the virulence genes. This has contributed to the emergence of a system of classifying transcription regulators in so-called regulatory 'families', including the 'two-component family' which consists, as the name suggests, of pairs of functionally coupled regulatory proteins. Other groups that contain significant numbers of virulence gene regulators include the AraC and LysR families, named after their founder members, each of which is a regulator of housekeeping gene expression.

Negative regulation of transcription: the *lac* repressor

Initiation of transcription may be prevented by specific repressor proteins. The system that negatively regulates expression of the *lac* operon is understood in considerable detail (Reznikoff 1992b) (Fig. 6.1). Here, the Lac repressor, the product of the *lacI* gene, binds to an operator site which overlaps the *lac* promoter. Binding of repressor to the operator is stabilized by DNA looping and loop formation is promoted by negative supercoiling of the DNA (Borowiec *et al.*, 1987; Whitson et al 1987).

The repressor and RNA polymerase coexist at the promoter in a non-

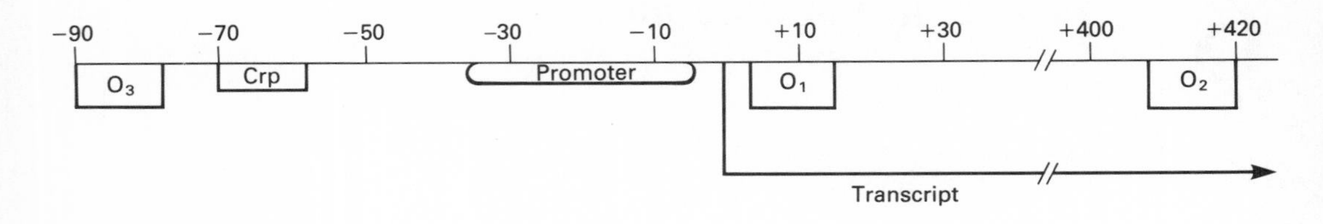

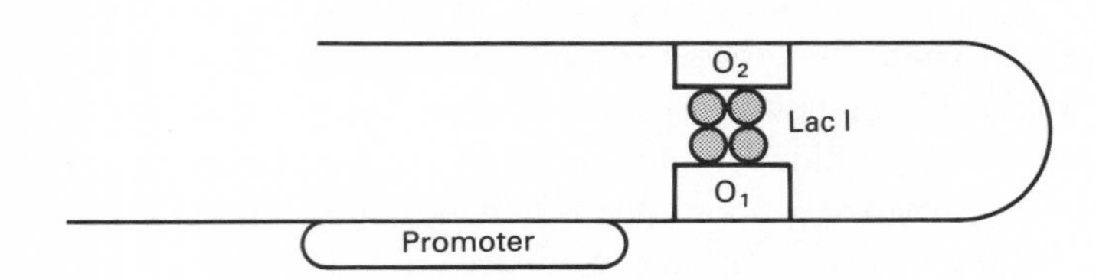

Fig. 6.1 The regulatory region of the *Escherichia coli lac* operon. The promoter region of *lacZ*, the first structural gene in the *lac* operon is illustrated. Distances are calibrated in bp. There are three operator-like sites (O_1, O_2 and O_3) and one binding site for cAMP-Crp (Crp). Binding of the tetrameric Lac 1 repressor cooperatively to O_1 and O_2 results in looping of the DNA between the operator sites; similar interactions may occur involving O_1 and O_3.

productive complex which can be induced by the gratuitous inducer isopropyl thiogalactoside (IPTG) to make mRNA (Strancy & Crothers, 1987). Repressor prevents RNA polymerase from escaping from the initial transcribing complex (ITC) into RNA elongation. The *lac* initial transcribed sequence (ITS) contains a high K_{NTP} site which requires an optimal level of nucleoside triphosphates (NTP) for elongation to begin. Under non-repressed conditions, suboptimal levels of NTP result in pausing at this site. Repressor acts on RNA polymerase to increase the K_{NTP} for the substrate at each stage of RNA chain extension. Thus, repressor prevents RNA polymerase from extending past the high K_{NTP} site even under conditions of abundant NTP. The site on RNA polymerase at which repressor exerts its effect may have evolved as a site for communication with heterologous transcription regulators. For example, it is thought to be the same site at which transcription terminator functions such as NusA have their input during the termination of elongation (Lee & Goldfarb, 1991; see section on Transcription termination/antitermination in this chapter).

Positive control of transcription: cAMP-Crp

A productive interaction between RNA polymerase and the promoter may require the assistance of an activator protein or of a special DNA structural feature such as a bend or loop or a particular degree of supercoiling. It may require more than one of these. Of course, all of these can also contribute to negative regulation of transcription. A classic example of a positive control protein is the cyclic AMP receptor protein, Crp (also called catabolite activator protein, CAP). In most Crp-regulated promoters, Crp acts as an activator. However, there are some examples of promoters that are repressed by Crp, including that of *cya*, the gene coding for adenylate cyclase, whose activity is required to synthesize the Crp cofactor, cAMP (Aiba, 1985).

The Crp protein is composed of two identical subunits of 209 amino acids, with a large aminoterminal domain concerned with cAMP binding and dimerization and a smaller carboxyterminal domain containing the helix-turn-helix motif which binds to DNA. Crp binds to a consensus sequence (5′-TGTGAnntngnTCACA-3′) usually found within 70 bp of the promoter (Botsford & Harman, 1992; reviewed in Busby, 1986). Binding of cAMP to Crp is necessary in order to introduce the conformational change that permits DNA binding. Thus, DNA binding is modulated by the intracellular concentration of cAMP, which is itself modulated by the nature of the carbon source on which the cells grow (Chapter 7). A contact site for Crp has been mapped within the carboxyterminal domain of the RpoA (or α subunit) protein of *E. coli* RNA polymerase, showing that these proteins interact physically at Crp-regulated promoters (Zou *et al.*, 1992).

Crp binding distorts the path of the DNA helix, introducing kinks at both the 3′ and the 5′ side of its binding site which act additively to bend the DNA at Crp by more than 90° (Schultz *et al.*, 1991). Experiments in which bends produced by other proteins or intrinsically bent DNA sequences were substituted for the Crp-induced bend have shown that these bends are essential for the effects of Crp on transcription of the *E. coli gal* operon, *in vivo* (Bracco *et al.*, 1989; reviewed in Lilley, 1991). Crp makes physical contact with RNA polymerase and the bending of DNA is thought to promote this protein–protein interaction (Fig. 6.2). Furthermore, when both Crp and RNA polymerase are present on the

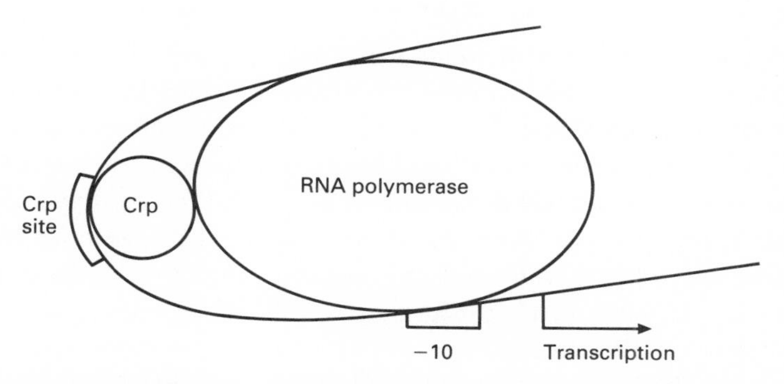

Fig. 6.2 Model of the interaction of Crp with RNA polymerase. The cAMP–Crp complex is shown bound to DNA upstream of a promoter. Crp has bent the DNA, making an additional contact between DNA and RNA polymerase. Crp is also in contact with polymerase.

DNA, the helix is bent by up to 180°. This could bring DNA far upstream into play in the control of transcription initiation. Loop formation is certainly important in the activation of σ^{54}-dependent promoters (see later) where it brings to the promoter transcription activators bound at distant sites, such as NifA which activates the *nif* genes in *Klebsiella pneumoniae* and other species. Here, the integration host factor (IHF) frequently plays a key role in bending the intervening DNA to facilitate loop formation (Hoover *et al.*, 1990). Bending seems to be less important in other Crp-dependent systems; Crp positive control mutants which still bind and bend DNA normally fail to activate *E. coli lac* expression (Bell *et al.*, 1990; Reznikoff, 1992a). Other data indicate that bringing upstream sequences into contact with RNA polymerase may not be important for Crp activation of *lac*: deletion of sequences upstream of position −83 has little effect on the ability of Crp to stimulate transcription from the *lac* promoter (Straney *et al.*, 1989).

Thus, the precise nature of the mechanism by which Crp activates transcription remains unclear; it is not certain which step in transcription initiation is enhanced by Crp. In the case of the *E. coli malT* gene, Crp activates transcription at the promoter clearance step (Menendez *et al.*, 1987). Crp can cooperate with other positive activators to induce transcription. For example, Crp and MalT, the maltose regulon activator, act in concert to activate the *E. coli malK* promoter. Here, Crp appears to redirect MalT to sites where a productive interaction with RNA polymerase becomes possible (Richet *et al.*, 1991).

Since Crp requires cAMP to function, the level of cAMP in the cell regulates Crp-dependent gene expression indirectly. cAMP levels are carbon source-regulated; for example, they fall when bacteria are grown in the presence of glucose (Makman & Sutherland, 1965). Thus, all of the Crp-controlled genes may be coordinately regulated in response to cAMP levels and may be said to form a 'regulon' of very diverse membership (Magasanik & Neidhardt, 1987).

cAMP-Crp and bacterial virulence

Given the pleiotropic nature of mutations in the Crp system, it is unsurprising to learn that it appears to play a crucial role in *Salmonella* virulence; specifically, *S. typhimurium* mutants deficient in *crp* or *cya* function are attenuated for mouse virulence (Curtiss & Kelly, 1987). It is difficult to know precisely how cAMP-Crp contributes to bacterial virulence and it is likely that multiple effects are involved. cAMP-Crp has been shown to regulate expression of the *ompB* regulatory locus (in *E. coli*) and this locus contributes significantly to *S. typhimurium* virulence in mice (Dorman *et al.*, 1989a; Huang *et al.*, 1992). Furthermore, cAMP is a regulator of enterotoxin gene expression in *E. coli* (Gilbert *et al.*, 1989; Martinez-Cadena *et al.*, 1981) and cAMP-Crp is involved in the control of flagellar synthesis in *S. typhimurium* (as it is in *E. coli*) (Kutsukake *et al.*, 1990). It is possible that these structures may influence virulence, particularly during the attachment phase of the invasion process (Lee *et al.*,

1992). A regulatory overlap exists between the cAMP-Crp and the iron regulons in *E. coli*, where it has been shown that Crp also regulates transcription of *fur*, the autoregulated gene encoding the pleiotropic iron regulatory protein, Fur (deLorenzo *et al*., 1988). Fur is conserved in many bacterial species and regulates virulence factor expression, such as that of shiga-like toxin, in response to iron availability (see later).

The *algD* gene from *Pseudomonas aeruginosa*, which encodes GDP-mannose dehydrogenase in alginate synthesis, is positively regulated by cAMP-Crp when the *algD* gene is expressed in *E. coli*. Perhaps the *Ps. aeruginosa* homologue of Crp is involved in its regulation when the gene is in its native genetic background; its expression does show carbon-source control in *Ps. aeruginosa* (DeVault *et al*., 1991). In this way, the expression of alginate, an important virulence factor in *Ps. aeruginosa* infections of cystic fibrosis patients, could be controlled by cAMP-Crp.

Transcription regulation by different sigma factors

RNA polymerase holoenzyme composed of $\alpha_2\beta\beta'\sigma$ (or $E\sigma^{70}$), encoded by the *rpoA*, *rpoB*, *rpoC* and *rpoD* genes, respectively, is not solely responsible for transcription in *E. coli*. The discovery of the existence of additional sigma factors capable of interacting with core polymerase has led to the realization that bacteria possess a highly successful method for controlling gene expression at the level of transcription which involves the reprogramming of RNA polymerase with alternative sigma factors which then allows it to recognize non-standard (i.e. non-σ^{70}) promoters (Table 6.1). This type of control mechanism is used in stress response systems such as the heat shock response and in developmental cascades in sporulating Gram-positive bacteria (Georgopoulos *et al*., 1990; Hoopes & McClure, 1987; Lonetto *et al*., 1992; Smith *et al*., 1990). Expression of alternative sigma factors may be induced by the environmental stress which they are involved in resisting. For example, σ^{32}, the heat shock sigma factor of *Escherichia coli* is transcribed from a message that displays enhanced stability at increased temperatures (Straus *et al*., 1987). Alternative sigma factors may also be subject to negative regulation. For example, the σ^F factor required for expression of flagellum genes in *Salmonella typhimurium* can be inactivated by

Table 6.1 Summary of consensus promoter sequences recognized by alternative sigma factors

Sigma factor	Promoter consensus sequence
σ^{70}	TTGACA (17 bp) TATAAT (6 bp) + 1
σ^{54}	CTGGCAC (5 bp) TTGCA (6-11 bp) + 1
σ^{32}	CTTGAA (13–15 bp) CCCCAT-TA (7 bp) + 1
σ^{24}	GAACTT (5 bp) ATAAAA (5 bp) TCTGA (6 bp) + 1

binding the regulatory protein FlgM, an 'antisigma factor' (Ohnishi *et al.*, 1992).

σ^{70} promoters are numerically the most important class in the enteric bacteria; a recent database survey identified almost 110 molecularly defined promoters that are transcribed using this sigma factor (Collado-Vides *et al.*, 1991). A few, such as the *lac* promoter, have been studied in minute detail at the molecular level and most of them display complex regulation, in which transcription is controlled by multiple and overlapping regulatory elements. In the case of *lac*, the operon is controlled negatively by the Lac repressor and positively by the cAMP receptor protein (Crp) (Fig. 6.1). This type of regulation permits multiple control inputs which can fine-tune transcription to the needs of the bacterium at any particular time in any particular environmental niche. Because they constitute the majority of known bacterial promoters, the σ^{70} group will not be considered together under this heading, but rather under the headings of the more specific regulatory elements which control them.

The following is not an exhaustive listing of alternative sigma factors. Instead it serves to illustrate a number of those that have been characterized in some detail at the molecular level and that make actual or potential contributions to the control of virulence gene expression.

The RpoN (σ^{54}) promoter group

σ-54 is encoded by the *rpoN* (also called *ntrA* and *glnF*) gene of enteric bacteria and it can promote transcription by RNA polymerase core enzyme in the absence of σ^{70}. σ^{54} has a radically different structure to σ^{70} and σ^{54}-driven core polymerase recognizes a different promoter consensus sequence to that bound by σ^{70} (Table 6.1). In general, genes activated by σ^{54} frequently code for products required for the nitrogen starvation response (e.g. *glnA* which codes for glutamine synthetase). Thus, the σ^{54}-dependent genes may be regarded as a stress response regulon. However, genes other than those involved directly in the nitrogen limitation response are σ^{54}-dependent. These include genes involved in bacterial virulence.

In *Pseudomonas aeruginosa*, σ^{54} is involved in controlling expression of a number of virulence factors. It is required for expression of urease and the pilin virulence factor (Ishimoto & Lory, 1989; Totten *et al.*, 1990) and contributes to flagellin expression. The latter is significant because motility may assist the bacterium in colonization of the host. σ^{54} does not control flagellin transcription directly but may regulate its expression through an indirect mechanism; *Ps. aeruginosa* flagellin expression is transcriptionally regulated by yet another alternative sigma factor, σ^{28} (RpoF) (Starnbach & Lory, 1992).

Alginate is an important virulence characteristic of those *Ps. aeruginosa* strains that infect cystic fibrosis patients. It is unclear if σ^{54} regulates alginate synthesis in *Ps. aeruginosa*. Some workers have found no role for *rpoN* in alginate synthesis (Mohr *et al.*, 1990; Totten *et al.*, 1990) while others have suggested that *rpoN* may regulate transcription of both the *algD* structural gene and the *algR* regulatory gene of the alginate biosynthetic pathway (Kimbara & Chakrabarty, 1989).

In the plant pathogen, *Pseudomonas syringae* pv. phaseolicola, σ^{54} has been identified as being required for the regulation of expression of several virulence genes of the *hrp* class. These genes are involved in causing disease in susceptable hosts and in causing a 'hypersensitive response' in resistant plants (Fellay, R., Rahme, L.G. & Panopoulos, N.J., cited in Rahme *et al.*, 1992).

RpoH (σ^{32}) promoters and the heat shock response

Bacteria respond to an increase in growth temperature by increasing the rate of expression of a small number of proteins. This induction occurs at the level of transcription and requires an alternative sigma factor called σ^{32}, the product of the *rpoH* gene (also known as *htpR*). σ^{32} programmed RNA polymerase recognizes a non-standard promoter sequence (Table 6.1). Mutations in the *E. coli rpoH* gene are lethal above 20°C, implying that its product is essential for life above that temperature (Zhou *et al.*, 1988). The thermally induced proteins have been formally classified as heat-shock proteins but in some cases their expression is now known to be inducible by several other stimuli, including exposure of the cell to hydrogen peroxide, ethanol, bacteriophage infection, heavy metals, DNA-damaging agents, the presence of abnormally folded polypeptides (reviewed in Georgopoulos *et al.*, 1990).

A number of studies indicate a role for the heat shock response in bacterial virulence. Data implicating σ^{32} specifically are rarer. The cholera toxin regulatory gene *toxR* (a positive regulator of the *ctxAB* operon, see Chapter 8) is transcribed divergently from the *Vibrio cholerae htpG* gene and *toxR* expression is thermoregulated. It has been suggested that this control may be exerted by RNA polymerase under the control of the *V. cholerae* analogue of σ^{32}. In this model, transcription of *htpG* by RNA polymerase (σ^{32}) occludes the σ^{70} promoter of *toxR*, resulting in downregulation of *toxR* transcription at higher temperatures (Parsot & Mekalanos, 1990). Experiments carried out in *E. coli* have shown that overexpression of σ^{32} results in a decrease in *toxR* expression (Parsot & Mekalanos, 1990).

RpoS (σ^{38}), a sigma factor for gene activation in stationary phase

The *E. coli* gene *rpoS* (also known as *appR*, *csi-2*, *katF* or *nur*) encodes a sigma factor whose expression is triggered by entry of the cell into the stationary phase of the growth cycle (Hengge-Aronis, 1993; Kolter, 1992; Siegle & Kolter, 1992). Mutants deficient in RpoS (σ^{38}) fail to develop resistance to hydrogen peroxide (due to the dependence of the catalase gene, *katE* on RpoS for transcription) and do not express *xthA*-encoded exonuclease III. They do not produce glycogen nor express a stationary phase-specific acid phosphatase known as AppA. RpoS is required for stationary phase expression of the 'morphogene' *bolA*, whose product is concerned with controlling the shape of the cell (Aldea *et al.*, 1988; Lange & Hengge-Aronis, 1991a). It regulates expression of the gene coding for Dps, a non-specific DNA-binding protein which protects DNA from damage during stationary phase and is also a pleiotropic regulator of stationary phase gene expression (Almirón *et al.*, 1992). Thus, cells appear to possess a regulon of stationary phase genes, most of which remain to be identified, whose transcription is dependent on the RpoS sigma factor (Lange & Hengge-Aronis, 1991b; Mulvey and Loewen, 1989; Mulvey *et al.*, 1990). A list of known RpoS-dependent genes is given in Table 6.2.

Two-dimensional gel electrophoresis reveals that the RpoS-deficient mutants fail to synthesize many products normally expressed in stationary phase, that they also fail to express about 18 osmotically inducible proteins (Hengge-Aronis, 1993) and fail to express some 32 carbon starvation proteins. The mutants survive carbon and nitrogen starvation poorly and fail to acquire starvation-mediated cross-resistance to osmotic, oxidative and heat stresses (McCann *et al.*, 1991) and are thermo-osmotically sensitive.

The signal that heralds the onset of stationary phase and hence *rpoS* transcription is unknown but may consist of an accumulation of a low-molecular-weight molecule. Some weak organic acids (such as acetic acid, benzoic acid and propionic acid) induce the gene (Mulvey *et al.*, 1990; Schellhorn & Stones, 1992) while others (such as *o*-aminobenzoic acid) do not, suggesting that the effect is not due simply to changes in internal pH but could be due to a structural relationship between the inducing acid and the genuine inducer (Mulvey *et al.*, 1990). Data from experiments in which exponentially growing *E. coli* cells were resuspended in spent growth medium (presumably containing

Table 6.2 Genes under RpoS control in *Escherichia coli*

Gene	Map position (min)	Function	Reference
appY	13	Regulator of *appA cyxAB*	Atlung *et al.*, 1989
bolA	10	Morphogene	Aldea *et al.*, 1989
csgA	23.1	Curli subunit	Olsén *et al.*, 1993
dps	18	DNA protection	Almirón *et al.*, 1992
glgS	66.6	Glycogen synthesis	Hengge-Aronis & Fischer, 1992
katE	37.2	Hydrogen peroxide resistance	Mulvey *et al.*, 1990
mcc	Plasmid-linked	Microcin C7 expression	Diáz-Guerra *et al.*, 1989
osmB	28	Lipoprotein	Jung *et al.*, 1990
osmY (*csi-5*)	99.3	Periplasmic protein	Lange & Hengge-Aronis, 1991b Yim & Villarejo, 1992
otsBA	41.6	Trehalose synthesis	Kaasen *et al.*, 1992
treA	26	Trehalose uptake	Hengge-Aronis *et al.*, 1991
xthA	38	Exonuclease III	Sak *et al.*, 1989

accumulated inducer molecules) or in fresh medium devoid of carbon sources indicate that it is starvation that provides the primary signal for *katF* induction (McCann *et al.*, 1993).

Work with transcriptional and translational *lacZ* fusions to *katF* indicates that its expression is regulated post-transcriptionally (Loewen *et al.*, 1993; McCann *et al.*, 1993). Some workers have shown that during carbon starvation, *katF* expression is negatively affected by cAMP (Lange & Hengge-Aronis, 1991b). However, McCann *et al.* (1993) found this work difficult to reproduce.

A candidate for the RpoS-dependent promoter consensus has been suggested based on the transcription start sites for two RpoS-dependent genes and conservation of upstream sequences: GTTTAGC-15 bp-ACGTCC-6 bp-G (*katE*) and GGTAAGC-17 bp-CCATCC-4 bp-A (*xthA*) (von Ossowski *et al.*, 1991). The *bolA* promoter (GTTAAGC-19 bp-GCGGCT) shows some similarity to this consensus (Lange & Hengge-Aronis, 1991a). Like the promoters used by other non-standard sigma factors, this sequence differs considerably from the σ^{70}-dependent promoter consensus sequence (Table 6.1). However, results from *in vitro* transcription experiments show that *E. coli* promoters can be divided into three classes on the basis of their recognition by RpoS or RpoD (σ^{70}). Type 1 promoters are recognized by both RpoS and RpoD; type 2 promoters are recognized by RpoD only while type 3 promoters are recognized by RpoS only (Tanaka *et al.*, 1993). Interestingly, all three promoter types possess RpoD-type −10 boxes. Type 3 promoters lack sequences matching the −35 box of classical RpoD-driven promoters.

E. coli promoters dependent on RpoS are associated with regions of DNA which display intrinsic curvature (Espinosa-Urgel & Tormo, 1993). These curves occur upstream of the promoters and may contribute to promoter activation, by analogy with the proposed role of DNA curvature (intrinsic or induced) in activation of σ^{70} dependent promoters. The curves may attract the H-NS protein, which is known to bind to such structures, and this may provide an additional layer of regulation (see Chapter 7). In general, H-NS represses those promoters to which it binds. The genes coding for the curli adhesin in fibronectin-binding strains of *E. coli* are subject to negative control by H-NS and positive control by RpoS and are a good example of this type of multi-factorial control (see Chapter 8).

Transcriptional regulation by DNA supercoiling

The supercoiled nature of bacterial DNA was described in Chapter 2. Evidence for a role for DNA supercoiling in the control of transcription comes from both *in vivo* and *in vitro* studies. Research on promoter function in the *in vivo* situation has usually involved topoisomerase-inactivating antibiotics or mutants with genetic lesions in the topoisomerase genes. *In vitro* studies compare the efficiencies of the promoter of interest at a range of superhelical densities. Data from this research has shown that the issue of transcriptional regulation by DNA supercoiling is complex and that one needs to proceed with caution when drawing conclusions from the results. Some promoters (for example, the *E. coli trp* promoter) seem to be indifferent to changes in supercoiling, at least over the ranges achievable by inhibiting DNA gyrase with antibiotics *in vivo* (Sanzey, 1979). Some promoters display context-dependent sensitivity to changes in the global level of supercoiling, performing differently on plasmids and on the chromosome, or *in vivo* and *in vitro*. The *E. coli tyrT* promoter seems to be supercoiling-sensitive *in vitro* but not when carried on plasmids *in vivo* (Lamond, 1985). However, experiments with this promoter on the chromosome show that it is very sensitive to changes in DNA supercoiling when in its native context (A. Free & C.J. Dorman, unpublished results). The *leu500* promoter of *Salmonella typhimurium* is another promoter with context-dependent sensitivity to supercoiling, being supercoiling sensitive on the chromosome but not when carried on recombinant plasmids (Richardson *et al.*, 1988). The nature of this dichotomy is discussed in detail for the *leu500* promoter below.

In principle, superhelical torsion may affect thermodynamically the activation of a promoter at a number of levels. It has the potential to influence the binding of RNA polymerase to the promoter sequence during the formation of the closed complex. This could be done by rotating the polymerase-binding hexameric −10 and −35 sequences to different faces of the helix. Similar helical displacements could affect the binding of accessory proteins, such as enhancer-binding transcription activators. Even if activator and polymerase manage to bind to the DNA, their mutual interaction could be modulated by helical torsion. This is true both for the close range regulator–polymerase interactions associated with σ^{70}-driven promoters (Kramer *et al.*, 1988) and for the longer range DNA loop-dependent interactions of the σ^{54} promoters in which enhancer-bound activators must make contact physically with polymerase.

As has been discussed in Chapter 2, the changes in the linking number of DNA are partitioned between the twist and the writhe (i.e. $\Delta Lk = \Delta Tw + Wr$). In terms of exerting an influence on protein contact with DNA over short distances, as occurs between polymerase and the −10 and −35 sequences within the promoter itself, the twist component might be expected to play an important part. Here, twisting of the DNA could align or misalign the canonical hexameric repeats, influencing interaction with polymerase and hence the formation of the closed complex. It has been proposed that promoters with non-standard spacing between the −10 and −35 sequences (i.e. greater or less than 17 bp) may be particularly sensitive to changes in twist and that these may exploit the twist component of supercoiling to regulate their function (Borowiec & Gralla, 1985; Wang & Syvanen, 1992). The formation of the open complex (RPo) requires a substantial unwinding of the DNA helix and this process is assisted by negative supercoiling of the DNA. Supercoiling is not only required for the formation of the initiation complex, it is also required to stabilize it (Richet & Raibaud, 1991). It has been shown that the structure of the promoter as a whole (i.e. −10, −35 and spacer) is important in determining the degree to which promoter function is affected by negative supercoiling in the DNA (Borowiec & Gralla, 1987).

The negative supercoiling that influences promoter function may be generated by DNA gyrase or by distortions of the DNA helix in the neighbourhood of the promoter due to the transcription of nearby genes or the replication of the DNA. The twin supercoiling domain model of Liu & Wang (1987) predicts that a moving transcription complex has a domain of relaxed (or even positively supercoiled) DNA ahead of it and a domain of negatively supercoiled DNA behind. If a promoter lies upstream from an actively transcribing gene whose direction of transcription is opposite to its own, it and the intervening DNA will be within the negatively supercoiled domain (Fig. 2.6). The activities of one neighbouring gene may influence strongly the function of the other's promoter. Given the number of examples of divergent transcription units, it is possible that this represents an important form of transcription control. It has been shown that such localized changes in DNA supercoiling can influence structural transitions in neighbouring DNA sequences, such as the formation and stabilization of Z-DNA from B-DNA (Rahmouni & Wells, 1989). Local supercoiling can also activate promoters.

The *leu500* mutant promoter of the *Salmonella typhimurium* leucine biosynthetic operon carries a GC base pair in place of an AT base pair within the -10 box and is transcriptionally silent in cells possessing an active DNA topoisomerase I gene (*topA*). If *topA* is mutationally inactivated, the *leu500* promoter is activated. This has led to speculation that increases in the global level of negative supercoiling (due to the unrestrained activity of DNA gyrase in the absence of topoisomerase I) can melt the mutant -10 region of the *leu500* promoter. However, when compensatory mutations are introduced to the *topA* mutant strain which restore the global level of supercoiling to values equivalent to those in *topA*$^+$ wild-type strains, the *leu500* promoter remains resolutely active (Richardson *et al.*, 1984). Thus, no correlation is possible between global supercoiling levels and *leu500* promoter activity; the only obvious correlation is that the chromosomally located *leu500* promoter is active in the absence of topoisomerase I. However, even this correlation is found to be conditional. When the *leu500* mutant promoter is carried in a plasmid, it fails to function in *any* genetic background, regardless of the level of supercoiling or the status of the *topA* gene (Richardson *et al.*, 1988). This shows that the activity of this promoter is highly sensitive to its context.

An application of the twin supercoiled domain model offers some hope of resolving the *leu500* enigma. Here, a putative upstream promoter transcribing the opposite DNA strand provides it with the negative supercoiling levels required to activate it. But, in the presence of active topoisomerase I, this negatively supercoiled domain is relaxed as rapidly as it forms and *leu500* remains dormant. When the *topA* gene is mutationally inactivated, topoisomerase I can no longer provide this relaxing activity and *leu500* is activated (Fig. 6.3). The model also suggests an explanation for the failure of the *leu500*

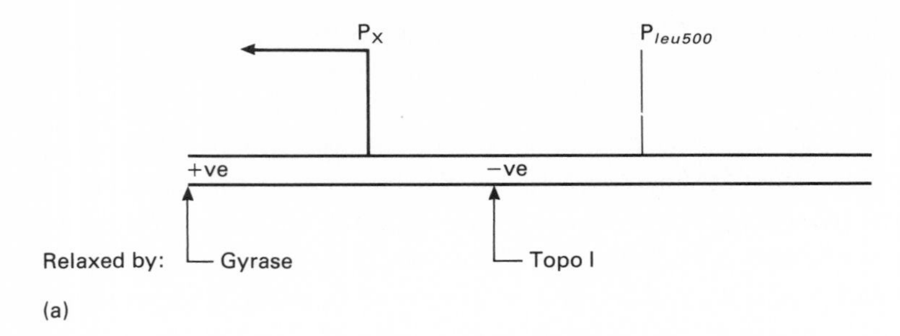

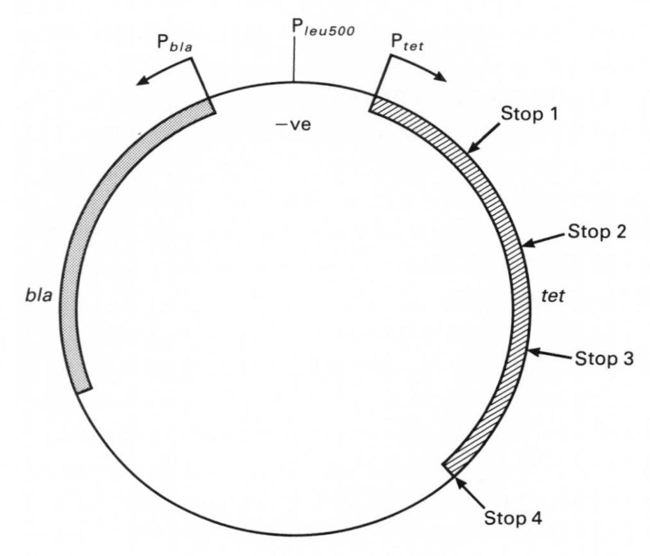

Extent of translation	Percentage *leu500* activation	
	Topo I$^+$	Topo I$^-$
No stop	0	100
Stop 1	0	10
Stop 2	0	35
Stop 3	0	60
Stop 4	0	80

(b)

Fig. 6.3 Effect of local DNA supercoiling on *leu500* promoter activity. (a) The activity of an upstream promoter (P_x) generates a domain of negative supercoiling at the *leu500* promoter, providing sufficient free energy to activate it. However, DNA topoisomerase I can relax this negatively supercoiled domain, suppressing transcription from the *leu500* promoter. (b) In a plasmid system, *leu500* activation depends on the transcription and the extent of translation of a neighbouring gene. The negative effects of progressive truncation of the *tet* gene product on *leu500* promoter activity are shown.

promoter to work when carried on a plasmid in a *topA* mutant cell; simply put, the negative supercoils provided by neighbouring gene activity escape from the vicinity to P_{leu500}. When a plasmid is designed which brackets *leu500* with divergently transcribing genes, trapping transcriptionally generated supercoils at P_{leu500}, the promoter is found to be active in the absence of topoisomerase I, as it is on the chromosome. The trapping of the negative supercoils is probably due to the inability of the transcription/translation complexes to rotate rapidly around the DNA template, due to torsional drag, as predicted by the model of Liu & Wang (1987) (Figs 2.6 & 6.3). This is supported by the demonstration that *leu500* activity depends not only on the transcription but also on the extent of translation of the neighbouring gene (Fig. 6.3) (Chen *et al.*, 1992).

Thus, active genes subdivide the DNA template into negatively or positively supercoiled domains and the topoisomerases restore the levels of supercoiling in these domains to wild-type levels (gyrase relaxing the positive supercoils and topoisomerase I relaxing the negative supercoils). These positive and negative domains may also extinguish each other if they come into contact. Fluctuations in the activities of the topoisomerases will have sudden and profound effects on topological activity and represent a potentially important level of modulation of gene expression.

An attraction of DNA supercoiling as a transcription regulator is its potential to affect promoters differentially and simultaneously. As we have seen, sensitivity to supercoiling depends upon several factors, not least of which is the primary structure of the promoter. This includes the degree of relatedness of the promoter to the canonical sequence and the size of the spacing between different parts of the promoter sequence (for example, an optimal spacing of 17 bp between the -10 and -35 hexamers of a σ^{70} promoter). These factors may mean that while one promoter is activated by increases in the supercoiling level, another promoter may find such increases to be inhibitory. The result is that promoters may vary in both the direction of their response to changes in supercoiling (that is, being either activated or repressed) and in the degree of their response, with some showing a much greater range of responsiveness than others to an equal change in supercoiling level. Experiments with random *lacZ* fusions to genes located around the chromosome of *Salmonella typhimurium* show that while most promoters show some level of sensitivity to changes in DNA supercoiling, the degree of change seen is usually modest and the direction of the change is not the same for all promoters (Jovanovich & Lebowitz, 1987).

In practice, sensitivity to changes in DNA supercoiling is rarely (if ever) seen as the only regulator of a promoter. In general, it forms just the basal layer of a hierarchy of regulatory influences which include more conventional transcriptional regulators, such as repressor or activator proteins. Its utility as a regulator lies in its ability to affect simultaneously the activities of so many promoters. This gives it the potential to link otherwise functionally or geographically unrelated genes within a regulatory network. The power of such a global control system to coordinate the expression of the genetic complement of the cell is obvious, particularly in the light of data indicating that DNA supercoiling levels vary in response to the very environmental stimuli that commensal and pathogenic bacteria encounter on moving from a free-living state to a host-associated state and back again. This topic is discussed in Chapters 7 and 8.

DNA looping and transcriptional regulation

Looping represents the application of yet another aspect of DNA structure to the control of DNA-dependent processes, including transcription. DNA looping occurs when a protein or protein complex binds simultaneously at two distinct positions in DNA (Fig. 6.4). Loop formation of this type offers several advantages in terms of promoting protein–DNA interactions. It allows new levels of regulation to be built in by involving two rather than one section of the DNA duplex with their associated proteins. It permits cooperativity in binding to the two sites at the neck of the loop. Thus, relatively low concentrations of the protein(s) are required to saturate the binding sites and these sites do not need to have very high affinities for the protein(s) (reviewed in Ptashne, 1986; Schleif,

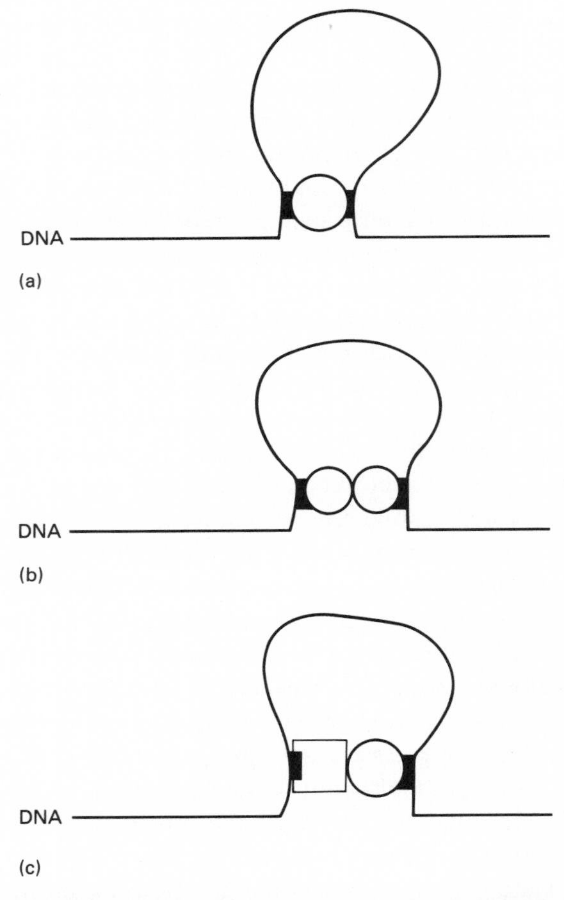

Fig. 6.4 Protein-induced DNA loop formation. (a) A loop generated by a monomeric protein with two DNA binding domains. (b) A loop generated by a dimeric protein in which each monomer possesses a single DNA binding domain. (c) A loop generated by a heterodimeric protein in which each monomer possesses a single (but distinct) DNA-binding domain.

1992; Wang & Giaever, 1988). Looping can be accomplished by a single protein with two DNA-binding sites, by a dimeric or tetrameric protein with one binding site per monomer or by heterologous DNA-binding proteins which associate with one another to form the loop (Fig. 6.4).

Protein-induced loop formation is sensitive to the phasing of the protein binding sites on the DNA. When the sites are on the same face of the helix, looping is favoured. Insertion of a half helical turn into the DNA of the loop will displace the sites to opposite faces of the helix and inhibit loop formation (Figs 6.5 & 6.6). Looping will also be influenced by the flexibility of the DNA or by the torsional stiffness of the DNA, as these will determine the ability of the protein to cope with slightly out-of-phase binding sites. DNA supercoiling will promote looping by bringing distant sites closer together. However, as the distance between the sites becomes very large (greater than several thousand base pairs) the benefits from supercoiling diminish. In bacterial systems, detailed studies have been carried out on the role of DNA looping in controlling the expression of the arabinose and lactose operons and the NtrC-dependent and NifA-dependent genes, among others. Some of these are discussed in Chapter 7.

DNA curving, bending and transcriptional regulation

To form the loops described in the last section, the DNA between the loop-forming protein binding sites must of necessity follow a curved or bent trajectory. Curved DNA can arise either as a consequence of the action of a protein upon a flexible DNA sequence or as a result of the presence of inherently curved sequences in the DNA itself, such as the conformationally rigid oligo-(dA)-(dT) tracts (Travers, 1989; Trifonov, 1985) (Fig. 6.7). Curvature is not restricted to sequences with poly(dA)-(dT) runs, although poly(dA)-(dT) tracts result in the greatest curvature (Harrington, 1992).

The presence of curved sequences upstream of bacterial promoters is believed to enhance their efficiency; a survey of 43 prokaryotic promoters established a correlation between the presence of such sequences and promoter strength (Plaston & Wartell, 1987). Furthermore, the direction of the curve appears to be very important. In experiments in which the curve found upstream of a *Bacillus subtilus* bacteriophage promoter was displaced further upstream by insertion of short DNA fragments, insertions of 11 or 21 base pairs (i.e. full helical turns) resulted in efficient promoter function whereas insertions of 15 or 25 base pairs (introducing half helical turn rotations) produced a reduction in promoter efficiency (McAllister & Achberger, 1989). These data are consistent with a role for DNA curvature in looping upstream sequences around to contact the promoter. In this model, a linear displacement of the curve to a position further upstream would not be expected to affect promoter function significantly, which is the observed experimental result. However, the data show that the rotational orientation of the curve with respect to the promoter is important and this is consistent with a situation in which the upstream

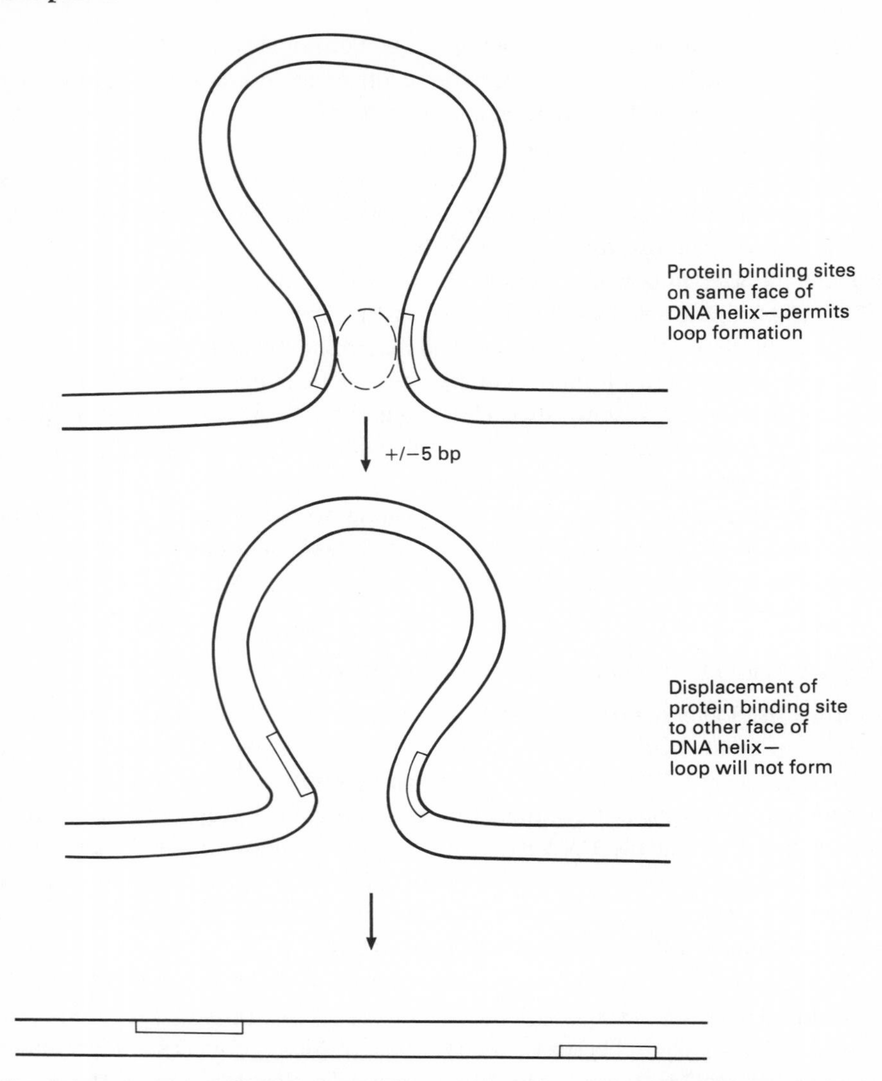

Fig. 6.5 Importance of helical phasing for DNA loop formation. When protein binding sites are on the same face of the helix, loop formation proceeds normally. If a half helical turn (5 bp for B-DNA) is added or removed from the sequence between the protein binding sites, the loop will not form because one site has been displaced to the back of the helix relative to the other.

sequences are required to make specific contact with RNA polymerase (Fig. 6.6).

Additional evidence that upstream curvature contributes to promoter efficiency comes from 'bend swap' experiments in which the binding site for the DNA bending cAMP-Crp complex in the *E. coli gal* promoter was replaced by an inherently bent DNA sequence (Bracco *et al.*, 1989). The results showed that DNA curvature activated the promoter and the extent of activation was

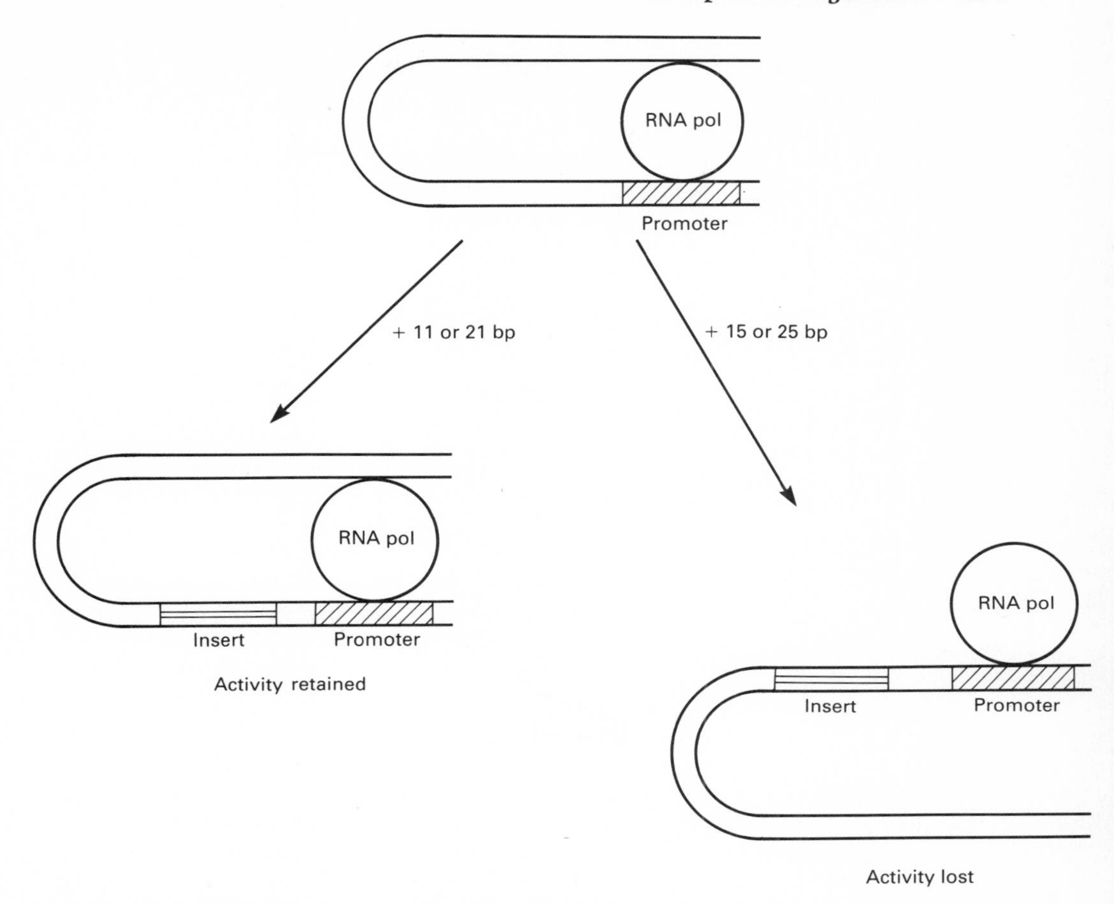

Fig. 6.6 Loop-dependent promoter activity. The effects of linear vs rotational displacement on DNA loop-dependent promoter function are shown. Adding DNA in increments equivalent to full helical turns (left of figure) increases the linear distance between the promoter and the upstream sequence but the loop trajectory is not altered and the promoter is still activated. Adding half-helical turns (right of figure) displaces the trajectory of the loop, abolishing contact between the upstream site and the promoter. In this situation promoter activation is lost.

dependent on the degree of curvature and the relative locations of the promoter and the centre of the curved sequence.

Transcriptional regulation by DNA methylation

Base-specific DNA methylation occurs throughout biology and has a number of important functions, including effects on gene expression (Dynan, 1989; Holliday, 1989; Razin & Cedar, 1991). It can alter DNA–protein interactions in a number of ways since the methyl group may provide a novel hydrophobic contact with the protein or it can induce a local change in the structure of the

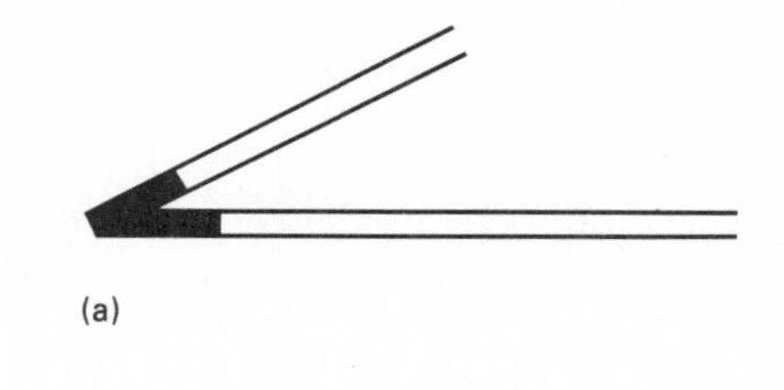

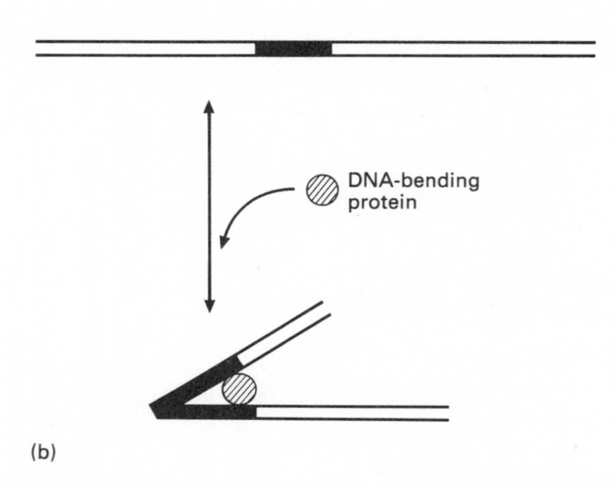

Fig. 6.7 Bends in DNA. A highly simplistic representation of DNA bending due to (a) the presence of an inherently bent sequence or (b) the presence of a binding site for a bend-inducing protein. In the latter case, the DNA remains unbent until the protein binds.

DNA which in turn modulates contact with the protein. A methyl group may have the ability to hinder sterically contacts between the protein and functional groups in the DNA. Such mechanisms have the potential to be highly specific in their effects and to affect contacts between particular DNA sites and heterologous proteins in different ways.

DNA in eukaryotes is methylated only on cytosines whereas bacterial DNA is methylated on cytosines and adenines. Chromosomal DNA in *E. coli* K-12 is methylated on 2% of adenines (N^6-methyladenine) and 1% of cytosines (5-methylcytosine) (Marinus, 1987). The detail of how gene expression is regulated by DNA methylation is best understood in *E. coli* K-12, where three types of DNA methylation have been described. One of these, catalysed by the HsdM methylase, is concerned with the protection of host DNA from cleavage by the K restriction system. This protection involves methylation at the second adenine in the sequence 5′-AAC(N_6)GTGC-3′, with the methylated DNA being recognized as 'self' and thus protected from attack by the restriction endonuclease (Mamelak & Boyer, 1970).

The Dcm system methylates DNA on cytosine residues. Specifically, it methylates the second cytosine in the sequence 5′-CC(A/T)GG-3′. Dcm methylation is not very widely distributed among the bacteria. It occurs in *E. coli* K-12 and *E. coli* C but not *E. coli* B. It is found in *Salmonella typhimurium, S. typhi, Shigella flexneri, S. sonnei, Citrobacter freundii, Klebsiella pneumoniae* and *K. oxytoca.* It is not found in *Aeromonas* species, *Agrobacterium tumefaciens, Bacillus* species, *Clostridium difficile, Lactobacillus casei, Proteus* species, *Pseudomonas aeruginosa, Rhizobium phaseoli, Serratia marcescans, Staphylococcus aureus, Streptomyces* or *Yersinia enterocolitica* (Gomez-Eichelmann *et al.*, 1991). The biological significance of the Dcm system is poorly understood; *dcm* mutants appear to be fully viable. Strains with defective Dcm activity may contain hemimethylated 5′-CC(A/T)GG-3′ sites and in *arl* mutants these can serve as recombination hot-spots (Korba & Hayes, 1982). It is possible that Dcm methylation serves to protect the cell from the effects of the *Eco*RII endonuclease (Schlagman *et al.*, 1976).

The Dam methylase methylates the adenine in the sequence 5′-GATC-3′ and this reaction affects a wide range of cellular processes, including gene expression (Barras & Marinus, 1989). The interaction of the Dam methylase with DNA is negatively modulated by methylation of the DNA. Dam methylation is thought to play an important role in chromosome replication since the absence of methylated adenines from the newly synthesized DNA strands permits discrimination between this and its methylated template strand. However, hemimethylation of the chromosome is probably not important for correct partitioning since this process proceeds normally in a *dam* mutant (Vinella *et al.*, 1992).

5′-GATC-3′ methylation is important in mismatch repair. Repair enzymes recognize the newly synthesized DNA strand behind the replication fork because this is unmethylated. Fully methylated DNA is not a substrate for repair. An early step in repair involves cleavage of the daughter strand DNA by the MutH protein at a position 5′ to the G residue in the 5′-GATC-3′ sequence. The state of methylation of the DNA influences the ability of the MutH protein to interact with it since it is only active on the unmethylated DNA strand. Using the methylated template strand as a guide, the mismatch repair system will remove any newly-incorportated base in the unmethylated strand that is mismatched to the template.

By influencing protein–DNA interactions, Dam methylation has the potential to affect transcription. There are now several examples of bacterial promoters that contain Dam methylation sites. These include the −35 regions of the *dnaA2P, mioC, sulA, trpS, trpR* and *tyrR* promoters and the −10 regions of the *glnS* promoter. Data from experiments with *dam* mutants show that the *dnaA* and *mioC* promoters are activated by methylation while the others are repressed (reviewed in Barras & Marinus, 1989). The *dnaA* and *mioC* gene products contribute to chromosome replication and both genes are located close to the origin of replication, *oriC*. Therefore, their promoters would be transiently hemimethylated during the early stages of replication and this would afford a mechanism of linking their expression to the cell cycle.

Transcription termination/antitermination

Once transcription elongation has been established, RNA polymerase is capable of synthesizing a very long transcript. It will cease to do so if it encounters a transcription terminator sequence in the DNA (which specifies an RNA sequence composed of a GC-rich inverted repeat followed immediately by a run of U residues) or if it pauses and then encounters the termination factor Rho, in the presence of NTPs. Transcription complexes paused during elongation can be restarted by the *E. coli* GreA protein, which interacts with RNA polymerase (Borukhov *et al.*, 1992).

Like the initiation step, termination provides a site for transcription regulation. Modulation of termination efficiency plays an important role in controlling gene expression in several bacteriophage and in some operons of *E. coli.* The *E. coli* NusA protein increases the efficiency of the bacteriophage lambda t_{R2} terminator, lying between the *P* and *Q* genes, by 90% and increases the efficiency of the t_1 terminator in the *E. coli rrnB* operon. NusA binds to RNA polymerase, occupying during the elongation phase the position occupied by sigma factor at initiation (reviewed in Friedman, 1988b; Yager & von Hippel, 1987).

The Q protein of bacteriophage λ functions as an antiterminator. It is effective at low efficiency alone but is highly effective in the presence of the host-encoded NusA protein; the role of the latter may be to load the Q protein on to the polymerase. The N protein of bacteriophage λ is also an antiterminator and binds to the NusA protein (Das, 1992). To be N-antitermination sensitive, a gene must have a *nut* site between its promoter and terminator. The terminator must be of the Rho-dependent type. The function of Rho is to release a paused transcription complex from the template. The Psu protein of bacteriophage P4 acts *in trans* to suppress Rho-dependent polarity in late gene expression of infecting bacteriophage P2, in chromosomal and in plasmid gene expression (Linderoth & Calendar, 1991).

The *bgl* operon of *E. coli* represents a bacterial chromosomal system in which the molecular detail of gene regulation through transcription antitermination in response to an environmental stimulus has been elucidated. This operon is required for the catabolism of aromatic β-glucosides and is transcribed from a promoter within the *bglR* regulatory sequences found upstream of *bglG.* A complication in studies of *bgl* expression is the fact that the operon is frequently cryptic and requires activation by point mutations or insertion sequences in *bglR* or by mutations at the unlinked *hns* locus (called *bglY* in the *bgl* literature). Full expression also requires a β-glucoside inducer. The *bglG* and *bglF* gene products play regulatory roles in expression of the operon.

Transcription from the *bgl* promoter is prone to termination at the t_1 terminator within the 130 bp leader sequence of the *bglG* gene and at the t_2

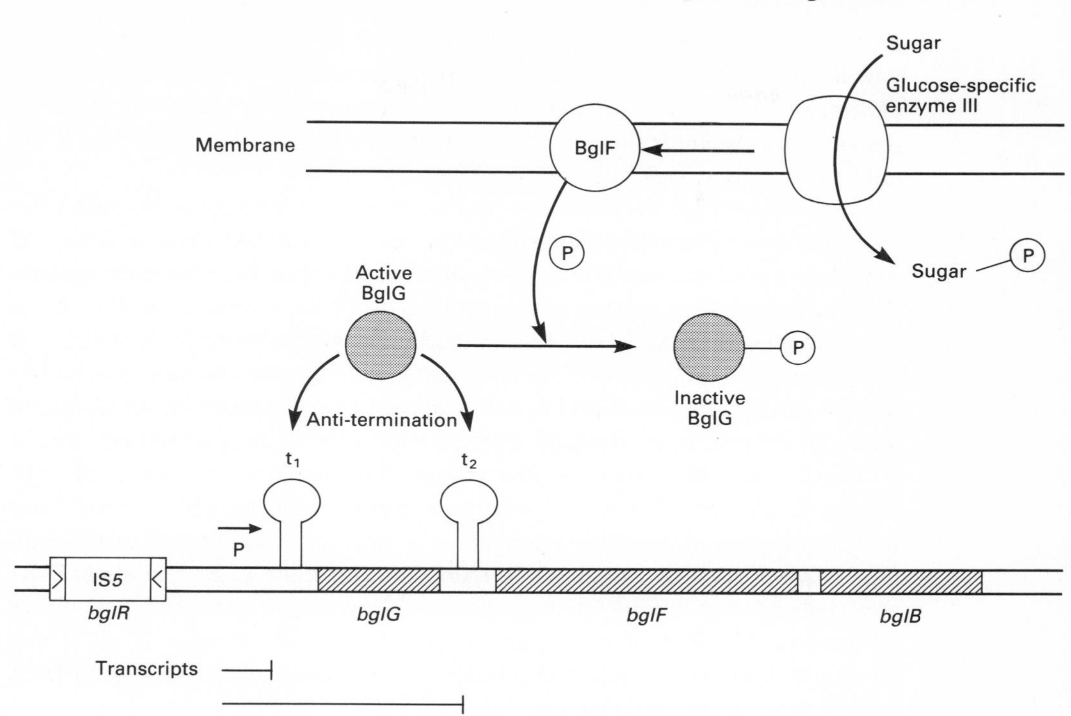

Fig. 6.8 Transcription antitermination and *bgl* operon regulation. The BglG protein is an antitermination factor whose activity is regulated by BglF. The kinase activity of the membrane-associated BglF protein is modulated by glucose-specific enzyme III. BglF transfers its phosphoryl group to BglG, inactivating it as a transcription antiterminator. Active BglG acts to prevent transcription termination at the t1 and t2 terminators within the *bgl* operon.

terminator located in the *bglG-bglF* intercistronic region (Fig. 6.8). Transcription termination at these sites is prevented by the BglG antiterminator protein (Mahadevan & Wright, 1987; Schnetz & Rak, 1988). Furthermore, the BglF protein (or Enzyme II Bgl) exerts negative control by phosphorylating BglG, inactivating its antitermination function. BglG is maintained in a phosphorylated state in the absence of β-glucosides but becomes dephosphorylated when β-glucosides are transported. The *bgl* promoter is also subject to catabolite repression and it has been proposed that the glucose-specific Enzyme III (involved in catabolite repression) may act in conjunction with Enzyme II Bgl to modulate the transport of β-glucosides and thus phosphorylation of BglG (Schnetz & Rak, 1990) (Fig. 6.8).

The *sac* operons of the Gram-positive bacterium *Bacillus subtilus* are also regulated by transcription antitermination. Sucrose induction of the levansucrase gene is dependent on a transcription terminator (*sacRt*) located upstream of the coding sequence, *sacB*. Termination here is antagonized by the SacY protein. The activity of SacY may be inhibited by a second regulator, SacX, which may be phosphorylated by the phosphoenolpyruvate-dependent phosphotransferase system, making SacX an inhibitor of SacY antitermination. In response to exogenous sucrose, SacX may become dephosphorylated, permitting antitermination at *sacRt* (Crutz *et al.*, 1990).

The *sacPA* operon of *B. subtilis* is also induced by sucrose and is thought to be regulated by transcription antitermination. SacA is an intracellular sucrase and SacP is a membrane-associated component of the phosphotransferase system. The SacT regulatory protein which controls *sacPA* expression is similar to the SacY antiterminator of *B. subtilis* and the BglG antiterminator protein of *E. coli* (Debarbouille *et al.*, 1990). The *glpD* gene of *B. subtilis* encodes glycerol-3-phosphate dehydrogenase and is also regulated by transcriptional attenutation, although the precise mechanism remains unclear (Holmberg & Rutberg, 1991).

Genetic data show that the Rho protein contributes to the control of DNA supercoiling in bacterial cells. *E. coli* mutants harbouring the *rho-15* lesion have reduced degrees of negative supercoiling in reporter plasmids and this phenotype is suppressed by mutations in *rpoB.* The capability of Rho to affect DNA superhelicity may result from an increase in the ability of transcription complex to unwind the DNA at transcription termination sites (Fassler *et al.*, 1986; Arnold & Tessman, 1988). The relationship of transcription pausing to DNA superhelicity is discussed next.

Transcriptional pausing and DNA supercoiling

During transcript elongation, RNA polymerase may pause transiently (as distinct from terminating) at different places along the DNA template. Pausing, influenced by the supply of NTPs, by NusA protein or by the secondary structure of the transcript can result in changes in DNA topology. Paused RNA polymerase can act as a topological barrier on the template, partitioning it into a negatively supercoiled domain which extends from itself downstream to the preceding, moving polymerase and a positively supercoiled domain between itself and the next approaching polymerase (Fig. 6.9). Furthermore, there is experimental evidence that some pause sites only function at particular super-

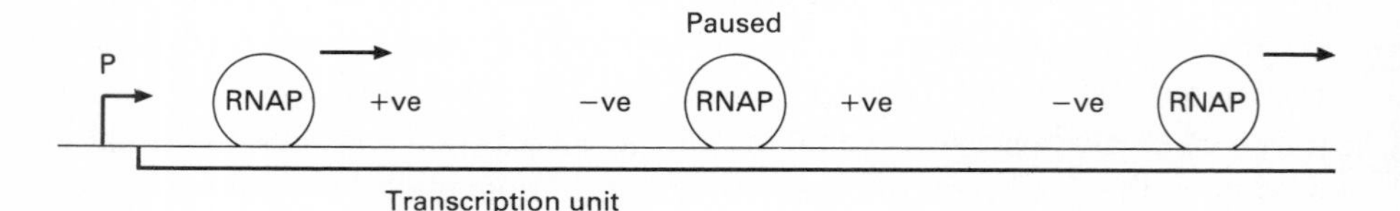

Fig. 6.9 Effects of RNA polymerase (RNAP) pausing on local topology of the DNA template. Paused transcription complexes may represent topological barriers within transcribed regions. Moving complexes can generate differentially supercoiled domains between themselves and the paused complexes which require the activity of topoisomerases to eliminate excessive overwinding or underwinding of the template.

helical densities (Krohn *et al.*, 1992). Thus, transcription pause sites must also be included when considering those factors that influence the topology of the genome.

Post-transcriptional regulation

The wide variety of mechanisms used by cells to regulate gene expression at the level of transcription indicates that this process is the target of choice for gene regulation. However, there are still opportunities to control gene expression once the transcript has been synthesized and these can involve modulation of the efficiency of translation initiation (reviewed in Gold, 1988).

One way to achieve this in a polycistronic operon is to place the translation start site for a promoter-distal gene within the coding sequence of the previous gene. The result is a coupling of the translation of the downstream gene to that of its upstream partner (Fig. 6.10). This mechanism can be particularly useful if the relative concentrations of the two gene products need to be maintained at fixed values. In the *ompB* operon, coding for the OmpR transcription factor and the EnvZ membrane-located histidine protein kinase (Chapter 7), there is an overlap of four base pairs between the OmpR and EnvZ open reading frames (Comeau *et al.*, 1985). The *envZ* protein lacks a good ribosome binding site of its own and probably depends on reintiation of translation following translation of the *ompR* message. This could explain why the cell appears to contain relatively higher levels of OmpR than EnvZ.

Other examples of translational coupling have been described in the *trp* operon of *E. coli*, the rIIb gene of bacteriophage T4 and the *E. coli gal* operon (Napoli *et al.*, 1981; Oppenheim & Yanofsky, 1980; Schumperli *et al.*, 1982). In cases where the downstream gene has a good ribosome binding site, the reinitiation achieved through translational coupling can result in an enhancement in downstream gene expression, contrary to the relatively poor level of expression observed in the case of *envZ* (Comeau *et al.*, 1985).

The *ompB* regulon provides a further illustration of post-transcriptional regulation. DNA sequence analysis of the region upstream of the *ompC* promoter reveals the existence of a small gene coding for a 174 base message which has some complementarity with the 5′ end of the *ompF* mRNA (Andersen *et al.*, 1987; Mizuno *et al.*, 1984) (Fig. 6.11). This small gene, *micF*, is transcribed on

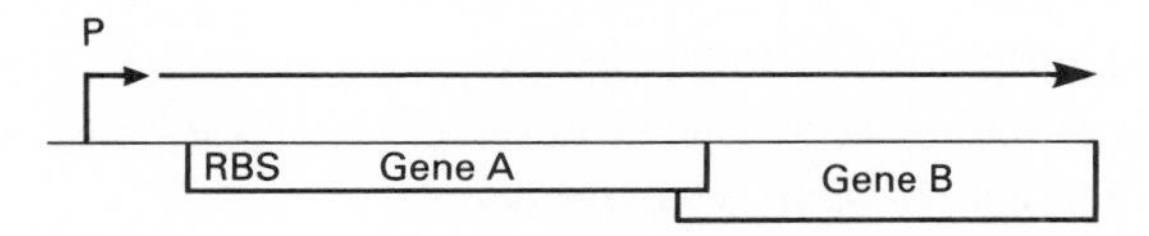

Fig. 6.10 Translational coupling. Genes A and B are transcribed from the same promoter but only gene A has a ribosome binding site (RBS). Therefore, translation of the gene B message requires a translational restart. Frequently, this results in poorer expression of the downstream gene.

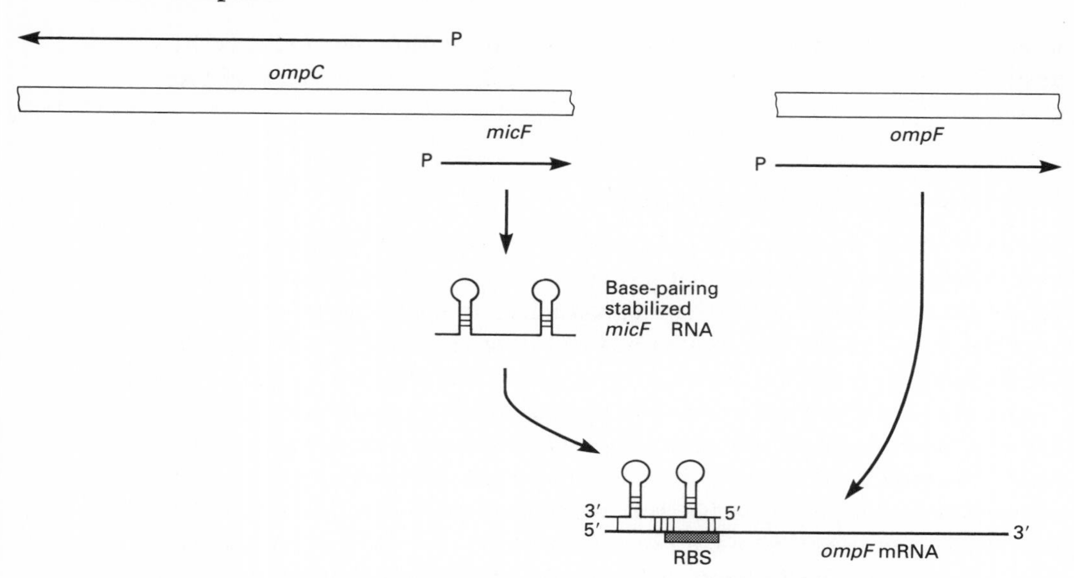

Fig. 6.11 Antisense RNA and gene regulation. The antisense RNA gene *micF* is transcribed divergently from the *ompC* porin gene on the *Escherichia coli* chromosome. Both genes are coregulated. The *micF* message shows sequence complementarity with the 5′ end of the *ompF* message. Base pairing between these sequesters the *ompF* ribosome binding site, preventing translational initiation.

the strand opposite to *ompC* and has a promoter that is coregulated with the *ompC* promoter. The *micF* message is an 'antisense' RNA and its interaction with the complementary region of the *ompF* mRNA appears to sequester the *ompF* ribosome-binding site by forming a region of double-stranded RNA, thus preventing translation initiation at that ribosome binding site. The precise *in vivo* role of *micF* has remained elusive, but it may be concerned with thermal regulation of *ompF* expression (Andersen *et al.*, 1989).

Antisense RNA is exploited as a post-transcriptional regulator in many systems in prokaryotes (reviewed in Simons & Kleckner, 1988). Most of these RNA molecules are small (around 100 bases in length) and capable of folding up into one or more stem loops, which may confer stability upon them. In addition to *micF*, another antisense RNA, *tic*, has been proposed as a regulator of chromosomal gene expression in *E. coli*. This molecule is thought to regulate negatively expression of the *crp* gene, coding for the cAMP receptor protein, Crp (Okamoto & Freundlich, 1986). *tic* displays complementarity to the 5′ end of the *crp* message and addition of *tic* RNA terminates transcription of *crp* prematurely *in vitro*. The *tic* promoter is regulated positively by cAMP-Crp, which is to be expected given the role of *tic* in *crp* expression.

Antisense RNA molecules also play important roles in controlling plasmid replication. In the case of plasmid ColE1, antisense RNA controls copy number through interaction with the precursor of the DNA replication primer molecule (Chapter 3). By inhibiting processing of the precursor, antisense control negatively regulates the initiation of DNA synthesis. Mutants defective in antisense control reach very high (and sometimes unsustainable) copy numbers (Moser *et al.*, 1984; Muesing *et al.*, 1981). Antisense control of plasmid replication can also work by inhibiting expression of a plasmid-encoded replication initiation protein, as in the case of IncFII plasmids (reviewed in Simons & Kleckner, 1988). The killer functions of F and plasmid R483, which kill plasmid-free cells which arise following failure to segregate the plasmid, are subject to antisense control (Nielsen *et al.*, 1991). Expression of an *E. coli*-encoded killer function, Gef, is also regulated by this sort of mechanism (Poulsen *et al.*, 1991).

Antisense control is exerted in the lifecycles of the lambdoid phages P22 (of *Salmonella typhimurium*) and λ (of *E. coli*). In the case of P22, the *sar* RNA sequesters the ribosome binding site of the *ant* gene which encodes an antirepressor protein responsible for inhibiting the binding of λ *c*I-like repressors to DNA (Liao *et al.*, 1987; Wu *et al.*, 1987). Antisense control of λ Q protein expression mediated by the *paq* RNA delays late gene expression in λ-infected cells. This RNA is transcribed from a *c*II promoter located within, and antisense to, the Q gene (Hoopes & McClure, 1985). In addition, the expression of the λ *c*II gene may be regulated negatively by the antisense RNA species *oop* which is expressed from a sequence within, and antisense to, the *c*II gene itself (Krinke & Wulff, 1987). The *oop* RNA results in cleavage of the *c*II mRNA (Krinke *et al.*, 1991).

Antisense control underlies the phenomenon of multicopy inhibition in IS*10* transposition regulation (see Chapter 4; Simons & Kleckner, 1988). Here, an antisense transcript pairs with and inhibits the message coding for the transposase protein. It is believed that this mechanism serves to reduce the rate of accumulation of IS*10* copies in the genome.

Translational pausing has long been recognized as a mechanism for regulating gene expression in bacteria. One of the most elegant examples is the role of pausing in the control of some *E. coli* and *S. typhimurium* amino acid biosynthetic operons, such as *his, ilv, leu, thr, trp* and the phenylalanine tRNA synthetase operon, *pheST* (Fig. 6.12). In these, starvation for a particular amino acid causes ribosomal pausing during translation of a leader sequence rich in codons specific for the missing amino acid. This results in the formation of a transcription antiterminator sequence in the leader RNA which allows RNA polymerase to travel into the sequences coding for the biosynthetic enzymes. In amino acid-replete growth conditions, translational pausing in the leader fails to occur and, instead of an antiterminator, a transcription terminator structure forms which prevents read-through by RNA polymerase to the structural genes of the operon (Fig. 6.12; reviewed in Landick & Yanofsky, 1987).

In the Gram-positive organism, *Bacillus subtilus, trp* expression is also controlled by alternative secondary structures in *trp* leader RNA. However, in this case, the adoption of the transcription terminator or the antiterminator structure is determined by an L-tryptophan-dependent RNA-binding protein called MtrB (or TRAP). In the presence of this amino acid, MtrB binds to the

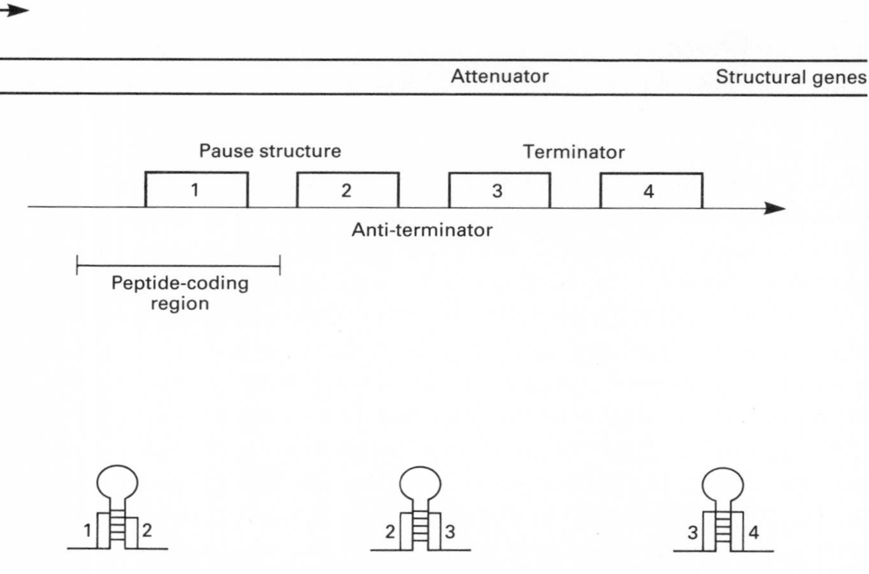

Fig. 6.12 Generalized transcription attenuation system. The attenuator region precedes the structural genes of an amino acid biosynthetic operon. The attenuator region transcript can form three mutually exclusive secondary structures. Which structure is formed is determined by levels of the charged tRNA specific for the particular amino acid synthesized by the products of this biosynthetic operon. Codons for this amino acid are found in tandem within the peptide-coding region of the attenuator sequence. The formation of the pause structure ensures close coupling of transcription and translation. In the absence of the appropriate charged tRNA, the ribosome stalls while translating the cognate amino acid codons, allowing the antiterminator structure to form. The genes of the biosynthetic operon will now be transcribed. If the charged tRNA is present, the ribosome will not stall. Instead, the terminator structure forms and the biosynthetic genes will not be transcribed.

sequence 5′-AGAAUGAGUU-3′, found in the antiterminator structure of the leader RNA, promoting the formation of the transcription terminator, and so preventing transcription of the *trpEDCFBA* biosynthetic operon (Babitzke & Yanofsky, 1993; Otridge & Gollnick, 1993).

Translational attenuation mechanisms have been described as controlling expression of some plasmid-linked antibiotic resistance genes in Gram-positive bacteria. Here, translation-inhibiting antibiotics such as chloramphenicol or erythromycin stall ribosomes in the act of translating leader sequences, preventing the mRNA from forming secondary structures which sequester the ribosome binding site of the gene (Fig. 6.13) (Bruckner *et al.*, 1987; Dubnau, 1985; Narayanan & Dubnau, 1987). In the case of the *cat-86* system of *Bacillus subtilis*, stalling of the ribosome brought about by amino acid starvation during translation of the leader sequence also induces expression of the resistance gene (Duvall *et al.*, 1987).

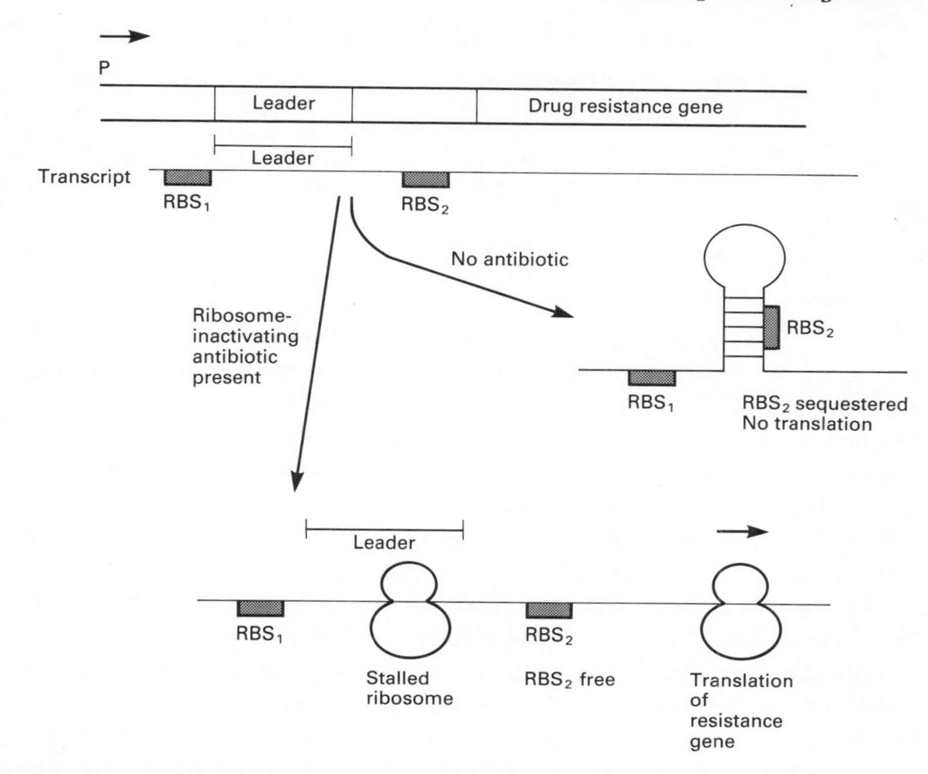

Fig. 6.13 Translational attenuation. A gene specifying resistance to a ribosome-inactivating antibiotic is prefaced by a leader peptide gene which is transcribed from the same promoter. The leader peptide and the structural gene have their own ribosome binding sites, RBS_1 and RBS_2, respectively. In the absence of antibiotic, the message adopts a secondary structure which sequesters the RBS for the structural gene, preventing its translation. In the presence of the ribosome-inactivating antibiotic, ribosomes become stalled while translating the peptide gene, preventing formation of the secondary structure. Under these conditions, the structural gene is translated and the product can protect the cell from the effects of the antibiotic.

Conclusion

In this chapter many of the best-characterized mechanisms by which bacteria regulate expression of their genes at the level of transcription have been described, together with specific examples. The reader has also been given access to the original literature on these systems so that the subject can be pursued in even greater depth, where appropriate. Of course, these systems do not function in isolation within the cell. Little attempt has been made in this chapter to show how the bacterium coordinates gene expression. In the following chapters, the principles of coordinated control will be described and their application to virulence gene regulation discussed.

7 Coordinated Control of Gene Expression

Introduction

Since unicellular prokaryotes are equipped with relatively small genomes and are devoid of specialized tissues, they face significant problems in interacting with the (frequently hostile) external environments they encounter during interaction with the host or most other environments. Environmental parameters may vary rapidly, imposing upon the bacterium a need to respond quickly. Adaptation to a dynamic environment or series of environments requires that the cell modulate its transcriptional profile so that the correct genes are expressed on cue at appropriate levels to ensure survival.

One solution to these problems has been the evolution of mechanisms for the coordinate control of gene expression in response to environmental change. This coordination is achieved at more than one level and reflects the manner in which genes are grouped together for regulatory purposes. Some genes are grouped physically, as in the case of operons. Here two or more genes may be transcribed as a polycistronic message from a single promoter, allowing all the genes to be switched on or off simultaneously as a function of the activity of that promoter. In other cases, coregulated genes may be located at physically distinct parts of the chromosome or even on separate replicons (for example, situations in which some genes are on a plasmid and others are on the chromosome). Such dispersed regulatory groups are called regulons and they represent a more complex regulatory unit than the operon. In fact, regulons may be built up of two or more operons.

Classical examples of these methods of coordinate regulation from bacterial molecular biology include the lactose operon and the maltose regulon. In the lactose operon, three genes, *lacZYA* whose products are concerned with the uptake and utilization of lactose, are expressed as a three-gene message which is then translated into three polypeptides. The transcription of the operon is controlled negatively by a repressor, encoded by the linked *lacI* gene, which (in the absence of the inducer molecule, lactose) binds to an operator site allowing it to interfere with the activity of RNA polymerase (Fig. 7.1; Chapter 6). In addition, transcription of *lacZYA* is positively regulated by cAMP-Crp (Fig. 7.1; Chapter 6).

The maltose regulon is genetically more complex than the *lac* operon. In *E. coli*, it is composed of four operons, one monocistronic (*malT*), one dicistronic (*malQP*) and two tricistronic (*malEFG* and *malKlamBmalM*) (Fig. 7.2). The regulon is divided between two chromosomal sites: *malT* and *malQP* are transcribed

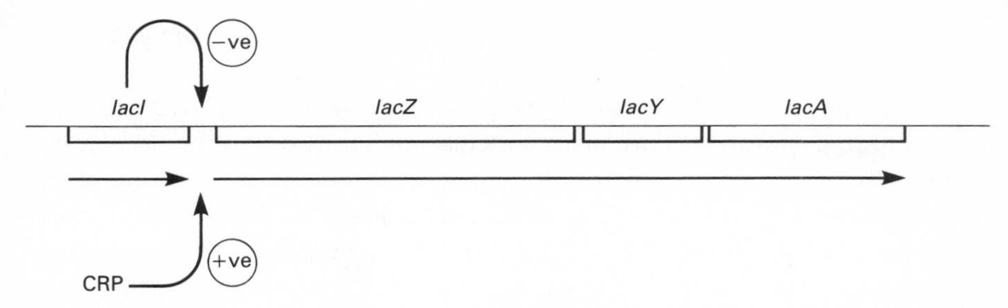

Fig. 7.1 Transcriptional regulation of the *Escherichia coli lac* operon. The *lac* repressor is encoded by the *lacI* gene and this regulates negatively transcription of the three structural genes *lacZYA.* In the absence of lactose, the repressor binds to the *lac* operator sequences; in the presence of this carbohydrate inducer, repressor binding is prevented and the structural genes are expressed. In addition, the operon is positively regulated by the cAMP-Crp complex.

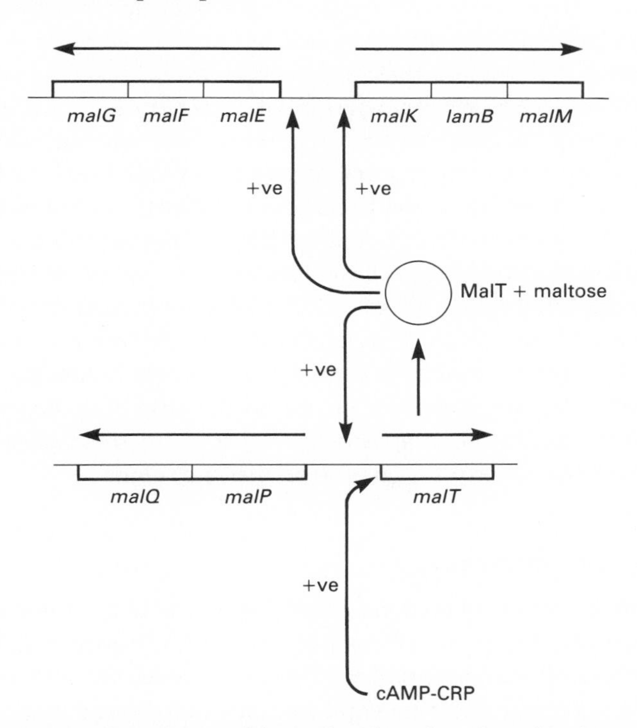

Fig. 7.2 Transcriptional regulation of the *Escherichia coli mal* regulon. Transcription of the operons of the maltose regulon is regulated positively by the MalT protein in the presence of maltose. The *malT* gene is itself regulated positively by the cAMP-Crp complex.

divergently at 75 min on the *E. coli* genetic map while *malEFG* and *malKlamB-malM* are transcribed divergently at 92 min (Bachmann, 1990). The products of the maltose regulon are concerned with the uptake and utilization of maltose and maltodextrins (Schwartz, 1987). Transcription of its constituent genes is positively regulated in the presence of maltose by the product of the *malT* gene (Fig. 7.2). Thus, this gene activator protein links together the constituent operons of the regulon. In addition, *mal* gene expression is regulated positively by the cAMP-Crp system. This forms a regulatory link between the distinct but functionally related operons within the maltose regulon and expression of the unrelated lactose operon, as well as other carbon source-regulated genes which are under cAMP-Crp control (Figs 7.1 & 7.2). The regulators just discussed may be ranked in terms of their pleiotropy such that MalT ranks above LacI whereas cAMP-Crp ranks above both MalT and LacI.

This type of regulatory hierarchy is commonplace in bacterial gene regulation and is frequently a feature of genes involved in bacterial virulence. In general, genes contributing to environmental adaptation or pathogenicity are not expressed constitutively but are activated in response to a signal with an external origin. These genes will possess a highly specific regulator, akin to the LacI repressor protein of the lactose operon or the MalT activator protein of the maltose regulon. In addition, their expression may be linked to other genetic systems via more pleiotropic regulators (analogous to cAMP-Crp), allowing the cell to operate its genetic complement as a series of overlapping gene regulatory networks. Placing genes under multiple regulatory influences makes a great deal of sense intuitively since the bacterium must adapt to a multicomponent environment. By taking multiple soundings through the operation of an array of signal transduction systems, the cell can fine-tune its response to environmental change by modulating gene expression in a highly sophisticated manner. Thus, the composition of the cell can be varied dramatically or subtly as circumstances demand. This survival strategy is essential to bacteria inhabiting a dynamic world whether they are pathogens or not. In the case of pathogens, it is not surprising to learn that they appear to exploit these same gene regulatory processes to control the expression of their virulence genes.

Regulons and stimulons

Regulons usually consist of operons under the control of a common regulatory protein. Frequently, this protein recognizes and binds to a particular sequence in the DNA close to the promoters of these operons. However, all the genes that respond to a particular environmental stimulus may not be coregulated by a common protein. Instead, these may belong to a number of regulons that respond independently to the same environmental signal. Such a group of regulons is called a 'stimulon' (Neidhardt, 1987). The osmotic stress response stimulon provides a good example of this sort of arrangement. Expression of the OmpC and OmpF porins is regulated transcriptionally by the DNA-binding protein OmpR in response to changes in osmotic pressure transmitted by the inner membrane protein kinase EnvZ. The osmoprotectant transport operon encoded at the *proU* locus of *E. coli* and *Salmonella typhimurium* is induced strongly by osmotic stress but no specific *trans*-acting regulatory protein has been identified which activates *proU* transcription at high osmolarity or represses it at low osmolarity. Thus, *ompC* and *proU* are both activated transcriptionally by osmotic stress (*ompF* is repressed under these conditions) but the mechanisms seem distinct (Csonka & Hanson, 1991).

Some operons are capable of belonging to more than one stimulon at a time. Thus, genes characterized as members of the heat shock response can be activated transcriptionally by oxidative stress. Here, the flexibility of response lies in the regulator since transcription activation requires the 'heat shock' sigma factor (σ^{32}) in both cases. In fact, a 'heat shock' response in *E. coli* can be elicited by many types of environmental stimuli, including exposure to alkylating and methylating agents, cadmium chloride, ethanol, hydrogen peroxide, nalidixic acid and amino acid starvation (Neidhardt & Van Bogelen, 1987). The heat shock response is now regarded as something of a misnomer since the induction of its constituent genes is really a response to many forms of stress.

Heat shock response-inducing stimuli such as ethanol treatment and amino acid starvation provoke increases in the levels of the 'alarmone' guanosine 3′-diphosphate-5′-diphosphate, which activates genes of the 'stringent response' (Cashel & Rudd, 1987; described later). Thus, these stimuli activate members of more than one stimulon. In fact, Van Bogelen *et al.* (1987) have detected overlaps in the induction patterns of proteins formally identified as belonging to the SOS response, to the oxidative stress response or to the heat shock response in *E. coli.* In the following sections, environmental stress response systems will be described and it will become apparent that most play direct or supporting roles in the pathogenic processes of bacteria.

Global regulation of gene expression

The existence of the regulatory hierarchies discussed earlier raises the possibility that the cell might possess control elements so pleiotropic that they have the potential to affect the expression of very large subsets of the genes in the cell. These regulators would possess truly global abilities, far beyond the rather restricted scope of elements controlling, for example, only the osmotically responsive genes or just the anaerobically inducible genes. In the bacterial gene regulation literature, many regulators are classed as 'global' but most have no effects beyond a few tens of genes at most. Clearly, to fulfil the role of global regulator, a control element must possess properties over and above those of classical, sequence-specific transcription factors. Recently, several candidates have emerged that have impressively pleiotropic effects on the transcription profile of the cell. One of these is the DNA-binding protein H-NS, which is a major component of bacterial chromatin (Chapter 2). Variations in H-NS levels due to growth phase or growth conditions (e.g. temperature) result in major alterations in the transcriptional profile of the cell, as well as effects on other reactions of DNA, such as mutation and recombination (see later and Chapters 2 & 8). Given that H-NS is an important component of bacterial chromatin and can influence the structure of DNA, it is not surprising that other aspects of DNA topology have emerged as potential global regulators of gene expression. DNA supercoiling is one of these, being a DNA topological parameter with the potential to influence every promoter in the cell and one which varies in response to environmental stimuli (see later and Chapters 2 & 8).

In this chapter, the major regulons and stimulons of bacteria will be described in order to show the range and variety of mechanisms by which bacteria (mainly enteric bacteria) coordinate the expression of their genes. Sections on the roles of highly pleiotropic effectors such as H-NS and DNA supercoiling are also included. In the following chapter, the relevance of these gene regulatory circuits to the control of virulence gene expression will be described.

The osmotic stress response stimulon

Water movement into and out of cells has profound consequences for intracellular physiology and many bacteria experience a highly dynamic situation from the point of view of external (and hence internal) water activity (i.e. water 'concentration'). Bacteria adapt to such changes in water activity partly through changes in gene expression. Csonka & Hanson (1991) have pointed out that gene regulation in response to changes in water activity are analogous to changes in response to variations in barometric pressure, viscosity or temperature in that no chemical signal is involved (as is the case with, for example, the lactose operon and lactose) but rather, the change is induced by an alteration in a physical parameter of the environment.

The osmotic pressure of the cytoplasm is maintained at a level (close to 0.6 MPa for *E. coli*) conducive to the metabolic functions which take place there and is set by homoeostatically balanced processes which govern pH levels, ion concentrations and levels of metabolites. Shifting the cell to a hyperosmotic medium results in a loss of water from the cytoplasm and a rapid response at the level of transcription of genes whose products are concerned with the uptake of non-toxic osmolytes which are accumulated within the cell to replace the lost water. There has been much debate over what constitutes the signal leading to the induction of the osmoresponsive genes. One possibility is that a change in turgor pressure might suffice. Another sensing mechanism envisages changes in the concentration of internal solutes such as potassium. Other possibilities include sensing of changes in extracellular solute concentrations or changes in the area of the cytoplasmic membrane.

Under hyperosmotic stress, bacteria accumulate a wide variety of compatible solutes. These include potassium, amino acids, *N*-methyl substituted amino acids, sugars, peptides and other compounds. The genes coding for these uptake systems are frequently members of the osmoresponsive stimulon, that is to say, they are induced under hyperosmotic conditions. In addition to accumulating compatible solutes from the external milieu, bacteria can manufacture them themselves. Once again, genes concerned with osmoprotectant synthesis are often members of the osmoresponsive stimulon. Bacteria do not regard all compatible solutes as equivalent and a hierarchy of solute acquisition may be discerned in cells under hyperosmotic stress. In *E. coli*, potassium is accumulated rapidly and glutamate is synthesized. Next, potassium is excreted and glutamate is broken down and trehalose (a disaccharide) is produced, usually an indication that osmotic adaptation is complete. Enteric bacteria appear to favour glycine betaine above all other compatible solutes and the genetic locus encoding its uptake system (*proU*) has one of the best-studied examples of an osmotically inducible promoter. When cells are subjected to an hypo-osmotic shock, water enters the cytoplasm and the compatible solutes are lost, either passively through membrane leakiness or actively through excretion or catabolism (Csonka & Hanson, 1991).

In the enteric bacteria, there is perhaps a score of genes whose transcription is affected significantly by osmolarity and the majority of these are concerned with osmotic adaptation (Csonka, 1989). Two of these genes code for the OmpC and OmpF porins whose expression is reciprocally regulated at the level of transcription by the regulatory protein OmpR in response to phosphotransfer from the inner membrane-located protein kinase EnvZ such that OmpC is expressed at high osmolarity and OmpF is expressed at low osmolarity. This 'two-component' regulatory system is described in detail later.

In *E. coli*, the *kdpABC* operon encodes a high-affinity ATP-dependent uptake system for potassium (Silver & Walderhaug, 1992) (Fig. 7.3). Transcription of *kdpABC* is induced only transiently by hyperosmotic stress and is positively regulated by a 'two-component' histidine protein kinase/response regulator system composed of KdpD and KdpE. KdpD is associated with the inner membrane, shows homology to histidine protein kinase family and has been shown to function as a kinase and to transfer its phosphate to KdpE *in vitro* (Nakashima *et al.*, 1992). Its aminoterminus may be involved in sensing changes in turgor pressure. KdpE is the transcription activator protein whose biological function is modulated by KdpD (Walderhaug *et al.*, 1992). At high turgor pressure, KdpE is activated by KdpD (presumably via phosphotransfer) and in turn activates transcription of the *kdpABC* operon from the *kdpA* promoter. This promoter is not fully functional in the absence of a *cis*-acting sequence located upstream of its −35 hexameric sequence. The upstream sequence is the binding site for the KdpE transcription activator (Sugiura *et al.*, 1992). The *kdpDE* genes are organized as an operon adjacent to and overlapping the *kdpABC* operon such that the *kdpD* promoter is within *kdpC*; transcription of *kdpABC* has an enhancing effect on expression of *kdpDE* (Polarek *et al.*, 1992).

Expression of the high-affinity glycine–betaine transport system encoded by *proU* in *E. coli* and *Salmonella typhimurium* is activated transcriptionally by high osmolarity. In *S. typhimurium* (but not *E. coli*, Lucht & Bremer, 1991) induction is enhanced by anaerobiosis, a phenomenon also seen with *ompC* porin regulation, at least in *E. coli* (Ní Bhriain *et al.*, 1989). The ProU system is composed of an

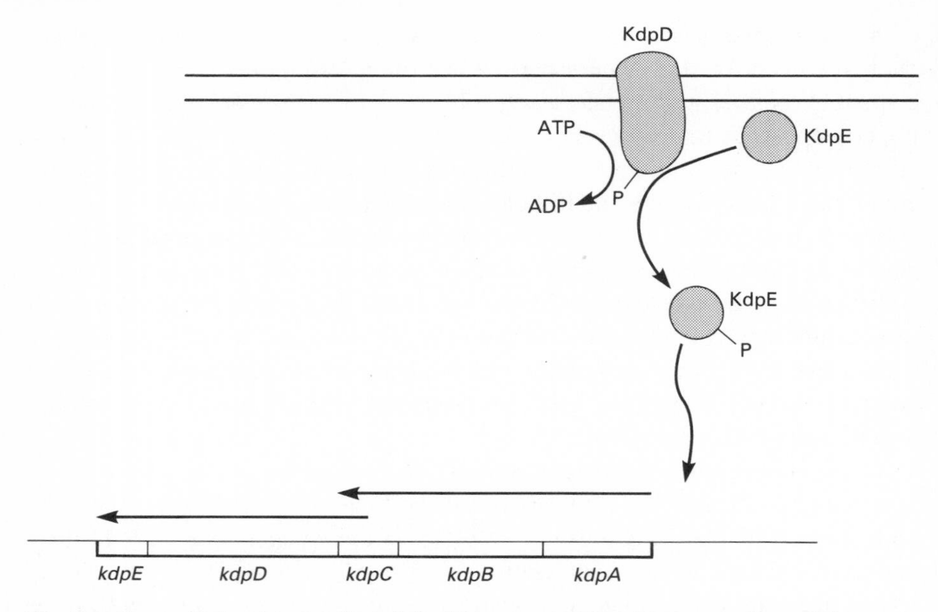

Fig. 7.3 Transcriptional activation of the *kdp* operon at low turgor pressure. This system consists of two overlapping operons. The *kdpABC* genes code for a potassium uptake system while *kdpD* and *kdpE* encode a histidine protein kinase and a transcription activator respectively. The promoter for *kdpDE* transcription is located within *kdpC* such that transcription of *kdpABC* has an enhancing effect on *kdpDE*.

operon of three genes *proVWX* coding for a periplasmic binding protein (ProX) and two inner membrane proteins (ProV and ProW) (Gowrishanker, 1989). The osmotic regulation of *proU* expression seems to be achieved by a complicated mechanism (Fig. 7.4). The promoters from the *E. coli* and *S. typhimurium* operons are almost identical in sequence, appear to be σ^{70} driven but have a non-standard 16 bp spacing between the -10 and -35 boxes (Overdier *et al.*, 1989; Stirling *et al.*, 1989b). This could make the promoter prone to activation by changes in DNA twist which could align the -10 and -35 boxes on the same face of the helix (Wang & Syvanen, 1992) and its activation correlates with changes in DNA topology produced in response to osmotic stress (Higgins *et al.*, 1988a). Deletion analysis has identified upstream sequences which are required for osmotic activation of the *E. coli* promoter (Lucht & Bremer, 1991) and other studies have shown that the operon's osmoresponsiveness is a feature of multiple mechanisms, including a second osmoresponsive promoter located in upstream sequences and a downstream sequence within the first structural gene (Dattananda *et al.*, 1991; Overdier & Csonka, 1992) (Fig. 7.4). This downstream sequence may undergo a structural transition in response to osmotic stress, altering the expression of *proU* in the process. In medium of low osmolarity, it may adopt a conformation that silences transcription of the operon (Overdier & Csonka, 1992). When it is carried on a plasmid, it alters the linking number *in vivo* (Dattanananda *et al.*, 1991). It has been suggested that this

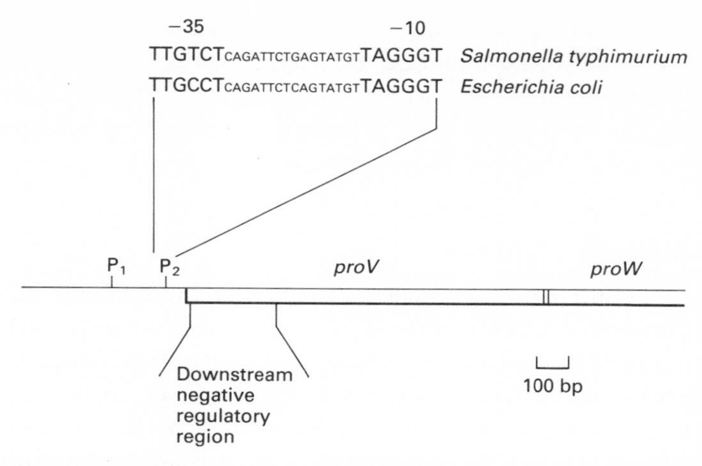

Fig. 7.4 Transcription control regions of the *proU* operon of *Salmonella typhimurium* and *Escherichia coli*. The structure of the P_2 promoter is highly conserved between the two species; the P_1 promoter has been identified in *E. coli*. The downstream negative regulatory region lies within the coding region of *proV*. The *proV* and *proW* genes overlap by 5 bp.

sequence may be responsible for the apparent sensitivity of *proU* to changes in DNA supercoiling brought about by osmotic stress, by inactivation of the *topA* gene or by inhibition of DNA gyrase by antibiotics (Higgins *et al.*, 1988a). The importance of DNA structure for *proU* function is illustrated by the discovery that mutations in the unlinked *osmZ* gene permit *proU* transcription at low osmolarity and the finding that *osmZ* codes for the nucleoid-associated protein H-NS (i.e. *osmZ* and *hns* are allelic). Thus, this DNA-binding protein which is a transcriptional silencer of several genes, negatively regulates *proU* (Higgins *et al.*, 1988a; Owen-Hughes *et al.*, 1992). Experiments with *fis* mutants have shown that the histone-like protein FIS is not required for *proU* osmoregulation and a slight effect seen in IHF-deficient cells is probably an indirect effect, perhaps reflecting an effect on DNA gyrase and hence *proU* via DNA supercoiling (Lucht & Bremer, 1991).

As with other stimulons, there are regulatory overlaps between the osmotic stress response and other stimulons. For example, the responsiveness of *E. coli* and *S. typhimurium ompC* and the *S. typhimurium proU* genes to anaerobiosis has already been referred to. In addition, the induction of some osmoregulated genes can also be achieved by growth to stationary phase (Jenkins *et al.*, 1990; Jung *et al.*, 1990). The *otsA* and *otsB* genes are required for synthesis of the osmoprotectant trehalose and their transcription is induced by osmotic stress and by the onset of stationary phase. Both forms of regulation require the RpoS stationary phase sigma factor (Hengge-Aronis, 1993; Kaasen *et al.*, 1992). Interestingly, *rpoS* mutants lose stationary phase thermotolerance and *otsA otsB* mutants are similarly sensitive to thermal damage in stationary phase, indicating that, in addition to its role in osmoprotection, trehalose also plays a role as a thermoprotectant in stationary phase (Hengge-Aronis *et al.*, 1991; reviewed in Hengge-Aronis, 1993). Furthermore, two-dimesional gel analysis reveals that up to 18 proteins are induced in an RpoS-dependent manner by osmotic stress, showing that this sigma factor plays a key role in controlling the expression of osmotically regulated genes in *E. coli* (Hengge-Aronis, 1993).

The heat shock response

As with many stimulons, the molecular detail of the bacterial 'heat shock response' has been worked out first in the enteric bacteria, and especially in *E. coli*. This is despite the fact that the response was recognized initially in the fruit fly, *Drosophila melanogaster* (Ritossa, 1962; Tissieres *et al.*, 1974). Information about heat shock systems in other organisms (prokaryotic or eukaryotic) is frequently founded upon the recognition of similarities in the sequences of proteins induced by heat stress or allied stimuli and those of their *E. coli* counterparts.

Significant progress with understanding the heat shock response in *E. coli* began with the discovery of the *rpoH* gene which encodes the heat shock sigma factor, σ^{32}, which reprogrammes RNA polymerase to recognize and bind to heat shock promoters (Neidhardt & Van Bogelen, 1987). The proteins encoded by the heat shock response genes carry out diverse functions. Some are involved in the folding and unfolding of other proteins or the assembly and disassembly of protein complexes, others are proteases and some have (as yet) unknown functions.

The principal transcriptional regulator of the heat shock response is the *rpoH* gene product, σ^{32}, and this is essential for life above 20°C (Zhou *et al.*, 1988). Transcription of *rpoH* involves multiple promoters, one of which is recognized by yet another sigma factor, σ^{24} (Erickson & Gross, 1989; Fujita & Ishihama, 1987). Once synthesized, the *rpoH* transcript is unstable unless the cell is growing at 42°C or higher, affording the system an element of post-transcriptional control (Straus *et al.*, 1987). Once translated, the σ^{32} polypeptide has a short half-life, introducing control of *rpoH* expression at the post-translational level (Straus *et al.*, 1987). This post-translational instability may be a feature of an interaction between σ^{32} and the 'chaperone' heat shock proteins DnaJ, DnaK and GrpE (Gamer *et al.*, 1992). Significantly, the 'standard' σ^{70} sigma factor is also a heat shock protein since its gene, *rpoD*, is transcribed from a heat shock promoter under heat shock conditions (Cowing *et al.*, 1985). The nitrogen limitation response sigma factor, σ^{54}, has also been implicated in the heat shock response. Specifically, the heat shock induction of the phage shock protein in *E. coli*, encoded by *psp*, requires the participation of σ^{54} (Weiner *et al.*, 1991).

The genes coding for DnaJ, DnaK and GrpE were discovered originally due to the contributions their products make to the life cycle of bacteriophage λ. Subsequently, their contributions to host cellular metabolism were also recognized. DnaK is the bacterial equivalent of the eukaryotic Hsp70 heat shock protein and DnaJ is highly conserved among prokaryotes (Bardwell & Craig,

1987; Georgopoulos *et al.*, 1990). The GrpE protein associates with DnaK and this association is dissolved in the presence of ATP. DnaJ, DnaK and GrpE all associate with the λ origin of replication and with proteins concerned with the replication of the virus (described in Georgopoulos *et al.*, 1990). These proteins also associate with σ^{32} (see earlier).

The *lon* gene product is a heat shock protein and an ATP-dependent protease with highly pleiotropic effects. These include a role in cell division where it degrades the SulA protein following an SOS response, rescuing the cell from the cell division arrest associated with the presence of SulA (Mizusawa & Gottesman, 1983). It also has a role in the regulation of colonic acid (extracellular polysaccharide) biosynthesis where it degrades the RcsA regulatory protein (Torres-Cabassa & Gottesman, 1987). The ClpB protein of *E. coli* is also a heat shock protein and may play a role in stabilizing ribosomal proteins when bacteria experience a heat shock (Squires & Squires, 1992).

Enteric bacteria possess a second heat shock regulon whose best-characterized member is the *degP* (or *htrA*) gene of *E. coli*. This gene codes for a periplasmic serine protease which degrades abnormally folded proteins (Lipinska *et al.*, 1990; Strauch & Beckwith 1988). Transcription of *htrA* is initiated exclusively by σ^{24}-programmed RNA polymerase, linking its expression to that of *rpoH*, which also possesses a σ^{24}-dependent promoter (Erickson & Gross, 1989; Georgopoulos *et al.*, 1990). Thus, at least two distinct but overlapping heat shock regulons exist in enteric bacteria and both play crucial roles in environmental adaptation.

Cold shock regulation of gene expression

The cold shock response in *E. coli* is induced by an abrupt temperature shift from 37 to 10°C (Jones *et al.*, 1987). The cold shock stops cell growth and results in a lag of several hours during which the synthesis of certain proteins (the cold shock proteins) increases. These proteins include H-NS, GyrA, NusA, RecA, polynucleotide phosphorylase, translation factors 2-alpha and 2-beta, pyruvate dehydrogenase (lipoamide), dihydrolipoamide acetyltransferase of pyruvate dehydrogenase and a protein called CS7.4 (Goldstein *et al.*, 1990; Jones *et al.*, 1987; 1992b; La Teana *et al.*, 1991).

When cultures of *E. coli* are shifted from 37 to 10°C, the cells synthesize most of the cold shock proteins at levels 2- to 10-fold higher than the non-induced level. However, a 7.4 kDa protein (CS7.4) is induced to levels 10- to 100-fold higher than the basal level, accumulating to levels equivalent to 13% of total cell protein within 90 min. The CS7.4 protein is encoded by the *cspA* gene and shows homology to human DNA-binding proteins such as DbpA, DbpB and YB-1. An upstream binding site for a cold-specific factor has been identified and this factor has been found to be expressed only during cold shock. If CspA really is a DNA-binding protein, it may be an important regulator of the cellular response to cold shock in *E. coli* (Tanabe *et al.*, 1992). In *Bacillus subtilis*, a

7365 Da cold shock protein, CspB, has been identified which is 61% identical to CspA/CS7.4 (Willimsky *et al.*, 1992).

Evidence that CS7.4 is a DNA-binding protein comes from the demonstration that it binds specifically to the promoter region of the *E. coli gyrA* gene (Jones *et al.*, 1992b). This suggests that CS7.4 may act as a transcription activator of *gyrA* during the cold shock response. Furthermore, the finding that *gyrB* expression is maintained during cold shock indicates that an increase in DNA gyrase may assist the cell in adapting to the low temperature shift. The *gyrA* promoter contains three copies of the motif 5′-ATTGG-3′, which is homologous to the sequence recognized by human Y box proteins. A reduction in the number of the repeats results in reduced binding of CS7.4 and when these sequences are eliminated, binding of CS7.4 to the promoter is abolished (Jones *et al.*, 1992b). CS7.4 has been shown to be a cold shock transcription activator of *hns* (La Teana *et al.*, 1991) and CS7.4 binding site homologues exist in the promoter of this gene. Similar sites occur in the promoter regions of the cold shock genes *cspA*, *nusA* and *recA* (cited in Jones *et al.*, 1992b). The existence of such sites in the *cspA* promoter suggests that CS7.4 expression may be subject to autoregulation.

In *E. coli*, cold shock results in the repression of heat shock protein synthesis as well as the induction of the cold shock proteins; proteins involved in transcription and translation are synthesized continuously during the shock. Temperature downshifts are also accompanied by a decrease in the levels of (p)ppGpp (see section below on The stringent response) and this has been shown to be physiologically significant in terms of its consequences for gene expression and adaptation to low temperature growth (Jones *et al.*, 1992a).

Thermoregulation of gene expression

Thermoregulation of gene expression may be distinguished from the 'heat shock response' in two ways. First, it involves genes not necessarily transcribed from classic heat shock promoters and, secondly, these genes are induced specifically by temperature changes and not by the myriad of other environmental stimuli that have been shown to activate the heat shock regulon members. In general, thermoregulated genes have promoters recognized by σ^{70} and their thermoresponsiveness is determined by *trans*-acting transcription factors which can be highly specific or, in some cases, highly pleiotropic. Many of the best-understood examples of thermal regulation involve virulence genes under the positive control of members of the AraC family of transcription factors (described later). Several of these genes are also regulated (negatively) by the nucleoid-associated protein, H-NS (see later and Chapter 8).

The stringent response

Starvation for amino acids results in an inability to supply sufficient quantities of aminoacylated tRNAs to meet the demands of translation. This activates a complex set of regulatory events known as the stringent response. The name comes from the recognition of a linkage between stable RNA synthesis and amino acid levels which implied the existence of a 'stringent requirement' for RNA control (reviewed in Cashel & Rudd, 1987). The signal initiating the stringent response is the presence of uncharged tRNAs in the cell. When an uncharged tRNA is located in the A site of a ribosome it can trigger the activity of the RelA protein (which is associated with a subset of ribosomes). This protein has (p)ppGpp synthetase I activity and is encoded by the *relA* gene in *E. coli* and *S. typhimurium*. It catalyses the conversion of GTP to guanosine-5′-diphosphate-3′-diphosphate (ppGpp) and guanosine-5′-triphosphate-3′-diphosphate (pppGpp). The *relA* gene is so-called because mutations there render the cells 'relaxed' rather than 'stringent' from the point of view of their response to amino acid starvation.

The unusual nucleotides ppGpp and pppGpp inhibit transcription of stable RNA genes by binding to RNA polymerase. In turn, inhibition of rRNA synthesis reduces production of ribosomal proteins. In addition to its negative effects on gene expression, ppGpp also activates transcription of some genes, such as the histidine biosynthetic operon in *S. typhimurium* (Riccio *et al.*, 1985; Shand *et al.*, 1989; Stephens *et al.*, 1975; Winkler *et al.*, 1978). The stringent response also induces the expression of genes of the heat shock response (Grossman *et al.*, 1985). The cellular pool of (p)ppGpp is degraded by the product of the *spoT* gene, (p)ppGpp 3′-pyrophosphohydrolase (Cashel & Rudd, 1987). Confusingly, this enzyme is now known to be bifunctional and to be responsible for (p)ppGpp synthetase II activity, i.e. it plays a role in (p)ppGpp synthesis as well as in its degradation (Hernandez & Bremer, 1991; Xiao *et al.*, 1991).

The precise mechanism by which ppGpp affects stable RNA synthesis is not completely clear. One suggestion is that ppGpp binding to RNA polymerase alters its promoter specificity such that the promoters of the stable RNA genes are no longer recognized (Baracchini & Bremer, 1988). In an alternative model, it has been proposed that ppGpp binding to RNA polymerase slows the rate of transcript elongation, resulting in a sequestration of RNA polymerase in the elongation phase. In this way the cellular concentration of free RNA polymerase is reduced, which is particularly detrimental to those promoters that require high concentrations of RNA polymerase for maximal operation (Jensen & Pedersen, 1990). The stringently controlled promoters of the stable RNA operons are of this type. Mutational analysis of stringently regulated promoters has permitted identification of residues required for sensitivity to this form of regulation. For example, a 4 bp substitution adjacent to the transcription start site of the *E. coli tyrT* gene results in loss of stringent regulation (Lamond & Travers, 1985). This mutation occurs within a GC-rich sequence, known as the 'discriminator', which is found between the −10 box and the transcription start site which is conserved among many stable RNA promoters (Travers, 1984). Similarly, the promoter from which the *E. coli fis* gene (coding for histone-like protein FIS, a regulator of stable RNA operons) is transcribed is itself subject to stringent regulation and possesses such a GC-rich region between the −10 box and the transcription start site (Nilsson *et al.*, 1990; Ninnemann *et al.*, 1992).

The SOS response regulon

The SOS response in *E. coli* is triggered by DNA damage caused by radiation or by chemical agents (reviewed in Thliveris *et al.*, 1990; Walker, 1987). During the response, cell division is delayed, the rate of DNA repair is accelerated, the DNA is subjected to elevated rates of mutation, and integrated prophages may excise from the chromosome. The SOS response is driven by the products of a group of unlinked genes and operons which constitute the SOS regulon. The expression of these transcription units (which code for about 20 proteins) is regulated negatively by the LexA repressor. This protein recognizes and binds as a dimer to a specific sequence (consensus: 5′-CTGTnnnnnnnnnnCAG-3′) found in the vicinity of the SOS regulon gene promoters. Following DNA damage, the RecA protein is reprogrammed to act as a co-protease which promotes self-cleavage of the LexA repressor, leading to induction of the SOS regulon genes (Little, 1993). In addition to acting as a co-protease in cleavage of LexA, activated RecA is involved in the cleavage of other proteins. These include the UmuD protein, whose cleaved carboxyterminus plays an important role in mutagenesis (enhanced mutagenesis is a feature of the SOS response). Other proteins subject to RecA-promoted self-cleavage include the bacteriophage λ repressor, whose destruction is an early step in the induction of the λ lytic pathway (see Chapter 4) (Little, 1993). Thus, the SOS response is highly pleiotropic in its effects and it overlaps with many other major cellular processes and stimulus-response pathways.

Anaerobic gene regulation

Facultative anaerobes such as *E. coli* have evolved complex mechanisms for energy generation in fluctuating environmental circumstances. In fermentation, energy is derived by substrate-level phosphorylation in which some metabolites have to be excreted in order to maintain the internal redox balance. Thus, when *E. coli* grows anaerobically using glucose as carbon and energy source, the by-products include succinate, acetate, formate, ethanol, carbon dioxide and hydrogen. The bacterium will switch from fermentation to respiration (oxidative phosphorylation) should an externally provided electron acceptor become available. When respiratory processes operate, proton motive force is generated as protons are driven out of the cell across the cytoplasmic membrane by redox reactions catalysed by membrane-associated protein complexes. The resulting proton gradient is used to drive processes such as ATP synthesis. Bacteria will always choose the electron acceptor offering the most positive midpoint potential (E'^{o}). Oxygen ($E'^{o} = +820$) is the most attractive and will be used in preference to nitrate ($E'^{o} = +430$) which is preferred to fumarate ($E'^{o} = +30$).

Exploitation of different terminal electron acceptors requires different fermentative pathways and the switches that govern the expression of these are primarily transcriptional. Three sets of pleiotropic gene regulators have been described, encoded by *fnr*, *narX*/*narL* and *arcB*/*arcA* (Iuchi & Lin, 1991).

The *fnr* gene encodes a DNA-binding protein that is required for the anaerobic activation of transcription of genes coding for fumarate reductase, nitrate reductase and nitrite reductase as well as formate dehydrogenase-N and anaerobic glycerol-3-phosphate dehydrogenase (Spiro & Guest, 1990). It is also required for the anaerobic expression of the *cea* gene coding for colicin E1 on the ColE1 plasmid (Eraso & Weinstock, 1992; Chapter 3). Fnr can also act as a repressor, negatively regulating transcription of its own gene as well as others, such as *ndh*, coding for a membrane-bound NADH dehydrogenase (Spiro *et al.*, 1989). Promoters subject to regulation by Fnr have a characteristic inverted repeat sequence (5′-TTGAT----ATCAA-3′) and genetic evidence indicates that Fnr interacts physically with RNA polymerase (Lombardo *et al.*, 1991). Transcription of *fnr* is negatively regulated by osmolarity and several Fnr-dependent genes also show negative regulation by osmolarity. This latter observation is complicated by the fact that this input appears to be Fnr-independent (Gouesbet *et al.*, 1993).

Structurally, Fnr resembles the cyclic-AMP receptor protein (Crp) but has a unique aminoterminal domain containing an iron-binding motif made up of a Cys-rich region (Cys-Ala-Ile-His-Cys-Gln-Asp-Cys) (Appendix I). Mutations that delete the Cys amino acids inactivate Fnr (Spiro & Guest, 1988). It has been proposed that the redox state of the bound iron determines the biological activity of Fnr; under aerobic conditions, the coordination state of the bound Fe^{3+} may render Fnr inactive whereas under anaerobic conditions, the altered coordination state of the reduced Fe^{2+} may allow a change in protein conformation to an active form. Alternatively, changes in the metal ion redox state may simply affect its binding to the Fnr protein (Spiro *et al.*, 1989). Kiley & Reznikoff (1991) have isolated mutant derivatives of Fnr (designated Fnr*) which activate an Fnr-dependent promoter under aerobic conditions. These mutations map to the Cys-rich aminoterminal domain or to regions of the protein which, by analogy with Crp, may be involved in dimer formation and therefore in conformational changes associated with the biological activation of the protein.

The Fnr protein has become the prototypic member of yet another 'family' of bacterial gene regulators. Besides Crp, other members of this family include: HlyX, a regulator of haemolysin expression in *Actinobacillus pleuropneumoniae* (MacInnes *et al.*, 1990); FnrN, a regulator of microaerophilic nitrogen fixation in *Rhizobium leguminosarum* (Colonna-Romano *et al.*, 1990); FixK, an activator of genes required for both the microaerophilic and symbiotic expression of the *Rhizobium meliloti fixN* operon (Batut *et al.*, 1989); CysR, a regulator of sulphur-responsive genes in a species of *Synechococcus* (Laudenbach & Grossman, 1991); NtcA, a regulator of nitrogen assimilation genes in the cyanobacterium *Synechococcus* (Vega-Palas *et al.*, 1992) and ANR, an activator of genes involved in aerobic growth in *Pseudomonas aeruginosa* (Sawers, 1991; Zimmermann *et al.*, 1991).

The NarX/NarL system is a member of the histidine protein kinase/response regulator group of environmental signal transducers, in which NarX is the sensor and NarL the response regulator, in this case, a DNA-binding protein (Nohno *et al.*, 1989). The NarX/NarL system is involved in the discrimination between different terminal electron acceptors. When nitrate and molybdenum are present, NarX/NarL activates the nitrate reductase operon (*narGHJI*) and represses the genes involved in the reduction of fumarate (*frdABCD*) and trimethylamine-*N*-oxide and dimethyl sulphoxide (*dmsABC*) (Kalman & Gunsalus, 1990). Regulatory overlaps exist between the Fnr-dependent genes and the NarX/NarL-dependent genes. For example, the *narGHJI* and the *frdABCD* operons are under the dual control of these regulatory systems (reviewed in Iuchi & Lin, 1991). A second sensor protein, NarQ, can activate NarL independently of NarX (Chiang *et al.*, 1992). This situation is analogous to that described for the independent activation of ArcA by either ArcB or CpxA (see later).

Aerobic gene regulation

Many genes in *E. coli* are highly expressed in the presence of oxygen but display high-level anaerobic expression only in mutants deficient in the ArcB/ArcA regulatory system. These proteins, encoded by unlinked genes, are homologous to the histidine protein kinase/response regulator 'two-component' signal transducer group (Iuchi & Lin, 1988; Iuchi *et al.*, 1990b). ArcB is a membrane-associated sensor protein and ArcA a DNA-binding protein that acts as a repressor of aerobically expressed genes following a shift to anaerobic conditions. ArcB autophosphorylates and transfers the phosphate group to an aspartate residue in ArcA (Iuchi & Lin, 1992). Exceptionally, it can activate transcription of some genes, such as *cyd*, the cytochrome d operon (encoding the high oxygen affinity cytochrome of *E. coli*) (Iuchi *et al.*, 1990a).

The ArcB/ArcA system is subject to regulatory cross-talk. For example, a second sensor protein, CpxA, can communicate with ArcA and direct it to regulate the synthesis of the F plasmid conjugation pilus. In this guise, the *arcA* gene was originally identified as *dye*, a pleiotropic effector of F plasmid-mediated conjugation and of sensitivity to redox dyes (Buxton & Drury, 1984). Some ArcA-dependent genes also form part of the Fnr regulon (*cydAB*, *cyoABCDE*) and one ArcA-regulated gene, *sodA* (coding for the manganese superoxide dismutase) is also under the control of the superoxide response regulator SoxRS and the iron regulator Fur. The *sodA* gene is regulated positively by SoxRS in response to superoxide radicals and by SoxQ in response to an unidentified signal (Tsaneva & Weiss, 1990; Demple & Amábile-Cuevas, 1991; Greenberg *et al.*, 1991;). Expression of *sodA* is repressed by at least four regulators: ArcA, Fnr, Fur and IHF. Fnr and IHF act to enhance the negative effect of ArcA on *sodA* while Fur links *sodA* transcription to the intracellular level of iron (Compan &

Touati, 1993). Fur, Fnr and ArcA all act to repress *sodA* anaerobically in a manner that is independent of SoxRS (Hassan & Sun, 1992). ArcA and Fur both bind directly to the promoter region of *sodA* and their binding is mutually exclusive. *In vitro*, ArcA can easily displace Fur from the promoter, but not vice versa. This may reflect hierarchical control by ArcA and Fur of *sodA* transcription (Tardat & Touati, 1993).

The succinate dehydrogenase operon (*sdhCDAB*) is negatively regulated by ArcA and positively controlled by cAMP-Crp. The cAMP-Crp complex also regulates the *cyoABCDE* operon. This operon and the *cydAB* operon are regulated by changes in pH, at least in mutants deficient in Fnr (Cotter *et al.*, 1990). Expression of *cydAB* is also subject to ArcB/ArcA-dependent heat shock regulation (Wall *et al.*, 1992). The Fnr-dependent *cea* gene of plasmid Col E1 is also regulated by the SOS response (anaerobically) and by catabolite repression (aerobically) (Eraso & Weinstock, 1992). Studies in *Salmonella typhimurium* have identified ArcB/ArcA as an anaerobic activator pf the cobalamin biosynthetic operon, *cob*, whose transcription is repressed aerobically and is induced by cAMP and propanediol and repressed by cobalamin (Andersson, 1992).

Thus in *E. coli*, the organism for which the most detailed data set is available, the regulation of transcription of genes concerned with aerobic and anaerobic respiration is governed by a complicated control network with links to regulons concerned with other central functions, such as response to low iron, to oxidative stress and to changes in carbon source.

Oxidative stress response

Bacteria can be damaged by active oxygen molecules generated by their own aerobic metabolism or by the external environment. Macromolecules such as DNA, RNA, proteins and lipids are all liable to suffer harm by oxidative stress and bacteria have evolved multiple mechanisms to protect themselves. Much of the research on bacterial oxidative stress responses has been carried out in *E. coli* and *Salmonella typhimurium*. The results have identified two distinct forms of response, one protecting against peroxide stress and the other protecting against superoxide stress (reviewed in Farr & Kogoma, 1991). About 30 H_2O_2-inducible proteins have been detected by two-dimensional gel analysis in *E. coli* and *S. typhimurium* of which nine are positively regulated by the OxyR protein. This transcription activator is a member of the LysR family of DNA-binding proteins (see later). The *oxyR* gene is negatively autoregulated. Genes subject to OxyR activation include *katG* (coding for catalase HPI) and *aphCF* (coding for alkyl hydroperoxide reductase). OxyR appears to bind to a region upstream of and overlapping with the -35 box of the σ^{70} promoters of the genes under its control, suggesting that it may interact physically with RNA polymerase (Tartaglia *et al.*, 1989). Recent evidence indicates that OxyR interacts with the α subunit of RNA polymerase (Tao *et al.*, 1993). In addition to repressing its own transcription, OxyR can repress transcription of the *mom* gene of bacteriophage

Mu. Here, it binds to the Mu DNA only when the latter is unmethylated (the target site includes the Dam methylase substrate sequence, 5′-GATC-3′) showing that this protein can discriminate between methylated and unmethylated DNA, at least while acting as a repressor (Bölker & Kahmann, 1989). The OxyR protein appears to be both the sensor and regulator of the H_2O_2 response. It is thought that the protein itself undergoes a conformational change when oxidized by H_2O_2 which alters its interaction with DNA and RNA polymerase (Storz *et al.*, 1990). OxyR may sit on the regulated promoter at all times and then activate it in response to oxidation by H_2O_2, allowing the cell to respond very rapidly to this form of oxidative stress.

E. coli possesses a distinct regulon of genes for responding to superoxide stress. Approximately nine of these have been found to be under the control of SoxRS. SoxR is a 17.1 kDa protein which shares homology with MerR, the regulator of mercury resistance genes. SoxS is a 12.9 kDa protein with homology to the DNA-binding domains of the AraC family of gene regulators (Amábile-Cuevas & Demple, 1991). The *soxR* and *soxS* genes are divergently transcribed with the *soxR* promoter lying within the *soxS* gene (Wu & Weiss, 1991). The SoxS protein has been shown to be able to induce transcription of the SoxRS regulon genes in the absence of SoxR, leading to the suggestion that the *soxRS* gene products act sequentially at the promoters they regulate, with the SoxS protein acting first (Amábile & Demple, 1991; Wu & Weiss, 1992). The relationship of SoxR to MerR, a metal-binding protein, and the observation that SoxR possesses four cysteines grouped near the carboxyterminus, has been used as the basis of a model in which SoxR binds a metal whose redox state determines the biological activity of the protein. Furthermore, like MerR, SoxR binds to the regulated promoter between the −10 and −35 boxes, and it has been suggested that, like MerR, SoxR may activate transcription through a torsional mechanism (Ansari *et al.*, 1992; Nunoshiba *et al.*, 1992). Instead of being triggered by mercury binding as in the case of MerR, SoxR is thought to be activated by redox. An analogous mechanism has been proposed for OxyR (see earlier; Storz *et al.*, 1990).

Mutations in *soxR* have been isolated which result in broad-spectrum resistance to structurally unrelated antibiotics such as ampicillin, bleomycin, chloramphenicol, nalidixic acid and tetracycline (Greenberg *et al.*, 1990). These act in part by altering the expression of the *rimK* gene, whose product adds glutamic acid residues to a ribosomal protein, and in part by altering the expression of the *ompF* gene coding for the outer membrane porin protein OmpF. The latter effect is achieved through an alteration in the expression of the antisense RNA gene *micF* whose product sequesters the ribosome-binding site in *ompF* mRNA and precludes its translation (Farr & Kogoma, 1991; Mizuno *et al.*, 1984). The effect of the *soxR* mutation on the ribosome protein would alter the target for several antibiotics and the change in OmpF expression would reduce entry of antibiotics into the cell.

Genes under the control of SoxRS include *sodA* (coding for the manganese-containing superoxide dismutase which catalyses the conversion of the superoxide radical to hydrogen peroxide and oxygen), *nfo* (coding for Endonuclease IV, required for the repair of oxidatively damaged DNA), *zwf* (coding for glucose-6-P dehydrogenase, required for NADPH synthesis) and the *soxRS* genes themselves. The expression of *sodA* is also under the control of the anaerobic regulator Fnr, the aerobic regulator ArcA and the iron limitation response regulator, Fur (see earlier). In addition to SodA (the SoxRS-dependent, inducible Mn-containing superoxide dismutase) bacteria also possess a constitutive enzyme, the Fe-containing superoxide dismutase, encoded by the SoxRS-independent *sodB* gene.

In addition to the HPI catalase (KatG) which is under OxyR control, *E. coli* possesses a second catalase, HPII, which is under the control of a separate gene activator. This enzyme, catalase HPII (or KatE) is inducible not by H_2O_2, but by entry into stationary phase. Transcription of *katE*, the gene for HPII, is positively regulated by KatF. This regulatory protein is believed to be an alternative sigma factor (called RpoS) for RNA polymerase (Mulvey & Loewen, 1989). In addition to *katE*, the KatF sigma factor is required for the expression of *xthA*, the gene coding for exonuclease III, an important enzyme for the repair of oxidatively damaged DNA (Demple *et al.*, 1986). Other KatF-dependent genes identified to date include *appA* (coding for an acid phosphatase) and the morphogene *bolA* whose product controls the transition of the rod-shaped exponential growth-phase cell to the spherical form characteristic of cells in stationary phase (Aldea *et al.*, 1989; Bohannon *et al.*, 1991; Lange & Hengge-Aronis, 1991a).

Overlaps have been detected between the oxidative stress regulons and other stress responses. For example, expression of the heat shock protein DnaK has been found to be inducible by H_2O_2 in both *E. coli* and *S. typhimurium* (Morgan *et al.*, 1986; Van Bogelen *et al.*, 1987). Expression of *E. coli* KatG, the HPI catalase, is inducible not only by H_2O_2, but also by nalidixic acid, an inducer of the SOS response; treatment with H_2O_2 induces expression of the *E. coli* RecA protein (Van Bogelen *et al.*, 1987). There may also be a link with the stringent response since treatment of cells with H_2O_2 results in an increase in the synthesis of ppGpp (Van Bogelen *et al.*, 1987). Oxidative stress has been reported to result in a relaxation of supercoiled reporter plasmids in *E. coli* (Horiuchi *et al.*, 1984). If such changes in DNA superhelicity also affect the chromosome, they might be expected to contribute to the networking of expression of heterologous genes in response to this form of stress.

Response to changes in carbon source

Glucose is the preferred carbon source for enteric bacteria in which it serves as a material for the construction of cellular components and as a source of energy. In the absence of glucose, the bacteria can 'make do' with alternatives, but should glucose once again become available, a series of mechanisms is called into play to switch from the less-preferred carbon source to glucose. Not all of these mechanisms are concerned with the regulation of transcription. For example, glucose can simply prevent entry of the alternative carbon source to the cell by inactivating its specific permease. This is achieved through the agency of a protein called III^{Glc}, which is involved in the uptake and phosphorylation of glucose. In its unphosphorylated state, this protein will inhibit permeases for non-glucose carbon sources. III^{Glc} becomes dephosphorylated by donating its phosphate to glucose; in the absence of this carbohydrate, the phosphorylated III^{Glc} will not inhibit the permeases specific for the alternative carbon sources (reviewed in Magasanik & Neidhardt, 1987).

Protein III^{Glc}-phosphate is an activator of adenylate cyclase, the enzyme that catalyses the synthesis of cAMP. As was stated earlier, passage of glucose into the cell depletes the level of III^{Glc}-phosphate and this in turn causes a reduction in the concentration of cAMP (through a loss of activation of adenylate cyclase). The loss of cAMP leads to a decline in the level of cAMP-Crp, the active form of Crp. Thus, operons whose promoters are cAMP-Crp-dependent are down-regulated. So, there is a link between glucose uptake and the activation of cAMP-Crp-dependent promoters (such as those of the arabinose, galactose or lactose operons).

Other carbon sources entering the cell may exert analogous effects on the expression of cAMP-Crp-dependent genes. III^{Glc} can only bind to permeases and block them while they are being used by their cognate compounds. Entry of such compounds via their permeases could take III^{Glc} out of circulation, leading to a reduction in the intracellular pool available for phosphorylation and thence activation of adenylate cyclase. For a discussion of the DNA-binding and transcription-activating properties of Crp, see Chapter 6.

Response to changes in nitrogen source

The preferred nitrogen source for enteric bacteria is ammonia. If this compound becomes limiting, the cell activates expression of several genes whose products facilitate the scavenging of low levels of ammonia or, if this is completely absent, permit the use of alternative nitrogen sources. Ammonia levels are sensed by glutamine synthetase, an enzyme present in very low amounts in cells replete with ammonia. When the ammonia level drops below 1 mM, the result is a decline in the glutamine concentration. A consequence of this is a rise in the levels of 2-ketoglutarate (conversion of 2-ketoglutarate to glutamate requires both glutamine and ammonia). These intracellular signals activate the products of the *glnB* gene (encoding protein P_{II}), the *glnD* gene (uridylyl-transferase/uridylyl-removing protein), the *ntrB* gene (the histidine protein kinase, NtrB) and the *ntrC* gene (the transcription factor and NtrB partner protein, NtrC).

In conditions of low glutamine and high 2-ketoglutarate, protein P_{II} is converted to its uridylylated form, P_{II}-UMP. P_{II} interacts with NtrB and stimulates its phosphatase activity. When P_{II} is uridylylated, NtrB stops being a phosphatase and becomes a kinase, transferring its phosphate to NtrC. In its phosphorylated form, NtrC is an activated transcription factor. Activated NtrC then stimulates transcription from RpoN-dependent promoters (i.e. σ^{54}-dependent promoters) such as that involved in high-level expression of the *glnA* gene, coding for glutamine synthetase. The details of the NtrB/NtrC 'two-component' system and of the interaction of NtrC with promoters are described later.

The histidine protein kinase/response regulator superfamily

The biological activities of transcription regulators may be modulated by covalent modification. An important example of a bacterial system using this form of control is the group of bipartite regulators known as the two-component family (Ronson *et al.*, 1987). Here, one partner is an environmental sensor and the other a response regulator (Appendices I & II). Not all response regulators are concerned with controlling transcription (e.g. a system of this type controls bacterial chemotaxis wholly through protein–protein interactions). Sensors are protein kinases which phosphorylate on a conserved histidine residue in response to a stimulus. The phosphate group is then transferred to an aspartate residue in the response regulator. If this is a transcription regulator, phosphotransfer will enable it to activate subservient promoters. In some cases, the sensor is associated with the cytoplasmic membrane, usually with its stimulus receiver domain on the outer surface of the membrane. In other cases, the sensor is wholly cytoplasmic. Although this regulatory family consists of very many members (reviewed in Stock *et al.*, 1989), only a few have been studied in detail at the molecular and biochemical levels. Frequently, the characteristics of the other members have simply been inferred from data obtained from amino acid sequence homologies with the best-studied examples. These systems participate in the control of responses to a wide range of stimuli. The conservation of the kinase domains and phosphoacceptor domains throughout the family means that any histidine protein kinase has the potential to donate its phosphate to any response regulator. This cross-talk between systems has been genetically and biochemically demonstrated to occur and may serve to integrate the response of the cell to its environment as a whole (Igo *et al.*, 1989; Ninfa *et al.*, 1988; Wanner, 1983, 1992). Several of these systems are directly involved in the control of virulence gene expression and, given their central role in adaptation, all must contribute to the overall 'fitness' of the bacterium that possesses them.

Recently, it has been shown that low molecular weight phosphate donors (such as acetyl phosphate) can phosphorylate regulator molecules. Since no protein kinase is involved, the regulators must be capable of autophosphorylation (McCleary *et al.*, 1993). Thus, the phosphorylated sensor kinase may serve simply as a substrate for dephosphorylation by the response regulator. McCleary *et al.* (1993) have suggested that the presence of acetyl phosphate may signal to enteric bacteria that they are in the intestine, since the vast population of gut anaerobes produce large amounts of acetate during the fermentation of carbohydrates. Since acetate cannot be metabolized in the absence of respiration, large amounts of acetyl phosphate might be expected to accumulate, thus signalling to the enterics that they are in the gut.

The EnvZ/OmpR system

The primary role of the EnvZ/OmpR system is to regulate the expression of the OmpC and OmpF outer membrane porin proteins of *E. coli* and *Salmonella typhimurium* in response to changes in osmolarity. These enteric bacteria regulate the expression of OmpC and OmpF reciprocally such that the relative abundance of OmpC increases as osmolarity increases, with a concomitant decrease in OmpF levels. If osmolarity decreases, the level of OmpF is increased with a compensating reduction in OmpC. Under all growth conditions, the overall level of OmpC/OmpF porin in the outer membrane remains constant at about 10^5 per cell (Nikaido & Vaara, 1987). The non-specific pores produced by the porin trimers in the outer membrane admit small hydrophilic molecules to the periplasm. The estimated pore diameter of the *E. coli* K-12 OmpF porin is 1.16 nm while that of OmpC is 1.08 nm (Nikaido & Rosenberg, 1983). This difference in pore size is believed to have important consequences for the survival of the cells and to underly the need for the reciprocal regulation of OmpF and OmpC expression. OmpC is a much more effective molecular sieve than OmpF and more efficiently excludes compounds above 200 Da, particularly if these are negatively charged. In the gut, the high osmolarity conditions should favour OmpC expression, preserving the bacterium from the harmful effects of detergent molecules such as bile salts. Conversely, OmpF expression in low-osmolarity environments such as those found outside the host will assist the organism in scavenging for nutrients. In addition to osmolarity, OmpC and OmpF porin expression is regulated in response to changes in temperature, oxygen availability, pH and carbon source (Mizuno & Mizushima, 1990; Stock *et al.*, 1989). Some of these environmental inputs are routed through EnvZ/OmpR while others are transmitted through other regulatory circuits. As will be seen, the control of OmpC and OmpF expression is achieved through multiple and overlapping regulators (Fig. 7.5).

EnvZ is the histidine protein kinase member of the partnership (Appendix II). It is an inner membrane protein of 50.3 kDa and with two membrane-spanning sequences in its aminoterminal domain (Comeau *et al.*, 1985) (Figs 7.5 & 8.3). His-243 is the site of ATP-dependent phosphorylation and mutations that alter the amino acid at this position abolish EnvZ phosphorylation (Forst *et al.*, 1989). This histidine is within the cytoplasmic domain of the protein. The cytoplasmic domain of EnvZ also has phosphatase activity and can dephosphorylate OmpR-P in an ATP-independent manner (Aiba *et al.*, 1989; Igo *et al.*, 1989). The mechanisms by which changes in osmolarity trigger these events remain obscure. Presumably a conformational change occurs in the osmosensing

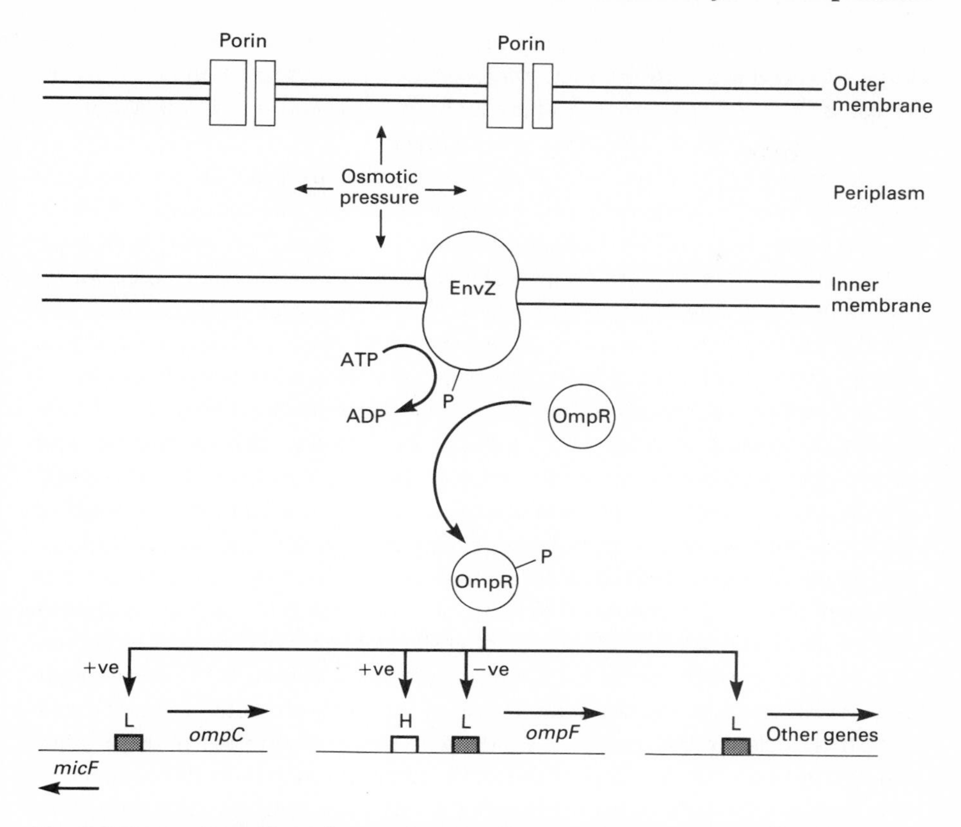

Fig. 7.5 Transcriptional regulation of porin gene expression by the EnvZ/OmpR two-component system. Osmotic pressure in the periplasm is thought to trigger phosphorylation of EnvZ. Phosphotransfer from EnvZ enhances the DNA-binding ability of OmpR. L, low-affinity binding site for phospho-OmpR; H, high-affinity site. *micF* encodes an antisense RNA which hybridizes to the 5′ end of *ompF* mRNA, preventing its translation (see Fig. 6.11). 'Other genes' subject to OmpR regulation may be controlled positively, negatively or both.

aminoterminal periplasmic domain and this activates the kinase in the cytoplasmic domain (Fig. 7.5).

OmpR is the response regulator and is a transcriptional activator of the *ompC* and *ompF* genes (Appendix I). OmpR is a 27.3 kDa cytoplasmic protein and is phosphorylated by EnvZ in its aminoterminal domain. The DNA-binding region is located in the carboxyterminal domain and DNA binding is stimulated by phosphorylation of the aminoterminal domain (Aiba *et al.*, 1989). The genes *envZ* and *ompR* form an operon, *ompB*, at 75 min on the *E. coli* and 74 min on the *S. typhimurium* genetic maps (Bachmann, 1990; Comeau *et al.*, 1985; Liljestrom *et al.*, 1988; Rudd; 1992; Sanderson & Roth, 1988). The relative abundance of the *ompR* and *envZ* gene products may be controlled by translational coupling;

the translational start site for EnvZ overlaps with the termination codon of OmpR (Comeau *et al.*, 1985). In *E. coli* K-12, EnvZ is present at about 10 copies per cell and OmpR is present at approximately 1000 copies per cell (Mizuno *et al.*, 1982; Wurtzel *et al.*, 1982).

The regulation of *ompC* and *ompF* transcription by EnvZ/OmpR has been studied in detail at the genetic and biochemical levels. The regulatory proteins are absolutely required for transcription activation from the *ompC* and *ompF* promoters yet the genes are reciprocally controlled in response to osmolarity. This differential control is a function of the nature and distribution of the OmpR-P binding sites upstream of the *ompC* and *ompF* promoters (Slauch & Silhavy, 1989; Stock *et al.*, 1989). The *ompF* promoter possesses a high-affinity site for OmpR-P. At low osmolarity the cellular concentration of OmpR-P is low. This small amount of activated OmpR can bind to the high-affinity site and activate the *ompF* promoter. The OmpR-P binding site prefacing the *ompC* promoter is a low-affinity site and does not bind the low-abundance OmpR-P present in the cell at low osmolarity; under these environmental conditions, *ompC* is poorly transcribed, if at all. At high osmolarity, the level of OmpR-P is increased due to the enhanced activity of EnvZ and there is now sufficient OmpR-P to occupy the low-affinity binding site upstream of the *ompC* promoter. This activates *ompC* transcription. Conversely, at high osmolarity, OmpR-P binds to low-affinity sites in the upstream region of the *ompF* promoter and is thought to form a nucleoprotein complex which represses transcription of *ompF*. This complex involves the bending of the DNA upstream of the *ompF* promoter to bring together the fully occupied OmpR-P sites in the complex. Mutations that 'stiffen' the sequences involved in the bend prevent formation of this nucleoprotein complex and prevent *ompF* repression at high osmolarity (Slauch & Silhavy, 1991).

Genetic analysis has revealed that the ability of phosphorylated OmpR to activate transcription depends on direct contact with RNA polymerase. Specifically, mutations in *rpoA*, encoding the α subunit of RNA polymerase, alter transcriptional control of porin expression by OmpR (Garrett & Silhavy, 1987; Matsuyama & Mizushima, 1987; Slauch *et al.*, 1991). These mutations map to the carboxyterminal portion of the α subunit protein, suggesting that this is the point of contact between polymerase and the gene activator (Slauch *et al.*, 1991).

Mutations in *ompB* that affect expression of OmpR and EnvZ have pleiotropic effects beyond *ompC* and *ompF*. In *S. typhimurium*, the *tppB* operon, encoding the tripeptide permease, is *ompB*-dependent (Gibson *et al.*, 1987). The *E. coli malT* (positive regulator of the maltose regulon), *phoA*, *phoE* (components of the phosphate utilization system) and *opr* (outer membrane protease) genes are also regulated by *ompB* (Case *et al.*, 1986; Sarma & Reeves, 1977; Wandersman *et al.*, 1980). Iron-regulated gene expression is also affected in *E. coli* (Lundrigan & Earhardt, 1981).

The NtrB/NtrC system

The NtrB/NtrC two-component system is a positive regulator of genes required to respond to nitrogen limitation (Appendix I & II). At the centre of the nitrogen limitation response is the *glnA* gene, encoding glutamine synthetase, which is required for the assimilation of fixed nitrogen. In the chromosomes of enteric bacteria, *glnA* lies immediately upstream of the *ntrBC* operon, which codes for the NtrB/NtrC two-component system. Under nitrogen-limiting conditions, all three genes are cotranscribed as an operon (Fig. 7.6). When nitrogen levels are high, the *glnA* gene and the *ntrBC* operon are transcribed independently. There are three promoters regulating their expression. Two are upstream of *glnA*. Of these, P_1 is a σ^{70} driven promoter which maintains a basal level of *glnA* transcript in the cell under nitrogen-replete conditions and the *glnA* P_2 promoter is a powerful σ^{54}-dependent promoter which can transcribe all three genes when nitrogen becomes limiting (Fig. 7.6). The *ntrBC* operon is prefaced with its own σ^{70}-dependent promoter which is important for transcription of the operon at a low level during periods when nitrogen is plentiful. NtrC negatively regulates transcription from the *glnA* P_1 promoter and the *ntrBC* promoter.

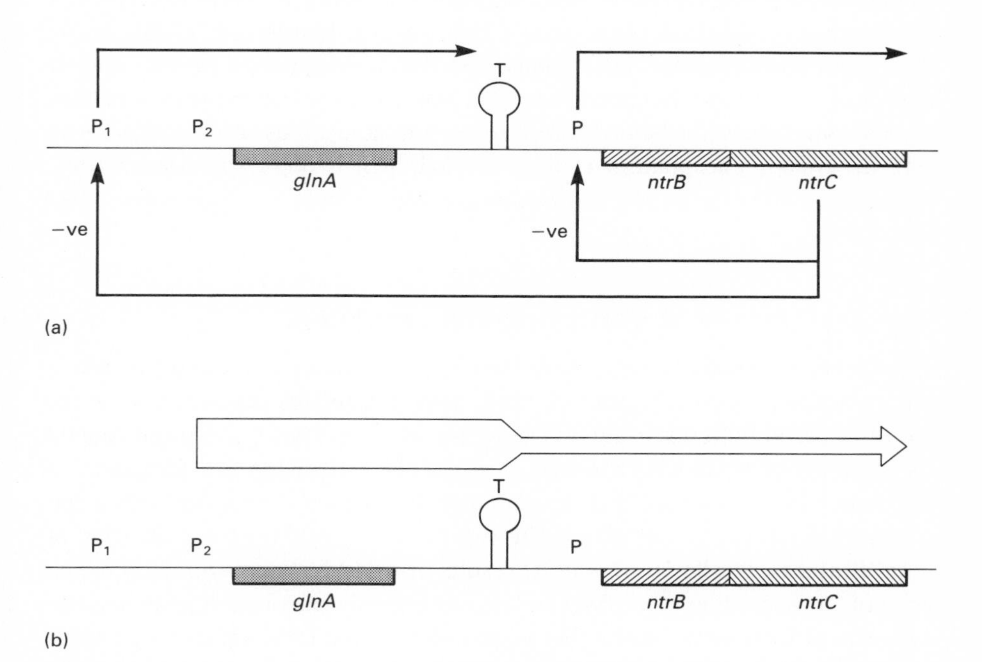

Fig. 7.6 Transcriptional regulation of the *glnAntrBC* operon. (a) Under high nitrogen conditions, NtrC negatively regulates the σ^{70} promoters in front of *ntrB* and *glnA* (P_1). Transcription from P_1 is terminated at the terminator site (T) between *glnA* and *ntrB*. (b) Under nitrogen-limiting conditions, the σ^{54} promoter P_2 is activated, greatly increasing transcription of *glnA* and allowing read-through of *ntrBC*.

σ^{54}-programmed RNA polymerase will bind to the *glnA* P_2 promoter and form a closed complex but it cannot form an open complex in the absence of NtrC. The formation of the open complex is ATP-dependent. In order to catalyse the formation of an open complex, NtrC must itself be phosphorylated. Phosphorylated NtrC has an ATPase activity that is essential for open complex formation (Weiss *et al.*, 1991). Thus, NtrC differs radically from the OmpR transcription activator in the detail of the mechanism by which it activates promoters. The sites to which NtrC binds are often far upstream of the promoters that it regulates. Contact between the transcription activator and the promoter-bound polymerase is achieved by looping of the intervening DNA sequences (Su *et al.*, 1990). In distinct but related σ^{54}-dependent systems, such as the NifA-dependent *nifH* promoter of *Klebsiella pneumoniae*, the bending of the DNA between the upstream transcription enhancer with its bound activator (NifA) and the promoter with its bound σ^{54} RNA polymerase holoenzyme is assisted by the DNA-bending histone-like protein IHF (Hoover *et al.*, 1990; Santero *et al.*, 1992) (Fig. 7.7; and see Chapter 2).

NtrB is a histidine protein kinase which autophosphorylates and transfers the phosphate to an aspartic acid residue (Asp-54, Sanders *et al.*, 1992) within its partner response regulator, NtrC, activating the latter as a positive regulator of transcription (Ninfa & Magasanik, 1986; Weiss & Magasanik, 1988). Unlike the membrane-associated EnvZ kinase, NtrB is a cytoplasmic protein and its phosphorylated state is controlled by a complex protein–protein signalling cascade which heralds the onset of nitrogen limitation. This cascade triggers an ATP-dependent phosphatase activity in NtrB which dephosphorylates NtrC, rendering it inactive as a transcription activator (Ninfa & Magasanik, 1986) (Fig. 7.8).

The AraC family of transcriptional regulators

This family of regulatory proteins derives its name from its prototypic member, the L-arabinose operon regulator, AraC, which regulates transcription of the *araBAD* operon both positively (in the presence of arabinose) and negatively (in the absence of arabinose) through a mechanism involving the formation of alternative loops in the DNA 5′ to *araB* (Lobell & Schleif, 1990) (Fig. 7.9). AraC can bind as a dimer to four sites in this region, O_1, O_2, and the two 'half-sites' of I, i.e. I_1 and I_2. Binding to O_2 and I_1 results in a 211 bp repression loop which prevents transcription of *araBAD* and the divergently transcribed gene *araC. In vitro* experiments demonstrate that supercoiling of the DNA assists loop formation (Hahn *et al.*, 1986; Lobell & Schleif, 1990). Binding of AraC to O_1 and O_2 represses *araC* expression and permits occupation of I_1 and I_2 which breaks the repression loop and stimulates transcription of *araBAD*. The disruption of the O_2–I_1 repression loop is caused by the interaction of L-arabinose with the AraC protein in the loop and is believed to be due to a conformational change in the AraC dimer resulting from sugar binding (Lobell & Schleif, 1990). Binding of

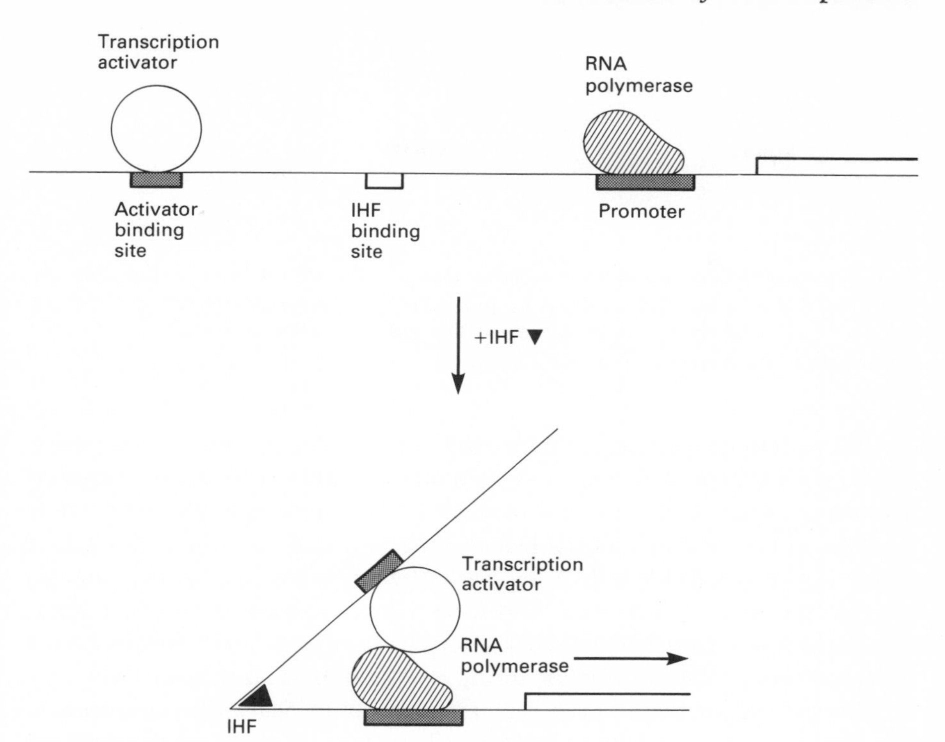

Fig. 7.7 Role of integration host factor (IHF)-induced DNA bending in activation of a σ^{54}-dependent promoter. RNA polymerase cannot form an open complex at a σ^{54}-dependent promoter unless an upstream-bound activator protein is brought into contact with it. Binding of IHF to a position approximately midway between the promoter and the upstream activator site causes the DNA to bend, bringing the transcription activator into contact with polymerase. This allows transcription to be initiated.

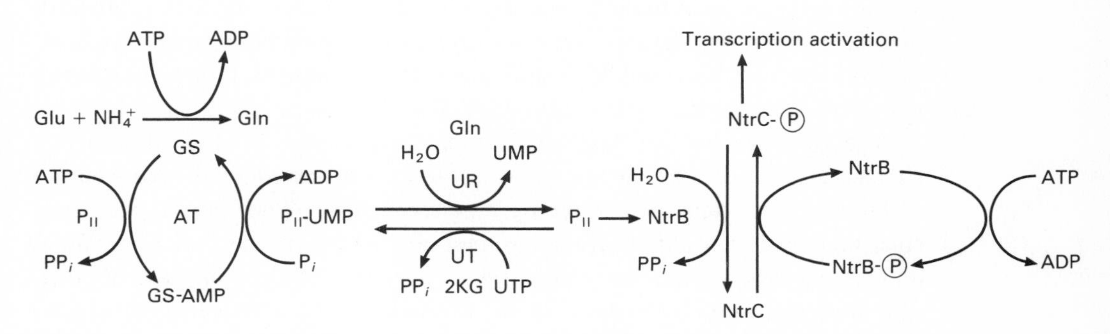

Fig. 7.8 Regulatory system for nitrogen utilization in enteric bacteria. The cascade by which combined nitrogen is assimilated is illustrated. GS, glutamine synthetase; AT, adenylyl transferase; UR, uridylyl-removing activity (glutamine-stimulated); UT, uridylyl transferase (2-ketoglutarate-stimulated); 2KG, 2-ketoglutarate.

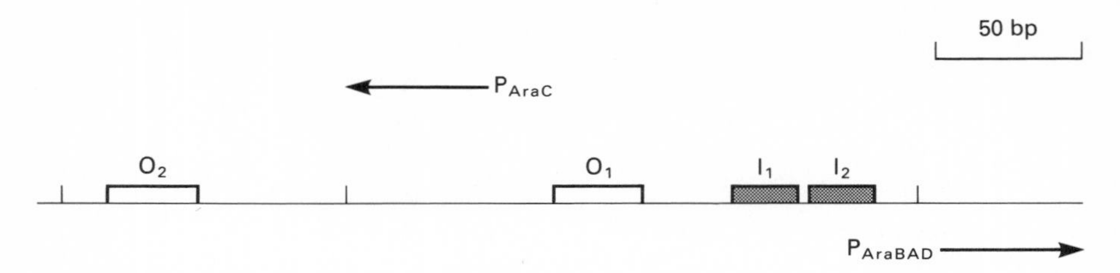

Fig. 7.9 AraC-binding sites in the regulatory region of the *ara* operon of *Escherichia coli*. There are four binding sites for the AraC protein within the regulatory region of the *ara* operon: O_1, O_2 and the I_1 and I_2 'half sites'. The positions of the promoters for transcription of *araBAD* and *araC* are indicated.

cAMP-Crp to its recognition site within the loop provides a second loop-breaking activity (Lobell & Schleif, 1991). An important feature of this loop-mediated regulation is the relative positions of the AraC binding sites on the DNA; these must be on the same face of the helix for the loop to form. The *in vivo* helical repeat of the loop DNA is 11.1 bp. If half helical turns of DNA are inserted between the AraC binding sites, displacing them to opposite faces of the helix, loop formation is strongly impaired; conversely, the insertion of full helical turns is not inhibitory to loop formation (Dunn *et al.*, 1984; Lee & Schleif, 1989).

The *araC* genes from *E. coli* B/r and K-12, from *Salmonella typhimurium*, *Citrobacter freundii* and *Erwinia carotovora* have been cloned and the deduced amino acid sequences show that they are highly related. Indeed, AraC proteins from all four species regulate the *araCBAD* genes in *E. coli* and *S. typhimurium* (Burke & Wilcox, 1987). AraC consists of two functionally distinct domains; the aminoterminal domain is involved in the binding of L-arabinose while the carboxyterminal domain contains the DNA-binding region (Brunelle & Schleif, 1989; Cass & Wilcox, 1986; Lauble *et al.*, 1989).

It has become apparent that bacteria possess a family of proteins related in amino acid sequence to AraC. The relevance of the detail concerning AraC interactions with DNA to these proteins is unclear. However, since the regions of homology to AraC always involve the carboxyterminal domain, where the AraC DNA-binding site is located (Fig. 7.10), it is possible that these molecules interact with DNA in a manner similar to that of AraC.

The members of the AraC-like protein family may be divided into two subclasses, one containing regulators of genes concerned with carbohydrate metabolism and another containing regulators of virulence genes. The most conserved residues among these proteins are found in the carboxyterminal domain, i.e. the region corresponding to the DNA-binding domain of AraC (Fig. 7.10; Appendix I). Presumably, the variability of the aminoterminal domain reflects differences in sugar binding in the carbohydrate metabolizing subclass and (perhaps) in environmental signal reception in the virulence gene regulator subclass (see later).

The carbohydrate metabolism regulators include CelD (cellobiose, *E. coli*)

```
E. chry AraC 201 RIDEVARHVCLSPSRLAHLFREQVGINILRWREDQRVIRAKLLLQTTQESIANIGRVVGYDDQLYFSRVFRKRVGVSPSDFRRRSSEINYPAAKTLPVAWGEQIPHAVSS
E. coli AraC 196 DIASVAQHVCLSPSRLSHLFRQQLGISVLSWREDQRISQAKLLLSTTRMPIATVGRNVGFDDQLYFSRVFKKCTGASPSEFRAGCEEKVNDVAVKLS
S. typh AraC 196 DIASVAQHVCLSPSRLSHLFRQQLGISVLSWREDQRISQAKLLLSTTRMPIATVGRNVGFDDQLYFSRVFKKCTGASPSEFRAGCE
E. coli MelR 208 TINDVAEHVKLNANYAMGIFQRVMQLTMKQYITAMRINHVRALLSDTDKSILDIALTAGFRSSSRFYSTFGKYVGMSPQQYRKLSQQRRQTFPG
E. coli RhaR 224 ALDKFCDEASCSERVLRQQFRQQTGMTINQYLRQVRVCHAQYLLQHSRLLISDISTECGFEDSNYFSVVFTRETGMTPSQWRHLNSQKD
E. coli RhaS 188 NWDAVADQFSLSLRTLHRQLKQQTGLTPQRYLNRLRLMKARHLLRHSEASVTDIAYRCGFSDSNHFSTLFRREFNWSPRDIRQGRDGFLQ
P. aeru XylS 229 SLERLAELAMMSLYNLFEKHAGTTPKNYIRNRKLESIRACLNDPSANVRSITEIALDYGFLHLGRFAENYRSAFGELPSDTLRQCKKEVA
Y. ent  VirF 182 KLSKFAREFGMGLTTFKELFGTVYGISPRAWISERRILYAHQLLLNGKMSIVDIAMEAGFSSQSYFTESYRRRFGCTPSQARLTKIATTG
Y. pest LcrF 182 KLSKFAREFGMGLTTFKELFGTVYGISPRAWISERRILYAHQLLLNGKMSIVDIAMEAGFSSQSYFTQSYRRRFGCTPSQARLTKIATTG
P. aeru ExsA 207 KLSDFSREFGMGLTTFKELFGSVYGVSPRAWISERRILYAHQLLLNSDMSIVDIAMEAGFSSQSYFTQSYRRRFGCTPSRSRQGKDECRAKNN
E. coli Rns  178 WTLGIIASAFNASEITIRKRLESENTNFNQILMQLRMSKAALLLLENSYQISQISNMIGISSASYFIRIFNKHYGVTPKQFFTYFKGG
E. coli CfaD 178 WTLGIIADAFNVSEITIRKRLESENTNFNQILMQLRMSKAALLLLENSYQISQISNMIGISSASYFIRVFNKHYGVTPKQFFTYFKGG
S. flex VirF 175 WRLSSISNNLNLSEIAVRKRLESEKLTFQQILLDIRMHHAAKLLLNSNSYINDVSRLIGISSPSYFIRKFNEYYGITPKKFYLYHKKF
S. sonn VirF 175 WRLSSISNNLNLSEIAVRKRLESEKLTFQQILLDIRMHHAAKLLLNSNSYINDVSRLIGISSPSYFIRKFNEYYGITPKKFYLYHKKF
E. coli AppY 147 WHLKDIAELIYTSESLIKKRLRDEGTSFTEILRDTRMRYAKKLITSNSYSINVVAQKCGYNSTSYFICAFKDYYGVTPSHYFEKIIGVTDGINKTID
E. coli EnvY 162 WNLRIVASSLCLSPSLLKKKLKNENTSYSQIVTECRMRYAVQMLLMDNKNITQVAQLCGYSSTSYFISVFKAFYGLTPLNYLAKQRQKVMW
E. coli FapR 168 WKLSDIAEEMHISEISVRKRLEQECLNFNQLILDVRLMKARHLLRHSEASVTDIAYRCGFSDSNHFSTLFRREFNWSPKKFEIGIKENLRCNR
V. chol ToxT 218 WRWADICGELRTNRMILKKELESRGVKFRELINSIRISYSISLMKTGEFKIKQIAYQSGFASVSYFSTVFKSTMNVAPSEYLFMLTGVAEK
P. sola HrpB 386 EVAAHINVTERALQLAFKSAVGMSPSSVIRRMRLEGIRSDLLDSERNPSNIIDTASRWGIRSRSALVKGYRKQFNEAPSETIWR
E. coli SoxS  23 NIDVVAKKSGYSKWYLQRMFRTVTHQTLGDYIRQRRLLLAAVELRTTERPIFDIAMDLGYVSQQTFSRVFRRQFDRTPSDYRHRL
Tn10    TetD  23 LLDDVANKAGYTKWYFQRLFKKVTGVTLASYIRARRLTKAAVELRLTKKTILEIALKYQFDSQQSFTRRFKYIFKVTPSYYRRNKLWELEAMH
E. coli CelD 190 ALENMVALSAKSQEYLTRATQRYYGKTPMQIINEIRINFAKKQLEMTNYSVTDIAFEAGYSSPSLFIKTFKKLTSFTPKSYRKKLTEFNQ
Consensus             A         L   F    G          R   A  LL      I DIA   GF S  YF   F    G TPS  R
```

Fig. 7.10 The AraC family of transcription activators. An alignment of the carboxyterminal domains of the AraC family members is shown. Work with the prototypic AraC protein of *E. coli* indicates that the DNA-binding domain is located within this region. In addition to proteins from *E. coli*, sequences from *Erwinia chrysamthemi* (*E. chry*), *Salmonella typhimurium* (*S. typh*), *Pseudomonas aeruginosa* (*P. aeru*), *Yersinia enterocolitica* (*Y. ent*), *Yersinia pestis* (*Y. pest*), *Shigella flexneri* (*S. flex*), *Shigella sonnei* (*S. sonn*), *Vibrio cholerae* (*V. chol*), *Pseudomonas solanaceum* (*P. sola*) and transposon Tn*10* are shown. These proteins are involved in regulating the expression of carbohydrate utilization systems (AraC, CelD, MelR, RhaR, RhaS, XylS), bacterial virulence networks (CfaD, ExsA, FapR, HrpB, LcrF, Rns, ToxT, VirF), oxidative stress response genes (SoxS), growth phase-dependent genes (AppY), bacterial surface proteins and antibiotic resistance (TetD). A consensus sequence for the family is shown at the base of the alignment.

(Parker & Hall, 1990), MelR (melibiose, *E. coli*), RhaR and RhaS (rhamnose, *E. coli*), XylS (benzoate and alkylbenzoates, *Pseudomonas putida*) (Gallegos *et al*., 1993; Ramos *et al*., 1990; Tobin & Schleif, 1987; Webster *et al*., 1987) (Fig. 7.10). The EnvY thermoregulatory protein of *E. coli* which is involved in controlling porin gene expression is also an AraC family member (Lundrigan *et al*., 1989). The *soxS* gene product, which is involved in regulating the expression of genes for defence against superoxide stress, is a member of the AraC family. It is transcribed divergently from the *soxR* gene (whose product is related to the mercury resistance regulator, MerR). SoxR and SoxS may act sequentially to activate the genes of the oxidative stress regulon (Amábile-Cuevas & Demple, 1991; see earlier).

The LysR family of transcriptional regulators

Bacteria possess a family of transcription regulators with an aminoterminally located DNA-binding domain of the 'helix-turn-helix' (HTH) type called the LysR family, after the prototypic member, LysR (Henikoff *et al.*, 1988). Members of the family include regulators of amino acid biosynthesis such as CysB, IlvY, LysR, MetR and TrpI (Chang *et al.*, 1989; Henikoff *et al.*, 1988; Plamann & Stauffer, 1987; Stragier & Patte, 1983; Wek & Hatfield, 1986), antibiotic resistance gene expression in *Citrobacter freundii* and *Enterobacter cloacae* (AmpR) (Honoré *et al.*, 1986; Lindquist *et al.*, 1989), oxidative stress response and phage Mu *mom* gene regulation (OxyR, Christman *et al.*, 1989; the OxyR analogue MomR, Bölker & Kahmann, 1989), hydrocarbon degradation in *Pseudomonas* (NahR) (Schell & Sukordhaman, 1989), nodulation gene expression in *Rhizobium* (NodD, NolR, SyrM) (Barnett & Long, 1990; Henikoff *et al.*, 1988; Kondorosi *et al.*, 1991), initiation of chromosome replication in *E. coli* (IciA) (Thöny *et al.*, 1991), activation of iron-regulated virulence gene expression in *Vibrio cholerae* (IrgB) (Goldberg *et al.*, 1991; Chapter 8), cyanate detoxification in *E. coli* (CynR) (Sung & Fuchs, 1992), regulation of plasmid-encoded virulence genes in several serovars of *Salmonella* (SpvR) (Caldwell & Gulig, 1991; Gulig *et al.*, 1993; Krause *et al.*, 1992; Pullinger *et al.*, 1989; Chapter 8) (Fig. 7.11).

Gene regulation by *N*-acyl-L-homoserine lactones

The prototypic example of gene regulation by the small molecule inducer *N*-(3-oxohexanoyl) homoserine lactone (HSL) is that of the *lux* bioluminescence system of the marine bacterium *Vibrio fischeri*. Here, a regulatory gene, *luxR*, is transcribed divergently from an operon that includes a second regulatory gene (*luxI*) and structural genes concerned with specifying factors required for the bioluminescent phenotype. Transcription of the major operon is induced by the accumulation of the autoinducer HSL, which is synthesized by the product of the *luxI* gene (Silverman *et al.*, 1989). This activation requires LuxR, whose ability to activate transcription depends on its binding HSL. Since HSL synthesis depends on *luxI* expression, the more LuxR activates *luxI*, the more HSL the cell produces. Once the level of autoinducer reaches a critical threshold value, the bioluminescent phenotype is expressed.

Recently, the HSL production has been detected in many other bacteria suggesting that these encode a LuxI homologue (and presumably a LuxR analogue) (reviewed in Williams *et al.*, 1992). In terms of specific systems, HSL activation has been described for carbapenem antibiotic biosynthesis in the plant pathogen *Erwinia carotovora*, for regulation of conjugation in the plant pathogen *Agrobacterium tumefaciens* and a LuxR homologue has been found to regulate elastase expression in *Pseudomonas aeruginosa* (Gambello & Iglewski, 1991; Piper *et al.*, 1993; Williams *et al.*, 1992; Zhang *et al.*, 1993; Chapter 8).

```
                                    Helix-Turn-Helix
                                  *********************

E. cloc AmpR MTRSYLPLNSLRAFEAAARHLSTFHAAIELNVTHSAISQHVKTLEQHLNCQLFVRVSRGLM-LTTEGENLLPVLNDSFDRIAGMLDRFANH--RAQEKLKIGVVGTFATGVLFS
E. coli CysB     MKLQQLRYIVEVVNHNLNVSSTAEGLYTSQPGISKQVRMLEDELGIQIFSRSGKHLTQVTPAGQEIIRIAREVLSKVDAIKSVAGEHTWPDKGSLYIATTHTQARYALPN
E. coli IlvY      MDLRDLKTFLHLAESRHFGRSARAMHVSPSTLSRQIQRLEEDLGQPLFVRDNRTVT-LTEAGEELRVFAQQTLLQYQQLRHTIDQQGPSLSGELHIFCSVTAAYSHLPP
E. coli LysR    MAAVNLRHIEIFHAVMTAGSLTEAAHLLHTSQPTVSRELARFEKVIGLKLFERVRGRLH-PTVQGLRLFEEVQRSWYGLDRIVSAAESLREFRQGELSIACLPVFSQSFLPQ
S. typh MetR     MIEIKHLKTLQALRNSGSLAAAAAVLHQTQSALSHQFSDLEQRLGFRLFVRKSQPLR-FTPQGEVLLQLANQVLPQISRALQACNE---PQQTRLRIAIECHSCIQWLTP
P. aeru TrpI MSRDLPSINALRAFEAAARLHSISLAAEELHVTHGAVSRQVRLLEEDLGVALFGRDGRGVK-LTDSGVRLRDACGDAFERLRGVCAELRRQTAEAPFVLGVPGSLLA--RWFIP
E. coli OxyR    MNIRDLEYLVA-LAEHR-H-FRRAADSCHVSQPTLSGQIRKLEDELGVMLLERTSRKVLFTQAGMLLVDQARTVLREVKVLKEMA-SQQGETMSGPLHIGLIPTVGPYLLPH
P. puti NahR MELRDLDLNLLVVFNQLLVDRRVSITAENLGLTQPAVSNALKRLRTSLQDPLFVRTHQGMEPTPYAAHLAEPVTSAMHALRNALQHHESFDPLTSERTFTLAMTDIGEIYFMPR
R. meli NolR  MNFRMEHTMQPLPPEKHEDAEIAAGFLSAMANPKRLLILDSLVKEEMAVGALAHKVGLSQSALSQHLSKLRAQNLVSTRRDAQTIYYSSSSDAVLKILGALSDIYGDDTDAVL
R. meli NodD MRFRGLDLNLLVALDALMTERKLTAAARRINLSQPAMSAAIARLRTYFGDELFSMQGRELIPTPRAEALAPAVRDALLHIQLSVIAWDPLNPAQSDRRFRIILSDFMILVFFAR
V. chol IrgB     MQDLSAVKAFHALCQHKSLTAAAKALEQPKSTLSRRLAQLEEDLGQSLLMRQGNRLTLTKAGEVFAVYSEQLLELANKSQEALQELNNQVTGELTLVVHPNLIRG-WLSQ
E. coli CynR      MLSRHINYFLAVAEHGSFTRAASALHVSQPALSQQIRQLEESLGVPLFDRSGRTIRLTDAGEVWRNYASRALQELGAGKRAIHDVADLTRGSLRIAVTPTFTS-YFIGP
E. coli IciA MKRP-DYRTLQALDAVIRERG-FERAAQKLCITQSAVSQRIKQLENMFGQPLLVRTVPP-RPTEQGQKLLALLRQVELLEEEWLGDEQTGSTPLLLS-LAVNADSLAT--WLLP
S. typh SpvR    MDFLINKKLKIFITLMETGSFSIATSVLYITRTPLSRVISDLERELKQRLFIRKNGTLIPTEFAQTIYRKVKSHYIFLHALEQEIGPTGKTKQLEIIFDEIYPGSLKNLIIS
Consensus               L  F A     S    AA  L     Q ALS     LE  LG  LF R          T                                     L I
```

Fig. 7.11 The LysR family of transcription activators. An alignment of the aminoterminal domains of the LysR family of transcription regulators is shown. This region includes the helix-turn-helix domain which is required for DNA binding. The proteins come from *Enterobacter cloacae* (*E. cloc*), *Escherichia coli* (*E. coli*), *Rhizobium meliloti* (*R. meli*), *Pseudomonas aeruginosa* (*P. aeru*), *Pseudomonas putida* (*P. puti*), *Salmonella typhimurium* (*S. typh*) and *Vibrio cholerae* (*V. chol*). The functions of these proteins are described in the text. A consensus sequence for the family is shown at the base of the alignment.

The Lrp regulon of enteric bacteria

E. coli possesses a regulon of genes whose expression is under the control of leucine. This group is known as the leucine regulon (Lin *et al.*, 1990) and its constituent genes are under the control of the 19 kDa Lrp (Leucine-responsive regulatory protein), which exists as a dimer in solution (Willins *et al.*, 1991) and was formerly known as Rbl. Other genes have been found which, though not regulated by leucine, are also under Lrp control. This shows that Lrp has wide-ranging effects on transcription (Lin *et al.*, 1992). The protein binds to and bends DNA, by angles of about 52° for single sites or 135° for adjacent sites (Wang & Calvo, 1993).

Lrp can activate transcription of some genes while repressing others. In some cases, it acts antagonistically with leucine to activate transcription and in other cases it acts antagonistically with leucine to repress transcription. In yet other cases, Lrp and leucine act cooperatively to exert their effects (either positive or negative) on gene expression (Lin *et al.*, 1992; Newman *et al.*, 1992) (Table 7.1). It is estimated that the cell contains 3000 copies of Lrp and that some 75 genes may be under its control (Newman *et al.*, 1992). Mutations in the *lrp* gene do not compromise seriously the growth of the cell. Even in minimal medium, growth is normal provided that the cell is supplied with ammonium, leucine and a suitable carbon source. These must be supplied because the leucine biosynthetic operon is poorly expressed in *lrp* strains and *lrp* mutants are unable to use glycine as a nitrogen source (Gcv^- phenotype) or

Table 7.1 Genes under Lrp*-leucine control

Bacterium	Gene	Map position (min)	Function	Effect of Lrp	Effect of leucine
E. coli K-12	*fim*	98	Type 1 pili	?	?
	gcv	63	Gly cleavage	Activator	None
	ilvIH	1.85	Ilv, Leu, Val biosynthesis	Activator	Antagonist
	leu	2	Leu biosynthesis	Activator	Antagonist
	livJH	76	Ile, Val uptake	Repressor	Compressor
	lrp	20	Leucine-responsive regulatory protein	Repressor	None
	lysU	94	Lys tRNA synthetase II	Repressor	Antagonist
	ompC	48	Porin	Repressor	?
	ompF	21	Porin	Activator	?
	oppABCD	28	Oligopeptide uptake	Repressor	Antagonist
	sdaA	40.2	L-serine deaminase I	Repressor	Antagonist
	sdaB	60.5	L-serine deaminase II	Activator	Coactivator
	serA	63	Ser biosynthesis	Activator	Antagonist
		91	Threonine dehydrogenase	Repressor	Antagonist
Non-K-12 strains	*fan*		K99 pilus	Activator	Antagonist
	pap		Pap pilus	Activator	None

*Lrp, leucine-responsive regulatory protein.

several organic nitrogen sources at low concentrations (Glt$^-$ phenotype). As yet, the physiological significance of the leucine-Lrp regulon (as it has been renamed) is unclear.

H-NS and transcriptional control

The H-NS protein is encoded by the *hns* gene of enteric bacteria and has been shown to contribute to the (usually negative) regulation of expression of a diverse group of genes (reviewed in Higgins *et al.*, 1990). Mutations in *hns* correlate with changes in DNA supercoiling (Higgins *et al.*, 1988a) suggesting a link between the biological function of H-NS and DNA structure (Owen-Hughes *et al.*, 1992). This suggestion is supported by the finding that H-NS is a nucleoid protein (Chapter 2) and is capable of compacting DNA *in vitro* (Spassky *et al.*, 1984; Varshavsky *et al.*, 1977). The structure of H-NS is conserved in many bacteria, including *E. coli*, *Proteus vulgaris*, *Salmonella typhimurium*, *Serratia marcescens* and *Shigella flexneri* (Hromockyj *et al.*, 1992; Hulton *et al.*, 1990; La Teana *et al.*, 1989; Marsh & Hillyard, 1990) (Appendix I). The protein contributes to the control of transcription, to the control of site-specific recombination

Table 7.2 Phenotypes associated with *hns* mutations in enteric bacteria

System	Effect of *hns* mutation
Escherichia coli	
Inducible *bgl* operon	Derepression of transcription
Expression of type 1 fimbriae	Accelerated site-specific recombination
Expression of type 1 fimbriae	Derepression of *fimA* transcription at low temperatures
Osmotically regulated *ompC ompF* genes	Altered osmoregulation
Thermoregulated Pap pilus expression	Derepression of transcription at low temperatures
Mutation rate	Increased mutation rate
Osmotically inducible transcription of *proU*	Derepression at low osmolarity
Extracellular polysaccharide synthesis	Increased production of extracellular polysaccharide at low temperatures
Thermoregulated plasmid-linked invasion genes in enteroinvasive *E. coli*	Increased transcription at low temperatures
Cfa/I adhesin expression	Increased transcription at low temperatures
Expression of curli adhesins	Activation of cryptic curlin gene
Motility	Loss of motility
Shigella flexneri	
Thermoregulated plasmid-linked invasion genes	Increased transcription at low temperatures
Salmonella typhimurium	
Balb/C mouse virulence	Loss of virulence
Osmotically inducible transcription of *proU*	Derepression at low osmolarity

systems, to the control of homologous (i.e. general) recombination and to the control of the mutation rate in bacteria (Barr *et al.*, 1992; Dri *et al.*, 1992; Higgins *et al.*, 1990; Lejeune & Danchin, 1990) (Table 7.2). It also has the ability to restore growth and to export protein when introduced in multicopy to an *E. coli* mutant deficient in *secY*, whose product is required to move secreted polypeptides across the bacterial cytoplasmic membrane (Ueguchi & Ito, 1992).

In cases where the effect of *hns* mutations on specific promoters has been investigated, the role of H-NS has generally been that of a repressor. However, two-dimensional gel analysis of H-NS$^+$ and H-NS$^-$ strains of *E. coli* has shown that, while many proteins are certainly induced in the *hns* deletion mutant, others are repressed strongly (Yamada *et al.*, 1991). Such effects are not, of course, necessarily due to direct effects of H-NS on the genes coding for these proteins.

The precise mechanism by which H-NS alters transcription is unknown. It has been found that the protein has a preference for curved DNA sequences, and since these are found frequently upstream of strong promoters (Chapter 6) it has been suggested that H-NS could repress transcription by altering the topology of these sequences (Owen-Hughes *et al.*, 1992; Yamada *et al.*, 1990). The observation that mutations in *hns* result in changes in the supercoiling of reporter plasmids suggests another mechanism by which the protein may affect transcription (Dorman *et al.*, 1990; Higgins *et al.*, 1988a). The protein may alter the local supercoiling of the promoter and so modulate transcription initiation, either by affecting polymerase binding or open complex formation. It may also influence the binding of specific regulators of transcription, be they activators or repressors. To do these things, H-NS may act alone, or act cooperatively with other general DNA-binding proteins (for example, the HU histone-like protein) or with a DNA topoisomerase.

Transcriptional regulation by the integration host factor

The integration host factor (IHF) has been described in Chapter 2. This site-specific DNA-binding and DNA-bending protein has been implicated as a transcriptional modulator of several genes in *E. coli* (reviewed in Freundlich *et al.*, 1992). Its potential as a transcription factor is made all the more compelling by its similarities to the eukaryotic transcription factor TFIID. Not only does TFIID bind to DNA in a similar manner to IHF (both interact with the minor groove of the helix), it also bends DNA when bound at the TATA boxes of promoters, perhaps bringing upstream-bound regulatory proteins into contact with RNA polymerase (Horikoshi *et al.*, 1992; Lee *et al.*, 1991; Nash & Granston, 1991). IHF seems to perform an analogous function for prokaryotic promoters recognized by the σ^{54} form of RNA polymerase. Here, an upstream-bound transcription activator is brought into contact with RNA polymerase when IHF binds and bends the intervening DNA sequence (Collado-Vides *et al.*, 1991) (Fig. 7.7). For example, IHF strongly enhances transcription from the nitrogen-regulated *glnH* P2 promoter by binding between the sites occupied by σ^{54} RNA polymerase and the NtrC transcription factor (Claverie-Martin & Magasanik, 1991). A similar mechanism has been proposed to explain the ability of IHF to enhance open complex formation *in vitro* at the NifA-dependent *nifH* promoter of *Klebsiella pneumoniae* (Hoover *et al.*, 1990; Santero *et al.*, 1992).

IHF can also modulate transcription from σ^{70} promoters, including those of the *ilvBN*, *ilvGMEDA*, *ompB*, *ompC*, *ompF*, and *topA* genes of *E. coli* (reviewed in Freundlich *et al.*, 1992) (Table 7.3). Thus it has the ability to link together the expression of many (unrelated) genes. In addition to effects on gene expression at the level of transcription initiation, IHF can also affect transcript elongation (as in the case of *ilvBN*; Tsui & Freundlich, 1989, cited in Freundlich *et al.*, 1992) and translation initiation (as in the case of the bacteriophage λ *c*II gene; Mahajna *et al.*, 1986).

Table 7.3 Integration host factor (IHF)-dependent systems

System	Function	Role of IHF
Site-specific recombination		
Lambda	Integration/excision	Positive
Type 1 fimbriae	Promoter fragment inversion	Positive
Transposable elements		
IS*10*	Transposition	Positive
Tn*10*	Transposition	Positive
Bacteriophage Mu	Transposition	Positive
Plasmids		
pSC101	Replication	Positive
Phage P1 plasmid lysogen	Partitioning	Positive
F	Conjugal transfer	Positive
R100	Conjugal transfer	Positive
Genes		
aceB	Acetate utilization	Positive
glnH (P_2 promoter)	Nitrogen assimilation	Positive
ilvBN	Ile, Val synthesis	Positive
ilvGMEDA	Ile, Val synthesis	Positive
nifH	Nitrogen assimilation	Positive
ompB	Porin gene regulation	Negative
ompC	Porin	Negative
ompF	Porin	Negative
pf1	Pyruvate formate lyase	Positive
pyrBI	Pyrimidine biosynthesis	Negative
sodA	Superoxide dismutase	Negative
tdcABC	Threonine deaminase	Positive
topA	Topoisomerase I	Negative

The Fur regulon

A genetic screen for mutations leading to the constitutive expression of the *fhuA* gene of *E. coli*, which encodes the outer membrane receptor for uptake of ferrichrome-like siderophores, led to the discovery of the genetic locus *fur* (ferric uptake regulation) (Hantke, 1981). *fur* mutants are derepressed for all of the iron uptake systems and the *fur* locus encodes a 17 kDa (148 amino acid protein), Fur, which acts as a transcriptional repressor of the genes of the iron assimilation pathways (Hantke, 1987; Schäffer *et al.*, 1985) (Appendix I). The Fur repressor acts with ferrous iron (Fe^{II}) as a corepressor and it binds to a specific operator sequence, the 'iron box', found within the promoter regions of Fur-regulated genes (Calderwood & Mekalanos, 1987; De Lorenzo *et al.*, 1987) (Fig. 7.12). Confirmation that the 'iron box' functions as an operator site for Fur came from an experiment in which a synthetic oligonucleotide corresponding to the iron box sequence was inserted into the promoter region of the *ompF* gene, causing expression of OmpF to become iron-regulated and Fur-dependent (Calderwood & Mekalanos, 1988). The iron-binding pocket of the Fur protein includes four metal-binding histidine residues in the carboxyterminal domain of the protein (Saito *et al.*, 1991). Iron binding here results in a conformational change which permits the aminoterminal domain to interact with the DNA of the iron box (Coy & Neilands, 1991; Saito & Williams, 1991).

pH regulation of gene expression

pH-regulated gene expression has been studied extensively in enteric bacteria. Results indicate that these organisms possess a number of pH-sensitive promoters and that several of these are also sensitive to changes in other environmental parameters. In *E. coli*, the *cad* operon (coding for lysine decarboxylase) is transcriptionally induced by external acidification via the product of the linked *cadC* gene. This gene encodes a *trans*-acting transcription activator which shows homology to the response regulator members of the histidine protein kinase/ response regulator two-component family of environmental signal transducers, and to ToxR, the transcription regulator of virulence gene expression in *Vibrio cholerae*, which is itself a transducer of pH stimuli (Watson *et al.*, 1992; Chapter 8). Another member of this family of signal transduction systems, EnvZ/OmpR, has also been described as regulating transcription in enteric bacteria in response to changes in external pH (see Chapter 8). It has been suggested that, like ToxR, CadC is a membrane-attached DNA-binding protein and that CadC undergoes a conformational change in response to external pH changes which

5′-GATAATGATAATCATTATC-3′

Fig. 7.12 The iron box. Consensus sequence of the binding site for the iron regulatory protein, Fur. The sequence is an inverted repeat (as indicated by the arrows).

results in modulation of its transcriptional activation capacity (Watson *et al.*, 1992). Expression of *cadBA* is also controlled in response to lysine by the unlinked regulatory locus *cadR* and another locus, *exaR*, is required for aerobic expression of *cad* to anaerobic levels. Mutations in *exaR* do not alter the acid induction of the locus. Furthermore, *exaR* is required for aerobic growth of *E. coli* below pH 6 (Slonczewski, 1992).

The CadC regulator is an example of a system for responding to changes in external pH. Systems for dealing with internal pH change have also been described. The inducible acid tolerance response of *Salmonella typhimurium* is one of these (described in Chapter 8). The sodium antiporter of *E. coli* is encoded by *nhaA* and is induced at high external pH in the presence of high levels of sodium (Karpel *et al.*, 1991). NhaA activity acidifies the cytoplasm, contributing to internal pH homoeostasis and this process may be regulated at the level of transcription by the LysR-like protein NhaR (not to be confused with another LysR-like regulator, NahR, of *Pseudomonas putida*) (S. Schuldiner & E. Padan, cited in Slonczewski, 1992).

The acid-inducible genes *adi* (arginine decarboxylase) and *cadA* (lysine decarboxylase) in *E. coli* are regulated negatively by the nucleoid-associated protein, H-NS (Shi *et al.*, 1993) showing a dependence on this pleiotropic regulator with roles in many environmental response pathways.

DNA supercoiling as a global regulator of transcription

In Chapter 2, the supercoiled nature of bacterial DNA was described, as were the mechanisms by which supercoiling levels are regulated. In Chapter 6, the importance of DNA supercoiling for transcription initiation was discussed. When the fact that DNA supercoiling levels vary in response to changes in the environment is taken into account, this topological parameter of DNA has the potential to be a very important global regulator of gene expression. For example, several lines of evidence indicate that a shift from growth in low osmolarity medium to a high osmolarity medium result in elevated levels of negative supercoiling in bacterial DNA. This includes a considerable increase in the unconstrained (i.e. protein-free, plectonemically wound) supercoils that have the free energy to drive structural transitions in DNA, including promoter melting at transcription initiation. Simplistically put, the change to the external environment resets the transcription potential of the nucleoid via this DNA topological modulation.

The mechanisms by which this modulation is exerted remain obscure and are still a matter of intense study (see Chapter 2). Nevertheless, certain trends have emerged that tell us something of the extent to which supercoiling contributes to global control. In general, it tends not to be the sole regulator of any gene; other, more specific regulatory factors are usually superimposed. Such statements need to be made with great caution since there have not been attempts to examine the problem systematically for a very large sample of promoters. In most cases, the genes described as being sensitive to changes in DNA supercoiling were initially identified as being regulated by environmental factors (such as osmotic stress or anaerobiosis) now known to alter supercoiling, or code for topoisomerases which themselves regulate DNA supercoiling. Nevertheless, the emerging picture is one in which DNA supercoiling levels appear to provide the background against which more specific control elements have to act.

An interesting example of specificity concerns promoters that respond to multiple environmental stimuli. The *tppB* promoter of *Salmonella typhimurium* is induced by anaerobic growth, suggesting a correlation between increased levels of DNA supercoiling (a consequence of anaerobiosis) and increased *tppB* transcription. (The *tppB* operon codes for an uptake system for tripeptides.) Since osmotic stress also increases the level of negative supercoiling in DNA, one might expect a synergistic effect if the two environmental stimuli are combined; this is the case (Ní Bhriain *et al.*, 1989). However, exerting osmotic stress alone is not sufficient to induce the *tppB* promoter; anaerobic growth conditions are obligatory for this induction. How can this be if the *tppB* promoter responds to DNA supercoiling in a simple manner? The promoter ought to be induced regardless of the *source* of increased supercoiling.

The *proU* promoter of *S. typhimurium* is induced dramatically by osmotic stress, a condition that elevates DNA supercoiling levels. One might expect that anaerobic growth would induce this promoter, since this stimulus has a similar effect on supercoiling. A synergistic effect is indeed seen when the two stimuli are coimposed. However, osmotic stress is essential for *proU* induction; anaerobiosis alone cannot induce the promoter (Ní Bhriain *et al.*, 1989). Once again, the promoter displays a need for a specific stimulus; a simple correlation with increased supercoiling levels from *any* source is not observed. Furthermore, in *E. coli*, the anaerobic synergy is not detectable for the *proU* promoter; here the dependence on a specific stimulus is absolute.

How do we know that DNA supercoiling has any functional relationship with the expression of the *tppB* or the *proU* systems? This evidence comes from experiments with DNA gyrase-inhibiting drugs. In the case of *tppB*, the ability of osmolarity to enhance transcription is abolished in cells treated with novobiocin (Ní Bhriain *et al.*, 1989). Similarly, the osmotic induction of *proU* is prevented by novobiocin treatment (Higgins *et al.*, 1988a). This evidence supports the idea that the increase in DNA supercoiling levels seen in osmotically stressed cells requires an active DNA gyrase. Is gyrase required for anaerobically induced levels of supercoiling? Experiments with gyrase inhibitors in anaerobic cultures of *E. coli* suggest that it is (Dorman *et al.*, 1988). The data of Yamamoto & Droffner (1985) indicate that anaerobic growth increases DNA supercoiling levels by extinguishing topoisomerase I activity, giving DNA gyrase a free hand in the cell. It is not clear how DNA topoisomerase activity would be abolished anaerobically. However, an increase in gyrase activity and a decline in DNA topoisomerase I activity under anaerobic growth conditions could certainly account for the observed effects on DNA supercoiling. The work of Drlica and colleagues shows a correlation between increased gyrase activity under anaerobic conditions and an increase in the [ATP]/[ADP] ratio under these conditions. This could provide a mechanism for gyrase stimulation in anaerobic cultures (Hsieh *et al.*, 1991a).

A system with a sensitivity to DNA supercoiling but without any particular dependency on osmotic stress or anaerobiosis for expression can be used to learn more about how these environmental parameters affect gene expression in a global way. The *his* operon of *S. typhimurium* is induced when DNA gyrase is inhibited (Rudd & Menzel, 1987). This suggests that increasing the level of supercoiling by *any* means (gyrase stimulation or topoisomerase inhibition) ought to repress *his* expression. This is the case; inactivation of the gene coding for topoisomerase I (increases supercoiling) represses *his;* osmotic stress (increases supercoiling) represses *his;* anaerobic growth (increases supercoiling) represses *his* (O'Byrne *et al.*, 1992). This system, which exhibits a simple response to supercoiling *per se*, is repressed by any treatment that raises supercoiling levels. This is in contrast to true anaerobically induced promoters such as *tppB*, or genuine osmotically regulated promoters such as *proU* which require a topological input by a specific route. This specificity may have to do with the presence of binding sites for particular topoisomerases in the vicinity of the promoter, a further example of the importance of local supercoiling, as opposed to global supercoiling (as in the case of *his*). It is clear that a great deal of work remains to be done before this mechanism of gene regulation can be fully understood.

8 Virulence Gene Regulatory Networks

Introduction

Previous chapters have illustrated the array of mechanisms used by bacteria to govern the expression of their genetic material. In this chapter, the application of those mechanisms to the control of virulence gene expression will be described. It will become apparent that environmental responses form an important theme in their control. In most cases, gene expression will be seen to be controlled by multiple regulators and frequently, the expression of particular virulence genes will be seen to be networked to that of others, not all of which have a direct or obvious contribution to make to pathogenicity. This has intuitive appeal since the bacteria must deal with a complex environment and are required to optimize the responses of their housekeeping genes as well as their virulence genes if they are to be successful. Thus, it is probably more helpful to think of virulence gene regulation in the context of the general response to environmental change than to regard these genes as independent entities controlled only by their own private regulatory systems.

Environmental stimuli inform the bacterium about its surroundings in physical (temperature, pH, osmolarity, etc) and chemical (nutrient availability, presence of toxic substances, etc.) terms. The organism must mount a series of responses at the appropriate time and at the appropriate level if it is to survive. This involves modulating the gene expression programme within the cell. The control of virulence factor expression appears to be integrated with these responses. Many factors required for infection are under environmental control and are encoded by genes whose expression is modulated in response to changes in temperature, pH, osmolarity, anaerobiosis, the presence or absence of specific chemicals and other stimuli. Frequently, the genes respond to more than one stimulus, indicating that more than one regulatory circuit is operating. This would provide a useful way to optimize expression at a particular point in the infection process. Thus, if the cytoplasm of a host cell imposes a set of stresses not found outside, these would make excellent stimuli for the induction of genes needed to survive in the intracellular rather than the extracellular situation. If individual stresses from the 'intracellular set' are encountered singly, the virulence gene may not be induced, or at least not to the same level as is achievable under truly intracellular conditions. In this way, the metabolic drain of producing virulence factors at an inappropriate site may be diminished, if not avoided. Useful reviews on the topic of virulence gene regulation by environmental stimuli include Gross (1993); Mekalanos (1992); Miller *et al.* (1989a); Dorman, (1991).

Osmotically regulated virulence genes

Given the fluctuations in water activity characteristic of life within and movement between those environments that bacteria inhabit both in association with the host and when free-living, it is not surprising to discover that osmotic regulation of virulence gene expression is a commonly encountered theme among organisms of most genera. In many cases, the environmental signal transducer responsible for the regulation of the osmotically sensitive gene(s) has been found to be multifunctional. For example, ToxR of *Vibrio cholerae* is a versatile regulator, transducing not only osmotic signals but also information about pH, temperature and the availability of certain amino acids (see later). Similarly, the EnvZ/OmpR 'two-component' osmoregulator system of enteric bacteria, which has been implicated as playing an important role in the virulence of *Salmonella typhimurium* and *Shigela flexneri*, is also concerned with the transduction of other signals (see later). Osmolarity is an important signal in the highly complex regulatory cascade concerned with controlling expression of genes involved in alginate synthesis in *Pseudomonas aeruginosa*. Once again, a 'two-component' system has been described as contributing to this environmental control (see later).

Thermoregulated virulence genes

Temperature is one of the few environmental parameters that have been demonstrated as being significant *in vivo* as regulators of bacterial virulence (Smith, 1990). Genetic analysis, often using reporter gene fusions, has permitted temperature-regulated promoters (as opposed to 'heat shock'-regulated promoters) to be detected in many pathogens. This, in turn, has allowed regulatory genes to be identified and characterized. Despite the growing body of data on the subject, a definitive description of the precise mechanism by which a change in temperature causes a change in the profile of bacterial gene expression remains elusive. As with other environmental parameters believed on the basis of *in vitro* data to be important in gene regulation during infection, temperature changes can result in changes in DNA topology. This raises the possibility that temperature, like osmolarity, anaerobiosis, nutrient availability, growth phase, et cetera can contribute to the determination of the level of supercoiling in bacterial DNA with all the predictable consequences for promoter function. In keeping with the general theory of how global changes in supercoiling can affect gene expression (discussed in Chapters 2 and 7), one would expect to find promoter-specific regulators in the cell with roles in regulating transcription in response to temperature change. This is the case in many of the organisms in which temperature regulation has been studied at the molecular level.

Frequently, members of the AraC family of transcription factors have been found to be required for temperature-controlled expression of virulence genes. These genes include invasion functions in *Shigella* species, the Yop regulon genes of *Yersinia* species and colonization factor genes of *E. coli* (discussed later). In addition, a role for the nucleoid-associated protein H-NS (Chapter 2) or a histone-like analogue (such as YmoA in *Yersinia enterocolitica*) has been described in many of these systems. The porin genes *ompC* and *ompF*, which contribute to virulence in *Salmonella typhimurium*, are regulated in part by temperature. Their promoters are subject to regulation by H-NS and the thermal control has been shown genetically to require the EnvY protein. Analysis of the EnvY amino acid sequence reveals that it is related to AraC (Klaasen & De Graaf, 1990; Lundrigan *et al.*, 1989; Fig. 7.10). Despite some early suggestions that EnvY might be a membrane-associated protein (Lundrigan & Earhardt, 1984), the most likely situation is that it is a cytoplasmically located DNA-binding protein. H-NS has also been implicated in the control of other systems (such as *E. coli pap* and *fim* gene expression) where thermal regulation has been found but no AraC family member has yet been discovered.

In *Yersinia enterocolitica*, *Y. pestis* and *Y. pseudotuberculosis*, transcription of the plasmid-encoded Yop regulon is positively regulated by the VirF protein in response to increases in temperature (and negatively regulated by calcium via a VirF-independent mechanism) (reviewed in Cornelis, 1992). However, *virF* is itself thermoregulated at the level of transcription. This regulation is partly abolished by mutations in the gene coding for the histone-like protein YmoA. In the *ymoA* mutant, VirF is expressed at the non-permissive temperature, but its ability to activate the Yop regulon is still enhanced by increased temperature. Overexpression of VirF at low temperature does not result in Yop regulon expression, the temperature signal is also required. VirF binding to DNA does not appear to be thermoregulated; the protein binds at both the permissive and the non-permissive temperature (Lambert *et al.*, 1992). Thus, VirF and temperature probably act cooperatively yet distinctly in Yop regulon induction (Fig. 8.1). These observations are consistent with a model in which the topology of the DNA at the Yop regulon promoters is set by the temperature, allowing the DNA-bound transcription factor (VirF) to activate transcription. They are also consistent with a model in which the topology of the promoter region is set, at least in part, by the YmoA protein (Cornelis, 1992).

A similar situation obtains in *Shigella flexneri* in that invasion gene expression is thermally induced at the level of transcription through a mechanism that involves cooperation between chromatin structure (as determined by protein H-NS) and positive activators (another AraC homologue called VirF and a second transcription activator called VirB) (reviewed in Hale, 1991). Here, VirF is required for transcription of *virB* and the VirB protein activates the invasion regulon (Tobe *et al.*, 1991). Overexpression of *virB* using an IPTG-inducible promoter results in expression of invasion genes (such as *ipaB*, *ipaC* and *ipaD*) at either permissive or non-permissive temperatures. This indicates that activation

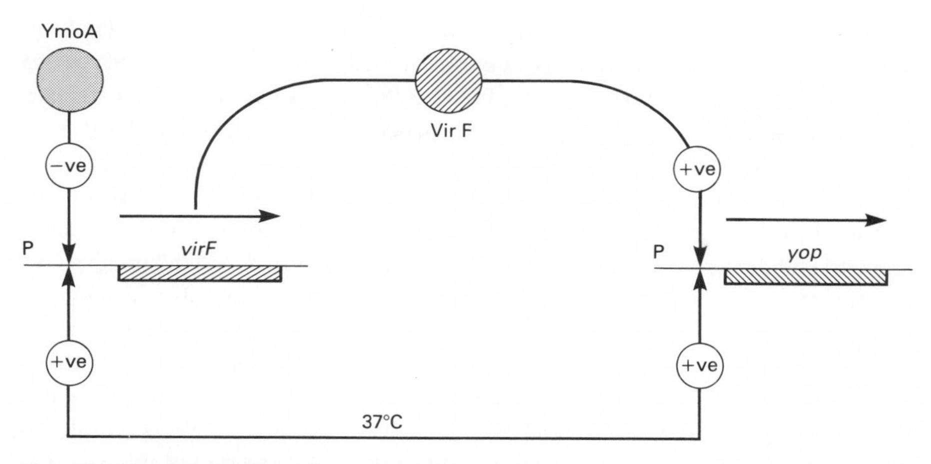

Fig. 8.1 Thermoregulation of Yop expression in *Yersinia enterocolitica*. The regulatory gene *virF* is transcriptionally activated at 37°C. This activation is modulated negatively by YmoA. The AraC-like protein VirF then activates transcription of the *yop* structural gene at 37°C.

of *virB* transcription by temperature and by VirF represents the main site at which the invasion regulon is controlled by changes in temperature (Fig. 8.2). The role of the H-NS protein in invasion gene thermoregulation is less clear. Mutants deficient in this protein express the invasion regulon at the non-permissive temperature but (as with *virF* gene expression in *ymoA* mutants of

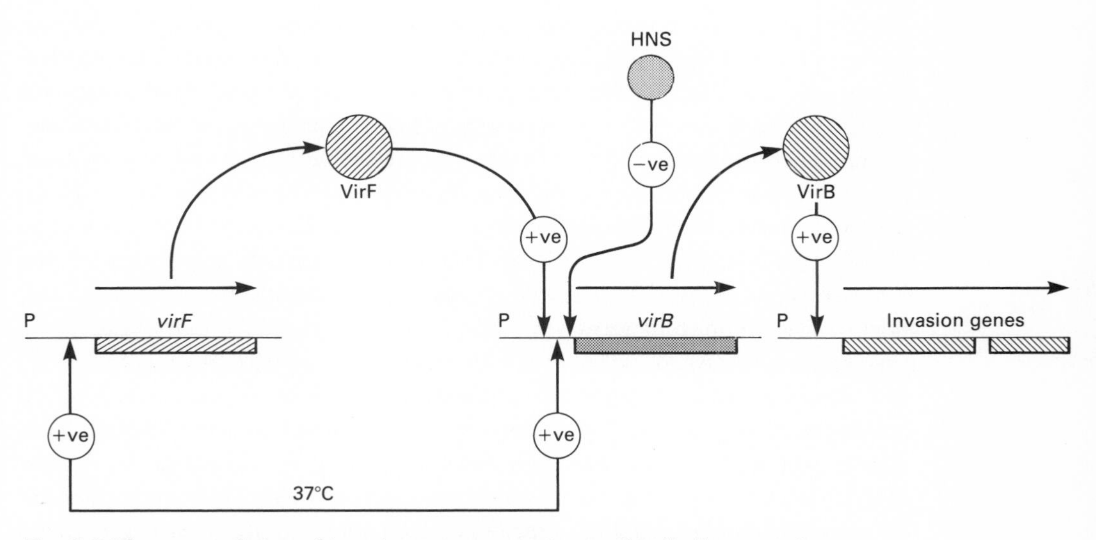

Fig. 8.2 Thermoregulation of invasion gene expression in *Shigella flexneri* and enteroinvasive *Escherichia coli*. The expression of the invasion gene regulon is subject to control by a regulatory cascade. The AraC-like protein VirF activates transcription of *virB* at 37°C. This activation is antagonized by the H-NS protein. Once expressed, VirB activates transcription of the invasion genes.

Yersinia) higher levels of expression are still achievable by growing the bacteria at the permissive temperature. Presumably, H-NS plays an important role in the determination of the topology of DNA at the promoters involved. An attractive possibility is that the *virB* promoter is sensitive to H-NS-induced changes in topology and that its activation is achieved by a combination of VirF acting in concert with DNA topology changes that are brought about by fluctuations in temperature. According to this model, mutations in *hns* mimic the thermally induced changes in the topology of the DNA (Dorman *et al.*, 1990). The invasion regulon of enteroinvasive *E. coli* (EIEC) is organized in a similar manner to that of *S. flexneri* (and includes a common nomenclature for the genes and their products) and here, changes in the intracellular levels of H-NS mimic the effects of temperature (Fig. 8.2). As with *S. flexneri*, the primary regulatory site for the H-NS input seems to be at the VirF-dependent *virB* promoter. This is consistent with the postive regulator VirF having a role antagonistic to that of the negative regulator H-NS at this promoter (Dagberg & Uhlin, 1992). Similar themes can be found in *E. coli* where colonization factor antigen expression is under temperature control at the level of transcription and is regulated by an AraC-like protein (CfaD) and the H-NS protein (see later).

Other mechanisms exist in pathogenic bacteria for regulating gene expression in response to temperature, although detailed molecular explanations are generally unavailable. Temperature control of virulence gene expression in *Bordetella* species is achieved via the BvgA/BvgS two-component signal transduction system, which also transduces other environmental signals such as changes in the concentrations of nicotinic acid and magnesium sulphate (see later). Temperature control of virulence factor expression in *Vibrio cholerae* involves the ToxR system (Miller *et al.*, 1989c). Again, this is an example of a system that is responsible for the transduction of several signals to the regulon. The *V. cholerae* system differs from others in that the genes under its control are optimally expressed below 37°C. In this way they resemble the low-temperature expression characteristic of the *inv* gene (encoding the 92 kDa protein 'invasin') in *Yersinia pseudotuberculosis* (Isberg *et al.*, 1988) and *Y. enterocolitica* (Pierson & Falkow, 1990; Young *et al.*, 1990). This low-temperature expression of *inv* correlates with the observation that *Y. enterocolitica* strains grown at 30°C are more invasive than those grown at 37°C. This is complicated by the existence of a second invasin, Ail (17 kDa), in *Y. enterocolitica* whose expression is higher at 37°C than at lower temperatures (Miller *et al.*, 1990b). Expression of *Y. enterocolitica* enterotoxin (Yst) is also thermoregulated in a counterintuitive manner; the toxin is only found in the supernatant of cultures grown below 30°C (Cornelis, 1992). Such apparently inappropriate expression may point to the existence of other regulatory factors which come into play *in vivo* but which are not observed in the *in vitro* conditions under which the gene expression experiments are carried out. This indicates the need to be cautious when attempting to apply data from such *in vitro* work to the *in vivo* situation.

Thermoregulation of transcription is not confined to virulence genes in

Gram-negative bacteria. In the Gram-positive pathogen *Listeria monocytogenes*, expression of genes required for the infection process is controlled by temperature via the transcription regulatory protein PrfA (Leimeister-Wächter *et al.*, 1992). For example, the *inl* locus, which encodes invasion functions, is activated at the level of transcription by PrfA in response to elevated temperature. It is also controlled by growth phase, illustrating yet again the very common theme of multifactorial control of bacterial virulence gene expression (Dramsi *et al.*, 1993).

The heat shock response and virulence gene expression

As stated in the section on Regulons and Stimulons in Chapter 7, 'heat shock response' is really a misnomer that has persisted, for historical reasons, into a period in which it has become recognized generally that this response is really concerned with protecting the cell from many forms of environmental stress. In addition, the response overlaps with other genetic networks concerned with survival in difficult circumstances, such as the SOS response and the stringent response. Since many pathogens pursue a way of life that exposes them to many forms of environmental stress, it comes as no surprise that the stress response systems turn out to be important for virulence in many bacteria. This is because of their general protective properties and/or because of their roles in regulating the expression of specific virulence genes.

Heat shock proteins are ubiquitous in prokaryotes and eukaryotes and several play important roles in bacterial virulence (Murray & Young, 1992). In *Coxiella burnetti*, the Q fever antigens HtpA and HtpB, encoded by the *htpAB* operon, are homologues of the *E. coli* GroES and GroEL proteins, respectively (Hemmingsen *et al.*, 1988; Vodkin & Williams, 1988). In *Mycobacterium tuberculosis* and *M. bovis*, the 12 kDa BCGa antigen is closely related to GroES (Young *et al.*, 1988). The 71 kDa antigen of *M. tuberculosis* is a target for the immune response during infection and the 70 kDa antigen of *M. leprae* cross-reacts immunologically with the *E. coli* DnaK protein (Mehlert & Young, 1989; Young *et al.*, 1988). Homologues of DnaJ have also been found in *Mycobacterium* species (Lathigra *et al.*, 1988). In *Borrelia burgdorferi* (the spirochaetal agent causing Lyme disease) genes coding for members of the Hsp60 and Hsp70 families of heat shock proteins have been characterized, including genes coding for DnaJ , DnaK and GrpE homologues (Tilly *et al.*, 1993). The *B. burgdorferi dnaJ* gene can be transcribed either alone or as part of a polycistronic message with the *hsp70* (*dnaK* homologue) gene (Anzola *et al.*, 1992a,b). This operon arrangement is reminiscent of the situation with *dnaK* and *dnaJ* in *E. coli*.

Expression of ToxR in *Vibrio cholerae* is modulated by the heat shock response (see later) and heat shock proteins are induced in *Salmonella typhimurium* as a result of invasion of eukaryotic cells, indicating that these may be important to the bacterium during intracellular life (Buchmeier & Heffron, 1990; Parsot & Mekalanos, 1991b). Similarly, a combination of heat shock and oxidative stress induces expression of the virulence gene coding for listeriolysin (which is necessary for intracellular growth) in *Listeria monocytogenes* (Sokolovic & Goebel, 1989).

As was noted in Chapter 7, enteric bacteria possess a second heat shock regulon whose best-characterized member is the *degP* (or *htrA*) gene of *E. coli*. This gene codes for a periplasmic serine protease which degrades abnormally folded proteins (Lipinska *et al.*, 1990; Strauch & Beckwith 1988). Interestingly, the *htrA* gene of *Salmonella typhimurium* has been shown to be essential for virulence in the BALB/c mouse model system (Johnson *et al.*, 1991).

Virulence gene regulation by carbon dioxide

Among the many environmental stresses that a host may exert on a bacterium is exposure to carbon dioxide. Several pathogens possess virulence genes whose expression is induced on encountering this stimulus. Presumably, it plays an important role in signalling to the organism that it has arrived in a niche where virulence factor production is appropriate. The Gram-positive pathogen *Bacillus anthracis*, the causative agent of anthrax, harbours two virulence plasmids which encode its major virulence determinants (Chapter 3). Growth medium composition influences the expression of the capsule (encoded by 90 kb plasmid pXO2) and the three toxin components (Cya, Lef and Pag, encoded by 174 kb plasmid pXO1). Toxin yields are low in rich media but high in minimal medium in the presence of CO_2 (Bartkus & Leppla, 1989). Similarly, the expression of the plasmid-linked *B. anthracis* capsule genes is induced by CO_2 (Makino *et al.*, 1988). CO_2 also plays a stimulatory role in the expression of genes involved in enterotoxin production in *Vibrio cholerae* (Shimamura *et al.*, 1985) and toxic shock syndrome toxin 1 (TSST-1) production in *Staphylococcus aureus* (Kass *et al.*, 1987).

Anaerobic regulation of gene expression

Anaerobic regulation of gene expression has been described in several pathogenic bacteria. In *Neisseria gonorrhoeae*, a 54 kDa lipoprotein called Pan1, with a possible role in colonization of anaerobic niches within the human host, is encoded by an anaerobically induced gene, *aniA* (Hoehn & Clark, 1992a, b). Transcript analysis reveals that *aniA* may be expressed from either a σ^{70}-dependent promoter or from a promoter of the 'gear-box' class. These promoter regions overlap in the upstream sequence of the *aniA* gene (Hoehn & Clark, 1992a). *N. gonorrhoeae* cannot grow anaerobically by fermentation and it uses nitrite as its sole terminal electron acceptor in anaerobic respiration, which led Clark (1990) to suggest that this organism is unlikely to possess an extensive regulon of anaerobically induced genes.

Anaerobic growth has been shown in independent studies to enhance the attachment and invasiveness of *Salmonella typhimurium* in mammalian cell lines *in vitro* (Ernst *et al.*, 1990; Francis *et al.*, 1992; Lee & Falkow, 1990; Lee *et al.*,

1992; Schiemann & Shope, 1991). This suggests that an anaerobic microenvironment at the site of invasion exerts environmental stimuli required to induce the invasive phenotype. Given that in *S. typhimurium*, anaerobic growth results in increased levels of DNA supercoiling, that osmotic stress alters invasion gene expression (Galán & Curtiss, 1990), that some enteric promoters display an overlap in sensitivity to both anaerobiosis and osmolarity (Ní Bhriain *et al.*, 1989), it is possible that anaerobic growth may collaborate with other environmental stimuli (such as osmolarity, pH and temperature, see the relevant sections in this chapter) to alter the transcription of genes required for invasion by changing the topology of the DNA at their promoters.

pH regulation of gene expression

Research by Foster and coworkers has elucidated some of the detail of pH-regulated gene expression in *Salmonella typhimurium*. This bacterium possesses an acid tolerance response which is triggered by growth at external pH values of 5.5–6. Induction of the response protects the cell against even lower pH values and may be regarded as an acid adaptation mechanism (Foster & Hall, 1990). This mechanism seems to be distinct from the oxidative stress response, the SOS response and the heat shock response. Studies carried out in *E. coli* have detected overlaps between the pH response and other stress response regulons. Among the proteins induced following a shift from pH 6.9 to 4.3 are the heat shock proteins DnaK, GroEL, HtpG and HtpM. Other acid-induced proteins in *E. coli* are also responsive to osmolarity or anaerobiosis or low temperature. One 'acid shock' protein is subject to control by RpoH (i.e. it is another, unidentified heat shock protein) (Heyde & Portalier, 1990).

Acid adaptation is likely to be very important to *Salmonella* species in both the free-living and pathogenic situations. In the latter case, the organism clearly possesses the means to survive in acid-stressful niches such as phagolysosomes. Interestingly, a *phoP* mutant of *S. typhimurium* is a 1000-fold less acid resistant than its *phoP*$^+$ parent, showing a link between a known virulence gene (*phoP*) and acid adaptation (Foster & Hall, 1990). Two-dimensional gel analysis shows that at least 18 proteins are altered in expression (6 repressed and 12 induced) as a result of an acid shift, indicating that the acid tolerance response involves a complex regulon. This has prompted a search for regulatory mutations. Strains deficient in the Mg^{2+}-dependent proton-translocating ATPase (*unc* mutants) cannot protect themselves against acid stress and in a constitutively acid tolerant mutant (*atr*), several acid tolerance proteins are overexpressed (Foster & Hall, 1990, 1991).

The genes coding for the major porins OmpC and OmpF in *E. coli* are subject to pH regulation at the level of transcription and this control requires the *ompB* locus (Heyde & Portalier, 1987). The two-dimensional electrophoretic analysis of Foster & Hall (1990) identified OmpC as one of the acid-inducible proteins in *S. typhimurium*; one of the acid-repressible proteins was identified as OmpF. Thus, as with osmolarity, pH regulation results in differential expression of

OmpC and OmpF. However, the acid tolerance response seems to be distinct from specific porin regulon control; an *ompR* : : Tn5 mutant displays the same enhanced acid tolerance as a wild-type strain following adaptation and acid challenge. This shows that *ompR* is not required for acid survival and suggests that the porins are not required either, at least under the experimental conditions used (Foster & Hall, 1990). Anaerobiosis has been identified as a regulator of porin gene expression in *S. typhimurium* (Ní Bhriain *et al.*, 1989) and other examples of genes subject to pH and anaerobic control have been reported. The *aniG* locus in *S. typhimurium*, which is transcriptionally regulated by the *earA* locus (Foster & Aliabadi, 1989), is anaerobically induced and is subject to control by pH (Aliabadi *et al.*, 1988). Interestingly, mutations in the *earA* locus result in a loss of pH control of *aniG* but not of anaerobic control, indicating that these environmental inputs are transmitted by distinct routes. Another regulatory overlap in *S. typhimurium* involves pH control of gene expression via the iron regulator, Fur. Mutants deficient in Fur exhibit an acid-sensitive phenotype and are unable to mount an acid tolerance response. However, altering the iron supply itself has no effect on acid tolerance response in a Fur^+ strain, suggesting that the role of Fur in the pH response is iron-independent (Foster & Hall, 1992). What this role might be is not clear.

pH has been identified as a regulator of virulence gene expression in plant pathogens. For instance, the *hrp* genes of phytopathogens influence the ability of the bacteria to cause disease in susceptible plants and to provoke a hypersensitive response in resistant plants (Lindgren *et al.*, 1986). Expression of some of these genes in *Pseudomonas syringae* pv. *phaseolicola* is regulated by pH (Rahme *et al.*, 1992). pH has also been identified as a regulator of *vir* gene expression in the plant pathogen *Agrobacterium tumefaciens* (Winans, 1990). However, in this case it is just one of many stimuli that exert an influence. *A. tumefaciens* seems to modulate *vir* transcription in response to a wide range of environmental inputs, reflecting the complexity of the plant–microbe interaction.

In the Gram-positive pathogen *Staphylococcus aureus*, expression of the pleiotropic regulator *agr* is decreased by alkaline pH. Using *agr*-encoded RNA III as a reporter, it can be shown that the locus is expressed maximally at pH 7.0 and is strongly repressed at pH 8.0. Expression of the *agr*-dependent enterotoxin C gene, *sec*, is also negatively regulated by alkaline pH, but only in strains wild-type for the *agr* locus, showing that the Agr system is required to transmit the pH signal to the *sec* gene promoter (Regassa & Betley, 1992). This shows that like the Vir 'two-component' regulatory system in *A. tumefaciens*, the *agr* 'two-component' system of *S. aureus* takes multiple environmental soundings and transmits them to its subservient genes. Both of these pleiotropic regulatory systems are discussed in detail elsewhere in this chapter.

Calcium regulation of gene expression in *Yersinia* species

Virulent Yersiniae exhibit a restricted growth phenomenon *in vitro* known as calcium dependency. In the absence of calcium in the millimolar range, Yersiniae will stop growing if shifted from 28 to 37°C. A return to the lower temperature or the addition of calcium to the medium will allow growth to resume. The *in vivo* relevance of calcium in *Yersinia* pathogenicity is difficult to assess. It has been pointed out that differential Ca^{2+} concentrations could enable the bacterium to distinguish between the extracellular mammalian environment (millimolar calcium levels) and the intracellular environment (micromolar calcium levels) (Brubaker, 1983). However, it has emerged that an intracellular niche is not essential for Yop production, i.e. these proteins are produced in a host environment that ought not to favour their expression (reviewed in Cornelis, 1992).

The presence of calcium ions has a repressive effect on transcription of the thermoregulated plasmid-encoded-*yop* regulon but how this is achieved is not really understood. The products of the plasmid-encoded-*car* operon are somehow involved. In the presence of Ca^{2+}, a product of the plasmid-encoded *car* operon (*lcrGVHyopBD*, coding for cytoplasmic regulatory functions — LcrG, LcrH — or secreted functions — LcrV, YopB, YopD) may inhibit the ability of the *lcrH* repressor gene or its product to repress *yop* transcription (Bergman *et al.*, 1991) (Fig. 3.16). The issue of how Ca^{2+} may influence genetic events is problematic since Yersiniae do not accumulate this cation (Perry & Brubaker, 1987). Models based on Ca^{2+} interaction with a surface receptor have been proposed. This candidate receptor is the secreted protein, YopN (Forsberg *et al.*, 1991). Unfortunately, YopN has yet to be shown to bind Ca^{2+} or to bind to the surface of the bacterium. Furthermore, the DNA-binding activity of the repressor LcrH remains to be established, as does a link between YopN and LcrH (see Cornelis, 1992, for a detailed discussion of these issues). Thus, despite much elegant research, the true physiological role of calcium in controlling virulence gene expression in the Yersiniae is still not fully understood.

Regulation of virulence gene expression during differentiation in *Proteus mirabilis*

Proteus mirabilis is a Gram-negative uropathogen which elaborates extracellular haemolysin and metalloprotease and intracellular urease as virulence determinants. This bacterium is unusual in that it undergoes a form of differentiation in which vegetative cells become filamentous, multinucleate and hyperflagellate and begin to exhibit a form of behaviour known as 'swarming'. Swarming involves not individual cells but a coordinated movement of a multicellular population across the surface of the substratum. It has been discovered that high-level expression of the virulence determinants correlates with the switch from the vegetative phase of life to the swarmer phase and mRNA analysis indicates that the virulence genes are regulated primarily at the level of transcription (Allison *et al.*, 1992b). Furthermore, invasion of urothelial cells by *P. mirabilis* is linked to motility and to swarming (Allison *et al.*, 1992a).

The nature of the regulatory mechanism and the signals involved remains obscure.

Histidine protein kinase/response regulator family members

These signal transduction systems were introduced in Chapter 7. They consist of a sensor protein which perceives fluctuations in one or more environmental parameters. This protein is a histidine protein kinase and it autophosphorylates in response to its cognate stimulus. These kinases are often, but not always, cytoplasmic membrane proteins (Fig. 8.3). The second part of the system is a response regulatory protein which, on receipt of a phosphate group from the kinase, alters a cellular function, which is usually (but not exclusively) gene transcription (Figs 8.4 & 8.5). Here, examples of these systems which regulate bacterial virulence genes will be described.

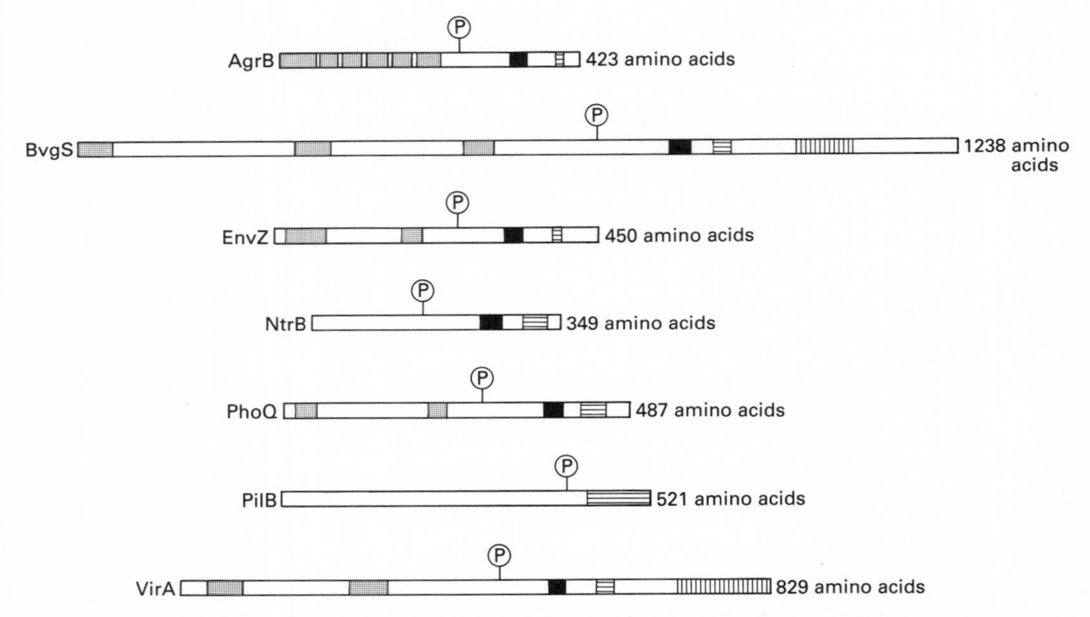

Fig. 8.3 The structures of some histidine protein kinases which have been characterized in great detail at the molecular level or which contribute to the control of bacterial virulence genes are summarized. These are AgrB (*Staphylococcus aureus*), BvgS (*Bordetella pertussis*), EnvZ (*Escherichia coli*, *Salmonella typhimurium*), NtrB (*Escherichia coli*), PhoQ (*Salmonella typhimurium*), PilB (*Neisseria gonorrhoeae*) and VirA (*Agrobacterium tumefaciens*). Shading in the aminoterminal domain indicates transmembrane sequences (AgrB, BvgS, EnvZ, PhoQ and VirA); the P symbol indicates the phosphorylation site (a conserved histidine residue); the filled and horizontally hatched boxes indicate further regions of conserved sequence; the vertical hatching in the carboxyterminal domains of BvgS and VirA show sequences with homology to the signal receiver domains found in response regulatory proteins (see Fig. 8.4 and Appendix II).

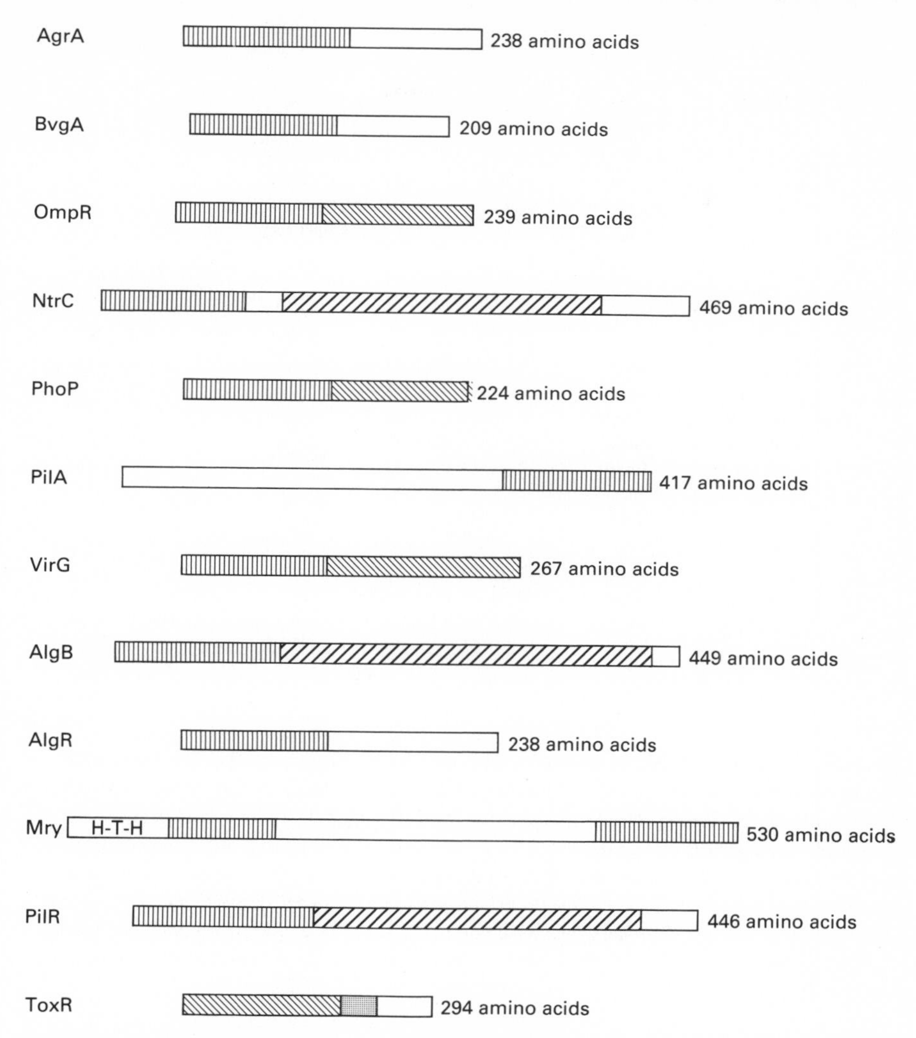

Fig. 8.4 Structures are summarized for AgrA (*Staphylococcus aureus*), BvgA (*Bordetella pertussis*), OmpR (*Escherichia coli, Salmonella typhimurium*), NtrC (*Escherichia coli*), PhoP (*Salmonella typhimurium*), PilA (*Neisseria gonorrhoeae*), VirG (*Agrobacterium tumefaciens*), AlgB, AlgR (*Pseudomonas aeruginosa*), Mry (*Streptococcus pyogenes*), PilR (*Pseudomonas aeruginosa*) and ToxR (*Vibrio cholerae*). The signal receiver domain is represented by vertical hatching and is found in all response regulators (except ToxR) at the aminoterminus (except PilA). Mry is exceptional in that it has two signal receiver domains, with the aminoterminal one being prefaced by a helix-turn-helix (H-T-H) DNA-binding motif. In general, the DNA-binding domains of the response regulators are found within the carboxytermini and differences among conserved sequences here allow the family to be further subdivided. Thus one subgroup consists of AgrA, BvgA and AlgR; another is composed of OmpR, PhoP, VirG and ToxR; while a third group contains NtrC, AlgB and PilR. ToxR is unique in possessing a transmembrane domain (shaded section). (See Appendix I).

(a)

S. aure	AgrB	KFRHDYVNILT
B. pert	BvgS	TMSHEIRTPMN
E. coli	EnvZ	GVSHDLRTPLT
K. pneu	NtrB	GLAHEIKNPLG
S. typh	PhoQ	DLTHSLKTALA
N. gono	PilB	AFSHEYDH–LF
A. tume	VirA	GIAHEFNNILG
		*

(b)

S. aure	AgrA	NAKNMNDIGCYF
B. pert	BvgA	FDVVITDCNMPG
E. coli	OmpR	FHLMVLDLMLPG
K. pneu	NtrC	GDLVITDVVMPD
S. typh	PhoP	PDIAIVDLGLPD
N. gono	PilA	VGEGIDDLRPFD
A. tume	VirG	VDVVVVDLNLGR
P. aeru	AlgB	FDLCFLDLRLGE
P. aeru	AlgR	PDIVLLDIRMPG
S. pyog	Mry	GYSAVYDNKKTS from N terminal domain
S. pyog	Mry	YDVIVTDVMVGK from C terminal domain
P. aeru	PilR	FDLCLTDMRLPD
		*

Fig. 8.5 Conserved domains in two component family members. (a) The conserved phosphorylation sites (*) (histidine residue underlined) in the histidine protein kinases. The structures of these proteins are summarized in Fig. 8.3. (b) The conserved phosphoacceptor sites (*) (aspartic acid residue underlined) in the response regulators. The structures of these proteins are summarized in Fig. 8.4. The sequences are from *Agrobacterium tumefaciens* (*A. tume*), *Bordetella pertussis* (*B. pert*), *Escherichia coli* (*E. coli*), *Klebsiella pneuomoniae* (*K. pneu*), *Neisseria gonorrhoeae* (*N. gono*), *Pseudomonas aeruginosa* (*P. aeru*), *Salmonella typhimurium* (*S. typh*), *Staphyloccus aureus* (*S. aure*) and *Streptococcus pyogenes* (*S. pyog*).

The OmpR/EnvZ system and virulence in enteric bacteria

The *ompB* locus, which encodes OmpR and EnvZ, is required for mouse virulence in *S. typhimurium*. It was originally characterized as a regulator of housekeeping gene expression and, for this reason, it is described in detail in Chapter 7. *S. typhimurium* strains with null mutations in *ompB* are attenuated in BALB/c mice and the inoculated mice become protected against oral challenge with high numbers (up to 10^{10}) of virulent, wild-type *S. typhimurium* cells (Dorman *et al.*, 1989a). Since *ompB* null mutants fail to express OmpC or OmpF porins, it might be expected that strains harbouring null mutations in both of the genes coding for these proteins would also be attenuated; this is the case (Chatfield *et al.*, 1991). However, since expression of OmpF should be environmentally repressed in the animal, it might be expected that a mutant carrying only a null allele of *ompC* would also be attenuated; this is not so. Presumably, OmpC is expressed at important interfaces between the bacterium and the pathogen during the infection and it is necessary to remove it genetically in order to achieve attenuation of virulence. Interestingly, the level of attenuation achieved with an *ompC ompF* double mutant is lower than that seen in the *ompB* mutant (Chatfield *et al.*, 1991). This points to the involvement of other *ompB*-dependent genes in *S. typhimurium* mouse virulence; the *ompB*-dependent *tppB* operon is not one of these (Chatfield *et al.*, 1991).

The *ompB* locus is also required for virulence in *Shigella flexneri* (Bernardini *et al.*, 1990). Here, plasmid-linked *vir* gene expression, which is enhanced by increases in growth medium osmolarity, is transcriptionally modulated by the *ompB* locus. Mutations in *ompB* result in a reduction in *S. flexneri* virulence.

PhoP/PhoQ and *Salmonella* virulence

The *phoP* regulatory gene controls expression of the *Salmonella typhimurium phoN* gene, encoding acid phosphatase (Kier *et al.*, 1977). Mutations in *phoP* attenuate *S. typhimurium* virulence in BALC/c mice (Miller *et al.*, 1989b). Analysis of the *phoP* gene reveals that it forms an operon with a second regulatory gene, *phoQ*, and that the predicted amino acid sequence of the *phoP* gene product displays extensive homology to OmpR (Appendix I) while the predicted amino acid sequence of the *phoQ* gene product resembles EnvZ (Miller *et al.*, 1989b) (Figs 8.3, 8.4 & 8.5). In the *phoP/phoQ* two-component system, PhoQ appears to be the sensor partner (Appendix II) (Figs 8.3 & 8.5). It has two potential membrane-spanning sequences and the conserved sequence of residues characteristic of the histidine protein kinase family, as judged from the inferred primary structure of the protein; PhoP is thought to be a DNA-binding protein (Miller *et al.*, 1989b) (Figs 8.4 & 8.5). Mutations in *phoN*, the gene originally identified as being under *phoP* control, do not affect *S. typhimurium* virulence. *phoP* mutants are sensitive to defensins, which are bacteriocidal peptides produced within macrophages and neutrophils (Fields *et al.*, 1989). Mutations in *phoN* do not affect defensin resistance. Thus, there must be a PhoP/PhoQ regulon of genes, some of which are concerned with defending the cell against specific antimicrobial agents produced intracellularly by the host.

Some progress has been made in identifying some of the genes controlled by the *phoP/phoQ* system which help *S. typhimurium* to survive in phagocytic cells. One of these is a gene called *pagC* (PhoP-activated gene), which is PhoP-dependent for expression and encodes an envelope protein required for phagolysosome survival inside macrophages (Miller *et al.*, 1989b). The PagC protein shows sequence homology to the Ail invasion protein of *Yersinia enterocolitica* and the Lom protein of bacteriophage λ (Pulkkinen & Miller, 1991).

The *phoP/phoQ* regulon is under multifactorial environmental control, responding to changes in carbon, nitrogen, phosphorus and pH (Kier *et al.*, 1977; Weppleman *et al.*, 1977; Miller *et al.*, 1989b). Low pH is probably an important signal in the intracellular environment. Bacteria are subjected to acid stress in the phagolysosome and this could signal the expression of protective bacterial proteins, such as PagC.

The VirG/VirA system of *Agrobacterium tumefaciens*

The *virA* and *virG* genes of the Gram-negative bacterium *Agrobacterium tumefaciens* are essential for virulence in all species of plant which this bacterium infects (Winans, 1992). These genes are located on the Ti plasmid (introduced in Chapter 3) and, unlike the other *vir* genes, they are transcribed during normal vegetative growth (Stachel & Zambryski, 1986). Expression of VirA and VirG is autoregulated. Expression is modulated by plant wound-derived phenolic compounds. In the presence of an inducer such as acetosyringone and active copies of *virA* and *virG* the expression of VirA and VirG is enhanced. VirA- and VirG-independent induction is also achievable through growth medium transitions and this is acetosyringone-independent. This growth medium-induced expression is due to a reduction in pH. Such induction is also achievable through starvation for phosphate. This is reminiscent of the induction of *pagC* expression via PhoP/PhoQ in *Salmonella typhimurium*, which is achieved in response to phosphate starvation and a reduction in pH (Miller *et al.*, 1989b). This alternative (i.e. VirA- and VirG-independent) agrobacterial regulatory pathway acts via a chromosomal locus called *chvD* (Stachel *et al.*, 1986; Veluthambi *et al.*, 1987; Winans *et al.*, 1988) whose product is a member of a large family of ATP-binding proteins and shows homology to the NodI transport protein of the nitrogen-fixing symbiont *Rhizobium leguminosarum* (Higgins *et al.*, 1986; Johnston, 1989; Winans *et al.*, 1988). *Agrobacterium* species and *Rhizobium* species are closely related plant-infecting bacteria.

VirA and VirG are responsible for activating the expression of the *vir* genes concerned with Ti plasmid DNA transfer to the host. They are members of the two-component family of sensor/regulators (Appendix I & II) (Figs 8.3, 8.4 & 8.5). VirA is a membrane-associated histidine protein kinase whose autophosphorylation activity is modulated by plant phenolics, such as acetosyringone. Autophosphorylation occurs on a specific histidine residue within the cytoplasmic domain (Jin *et al.*, 1990b) (Fig. 8.6). It transfers its phosphate group to the transcription activator VirG, which is a sequence-specific DNA-binding protein (Jin *et al.*, 1990a). The phosphate group is transferred from a phosphorylated histidine residue in VirA to an asparagine residue at position 54 in VirG (Jin *et al.*, 1990a). The aminoterminal domain of VirA is located in the periplasmic space and it is either to here or to the transmembrane sequences that the plant signal molecules are thought to bind (Melchers *et al.*, 1989; Winans *et al.*, 1989). VirG recognizes and binds to a 12 bp DNA sequence known as the *vir* box (5′-TNCAATTGAAAPy-3′, where Py is a pyrimidine and N is any residue) which is found upstream of all *vir* genes subject to VirA/VirG regulation (Jin *et al.*, 1990c). In some cases, interaction of VirG with its binding sites may be antagonized by the chromosomally encoded 15.5 kDa Ros protein, particularly in the cases of the *virC* and *virD* genes whose products play a key role in conjugal transfer of the Ti plasmid to the plant (Cooley *et al.*, 1991; Chapter 3).

The domain structure of VirG resembles that of OmpR (Figs 8.4 & 8.5) and the DNA-binding domain of VirG shows homology to that of OmpR (Stock *et al.*,

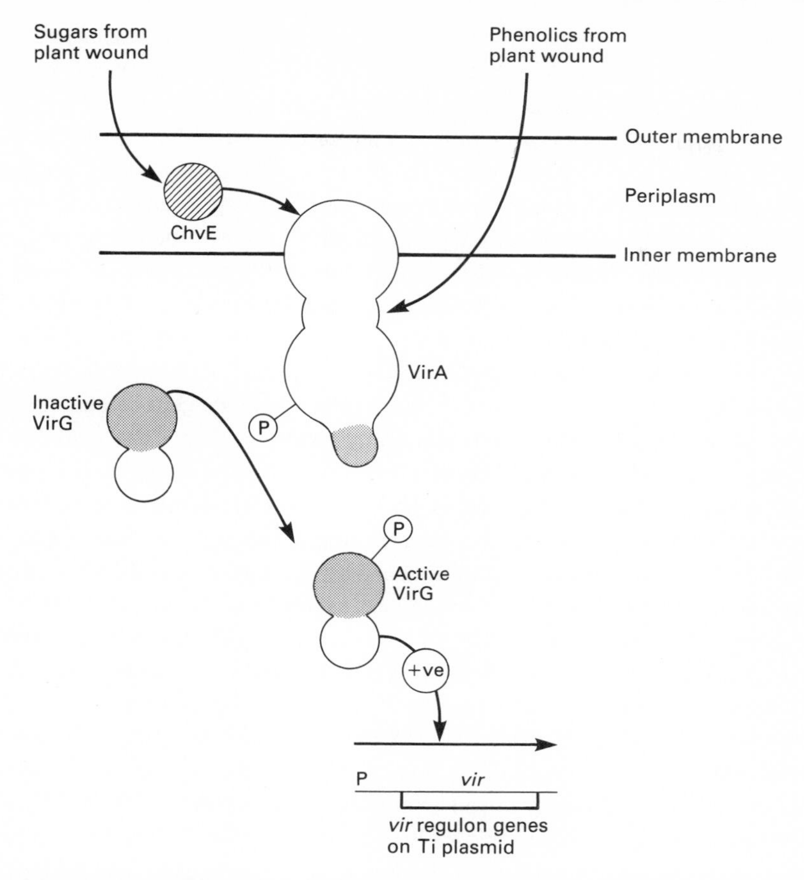

Fig. 8.6 Regulation of *vir* gene transcription by VirA/VirG in *Agrobacterium tumefaciens.* The membrane-located histidine protein kinase VirA detects the presence of sugars and phenolics from plant wounds and autophosphorylates in response. The phospho group is transferred to the transcription factor VirG which then activates the genes of the *vir* regulon located on the Ti plasmid. The receiver domain of VirG, which has a homologue in the carboxyterminus of VirA, is indicated by the shaded area.

1989; Winans *et al.*, 1986). The consensus binding sites of these transcription activators share up to seven nucleotides (Jin *et al.*, 1990c; Stock *et al.*, 1989). Unlike phospho-OmpR or phospho-NtrC, the phosphorylated form of VirG is stable for long periods of time even in the presence of VirA and ATP, indicating the lack of a phosphatase activity in this system (Jin *et al.*, 1990a). If this stability is not an experimental artefact, it may reflect a need for an ability to maintain the inducing signal over a long period in order to operate the *vir* system in cells growing under unfavourable conditions; certainly, the induction of the *vir* system is not rapid (Jin *et al.*, 1987).

The *bvg* (*vir*) locus of *Bordetella* species

Three closely related species of *Bordetella* have been investigated in terms of virulence gene regulation. These are *B. pertussis, B. bronchiseptica* and *B. parapertussis.* All adhere to ciliated cells in the upper respiratory tracts of mammals and cause disease by expressing and releasing several virulence factors. *B. pertussis* causes whooping cough in humans while *B. parapertussis* causes a milder form of the disease. *B. bronchiseptica* is an animal pathogen, causing atrophic rhinitis in pigs and kennel cough in dogs (reviewed in Aricó *et al.*, 1991). Virulence factors common to all three species include protactin (a non-pilus adhesin), filamentous haemagglutinin and pili (both of which are adhesins), haemolysin, necrotic toxin and adenylate cyclase. *B. pertussis* also expresses pertussis toxin.

The expression of many of the virulence genes of *Bordetella* species is regulated through the pleiotropic *bvg* (Bordetella virulence gene) locus, formerly known as *vir*. This locus controls the expression of subservient genes at the transcriptional level (Melton & Weiss, 1989) (Fig. 8.7). Genes under the positive control of *bvg* include those encoding pertussis toxin (in *B. pertussis*), adenylate cyclase toxin, haemolysin, dermonecrotic toxin, filamentous haemagglutinin and fimbriae. Their expression is modulated by changes in specific environmental parameters, including temperature (see section on Thermoregulated virulence genes, later) and levels of $MgSO_4$ and nicotinic acid (Armstrong & Parker, 1986; Lacey, 1960; Schneider & Parker, 1982). *Bordetella* species also possess genes whose expression is repressed by *bvg* (Knapp & Mekalanos, 1988).

The *bvg* locus has been cloned and its nucleotide sequence determined. The inferred amino acid sequences of the gene products indicates a close relationship to the two-component family of sensor/regulators (Aricò *et al.*, 1989). There are two Bvg proteins, BvgA (23 kDa) and BvgS (135 kDa). BvgA shows homology to the response regulator members of the family and can activate transcription of *bvg*-dependent genes (Roy *et al.*, 1989) (Figs 8.4 & 8.5; Appendix I). BvgA copy-number is crucial to the expression of *bvg*; when expressed *in trans* from a high copy-number plasmid, it represses the *bvg* system (Aricò *et al.*, 1991). The *bvg* operon appears to be autoregulated by BvgA (Roy *et al.*, 1990; Scarlato *et al.*, 1990). In this respect it shows similarity to the autogenously regulated *virA* and *virG* genes of *Agrobacterium tumefaciens* (see previous section).

The BvgS protein shows homology to the histidine protein kinase family and is located in the cytoplasmic membrane (Stibitz & Yang, 1991) (Figs 8.3 & 8.5; Appendix II). The original nucleotide sequence of the *bvg* locus indicated that it coded for three polypeptides, BvgA (cytoplasmic), BvgB (periplasmic) and BvgC (cytoplasmic membrane) (Aricò *et al.*, 1989; Gross *et al.*, 1989). The three-protein model is now known to be incorrect and to have arisen from a DNA sequencing error; proteins BvgB and BvgC are really one polypeptide, called BvgS (for 'Sensor' protein) (Stibitz & Yang, 1991).

The aminoterminal 541 amino acids of BvgS are in the periplasm. This domain includes the now-defunct 'BvgB' periplasmic protein and is presumed to include the environmental sensing region. Interestingly, a large part of the sensor sequence is recapitulated within the cytoplasmic domain of BvgS which

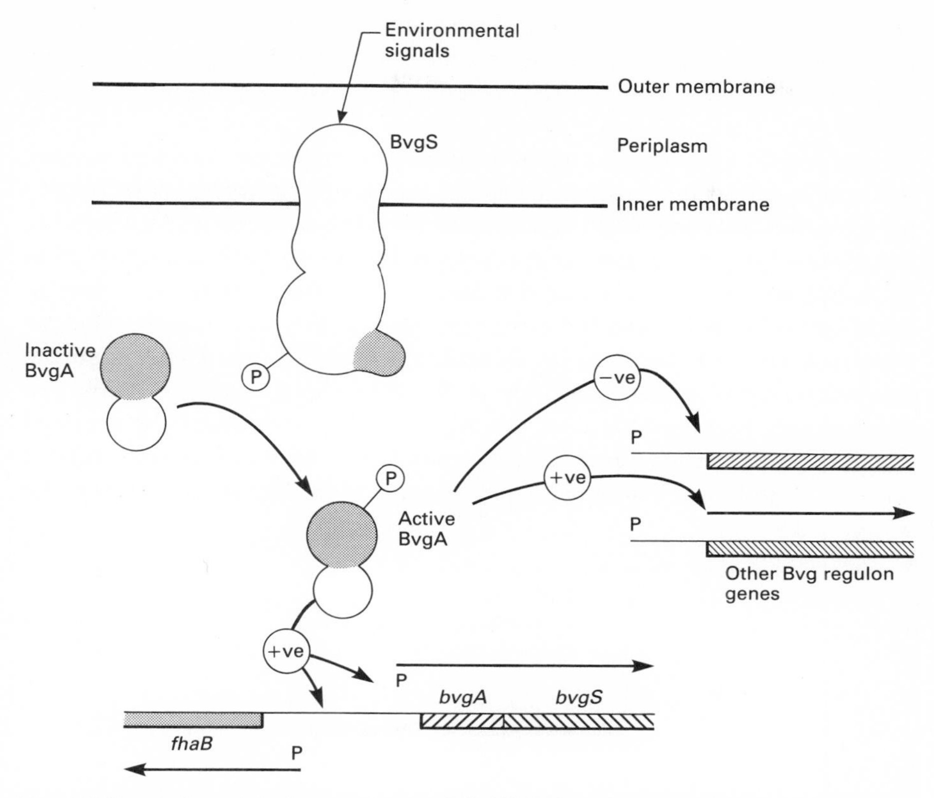

Fig. 8.7 Regulation of the *bvg* regulon in *Bordetella pertussis* by the BvgS/BvgA two-component regulatory system. The membrane-located sensor protein BvgS detects environmental signals and then transmits them to the genome via the transcription factor BvgA (presumably by phosphotransfer, as illustrated here). Once bvgA has been converted to an active state, it activates *fhaB* and the *bvgAS* operon. It also alters the transcription of other regulon members either positively or negatively, perhaps by acting through a regulatory cascade (see text). Note the presence in BvgS of a region (shaded) showing homology to the signal receiver domain of BvgA.

was formerly known as BvgC. Thus, this protein contains a large tandem repeat in the primary sequence of its aminoterminal domain which may be concerned with multimerization of the protein. In addition, the BvgS cytoplasmic domain contains a region of homology to response regulator proteins (Aricó *et al.*, 1989; Gross *et al.*, 1989). The presence of such a response regulator-like sequence in a sensor protein has also been detected in the case of the VirA protein of *Agrobacterium tumefaciens* (Stock *et al.*, 1989) (Fig. 8.3 and Fig. 8.6).

Although expression of BvgS is subject to autogenous control by BvgA at the transcriptional level, the control is not as stringent as in the case of the *bvgA* gene. This has been interpreted as being a means of ensuring a constant presence of the sensor in the cell under all environmental conditions and an ability to amplify the level of transcription regulator (BvgA) protein when this is necessary (Stibitz & Yang, 1991). The mechanism by which differential regulation of BvgA and BvgS expression is achieved is not clear, but may involve internal promoters within the *bvg* operon or a form of post-transcriptional control.

Bvg-dependent gene activation is achieved in a temporally regulated manner (see also Agr-dependent gene activation in *Staphylococcus aureus*, later). Temporal activation has been elucidated using time-course experiments in which the expression of Bvg-dependent genes is induced by shifting the culture from 25 to 37°C (Scarlato *et al.*, 1991). The first step in induction involves the transcriptional activation of *bvg* and the closely linked *fhaB* gene (encoding filamentous haemagglutinin). Before thermal induction, *bvg* is transcribed only from the Bvg-independent promoter, P_2 (Roy *et al.*, 1990; Scarlato *et al.*, 1990). Ten minutes after thermal induction, the Bvg-dependent promoter P_1 is activated (Fig. 8.8). As P_1 transcription increases, transcription from P_2 falls away. After 2 hours P_2 shuts down. Ten minutes after the shift to the higher temperature, the

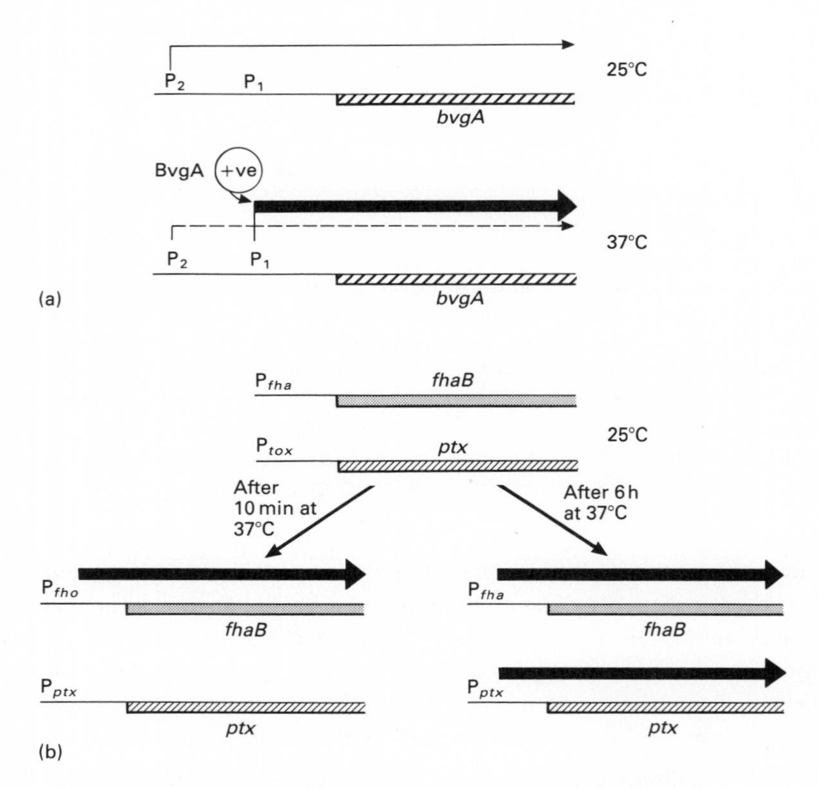

Fig. 8.8 Temperature and gene regulation by BvgA in *Bordetella pertussis.* (a) Activation of *bvgA* transcription by temperature. The *bvgA* gene has two promoters. P_2 is a weak constitutive promoter which is active at low temperature. At 37°C, activated BvgA protein binds to and activates the P_1 promoter. This strongly transcribed *bvgA.* (b) BvgA exerts differential temporal control on gene expression. At 25°C, *fhaB* and *ptx* are switched off. Following a shift to 37°C, *fhaB* is transcriptionally activated within 10 min, *ptx* remains unexpressed. After 6 hours, *ptx* is also expressed.

level of BvgA protein increases until at 6 hours post-induction, it is present at more than fifty times the uninduced level while the amount of BvgS protein increases fourfold in the same period. Two hours post-induction, the levels of pertussis toxin mRNA and adenylate cyclase mRNA increase in proportion with the BvgA level. This constitutes the second step of the induction process and concerns the *trans* activation of other *bvg*-dependent virulence genes. The effect of the thermal induction appears to be primarily one of increasing the intracellular levels of BvgA since the pertussis toxin and adenylate cyclase promoters can be induced artificially simply by overexpressing BvgA and BvgS in the absence of a thermal signal (Scarlato *et al.*, 1991). Presumably, the requisite level of phosphotransfer activity to BvgA is available under these conditions. These findings suggest that the induction of virulence factors in *B. pertussis* is sequential, with adhesin expression (FhaB) preceding expression of factors involved in tissue colonization and damage.

Huh & Weiss (1991) have been unable to demonstrate binding of BvgA protein to the promoter region of the pertussis toxin gene (*ptx*). Instead, they isolated a 23 kDa protein encoded by virulent *B. pertussis* which bound to the promoter region of the *cya* gene coding for adenylate cyclase, raising the possibility that the activation of *ptx* transcription by BvgA might be indirect, involving a regulatory cascade analogous to that described for virulence gene regulation in *Vibrio cholerae* (see later, this chapter). In this model, BvgA would first activate expression of the 23 kDa protein which in turn activates *ptx* and *cya* gene transcription. This DNA-binding protein has been called Act (Activator of toxin expression) (Huh & Weiss, 1991).

The *bvg* loci from *Bordetella pertussis, B. parapertussis* and *B. bronchiseptica* have been cloned and sequenced (Aricò *et al.*, 1991). They show a high degree of sequence conservation at the amino acid level and are functionally similar and interchangeable (Gross & Rappuoli, 1988; Monack *et al.*, 1989). The sequences are particularly well conserved in those regions likely to be involved in the histidine protein kinase and DNA-binding activities; BvgA is completely conserved between the three species. Differences occur in the periplasmic sensory domain of BvgS. The pattern of these amino acid changes confirms the proposed evolutionary relationship among these *Bordetella* species, i.e. that *B. bronchiseptica* and *B. parapertussis* are more closely related to each other than to *B. pertussis* (Aricò *et al.*, 1991).

Alginate synthesis in *Pseudomonas aeruginosa*

Pseudomonas aeruginosa is a significant respiratory pathogen in humans suffering from cystic fibrosis. Strains of *P. aeruginosa* found in the lungs of cystic fibrosis patients often have a mucoid phenotype, the appearance of which correlates strongly with a deterioration in the condition of the infected individual. This mucoid phenotype is due to the production by *P. aeruginosa* of an extracellular polysaccharide called alginate.

The genes involved in alginate biosynthesis in *Ps. aeruginosa* are subject to

complex control (Fig. 8.9). One level of control is exerted by proteins displaying homology to the two-component family of sensor/regulators (Deretic *et al.*, 1991). The *algD* structural gene encodes GDP-mannose dehydrogenase, is central to alginate biosynthesis and is subject to positive transcriptional control by the *algR* gene product (Deretic *et al.*, 1989; Deretic & Konyecsni, 1989). The AlgR protein shows homology to the transcription activators NtrC (at its aminoterminus) and OmpR (at its carboxyterminus) (Deretic *et al.*, 1989) (Figs 8.4 & 8.5; Appendix I). AlgR activates *algD* in response to osmotic stress and when the *algD* promoter is studied in *E. coli*, OmpR can replace AlgR as an *algD* activator at high osmolarity (Berry *et al.*, 1989). Three AlgR-binding sites have been mapped in the region upstream of the *algD* promoter, at positions

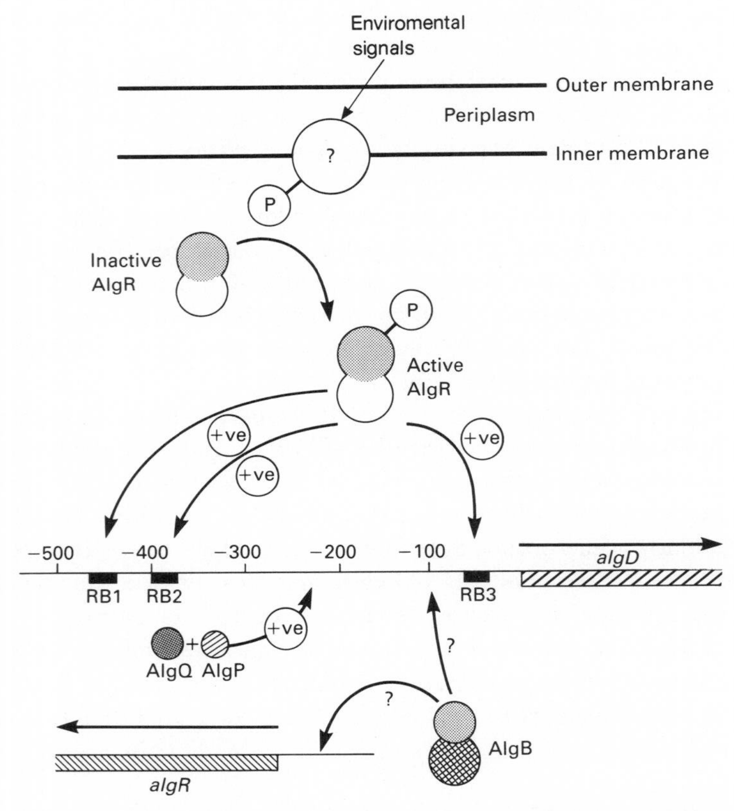

Fig. 8.9 Activation of *algD* transcription in *Pseudomonas aeruginosa*. The *algD* gene is subject to multicomponent control. The AlgR transcription activator shows homology to the response regulator family of proteins and may be environmentally modulated by a histine protein kinase (perhaps membrane-associated, as hypothesized here). Active AlgR has three binding sites upstream of the *algD* promoter (RB1, RB2, RB3). The AlgQ and AlgP regulatory proteins affect *algD* transcription positively, although the precise mechanism by which they do so is not known. The point at which AlgB makes its regulatory contribution is unclear; it might influence expression of *algD* directly or it may act via *algR*.

– 479 to – 457 (RB1), – 400 to – 380 (RB2) and – 50 to – 30 (RB3); RB1 and RB2 are referred to as the far upstream sites (FUS) (Mohr *et al.*, 1991a; 1992) (Fig. 8.9). *In vitro*, AlgR can be phosphorylated by the histidine protein kinase CheA, a regulator of chemotaxis in *Salmonella typhimurium*. This suggests that AlgR may be a substrate for phosphotransfer *in vivo* from its elusive sensor partner protein. Furthermore, phosphotransfer from CheA to AlgR is affected by carbamoyl phosphate and acetyl phosphate, suggesting that AlgR could interact with intermediary metabolites such as carbomyl phosphate and acetyl phosphate *in vivo* (Deretic *et al.*, 1992b). Low-molecular-weight phosphate donors have been shown to phosphorylate other bacterial response regulators such as CheY (Lukat *et al.*, 1992) and PhoB (Wanner & Wilmes-Reisenberg, 1992).

A further *algD* activator is encoded by the *algB* gene. The predicted AlgB amino acid sequence shows homology to NtrC (Figs 8.4 & 8.5; Appendix I). However, AlgB may not regulate *algD* directly; in *E. coli*, it fails to activate *algD* transcription in response to osmotic stress (Goldberg & Dahnke, 1992). Perhaps AlgB acts via AlgR or via one of the other regulators of *algD*. These include AlgP, a histone-like protein also called HP1, which is believed to be a component of the *Ps. aeruginosa* nucleoid (Appendix I). It has been postulated that AlgP may contribute to *algD* regulation by influencing the topology of the DNA within the regulatory region, possibly in cooperation with AlgR (Deretic *et al.*, 1992a). A very closely related histone-like protein (AlgR3) is thought to act in a similar manner in regulating *algD* expression (Kato *et al.*, 1990a). A role has been demonstrated for the *Ps. aeruginosa* homologue of the *E. coli* histone-like protein IHF in transcriptional control of *algB* (Wozniak & Ohman, 1993). IHF is thought to be important for the maintenance of a basal level of *algB* expression in non-mucoid cells. In mucoid cells, yet another positive regulator (AlgT) is required for high levels of *algB* transcription.

An additional regulatory protein, AlgQ (18 kDa), acts synergistically with AlgP in *algD* activation and the *algP* and *algQ* genes map close together on the *Ps. aeruginosa* chromosome, although they do not constitute an operon (Konyecsni & Deretic, 1990).

The PilA/PilB system of *Neisseria gonorrhoeae*

Neiserria gonorrhoeae is the causative agent of the sexually transmitted disease, gonorrhoea, in humans. Among its virulence factors are pili, which mediate attachment to the host cells. In *N. gonorrhoeae*, the *pilA* and *pilB* genes act *in trans* to regulate transcription of the *pilE* pilus genes. These genes are divergently transcribed and map downstream of the *pilE1* and *opaE1* loci in strain MS11 (Taha *et al.*, 1988). PilB shows homology to the histidine protein kinase family and PilA displays homology to the response regulators (Taha *et al.*, 1991) (Fig. 8.4; Appendices I, II). Data from analyses with Tn*phoA* fusions suggest that PilB is located in the cytoplasmic membrane and has both periplasmic and cytoplasmic domains; PilA is believed to be a DNA-binding protein and to be an activator of the pilin gene promoter. Since *pilA* null mutations are lethal to gonococci, this is probably an essential gene. Gonococcal *pilB* mutants are hyperpiliated. It is likely that the PilA/PilB system regulates gonococcal piliation in response to environmental cues (Taha *et al.*, 1991).

The PilR regulator of *Pseudomonas aeruginosa*

Like many other bacterial pathogens, *Pseudomonas aeruginosa* produces adhesin pili which assist the bacterium in attachment to its host. Expression of the pilin gene in *Ps. aeruginosa* depends on a transcriptional activator encoded by the *pilR* gene. This gene codes for a protein of 50 kDa showing homology to the nitrogen regulatory transcription activator protein NtrC (Fig. 8.4; Appendix I). The conservation in PilR of the aminoterminal aspartic acid residue at which NtrC receives a phosphate group from NtrB (Sanders *et al.*, 1992) has led to the suggestion that PilR may be activated by phosphotransfer from an (as yet unidentified) histidine protein kinase partner in response to environmental signals (Ishimoto & Lory, 1992).

The accessory gene regulator (*agr*) system of *Staphylococcus aureus*

The Gram-positive coccus, *Staphylococcus aureus*, produces an array of exported toxins. These include seven serologically distinct forms of staphylococcal enterotoxin (SE) which are classified under an alphanumerical system as SEA, SEB, SEC1, SEC2, SEC3, SED and SEE. These toxins are associated with several types of illness such as food poisoning, septicaemia, circulatory collapse, arthritis and heart disease. They are called enterotoxins because they stimulate the emetic and diarrhoeal response, possibly by affecting emetic receptors in the abdominal viscera. A distant relative of SE is toxic shock syndrome toxin-1, TSST-1. This is associated with tampon use and with symptoms such as diarrhoea, hypotension and renal failure, among others. All these toxins have been found to exert significant mitogenic effects on T-lymphocytes and they have been classified as 'superantigens' (reviewed in Hewitt *et al.*, 1992; Iandolo, 1990). Staphylococci also produce haemolytic toxins which damage membranes. These are classified under a greek alphabetical system. The alpha toxin inserts into membranes and causes fluid loss. The beta toxin has sphingomylinase activity and lyses cells whose membranes possess this lipid. The biology of the gamma toxin remains unclear. The delta toxin possesses a membrane-damaging detergent-like activity. Some strains of *S. aureus* are also capable of expressing epidermolytic toxins (A and B). These exfoliative toxins are responsible for the condition known as staphylococcal scalded skin syndrome. This is, by no means, an exhaustive list of *S. aureus* virulence factors.

These host-damaging toxins are clearly capable of carrying out more than one kind of function and are used at more than one level of host–pathogen interaction. Understanding how the bacterium controls toxin expression during the infection is a central issue in understanding the regulation of the entire pathogenic process. Expression of many exoproteins by the Gram-positive pathogen *Staphylococcus aureus* is regulated by the accessory gene regulator, *agr*. This should not be confused with another highly pleiotropic 'staphylococcal accessory regulator', *sar*, which controls the expression of many of the same exoprotein genes (Cheung *et al.*, 1992). The *agr*-regulated genes include the cytolytic toxins alpha, beta, gamma and delta haemolysin, enterotoxins, toxic shock syndrome toxin-1 (TSST-1), leucocidin, immunoglobulin-binding protein A, coagulase, and others. The expression of genes under *agr* control is differentially regulated such that those expressed during stationary phase (such as *hla*, encoding alpha haemolysin) are activated by *agr* while those expressed in exponential growth (such as *spa*, encoding protein A) are repressed by *agr*. Agr may also interact with the products of a poorly characterized locus, *xpr*, mutations in which alter the expression of the same set of virulence genes (Smeltzer *et al.*, 1993).

A transposon Tn*551* insertion mutation in *agr* was found to alter simultaneously the expression of exoprotein expression (Recsei *et al.*, 1985). Specifically, the *agr* : : Tn*551* mutant shows decreased expression of alpha toxin, beta haemolysin, delta lysin, TSST-1, enterotoxin B, epidermolytic toxins A and B, leucocidin, staphylokinase, nuclease, serine-protease and metalloprotease and acid phosphatase and increased expression of protein A and coagulase (Gaskill & Khan, 1988; Janzon *et al.*, 1989; O'Toole & Foster, 1986, 1987; Rescei *et al.*, 1985).

Cloning and nucleotide sequencing has revealed that the Tn*551* insertion is within an open reading frame whose gene product (AgrA) displays homology to the response regulator members of the histidine protein kinase/response regulator two-component family (Peng *et al.*, 1988; Stock *et al.*, 1989) (Figs 8.4 & 8.5; Appendix I). This gene is part of a polycistronic operon containing the additional genes *agrB*, *agrC* and *agrD* (Janzon *et al.*, 1989; Vandenesch *et al.*, 1991). The *agr* locus also includes the *hla* gene (encoding delta lysin) which is transcribed divergently from *agr* on the opposite DNA strand (Janzon *et al.*, 1989). The transcriptional organization of *agr* is complex (Fig. 8.10). *agrA* is transcribed as a monocistronic message (RNA I) from its own weak, constitutive promoter, P_1 while *agrA* and *agrBCD* are transcribed as a polycistronic message (RNA II) from the P_2 promoter. A third promoter, P_3, transcribes *hla* on the opposite strand (Fig. 8.10).

AgrB (nomenclature of Novick and coworkers) is related to the histidine protein kinase family of sensor proteins (Figs 8.3 & 8.5; Appendix II) and is believed to become phosphorylated in response to environmental stimuli and then to transfer the phosphate group to AgrA (Janzon & Arvidson, 1990). Furthermore, hydrophobicity analysis of the AgrB sequence indicates that it may be membrane-associated, in keeping with its proposed role as a sensor of environmental signals (Vandenesch *et al.*, 1991). AgrA is a positive regulator of

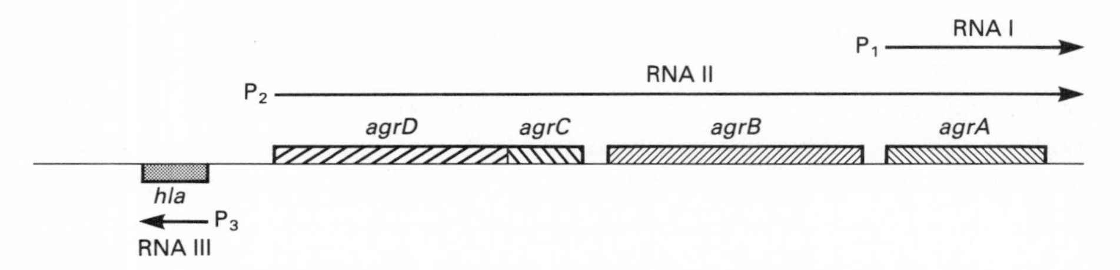

Fig. 8.10 Genetic organization of the *agr* locus of *Staphylococcus aureus*. The accessory gene regulator locus is composed of three operons. Promoter P_1 transcribes the *agrA* gene while promoter P_2 transcribes the *agrDCBA* operon. The P_3 promoter transcribes the *hla* gene on the opposite DNA strand.

agr, and thus shows functional similarites to BvgA in the *Bordetella bvg* system and VirG in the *Agrobacterium vir* system. Activation of AgrA results in an increase in *hla* transcription, indicating that *agr* and *hla* are transcriptionally coregulated (Janzon & Arvidson, 1990; Janzon *et al.*, 1989). The *hla* message is a 517-nucleotide transcript called RNA III and is translated to yield a secreted 26 amino acid polypeptide which forms channels in phospholipid bilayers, lysing cells or organelles in the process (Freer, 1988). The *hla* transcript appears to be a regulator of *agr*-dependent genes. Specifically, the *hla* transcript cooperates with *agr* in the transcriptional regulation of alpha toxin and protein A expression; the *hla* transcript and translation product cooperate with *agr* in the regulation of extracellular serineprotease and metalloprotease, although here the level at which this cooperative regulation is exerted is unclear (Janzon Arvidson, 1990).

The Tox regulatory system of *Vibrio cholerae*

The Gram-negative pathogen *Vibrio cholerae* induces severe diarrhoeae in humans and cholera toxin is a central virulence factor in this process. *V. cholerae* regulates the expression of its virulence genes in response to a wide range of environmental stimuli, including changes in osmolarity, temperature, pH, and the availability of certain amino acids (DiRita, 1992; DiRita *et al.*, 1990; Miller *et al.*, 1987). The genes coding for the A and B subunits of cholera toxin are arranged in an operon, *ctxAB*, an arrangement resembling that of the *E. coli* heat-labile toxin (LT) operon, *eltAB* (Pearson & Mekalanos, 1982). The open reading frames for the two subunits overlap by two base pairs and are expressed in the ratio A_1B_5. This differential expression is thought to be a function of the efficiencies of the ribosome-binding sites of each gene because when the *ctxB* gene is placed under the control of the *ctxA* transcription and translation signals, it is expressed nine times less efficiently than when it is under the control of its own expression signals (Mekalanos *et al.*, 1983). The promoter of the *ctxAB* operon is preceded by between three and eight copies of the sequence 5′-TTTTGAT-3′, depending on the isolate, and these repeat sequences are required by the *ctxAB* transcription activator ToxR. This activator is not encoded by *E. coli* and when the *ctxAB* genes were originally cloned in that species, their expression was greatly reduced in comparison with the situation in *V. cholerae* (Pearson & Mekalanos, 1982). The *V. cholerae toxR* gene was detected due to its ability to activate transcription of a *ctx–lacZ* fusion in *E. coli* when expressed *in trans* from a recombinant plasmid (Miller & Mekalanos, 1984). The 32 kDa ToxR protein is a sequence-specific DNA-binding protein located in the cytoplasmic membrane which activates *ctxAB* transcription via the 5′-TTTTGAT-3′ direct repeats located close to the promoter. *ctx* promoters with the repeats are hundreds of times more active in the presence of ToxR than those lacking them, even if the promoter sequence is otherwise unaffected (Miller *et al.*, 1987). The DNA-binding domain of ToxR is located in the aminoterminal region of the protein and displays homology to OmpR and other response regulator members of the two-component family of signal transducers (Miller *et al.*, 1987) (Fig. 8.4). Mutations in this domain of ToxR abolish DNA binding and gene activation (Ottemann, K. *et al.*, 1992). The observation that ToxR possesses a region of similarity to response regulator proteins such as OmpR led to the suggestion that it might constitute a 'one-component' version of the usual two-component system in which the membrane-associated sensor domain had become fused with the response regulator in a single polypeptide. Further analysis of the ToxR system has shown that it is actually in a rather different category of regulator.

On the *V. cholerae* chromosome, the *toxR* gene forms an operon with a second gene, *toxS* and this latter gene encodes a 19 kDa protein which is required by ToxR for activation of the *ctxAB* promoter (DiRita & Mekalanos, 1991; Miller *et al.*, 1989c). The active form of ToxR is believed to be a dimer and ToxS is thought to ensure dimerization when ToxR is present in the cell membrane at normal levels; overexpression of ToxR removes the need for ToxS, probably because the protein dimerizes spontaneously at high concentration (DiRita & Mekalanos, 1991; Miller *et al.*, 1989b). The interaction between ToxR and ToxS that leads to ToxR dimerization (and hence activation) is thought to occur in the periplasm and it is suggested that the transmembrane signalling may resemble mechanistically that of eukaryotic membrane receptor protein tyrosine kinases, with the important distinction that the event on the cytoplasmic side of the membrane is DNA binding and not phosphotransfer (DiRita & Mekalanos, 1991) (Fig. 8.11).

Vibrio cholerae possesses an array of genes whose expression is ultimately dependent upon ToxR. However, with the exception of *ctxAB*, the association of ToxR with these genes is not direct but often involves an intermediate regulator, ToxT. The *toxT* gene product is a 32 kDa protein with homology to the AraC family of transcription regulators (DiRita & Mekalanos, cited in DiRita, 1992; Higgins *et al.*, 1992; see later). It should be noted that ToxT has been discovered independently by Ogierman & Manning (1992) and named TcpN (Fig. 7.10; Appendix I).

Like ToxR, ToxT can activate the *ctxAB* operon in *E. coli*. However, ToxT can

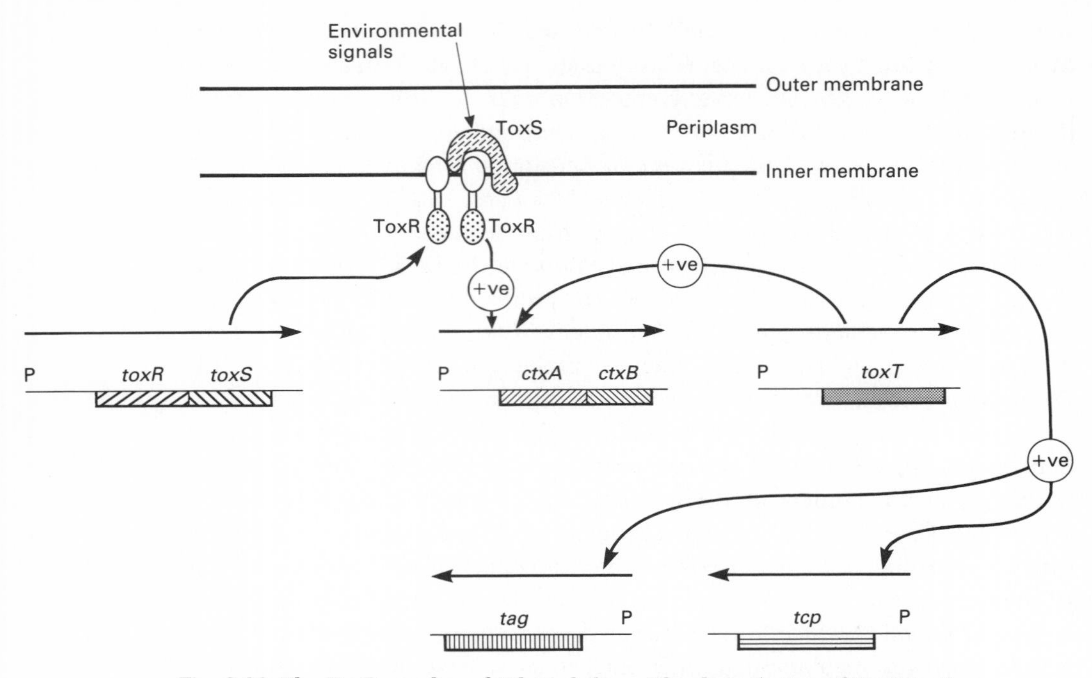

Fig. 8.11 The ToxR regulon of *Vibrio cholerae*. The dimerization of ToxR in the membrane is assisted by ToxS. This protein complex transmits environmental signals to the genome. The *toxR* and *toxS* genes form an operon. ToxR regulates transcription of the *ctxAB* operon directly. The ToxT protein is an AraC-like molecule which contributes to *ctxAB* regulation and also controls the expression of other members of the ToxR regulon. Expression of *toxT* is presumed to be under ToxR control. P, promoter.

also activate the other known ToxR-dependent genes when these are cloned in *E. coli* in the absence of ToxR (DiRita *et al.*, 1991). Many ToxR-activated genes (designated as *tag* genes) are concerned with the expression of the toxin coregulated pilus (TCP) or the accessory colonization factor (ACF) (Mekalanos *et al.*, 1988); others have no obvious role in virulence (e.g. *tagA*) and one (*aldA*, transcribed divergently from *tagA*) codes for aldehyde dehydrogenase, the first ToxR-dependent cytoplasmic protein to be characterized (Parsot & Mekalanos, 1991a). While some of these genes (e.g. *aldA, tagA, tcpA, tcpI*) can be activated by ToxT in the absence of ToxR (DiRita *et al.*, 1991), they cannot be activated by ToxR alone. The finding that ToxR activates the transcription of *toxT* is consistent with a regulatory cascade for the activation of virulence gene expression by ToxR acting through ToxT (Fig. 8.11). Whether this activation is direct or involves yet another intermediate regulator is unclear (DiRita, 1992). There is evidence that this cascade is even more extensive in the case of certain genes; for example, the ToxT-dependent *tcpI* gene is itself a regulator of *tcp* gene expression (Taylor, 1989). A further regulatory layer in ToxR-dependent gene expression is afforded by sensitivity on the part of some promoters to changes in DNA supercoiling. Specifically, the promoters of the *acfA* and the *acfD* genes require active DNA gyrase for normal function (Parsot & Mekalanos, 1992).

It should be noted that *V. cholerae* produces an additional, recently discovered, enterotoxin. This is the zonula occludens toxin (ZOT) which interferes with tight junctions between cells in the intestine (Fasano *et al.*, 1991). The genetics of ZOT toxin have not been worked out in detail and it is not clear if it forms a part of the ToxR cascade.

The Mry transcription activator of *Streptococcus pyogenes*

The M protein is a key virulence determinant of group A streptococci. These bacteria are responsible for acute streptococcal pharyngitis, acute rheumatic fever, skin infections, some forms of pneumonia and a condition that resembles toxic shock. M proteins are fibrillar in structure, are composed of two α- helical chains wrapped in a coiled-coil which extend from the surface and protect the bacterium from phagocytosis by polymorphonuclear leukocytes (reviewed in Fischetti, 1991). Expression of M protein is regulated at the level of transcription and details of the regulatory protein have become available, at least for some strains (Caparon & Scott, 1987).

Transcription of the *Streptococcus pyogenes emm6* gene, coding for the highly antigenically variable M surface protein is activated by the product of the closely linked *mry* gene (Perez-Casal *et al.*, 1991). The *mry* gene product is a *trans*-acting transcriptional activator and the amino acid sequence deduced from the primary sequence of the *mry* gene suggests that it has some resemblance to the conserved phosphorylation acceptor domains of the regulatory proteins of the histidine protein kinase/response regulator family (Perez-Casal *et al.*, 1991) (Fig. 8.4; Appendix I). These regions of similarity are found in small segments of both the aminoterminal and the carboxyterminal domains. The aminoterminal domain also has an amino acid sequence which resembles the helix-turn-helix motif found in some DNA-binding proteins, supporting the hypothesis that Mry acts on *emm* transcription by binding to DNA. As with other regulators of virulence gene expression, Mry appears to be involved in the transmission of environmental signals. In the case of the *emm* gene, an important signal appears to be carbon dioxide (Caparon *et al.*, 1992). Temperature, which had originally been thought to be a regulator of *emm* transcription (Perez-Casal *et al.*, 1991) does not seem to be relevant after all (Caparon *et al.*, 1992). The gene coding for Mry, *mry*, is autoregulated at the level of transcription in response to CO_2 (Okada *et al.*, 1993).

A regulator which is highly homologous to Mry, called VirR, is located upstream of the *emm12* gene in another serotype of *Streptococcus pyogenes*. This regulates coordinately the expression of the M12 protein and the C5a peptidase (ScpA) in that strain (Podbielski *et al.*, 1992).

AraC family members

The family of transcription activator proteins showing homology in their carboxyterminal domains to the DNA-binding region of the AraC protein of enteric bacteria was introduced in Chapter 7. Here, those members that control the transcription of specific virulence genes in response to environmental stimuli (typically changes in temperature) are described in more detail.

VirF/LcrF in *Yersinia* species

In *Yersinia pestis* (the agent responsible for bubonic plague in humans), *Y. pseudotuberculosis* and *Y. enterocolitica* (both of which cause enteritis), growth becomes restricted at 37°C on calcium-deficient medium. This calcium dependency correlates with the secretion of *Yersinia* outer membrane proteins (Yops) and with virulence. Despite the old designation, Yops are really secreted rather than outer membrane proteins. They are encoded by a 70 kb plasmid (Chapter 3) and *yop* transcription is thermally triggered through the plasmid-encoded VirF protein. This thermoactivation of *yop* genes by VirF is not calcium-dependent. VirF displays carboxyterminal domain homology to the other members of the AraC family, presumably reflecting a common DNA-binding region (Cornelis *et al.*, 1989b; Fig. 7.10; Appendix I).

In *Y. enterocolitica*, the expression of *virF* is thought to be autoregulated. *In vitro*, the *Y. enterocolitica* VirF protein binds *yop* promoter DNA even at 25°C. The discrimination between *yop* promoter activity at 37°C and inactivity at 25°C appears to be achieved through changes in the conformation of the promoter DNA and involves the *Y. enterocolitica* histone-like protein YmoA (Cornelis *et al.*, 1991; Lambert de Rouvroit *et al.*, 1992). VirF also activates *yadA* transcription at 37°C. YadA is a plasmid-encoded outer membrane protein involved in adherence and serum resistance in *Yersinia enterocolitica* and *Y. pseudotuberculosis* but not *Y. pestis* (Skurnik & Toivanen, 1992). The *Y. pseudotuberculosis* and *Y. pestis* counterparts of the *Y. enterocolitica* VirF protein are called LcrF. The *Y. pestis* LcrF protein is virtually identical in amino acid sequence to VirF. Of six sequence differences between LcrF and VirF, only one (Gly-95 in LcrF is Ala-95 in VirF) is a non-conservative change (Hoe *et al.*, 1992) (Fig. 7.10; Appendix I). Thus, the LcrF and VirF proteins almost certainly carry out identical biological tasks in these bacteria.

VirF in *Shigella* species

Shigellae are enteroinvasive pathogens and cause dysentery in humans. The molecular detail of the infection process has been best worked out in *Shigella flexneri* and *S. sonnei*. The plasmid-encoded invasion genes of *S. flexneri* and *S. sonnei* are thermally regulated at the level of transcription by a complex cascade that involves both plasmid- and chromosomally-encoded regulators (reviewed in Hale, 1991). Some of these genes may also be subject to post-transcriptional control via a chromosomal locus called *vacB* (Tobe *et al.*, 1992). Structural genes encoding the outer membrane-located invasion functions are transcriptionally activated by the VirB protein (called InvE in *S. sonnei*; Appendix I), which is structurally related to the ParB partition function of plasmid P1 and to the SopB partition function of plasmid F (Adler *et al.*, 1989; Watanabe *et al.*, 1990). Transcription of *virB* is activated by the VirF protein, a positive activator of transcription and a member of the AraC family (Cornelis, 1992; Dorman, 1992; Hoe *et al.*, 1992; Savelkoul *et al.*, 1990; Smyth *et al.*, 1991) (Fig. 7.10; Appendix I). The need for VirF can be circumvented by overexpressing VirB from a heterologous promoter (such as the IPTG-inducible *tac* promoter) (Adler *et al.*, 1989). As in the case of *yop* gene activation by the *Yersinia* VirF protein, a nucleoid protein (in this case, H-NS) is involved in modulating VirF-dependent virulence gene transcription in *S. flexneri* (Dorman *et al.*, 1990; Hale, 1991; Higgins *et al.*, 1990). It appears that H-NS acts antagonistically at VirF-dependent promoters. It is possible that the function of VirF is to overcome the negative influence of H-NS at these promoters.

The Rns and CfaD proteins of enterotoxigenic *Escherichia coli*

Enterotoxigenic *E. coli* (ETEC) strains elaborate serologically distinct surface adhesins. The first ETEC fimbrial antigen to be identified was CFA/I (colonization factor antigen I) (Evans *et al.*, 1975). Later, a second antigen, CFA/II, was identified and shown to consist of three surface-associated antigens, CS1, CS2 and CS3 (Cravioto *et al.*, 1982; Evans & Evans, 1978; Levine *et al.*, 1984; Mullany *et al.*, 1983; Smyth, 1982). ETEC strains harbour a plasmid that encodes a transcriptional activator for the expression of CS1 and CS2. This transcription activator is called Rns and its carboxyterminal domain shows considerable sequence homology to the corresponding region of the AraC proteins (Caron *et al.*, 1989) (Fig. 7.10; Appendix I). It is believed that the carboxyterminal domains of the VirF family members contains the DNA-binding domain. Furthermore, Rns shares amino acid sequence homology with the *S. flexneri* and *S. sonnei* VirF proteins throughout its length (Dorman, 1992). In addition, the *rns* gene, like the *Shigella virF* gene, has an abnormally low G + C content (28% for *rns*; 30% for *virF*, as opposed to about 50% for these enteric bacteria) suggesting that these genes may have been acquired from a source outside the enteric group (Caron *et al.*, 1989; Dorman, 1992).

In ETEC, expression of the thermally regulated plasmid-encoded fimbrial antigen CFA/I requires another AraC-like protein called CfaD. CfaD and Rns are highly homologous and can substitute for one another functionally (Savelkoul *et al.*, 1990; Fig. 7.10; Appendix I). The nucleotide sequence of the *cfaD* gene differs from that of *rns* at only 28 positions (Savelkoul *et al.*, 1990), suggesting that, like ETEC *rns* and *Shigella virF*, ETEC *cfaD* may have originated outside the enteric group.

Transcription of the regulatory gene *cfaD* (*rns*) is itself thermally regulated. Mutations that inactivate the gene coding for the H-NS nucleoid protein derepress transcription of CFA/I genes at low temperatures. In the absence of H-NS, the need for CfaD is abolished, although greater levels of CFA/I expression are seen in the presence of CfaD (Jordi *et al.*, 1992). Thus, it appears that a key role of this AraC-like regulatory protein is to overcome the negative effects of H-NS at the promoters it activates (Fig. 8.12). This is similar to the situation described for VirF in enteroinvasive *E. coli* (EIEC) and *Shigella* species.

The TcpN and ToxT proteins of *Vibrio cholerae*

The toxin coregulated pilus (Tcp) regulon of *Vibrio cholerae* includes the regulatory gene *tcpN* whose product, TcpN, is a member of the AraC family. As has been reported for several other AraC-like regulatory protein genes, *tcpN* has an abnormally low G + C content (28.3%, compared with an average for *V. cholerae* of 40–50%) (Ogierman & Manning, 1992). The *Vibrio cholerae* ToxT transcription activator involved in regulating many ToxR-dependent genes is the same as TcpN (Higgins *et al.*, 1992; Figs 7.10 & 8.11; Appendix I). Thus, TcpN is not simply a dedicated regulator of pilus genes but contributes in a general way to the control of *V. cholerae* virulence gene expression.

The FapR protein of *Escherichia coli*

The 987P pilus adhesin of enterotoxigenic *E. coli* (ETEC) is specific for attachment to the brush border cells of pig intestines. Infection of the pig by these bacteria can result in severe diarrhoeal disease. ETEC strains expressing 987P also produce heat-stable enterotoxin. A molecular explanation for this associa-

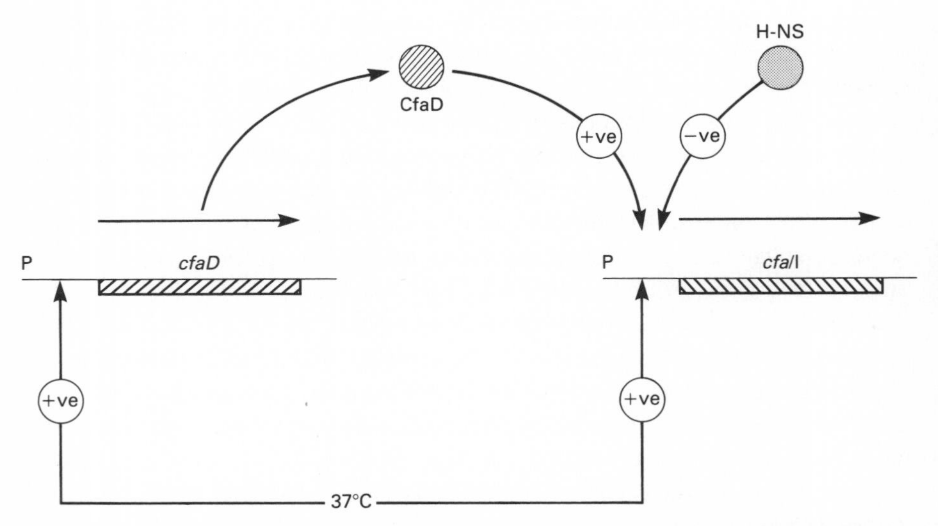

Fig. 8.12 Thermoregulation of Cfa/I expression in enterotoxigenic *Escherichia coli*. The regulatory gene *cfaD* is transcriptionally activated by temperature. The AraC-like DNA-binding protein CfaD acts in concert with temperature to activate transcription of the *cfa/I* structural gene. This activation is antagonized by H-NS.

tion has now been provided. The toxin gene is a component of a composite transposon (probably Tn*1681*) which consists of IS*1* elements flanking the gene coding for heat-stable enterotoxin ST_{pa}. The activities of the IS*1* elements which flank the composite transposon result in DNA rearrangements which can cause the 987P operon to be deleted. This provides a mechanisms for 987P pilus phase variation.

Insertion of the transposon into DNA adjacent to the 987P operon results in activation of a regulatory gene called *fapR*. The product of this gene, FapR, activates transcription of the 987P fimbrial gene cluster in ETEC. The mechanism by which the transposon turns on transcription of *fapR* is unclear but could involve a DNA topological influence similar to that by which transposable elements activate the cryptic *bgl* operon in *E. coli* K-12 (discussed in Klaasen *et al.*, 1990). Nucleotide sequence analysis of the *fapR* gene reveals that its product shows homology to the AraC family of transcription factors (Klaasen & De Graaf, 1990; Klaasen *et al.*, 1990; Fig. 7.10; Appendix I). Presumably, this protein interacts with DNA in a manner similar to that used by other members of the family. It is possible that this interaction is modulated by environmental factors.

The ExsA protein of *Pseudomonas aeruginosa*

Pseudomonas aeruginosa is an important opportunistic pathogen and is capable of expressing several extracellular proteins which assist it in infection. This bacterium can infect burnt tissue in burns patients and its spread to the bloodstream is helped by a virulence determinant called exoenzyme S. Consistent with this is the observation that if exoenzyme S expression is prevented, the bacterium shows reduced ability to spread and cause tissue damage. This enzyme is a secreted ADP-ribosyltransferase whose physiologically significant target in the host has not been identified unambiguously.

Exoenzyme S expression is environmentally modulated but is not iron-regulated. Thus, it is not coregulated with other major virulence determinants of *Ps. aeruginosa* such as elastase or exotoxin A whose expression is sensitive to iron levels. Regulation of exoenzyme S is complicated and involves up to three regulatory genes. One of these is a *trans*-acting factor called ExsA. The ExsA regulatory protein has been found to be a member of the AraC protein family and to be particularly similar in primary structure to the VirF protein of *Yersinia enterocolitica* (Frank & Iglewski, 1991; Fig. 7.10; Appendix I). For this reason, it is assumed that ExsA is a transcription factor. The environmental signal that activates ExsA or its gene has not been identified. The relationship between ExsA and VirF is made even more interesting by the discovery that a second regulator of exoenzyme S expression, ExsB shows homology to the *Y. enterocolitica* regulatory protein VirB (Frank & Iglewski, 1991). VirB plays a poorly characterized role in controlling expression of at least some *yop* genes in this bacterium.

The HrpB protein of *Pseudomonas solanacearum*

The plant pathogen *Pseudomonas solanacearum*, which causes bacterial wilt in solanaceous plant species, harbours a set of *hrp* genes which are required for plant virulence. Many of these genes are located on a high-molecular-weight plasmid and some are under transcriptional control by the positive regulator HrpB. The *hrpB* gene is partially autoregulated and induces *hrp* gene expression in response to unknown environmental signals which are mimicked by growth in minimal media but not by growth in rich media. Expression of *hrp* genes in *Pseudomonas syringae* pv. *phaseolicola* is regulated by environmental signals such as osmolarity, pH and carbon source (Rahme *et al.*, 1992). It would not be surprising if such signals were also significant in the *Ps. solanacearum* Hrp system. The predicted amino acid sequence of HrpB, derived from the nucleotide sequence of the *hrpB* gene, indicates that this protein is related to the AraC family of transcription factors (Genin *et al.*, 1992; Fig. 7.10; Appendix I).

Regulation of intimin expression in enteropathogenic *Escherichia coli*

Enteropathogenic *E. coli* (EPEC) strains are a major cause of human diarrhoea and have become the subject of intense genetic analysis (reviewed in Donnenberg & Kaper, 1992). The *eae* locus is located on the chromosome of EPEC strains and enterohaemorrhagic *E. coli* strains (EHEC). The EPEC locus encodes intimin, a 94 kDa outer membrane protein with homology to the invasin proteins of *Yersinia enterocolitica* and *Y. pseudotuberculosis* which promote *Yersinia* invasion by binding to integrin receptors on epithelial cells (Isberg & Leong, 1990; Chapter 1). Intimin is encoded by the *eaeA* gene. A second gene, *eaeB*, is located downstream of *eaeA* and encodes a protein showing homology to aminotransferase enzymes. The *eaeB* gene is required for normal intimate attachment of EPEC to epithelia, but its precise role in the process is unclear (Donnenberg *et al.*, 1993).

Expression of intimin is enhanced by the presence in the cell of the EAF plasmid. This molecule carries the *per* gene, which codes for a *trans*-acting regulatory protein showing homology to the EnvY thermoregulator of porin gene expression in *E. coli* K-12 (Donnenberg & Kaper, 1992; Lundrigan & Earhardt, 1984; Lundrigan *et al.*, 1989). Interestingly, the EnvY protein shows homology to the AraC family of transcription factors (Klaasen & De Graaf, 1990; Fig. 7.10; Appendix I), several of which are plasmid-encoded and control virulence gene expression in response to changes in temperature (e.g. the VirF protein of *Yersinia* species and the VirF protein of *Shigella* species). This is a further example of convergence occurring among bacterial regulatory circuits, both at the level of structure and of function. It will be very interesting to discover which environmental stimuli are involved in controlling the expression of the Per-dependent genes.

LysR family members

The LysR family of bacterial gene regulatory proteins was introduced in Chapter 7. These DNA-binding proteins share amino acid sequence homology in their aminoterminal domains, the region which contains the DNA-binding sequences (a helix-turn-helix motif). Here, those family members that contribute to bacterial virulence gene regulation are described in more detail.

IrgB and iron-regulated virulence genes in *Vibrio cholerae*

Iron plays an important role in *Vibrio cholerae* virulence. This bacterium encodes an outer membrane-located 77 kDa iron-regulated virulence protein called IrgA which has amino acid sequence homology to the FepA protein of *E. coli*. Mutations in the gene coding for this protein result in reduced virulence in a newborn mouse model. The *V. cholerae* protein may play a role in iron-vibriobactin uptake, by analogy with iron-enterochelin uptake via FepA in *E. coli* (Goldberg *et al.*, 1990a,b). However, since iron-vibriobactin does not appear to be essential for virulence in *V. cholerae* (Sigel *et al.*, 1985), loss of the IrgA protein must affect virulence in some other way.

Expression of IrgA is transcriptionally regulated. Negative control is thought to be exerted by the *V. cholerae* counterpart of Fur, the iron regulatory repressor protein of *E. coli*. A match to the consensus binding site for Fur overlaps the *irgA* promoter (Goldberg *et al.*, 1990a). In addition to negative regulation by a Fur-like protein, *irgA* transcription is positively controlled by the IrgB protein, which is a member of the LysR family of transcription regulators (Goldberg *et al.*, 1991; Appendix I). The *irgB* gene is transcribed divergently from *irgA* and their promoters are both subject to negative regulation by the Fur-like repressor (see later). Thus, in conditions of iron sufficiency, transcription of both genes is repressed. When iron becomes limiting, repression is relieved and the IrgB activator protein is synthesized. This then activates *irgA* transcription to the fully induced level, about 850-fold higher than the basal level of expression (Goldberg *et al.*, 1991). This represents an extremely impressive degree of transcriptional derepression.

Thus, the *V. cholerae irgA* gene is simultaneously a member of an iron regulon due to its control by the pleiotropic Fur regulator and also possesses a tightly linked, specific positive regulator, IrgB, permitting regulation of transcription over a very wide range.

SpvR and *Salmonella* virulence gene regulation

The high-molecular-weight virulence plasmids of the non-typhoid serovars of *Salmonella* carry a genetic operon whose products are essential for virulence in mouse model systems (Chapter 3). Specifically, the plasmid appears to be required for survival and proliferation in the tissues of the mouse reticuloendothelial system following either oral or intraperitoneal inoculation (Gulig *et al.*, 1992; Krause *et al.*, 1992). The *spv* system consists of four genes *spvABCD* whose role is presumed to be structural but whose functions are unknown and a regulatory gene, *spvR*, whose product has been assigned to the LysR family on the basis of amino acid sequence homology and whose regulatory function has been deduced from genetic analysis (Caldwell & Gulig, 1991; Fang *et al.*, 1991; Krause *et al.*, 1992; Pullinger *et al.*, 1989; Appendix I; Fig. 7.11). The introduction of non-polar translation termination mutations into the structural genes of the operon in *Salmonella dublin* indicates that SpvB is essential for virulence in mice, SpvA is not essential and SpvC and SpvD appear to make minor contributions to virulence (data from intraperitoneal inoculation of BALB/c mice; Roudier *et al.*, 1992).

SpvR acts as a transcriptional activator of the downstream genes, *spvABCD*, during stationary phase. It also positively regulates its own expression (Taira *et al.*, 1991) (Fig. 8.13). The four downstream genes are differentially transcribed in the order *spvA* > *spvB* > *spvC* > *spvD* (Krause *et al.*, 1992). The operon appears to be expressed in stationary phase as a series of discrete messages of increasing length of which the monocistronic *spvA* transcript is the most abundant and the polycistronic *spvAB*, *spvABC* and *spvABCD* messages are progressively less abundant. In fact, this pattern of expression remains constant throughout the growth cycle but the mRNA species (other than *spvA*) are difficult to detect prior to stationary phase.

The SpvR protein is thought to increase the rate of transcription initiation at the *spvA* promoters (there are two promoters) as the growth cycle advances. Thus, SpvR activity at the *spvA* promoters results in an amplification of the expression pattern. Presumably, the differential expression described earlier reflects transcription termination at successive intergenic sites in the *spv* operon

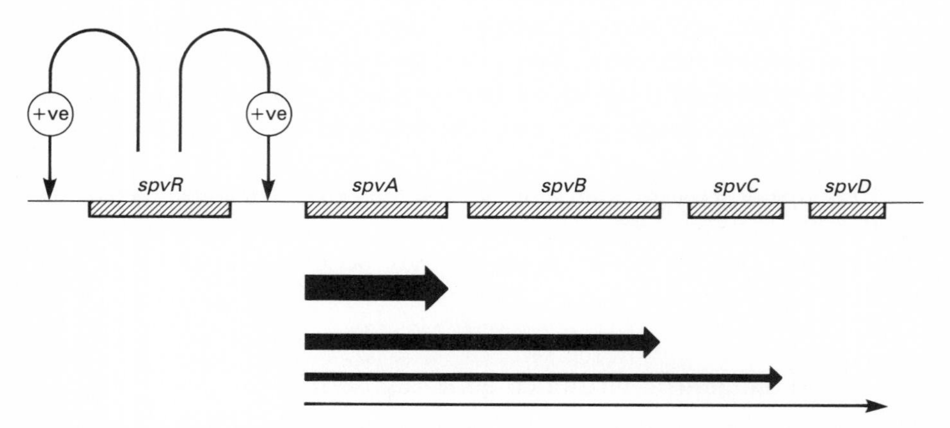

Fig. 8.13 Transcriptional regulation of *spv* gene expression in *Salmonella typhimurium*. The transcription of the four structural genes, *spvABCD* is under the positive control of the regulatory protein SpvR. SpvR also regulates its own gene positively. Transcript abundance is observed to follow the pattern *spvA* > *spvAB* > *spvABC* > *spvABCD*.

such that the longest message is the least abundant and the genes most distal to the *spvA* promoters are the least well expressed. Alternatively, the pattern may reflect differential message stability, with the shortest message being the most stable and the longest the most labile (Fig. 8.13).

Data from experiments with polyclonal antibodies indicate that glucose starvation results in induction of SpvA, SpvB and SpvC protein expression in exponential phase and starvation for glucose in combination with a heat shock in stationary phase results in enhanced expression of SpvA, but not SpvB or SpvC. Iron limitation or acid growth conditions induce these proteins in exponential growth even in the presence of glucose. These environmental inducers require SpvR in order to affect Spv protein expression (Valone *et al.*, 1993). Iron starvation and acid pH are both circumstances that *Salmonella* species would have to adapt to during host infection.

RpoS and virulence gene regulation in enteric bacteria

The *katF* gene of *Salmonella typhimurium* encoding the putative sigma factor RpoS has been identified and shown to be required for virulence (Fang *et al.*, 1992). An *S. typhimurium katF* mutant displays reduced resistance to oxidative stress in stationary phase due to its failure to express Rpos-dependent HPII catalase. The mutant is also impaired in resisting starvation and displays enhanced susceptibility to DNA damage by alkylating agents. It is also hypersensitive to acid pH. These data indicate that the *katF* mutant is unlikely to survive well within the hostile environment of the macrophage. Indeed, the mutant shows significantly reduced virulence in the mouse (Fang *et al.*, 1992). In terms of effects on the expression of specific virulence genes, the *katF* mutant shows reduced transcription of the *spvB* gene on the *S. typhimurium* virulence plasmid (see previous section).

Work with *Shigella flexneri* has also shown a requirement for *katF* for acid resistance and for virulence (Small & Falkow, 1992). Thus, the growth conditions experienced by some pathogens *in vivo* may approximate to those encountered during stationary phase growth *in vitro*. This would explain why genes required for survival and pathogenicity *in vivo* have been placed under stationary phase regulation.

Lrp-regulated genes

The Lrp regulon was introduced in Chapter 7. So far, it has been characterized in detail only in *E. coli* but already it has been shown to include several genes which contribute to virulence. These are discussed next.

The *Escherichia coli pap* operon

Most *E. coli* strains that cause pyelonephritis express P adhesin which binds to the P blood group antigen of host epithelial cells lining the intestinal and urinary tracts (reviewed in Tennent *et al.*, 1990). The pyelonephritis-associated pilus (*pap*) operon is made up of at least 11 genes coding for major (PapA) and minor pilin (PapH, PapE, PapF) proteins, an adhesin (PapG), a protein for pilin translocation across the bacterial membrane (PapC), a periplasmically located chaperone (PapD), a protein required for chaperone integrity (PapJ) and two transcription regulators (Pap I and PapB) (Fig. 8.14; reviewed in Smyth & Smith, 1992).

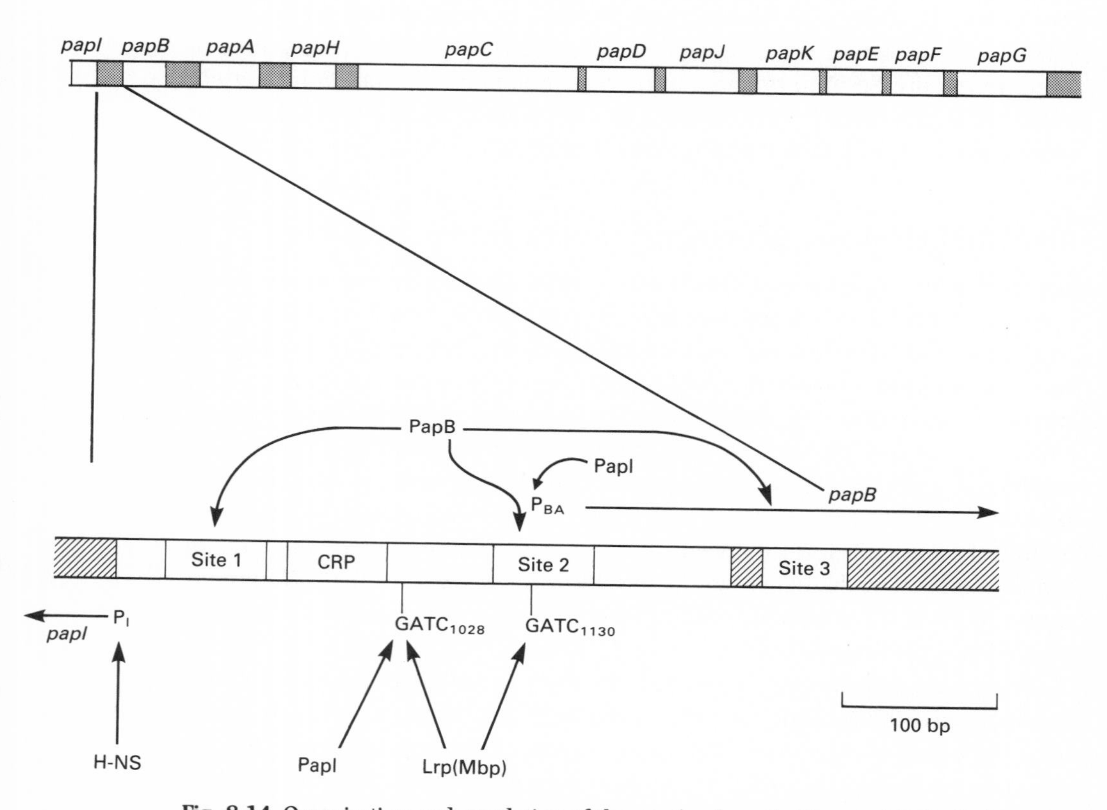

Fig. 8.14 Organization and regulation of the *pap* (pyelonephritis-associated pilus) operon of *Escherichia coli*. The upper part of the figure indictes the relative positions of the *pap* genes. Dark shading indicates intergenic spaces. An enlargement of the regulatory region between *papI* and *papB* is shown. There are three binding sites for PapB in this region (Sites 1, 2 and 3). Site 2 is also a target for PapI. There is a binding site for the cAMP-Crp complex adjacent to Site 1. The positions of the promoters for *papB* (P_{BA}) and *papI* (P_I) are shown. Promoter P_I is subject to negative regulation by H-NS. The GATC sequences through which *pap* phase variation is exerted through DNA Dam methylation are shown. The site at position 1028 is a target for PapI and Lrp (Mbp): the leucine regulatory protein (methylation blocking protein). The site at position 1130 is a target for Lrp (Mbp). The *pap*-encoded functions are: PapA, pilus major subunit; PapB, regulation; PapC, outer membrane translocation and assembly; PapD, chaperone/protein transport; PapE, minor pilin adaptor; PapF, minor pilin; PapG, adhesin; PapH, minor pilin; PapI, regulation; PapJ, chaperone/protein integrity; PapK, minor pilin.

Electron microscopic analysis of purified pili reveals that these consist of a two-stage pylon structure with a wide stem supporting a narrow tip. The main stem is made up of PapA pilin and the tip is made up of PapE. The adhesin is PapG and this is connected to the tip of the pilus by PapF. PapG makes a specific contact with Gal(α-4)Gal carbohydrate, its host surface receptor. While the structure of the pilus on the surface of the bacterium is certainly complex, the mechanism by which it is sent there is equally impressive. Components are transported first to the periplasm where PapD acts as a chaperone to ensure correct protein folding and assembly into the macromolecular structure. PapD is assisted by PapJ. PapC provides a gate in the outer membrane through which the growing pilus is extruded, beginning with the tip protein. Pilus growth is regulated negatively by PapH. Thus, the final length of the pilus is thought to be determined by the relative abundances of PapA and PapH. Evidence in support of this comes from experiments in which the relative amounts of PapA and PapH are varied. If PapH is overexpressed, the cell produces unusually short pili; if PapA is overexpressed, the cell produces unusually long pili.

Pap pilus operon transcription is under phase-variable control such that individual bacteria within the population either express Pap pili (phase ON) or do not express Pap pili (phase OFF). Phase variation is controlled at the level of transcription by Dam methylation at two 5′-GATC-3′ sites within the *papI-papB* intergenic region (see Chapter 7 for a discussion of Dam methylation). These 5′-GATC-3′ sites are positioned at locations 1028 and 1130 in the regulatory region (Fig. 8.14) (Blyn *et al.*, 1990). In phase ON cells, $GATC_{1028}$ is unmethylated; in phase OFF cells, $GATC_{1130}$ is unmethylated. This suggests that transcription of the *pap* genes is regulated in part by the differential protection of these sites. The analysis of this system was facilitated by the differential sensitivity of Dam-methylated DNA to digestion by restriction endonucleases recognizing the tetrameric sequence 5′-GATC-3′. Specifically, the enzyme DpnI cleaves methylated 5′-GATC-3′ sites while the enzyme *Mbo*I cleaves unmethylated 5′-GATC-3′ sites. Protection of $GATC_{1130}$ against methylation requires the *E. coli*-encoded Mbf (methylation blocking factor) protein; protection of $GATC_{1028}$ requires the Mbf and the *pap* operon-encoded PapI protein (Braaten *et al.*, 1991). Methylation of $GATC_{1028}$ inhibits binding of Mbf/PapI here and alters binding of Mbf at $GATC_{1130}$. Thus, the regulatory mechanism centres around the competition of Mbf/PapI and Dam methylase for binding near $GATC_{1028}$, with the methylation/non-methylation of this site regulating *pap* transcription (Nou *et al.*, 1993). The Mbf protein is also a postive regulator of *pap* transcription. Thus, *pap* transcription is under complex, multicomponent control since it is also regulated positively by the *papB* gene product and by cAMP-Crp and is regulated negatively by the H-NS protein (reviewed in Smyth & Smith, 1992) (Fig. 8.14).

The Mbf protein is identical with the the earlier-described Lrp protein, the regulator of the 'Lrp-leucine regulon', which is now known also to include genes not controlled by leucine (Braaten *et al.*, 1992; Lin *et al.*, 1992; van der Woude *et al.*, 1992; Chapter 7). The *pap* operon has been found to belong to the class of genes that is unresponsive to leucine (Braaten *et al.*, 1992).

The K99 (Fan) operon of *Escherichia coli*

The K99 pilus is encoded by the *fan* operon of enterotoxigenic *E. coli* and is required for attachment to the small intestines of lambs, calves and piglets (De Graaf, 1988; Tennent *et al.*, 1990). The major pilin subunit is encoded by the *fanC* gene and the *fanA* and *fanB* genes encode transcriptional regulators (Fig. 8.15). Expression of the *fan* operon is regulated positively by Lrp. Compared with the relatively minor contributions made by FanA and FanB to *fan* operon control, Lrp is a major regulator. Mutations in both *fanA* and *fanB* result in only a twofold decrease in *fan* expression whereas a mutation in *lrp* results in a 70-fold reduction (Braaten *et al.*, 1992). Thus, the Fan regulatory system resembles that of Pap in terms of Lrp involvement. In addition, FanA and FanB show homology to PapB (but not to PapI) (Roosendaal *et al.*, 1987). However, *fan* differs from *pap* in that the expression of *fan* is leucine-sensitive. Addition of L-leucine or L-alanine to the growth medium represses expression of K99 pili by between 25- and 50-fold and this effect is, at least in part, caused by a repression of *fan* transcription (Braaten *et al.*, 1992). The *fan* operon also differs from *pap* in that it does not appear to undergo phase variation. Interestingly, no 5′-GATC-3′ sequences are found within the *fan* regulatory region (van der Woude *et al.*, 1992) suggesting that the mechanism by which Lrp regulates *fan* is different to that by which it controls *pap*.

Lrp and other adhesin operons

The Lrp protein has been implicated in the control of expression of other fimbrial systems in different strains of *E. coli*. These include the type 1 fimbriae, where Lrp stimulates the rate of the site-specific recombination event controlling the orientation of the invertable promoter fragment (Blomfield *et al.*, 1993; Chapter 5). 5′-GATC-3′ boxes are located in the regulatory regions of the *sfa* operon (coding for S pili), the *daa* operon (coding for F1845 pili) and the *fae* operon (coding for K88 pili). Furthermore, the number (two) and spacing of the boxes (102–103 bp) are highly conserved among these operons even though the nucleotide sequences of the spacer regions are quite different (van der Woude *et al.*, 1992). The *sfa* operon is phase variable in its expression and the PapI protein

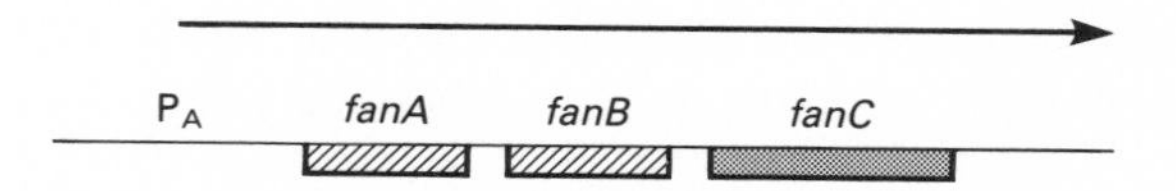

Fig. 8.15 The K99 operon of *Escherichia coli*. The 5′ end of the K99 operon is shown, indicating the *fanA* and *fanB* regulatory genes and the *fanC* gene coding for the K99 subunit. P_A, promoter.

can complement mutations in *sfaC*, a regulatory gene for *sfa* expression (Göransson *et al.*, 1988). There are also some experimental data indicating that Lrp regulates expression of *daa* (Moseley, S., cited in van der Woude *et al.*, 1992).

H-NS-regulated genes

The nucleoid-associated protein H-NS (also known as H1) was introduced in Chapter 2 and discussed in terms of its contributions to transcriptional regulation in Chapter 7. Several of the systems in which H-NS has been shown to play a regulatory role contribute to bacterial virulence and are frequently regulated in response to changes in environmental parameters (especially temperature). Here, those virulence systems in which H-NS regulation has been detected are described.

H-NS and the *pap* operon of *Escherichia coli*

The *pap* pilus operon of uropathogenic strains of *E. coli* is thermoregulated at the level of transcription. The genes are expressed at 37°C but not at lower temperatures, such as 25°C. The thermal control input affects both parts of the divergently transcribed *pap* system, that is both the monocistronic *papI* operon and the polycistronic *papB* operon are coordinately regulated by this environmental signal. An unlinked chromosomal mutation at the *drdX* locus in *E. coli* permits expression of the *pap* pilus operon at 25°C (Göransson *et al.*, 1990) (Fig. 8.14). *drdX* is allelic with *hns* and it has been suggested that it achieves its effect on *pap* gene expression by 'transcriptional silencing', a mechanism in which the H-NS protein makes the *pap* genes unavailable for transcription by sequestering them (or just their promoters) within a chromatin-like structure. This indicates a role for DNA structure in the control of this system, which is consistent with the known role of H-NS in organizing the local architecture of DNA.

H-NS and the *fimA* gene of *Escherichia coli*

The *fimA* gene of *E. coli* encodes the type 1 fimbrial subunit and its promoter is carried within a 314 bp sequence of invertible DNA. The inversion of the promoter fragment makes expression of *fimA* phase variable (Chapter 5). An unlinked mutation on the *E. coli* chromosome at *pilG* was found to result in an acceleration of the rate of *fimA* promoter fragment switching (Spears *et al.*, 1986). A mutation in *osmZ*, an allele of *hns*, was found to produce an identical effect (Higgins *et al.*, 1988a). The *pilG* gene is now known to be allelic with *hns* (Kawula & Orndorff, 1991). In addition to its role in controlling the rate of *fimA* promoter fragment switching, *hns* also regulates transcription negatively from the *fimA* promoter (Dorman & Ní Bhriain, 1992b). This indicates that local DNA structure is crucial to the control of *fim* expression.

H-NS and the porin genes of enteric bacteria

The genes coding for OmpC and OmpF porins in *E. coli* and *Salmonella typhimurium* are subject to transcriptional and post-transcriptional regulation at multiple levels (reviewed in Dorman & Ní Bhriain, 1992a). Transcription of *ompC* and *ompF* is regulated primarily by the EnvZ/OmpR histidine protein kinase/ response regulator system (see earlier). In addition, these promoters are dependent on the histone-like integration host factor (IHF) protein for normal expression and are sensitive to changes in DNA supercoiling (Graeme-Cook *et al.*, 1989; Huang *et al.*, 1990; Tsui *et al.*, 1988). In *E. coli*, mutations in the *hns* gene (called *osmZ* in the porin literature) increase expression of OmpC and decrease expression of OmpF, and these effects are transcriptional (Graeme-Cook *et al.*, 1989). Thus, expression of these important genes is coregulated with other factors contributing to environmental adaptation and virulence through the H-NS protein.

H-NS and the invasion genes of *Shigella flexneri* and enteroinvasive *Escherichia coli*

The structural genes required for invasion and intracellular and intercellular spread in *Shigella* species and enteroinvasive *Escherichia coli* are plasmid-located (Chapter 3) and subject to a complex regulatory cascade involving both plasmid-encoded and chromosomally encoded regulatory proteins. These genes are environmentally regulated. Specifically, expression of the plasmid-linked invasion regulon of *Shigella flexneri* is induced by growth at 37°C and is repressed at 30°C (reviewed in Hale, 1991; Maurelli, 1990). A chromosomal mutation in the *virR* locus of *S. flexneri* abolishes thermoregulation of these genes, such that they are expressed constitutively at both temperatures (Maurelli & Sansonetti, 1988b). The *virR* locus maps to a location on the *S. flexneri* chromosome equivalent to that occupied by *hns* in *E. coli* and when a *virR* : : Tn*10* mutation is introduced to *E. coli* by bacteriophage P1 transduction, it exerts an *hns*-like effect on H-NS-dependent gene expression, showing that *hns* and *virR* are allelic (Dorman *et al.*, 1990). Cloning and sequencing the *virR* locus from *S. flexneri* has confirmed its allelic relationship with *hns* (Hromockyi *et al.*, 1992).

The regulatory cascade which controls virulence gene expression has been described elsewhere in this chapter. A plasmid-encoded AraC-like transcription factor called VirF activates transcription of the plasmid-linked *virB* gene in response to increased temperature. VirB then activates transcription of the virulence genes. The most likely point at which H-NS might influence the expression of the invasion gene regulon is at the *virB* promoter. Overexpression of VirB allows the virulence genes to be expressed, even at the non-permissive temperature. Therefore, the H-NS effect is unlikely to be a factor at the structural gene promoters. The *virB* gene is thermoregulated and requires the AraC-like protein VirF for expression. By analogy with *Yersinia*, where the coincidentally named VirF protein exerts thermal control of transcription of the *yop* genes, temperature probably alters the topology of the DNA at the promoter, allowing VirF to activate transcription (Cornelis, 1992). A situation may be envisaged in which H-NS and VirF act antagonistically at the *S. flexneri virB* promoter, with the former protein achieving its negative regulatory effects at the non-permissive temperature through a change in DNA topology (see, for example, Dorman *et al.*, 1990). When the negative effector (H-NS) is removed genetically, the need for VirF should be removed. This is the case with Cfa/I expression in *E. coli*. Here, expression of this adhesin is positively regulated by an AraC-like protein (CfaD) but the requirement for this positive activator is diminished if H-NS is removed genetically from the cell (Jordi *et al.*, 1992).

In enteroinvasive *E. coli* (EIEC), invasion genes are maintained on a high-molecular-weight virulence plasmid with an operon structure and regulatory circuit which matches closely that seen in *S. flexneri*. Experiments in which H-NS was overproduced or inactivated in EIEC strains have allowed understanding of the role of this nucleoid-assocated protein in thermoregulation to be refined. Specifically, increases in the intracellular concentration of H-NS mimic the effects of reduced temperature and these effects are detected at the *virB* and the *virG* promoters, both of which are subject to positive regulation by the AraC-like protein, VirF. (VirG is a virulence protein required for intercellular and intracellular spread of *Shigella*.) Changes in the level of H-NS do not affect significantly transcription from the invasion gene promoters themselves (Dagberg & Uhlin, 1992). This is consistent with a regulatory cascade in which the thermal signal is manifested in terms of a change in H-NS levels, altering expression of the VirB regulator, whose intracellular concentration is a determinant of invasion gene transcription. The primary role of VirF is to act as a facilitator of *virB* (and *virG*) transcription, which includes a role as an antagonist of the negative input from H-NS.

H-NS and mouse virulence in *Salmonella typhimurium*

Given the pleiotropic nature of *hns* lesions and the demonstration that they can alter the expression of virulence genes, experiments have been carried out to examine the possibility that the *hns* locus contributes to virulence in *Salmonella typhimurium*. Work with *hns* mutants of mouse virulent strains has shown that insertional inactivation of the *hns* gene results in an attenuation of virulence (Harrison, J., Pickard, D., Khan, A., Chatfield, S.N., Dorman, C.J., Hormaeche, C & Dougan, G., unpublished data). However, the role of *hns* in virulence is unclear and it is not known which virulence genes are affected by the mutation. Furthermore, the *hns* lesion appears to be just the first step in a cascade of mutagenic events leading to attenuation. Thus, loss of active H-NS may simply be an enabling event which results, secondarily, in the loss of virulence due to *hns*-promoted mutation. Certainly, both loss of and overexpression of H-NS can result in a loss of cAMP production in *E. coli* (Barr *et al.*, 1992; Lejeune &

Danchin, 1990). If similar events occur in *S. typhimurium*, this might lead to a loss of virulence since mutations in the cAMP-Crp and adenylate cyclase genes of *S. typhimurium* do result in a loss of mouse virulence (Curtiss & Kelly, 1987). Given the contribution of the *S. typhimurium* OmpC and OmpF porins to mouse virulence and the H-NS-dependence which the *ompC* and *ompF* genes show for normal regulation, it is possible that part of the *hns* effect lies in a change in the structure of the cell surface (Chatfield *et al.*, 1991; Graeme-Cook *et al.*, 1989). Another possibility is that the changes in DNA supercoiling which result from inactivation of *hns* could alter the expression of virulence genes with supercoiling-sensitive promoters; the *ompC* and *ompF* genes possess promoters in this category, and so do the *S. typhimurium* invasion genes (Galán & Curtiss, 1990; Graeme-Cook *et al.*, 1989).

H-NS and Cfa/I expression in *Escherichia coli*

The Cfa/I fimbriae of *E. coli* are thermoregulated at the level of transcription, being expressed at 37°C but not at 20°C. The genes coding for these adhesins are carried on a plasmid in enterotoxigenic *E. coli* and two distinct genetic regions (called Region 1 and Region 2), separated by 40 kbp, are required for Cfa/I expression (reviewed in Smyth & Smith, 1992). Region 2 encodes a transcription activator, CfaD, which is required for the expression of the genes in Region 1. CfaD is a member of the AraC family of transcription regulators (Savelkoul *et al.*, 1990; see earlier).

A mutation in the *E. coli hns* gene results in a loss of temperature control of Cfa/I expression; specifically the genes are expressed at both the permissive and the (normally) non-permissive temperature. Furthermore, in the *hns* mutant background, the CfaD protein is redundant. This shows that the function of CfaD is to overcome the negative influence of H-NS on Cfa/I gene expression. However, although the CfaD protein is not needed for Cfa/I activation in the *hns* mutant, its presence still results in greater Cfa/I gene transcription (Jordi *et al.*, 1992). This shows that CfaD is a true transcription activator and not simply an antagonizer of H-NS activity at the Cfa/I gene promoter.

H-NS and curli expression in *Escherichia coli*

Curli are thin, coiled fibres found on the surface of fibronectin- and lamin-binding strains of *E. coli* (Olsén *et al.*, 1989). Their expression is temperature-regulated and the curlin subunit protein is encoded by the *csgA* gene. This gene is transcriptionally activated by the *trans*-acting regulatory gene *crl* at 26°C but not at 37°C. Expression of the *crl* gene is not itself thermoregulated. In *E. coli* K-12 strain HB101, the *crl* gene is absent and the *csgA* gene is cryptic. Introduction of the *crl* gene permits curli to be expressed in this strain (Arnqvist *et al.*, 1992). Alternatively, the cryptic gene can be activated by mutating the *hns* gene, suggesting that DNA topology plays a role in *csgA* regulation (Olsén *et al.*,

1993). Expression of curli is also regulated by osmolarity and by anaerobiosis, two environmental parameters which alter DNA topology in bacteria (Provence & Curtiss, 1992). Furthermore, curli expression is regulated positively by RpoS (the putative sigma factor required for stationary phase gene expression), linking their expression to other stationary phase and osmotic stress-controlled genes (Hengge-Aronis, 1993; Olsén *et al.*, 1993). The function of RpoS appears to be to relieve H-NS-imposed transcriptional repression of *csgA* in the stationary phase of growth (Olsén *et al.*, 1993). Functionally, this antagonism is reminiscent of the roles of AraC-like transcription factors such as CfaD (*E. coli*) and VirF (*Shigella* and enteroinvasive *E. coli*) which relieve transcriptional repression of virulence factor genes imposed by H-NS (see elsewhere, this chapter).

YmoA and gene regulation in *Yersinia enterocolitica*

Expression of the *yop*, *yadA* and *vir* genes of *Yersinia* species is temperature regulated and the *yop* and *virC* operon (*ysc* genes) require the VirF protein for transcription (see under AraC family, earlier). Not all thermoregulated virulence genes in *Yersinia* are VirF-regulated; for example, the *virA* and *virB* genes are VirF-independent. VirF expression is itself controlled by growth temperature. In fact, transcription of *virF* remains thermoregulated in a *Y. enterocolitica* strain which lacks the pYV plasmid (Cornelis *et al.*, 1989b). This shows that a chromosomal gene probably regulates *virF*. A mutation in the *ymoA* locus of the *Y. enterocolitica* chromosome results in derepression of VirF-dependent *yop* gene expression at 28°C; it also results in increased expression of VirF (Cornelis *et al.*, 1991). Growth at 37°C results in even higher levels of gene expression in the *ymoA* mutant. This is indicative of thermal response modulation rather than abolition, hence the designation *ymoA*, for *Yersinia* modulator.

The *ymoA* gene codes for a 8064 Da protein which is rich in charged amino acids. It shows no homology to H-NS, IHF or HU but appears to be a histone-like protein based on its amino acid composition (Cornelis *et al.*, 1991; Appendix I). Mutants deficient in *ymoA* expression are characterized by having fragile chromatin and can display alterations in plasmid DNA supercoiling, although this seems to depend on the plasmid and on the growth conditions. Furthermore, the expression of the thermoregulated, VirF-dependent gene *yopH* becomes osmoregulated in a *ymoA* mutant (Cornelis, 1992; Cornelis *et al.*, 1991).

Y. enterocolitica expresses a heat-stable enterotoxin, Yst, from a chromosomal gene. This toxin shows some structural similarity to the ST_a enterotoxin of *E. coli* and it is expressed in broth cultures grown below 30°C. It has been noted that *Y. enterocolitica* strains appear to lose the ability to express Yst on storage. However, in one Yst-negative strain, enterotoxin production was restored when a mutation in *ymoA* was introduced, suggesting that the toxin gene is maintained in a cryptic form in the non-expressing strains and that this crypticity can be relieved by inactivating *ymoA* (Cornelis, 1992). This is reminiscent of cryptic gene activation as seen in *E. coli* mutants deficient in H-NS (Higgins *et al.*,

1988a). It is not yet known if YmoA can affect the expression of other genes, particularly those with promoters known to be sensitive to variations in DNA topology.

The Hha protein and haemolysin gene expression in *Escherichia coli*

Alpha haemolysin is an important virulence determinant in *E. coli* strains causing infections in humans and animals. Strains isolated from the intestines of infected animals possess plasmid-located haemolysin (*hly*) genes whereas those from human urinary tract infections have the *hly* genes on the chromosome (Müller *et al.*, 1983). In those *hly* operons originating on transmissible plasmids, high-level expression of the operon depends upon *cis*-acting upstream sequences, including a 600 bp region called *hlyR* (Koronakis *et al.*, 1989; Vogel *et al.*, 1988). This region affects both transcription initiation and the antitermination function required for synthesis of the full-length operon transcript (see separate section later). An analogous expression-enhancement system has been found upstream of chromosomally located *hly* operons (Cross *et al.*, 1990).

The *hha* chromosomal gene of *E. coli* codes for an 8.6 kDa protein, which shows homology to the YmoA histone-like protein of *Yersinia entercolitica* (Appendix I). Hha negatively regulates *hly* gene expression but only in the absence of the *hlyR* sequence. The main effect of an *hha* mutation is to increase the level of the *hlyCA* transcript, probably by increasing the rate of transcription of the *hly* genes. Hha also has effects on the expression of other, unrelated genes carried on recombinant plasmids (Nieto *et al.*, 1991; Bailey *et al.*, 1992; Carmona *et al.* 1993). Mutations in *hha* alter the level of DNA supercoiling, indicating a role for DNA topology in the regulation of *hly* gene expression (Carmona *et al.*, 1993).

Hc1 and developmental regulation in *Chlamydia trachomatis*

Biovars of *Chlamydia trachomatis* that are infectious to humans cause trachoma (the leading cause of preventable blindness in the world) and sexually transmitted diseases (reviewed in Bavoil, 1990). *C. trachomatis* is an invasive prokaryote which has a biphasic life cycle consisting of an extracellular, infectious phase called the elementary body and an intracellular, replicating phase known as the reticulate body phase. The elementary body (EB) is metabolically inert but once it has infected a susceptible cell it undergoes differentiation to form the metabolically active reticulate body (RB). The EB form is characterized by a dense chromatin structure, the RB form possesses a more diffuse nucleoid, which is genetically active. *C. trachomatis* expresses an 18 kDa histone-like protein called Hc1, which is encoded by the *hctA* gene (Appendix I) (Hackstadt *et al.*, 1991). When expressed in *E. coli*, Hc1 condenses the bacterial nucleoid (Barry *et al.*, 1992) and the purified protein has the ability to pack DNA into condensed spherical bodies; it shows a slight preference for supercoiled DNA (Christiansen *et al.*, 1993). The *hctA* gene is developmentally regulated and is expressed late in the RB phase of the life cycle at the onset of differentiation to the EB form. Hc1 seems to play a general role in condensing the *C. trachomatis* nucleoid (Hackstadt *et al.*, 1991). When the *hctA* gene is expressed in enteric bacteria, it alters the level of DNA supercoiling and the transcription of genes with DNA supercoiling-sensitive promoters such as *ompC*, *ompF* and *proU* in *E. coli* and *his* in *Salmonella typhimurium* (Barry *et al.*, 1993). The results indicate that Hc1 may alter transcription within *Chlamydia* prior to condensation of the nucleoid (Barry *et al.*, 1993).

The Fur repressor and iron regulation of virulence gene expression

Iron is frequently a limiting factor in the determination of the ability of bacteria to colonize any environment (Chapter 7). Under aerobic, neutral pH conditions, ferric ions are largely insoluble and so unavailable to bacteria. Within the human or animal host, the bacterium must compete with highly efficient iron-sequestering systems such as albumin, ferritin, lactoferrin, haptoglobin, haemopexin, haemoglobin and transferrin. Bacteria use iron-scavenging systems to overcome iron deficiency in the host and these systems are frequently multi-component and complex. Consequently, the expression of the genes coding for these is strictly regulated. The Fur repressor protein plays a key role in the transcriptional control of these iron uptake systems (Chapter 7; Appendix I). It ensures that the iron-scavenging apparatus is only activated during periods of severe iron deficiency (Neilands, 1990).

Iron-regulated proteins have been identified in a number of pathogenic bacteria and include iron-repressible outer membrane proteins in *E. coli*, *Enterobacter cloacae*, *Erwinia* species, *Klebsiella aerogenes*, *Neisseria gonorrhoeae*, *Pseudomonas aeruginosa*, *Ps. putida*, *Salmonella typhimurium*, *Shigella flexneri*, *Vibrio anguillarum*, *V. cholerae* and *Yersinia* species (Carniel *et al.*, 1987; Expert & Toussaint, 1985; McIntosh & Earhart, 1976; Neilands *et al.*, 1982; Oudega *et al.*, 1979; Prince *et al.*, 1991; Sikkema & Brubaker, 1989; Staggs & Perry, 1991; Weisbeek *et al.*, 1990; West & Sparling, 1985; Williams *et al.*, 1984). The expression of several bacterial toxins is iron-regulated, for example, shiga toxin of *Shigella dysenteriae* (Dubos & Geiger, 1946), shiga-like toxin of *E. coli* (Calderwood & Mekalanos, 1987), haemolysin of *Vibrio cholerae* (Stoebner & Payne, 1988) and diphtheria toxin of *Corynebacterium diphtheriae* (Boyd *et al.*, 1990; Cryz *et al.*, 1983). These observations have prompted searches for iron regulatory proteins in these pathogens which function in a manner analogous to that of Fur in *E. coli*. A polyclonal rabbit serum containing antibodies that recognize *E. coli* Fur detects a polypeptide of similar molecular weight in *Pseudomonas aeruginosa*, *Salmonella typhimurium*, *Shigella flexneri* and *Vibrio cholerae* (Prince *et al.*, 1991).

A Fur homologue in *Yersinia*

A *fur* regulatory gene has been identified and cloned from *Yersinia pestis* using the *E. coli fur* gene as a DNA probe (Staggs & Perry, 1991; Appendix I). The cloned gene is capable of complementing a mutation in the *E. coli fur* locus and may contribute to the iron-mediated regulation of *Y. pestis* genes involved in ferric iron uptake, as well as those concerned with utilization of haemin, haeme/haemopexin, haeme/albumin, ferritin, haemoglobulin and haemoglobin/haptoglobin (Staggs & Perry, 1991). The 16 kDa *Y. pestis* Fur protein displays 84% amino acid sequence homology to its *E. coli* counterpart and can complement a *Y. enterocolitica fur* mutation (Staggs & Perry, 1992).

The haemin receptor of *Y. enterocolitica*, HemR, has been identified as a 78 kDa iron regulated outer membrane protein whose expression is rendered constitutive in a *fur* mutant of *Y. enterocolitica* (Stojiljkovic & Hantke, 1992). Furthermore, as with iron uptake systems in other Gram-negative bacteria, haemin uptake in *Y. enterocolitica* appears to depend on the periplasm-spanning TonB protein. Analyses carried out in *E. coli* and *Salmonella typhimurium* have shown that the *tonB* gene is itself subject to negative regulation by Fur and is anaerobically regulated and possesses a promoter sensitive to changes in DNA supercoiling (Dorman *et al.*, 1988; Postle, 1990). TonB is believed to link outer membrane proteins to the energy-generating processes of the cytoplasm.

A Fur homologue in *Pseudomonas aeruginosa*

In *Pseudomonas aeruginosa*, iron plays a key role in controlling the expression of virulence factors. The *toxA* gene, which codes for the ADP-ribosylating exotoxin A, is regulated by iron at the level of transcription. It is expressed at its maximum levels under conditions of low iron and is repressed under high-iron conditions. Maximum expression of *toxA* requires the products of the *toxR* gene (now renamed *regA*) (Hedstrom *et al.*, 1986) and the *regB* gene, located downstream of *regA* (Wick *et al.*, 1990). Transcription of *regA* involves two promoters (P_1 and P_2) one of which, P_2, is tightly regulated by iron while the other, P_1, is not (Storey *et al.*, 1990). Furthermore, transcription from P_1 occurs early in the growth cycle regardless of the iron status of the culture whereas transcription from P_2 occurs late in the growth cycle but only under iron-limiting conditions (Frank & Iglewski, 1988; Storey *et al.*, 1990).

The introduction of a multicopy plasmid encoding the *E. coli fur* gene results in repression of a *lacZ* fusion to the *toxA* gene even under low-iron conditions. Transcription of *regA* is also repressed by multiple copies of *E. coli fur* (Prince *et al.*, 1991). Multicopy *E. coli fur* also has a negative effect on the expression of protease in *Pseudomonas aeuginosa* strain PA103C (Prince *et al.*, 1991). A polyclonal serum containing antibodies that recognize *E. coli* Fur detects a protein of similar molecular weight in *Ps. aeruginosa* (Prince *et al.*, 1991). These results suggested that *Ps. aeruginosa* possesses a Fur homologue and that it may play a role in regulating the expression of virulence determinants. This has now been shown to be correct. The *Ps. aeruginosa* Fur protein shows over 50% homology to the *E. coli* protein but is missing the carboxyterminal domain believed to be required for metal-binding (Prince *et al.*, 1993). This domain is present in the Fur proteins of *Vibrio cholerae* and *Yersinia pestis*, suggesting that the *Ps. aeruginosa* protein works in a different way to these. Perhaps it has no affinity for its DNA-binding site under conditions of low iron availability, in contrast to the *E. coli* protein which may still have some binding ability under such conditions. This would mean that the iron regulons of these two organisms were regulated across different ranges of available iron.

Fur and haemolysin expression in *Serratia marcescens*

Expression of *Serratia marcescens shlA*-encoded haemolysin is regulated by iron in *E. coli* but not in an *E. coli fur* mutant, implying that this gene may be Fur-dependent in *Serratia* (Poole & Braun, 1988). This represents yet a further example of a virulence factor in a Gram-negative bacterium which is iron-controlled and possibly regulated by a member of the ubiquitous Fur-like protein group at the level of transcription.

Fur and gene expression in *Vibrio cholerae*

Several pieces of evidence have pointed to the existence in *Vibrio cholerae* of a Fur protein. Firstly, expression of the IrgA iron-regulated outer membrane protein, which may be involved in iron uptake in *V. cholerae* and dependent on a TonB-like protein, is positively regulated by the IrgB protein (a member of the LysR family; see earlier) and negatively regulated by the *V. cholerae* counterpart of Fur (Goldberg *et al.*, 1990a, 1992). In fact, since *irgA* and *irgB* are divergently transcribed and appear to share the same 'iron box', their expression may be coordinately repressed by Fur under iron-replete conditions. Secondly, iron regulation can be restored to a *V. cholerae* mutant which expresses constitutively a haemolysin when a recombinant plasmid encoding the *E. coli* Fur protein is introduced (Stoebner & Payne, 1988). Thirdly, the gene coding for the vibriobactin receptor, ViuA, has been found to be regulated by Fur in *E. coli* and to have a match to the consensus Fur-binding site overlapping the −10 and −35 boxes of the *viuA* promoter (Butterton *et al.*, 1992). Fourthly, Prince *et al.* (1991) detected an analogue of *E. coli* Fur immunologically in *V. cholerae.*

More recently, the *fur* gene of *V. cholerae* has been cloned and sequenced and the predicted amino acid sequence of its product shows a high degree of homology to the Fur protein of *E. coli* (Litwin *et al.*, 1992). This homology suggests that the *V. cholerae* protein probably interacts with DNA in a manner similar to that of the *E. coli* Fur protein. This is in contrast to the situation with the *Ps. aeruginosa* Fur protein (see earlier).

Fur and expression of *Escherichia coli* shiga-like toxin

Shiga-like toxin (SLT or vero toxin) of *E. coli* is quite similar to shiga toxin from *Shigella dysenteriae* type 1 in biological activity, subunit structure and immunological cross-reactivity (Calderwood & Mekalanos, 1987). *E. coli* strains expressing the toxin cause diseases such as haemorrhagic colitis. SLT enters target cells by receptor-mediated endocytosis. It is composed of two subunits, A and B. The B subunit interacts with the receptor glycolipid while the A subunit is thought to undergo proteolytic cleavage. Activated A subunit is an *N*-glycosidase and it cleaves specifically 28S rRNA. This activity blocks translation in the target cell (reviewed in Wren, 1992)

The A and B subunits are encoded by the bacteriophage-encoded *sltA* and *sltB* genes. These are probably transcribed as an operon in bacteriophages H19B and 933J (Calderwood *et al.*, 1987; Jackson *et al.*, 1987). Like shiga toxin, the expression of shiga-like toxin is iron regulated, with the level of toxin being greatest in low iron conditions (Dubos & Geiger, 1946; O'Brien *et al.*, 1982). The *slt* promoter is regulated negatively by Fur (Calderwood & Mekalanos, 1987) and this discovery has led to the identification and confirmation of the consensus sequence for the 'iron box' (Calderwood & Mekalanos, 1988; Chapter 7). Thus, iron starvation prompts *E. coli* to express SLT, causing damage to host cells and perhaps releasing growth factors, including iron. By placing the bacteriophage-encoded *slt* genes under Fur control, the cell coregulates these with its other major iron starvation response systems.

DtxR and expression of diphtheria toxin

A *galK* fusion to the diphtheria toxin gene (*tox*) promoter from *Corynebacterium diphtheria* was found to be iron regulated in *E. coli* but not in an *E. coli* strain carrying a *fur* mutation (Tai & Holmes, 1988). Given that expression of the toxin is iron-controlled in *C. diphtheria*, it appeared likely that this bacterium employs a Fur analogue to exert this control. Subsequently, the regulator of diphtheria toxin gene expression has been found to be encoded by the *dtxR* gene and is a 25 kDa protein with some amino acid sequence homology to Fur (Boyd *et al.*, 1990; Appendix I). This iron-responsive protein also regulates expression of siderophores in *C. diphtheriae* (Boyd *et al.*, 1990; Schmitt & Holmes, 1991). Thus, DtxR regulates expression of a phage-encoded toxin (diphtheria toxin) and chromosomally encoded *C. diphtheriae* genes with iron-responsive expression. The gene for DxtR is chromosomally-located. This is directly analogous to the situation with shiga-like toxin in *E. coli* which is phage-encoded and regulated by Fur in concert with other iron-controlled genes (see previous section). Although DtxR can regulate *tox* expression in *E. coli* in response to iron, it cannot regulate Fur-dependent *E. coli* genes, indicating a mode of action distinct from Fur in terms of *E. coli* gene control (Boyd *et al.*, 1990).

The promoter region of the *tox* gene resembles that of σ^{70} driven promoters from *E. coli*. Its −10 and −35 boxes are overlapped by a region of dyad symmetry to which DtxR binds (Schmitt & Holmes, 1993). This inverted repeat differs in structure from the 'iron box' recognized and bound by Fur in *E. coli* (Tao & Murphy, 1992; Tao *et al.*, 1992). DtxR must bind Fe^{2+} in order to be proficient for DNA binding. Other divalent cations, such as Co^{2+}, Mn^{2+}, Ni^{2+}, and Zn^{2+} can substitute for Fe^{2+} (Tao & Murphy, 1992). Some of these metals will also activate DNA binding by Fur (Coy & Neilands, 1991). DtxR is thought to bind to its operator site as a dimer, possibly through a helix-turn-helix motif (Schmitt & Holmes, 1993).

Regulation of elastase gene expression in *Pseudomonas aeruginosa*

Elastase is believed to be required for maximum virulence in *Pseudomonas aeruginosa*. It is a 32 kDa zinc-metalloprotease which degrades elastin, collagen and other biologically important molecules, perhaps providing the bacterium with an infection route (reviewed in Galloway, 1991). Elastase displays expression kinetics which mirror those of exotoxin A, where there is an initial iron-independent phase followed by an iron-repressible phase in late exponential growth. This biphasic pattern of *toxA*-encoded exotoxin A expression is imposed, at least in part, by the 29 kDa RegA regulatory protein (previously known as ToxR) (Hindahl *et al.*, 1988; Wozniak *et al.*, 1987). On the other hand, transcription of the elastase gene, *lasB*, is positively regulated by LasR (27 kDa), which shows sequence homology to other regulators such as LuxR, the *Vibrio fischeri* bioluminescence regulator (Gambello & Iglewski, 1991; Appendix I). Since *lux* gene expression is regulated by *N*-(3-oxohexanoyl) homoserine lactone (HSL) in *V. fischeri*, and *Ps. aeruginosa* is known to produce HSL, this suggests that elastase expression could be regulated by this diffusible signal molecule (Williams *et al.*, 1992). The *lasI* gene is involved in the synthesis of this signalling molecule, which has been named *Pseudomonas* autoinducer or PAI (Passador *et al.*, 1993).

LasB expression seems also to be regulated at the level of translation in response to iron and zinc, although the precise mechanism of this post-transcriptional control remains unclear (Brumlik & Storey, 1992); zinc is not a regulator of exotoxin A (Blumenthals *et al.*, 1987). Little is known of the regulation of *lasA*, the second elastase gene found in *Ps. aeruginosa* (Galloway, 1991).

Gene regulation by *N*-acyl-L-homoserine lactones in bacterial pathogens

The role of *N*-acyl-L-homoserine lactone (HSL) molecules in regulating elastase expression in *Pseudomonas aeruginosa* has been referred to in the previous section. These autoinducer molecules have also been discovered as regulators of *tra* operon expression in the plant pathogen, *Agrobacterium tumefaciens*. Here, a transcription regulator, TraR, shows significant homology to LuxR, the bioluminescence gene regulator from *Vibrio fischeri* (Piper *et al.*, 1993). Furthermore, a small diffusable molecule inducer of this system, previously called CF (for conjugation factor) has been identified as a member of the HSL family (Zhang *et al.*, 1993). HSL production has also been detected in *Citrobacter, Erwinia, Enterobacter, Hafnia, Proteus, Rhanella* and *Serratia* (reviewed in Williams *et al.*, 1992) showing that this type of regulator is widely present among bacterial species. In *Erwinia carotovora*, the cause of soft rot disease in potatoes, expression of exoenzymes required for plant infectivity is regulated by HSL (Jones *et al.*, 1993; Pirhonen *et al.*, 1993).

Transcription antitermination and regulation of haemolysin expression in *Escherichia coli*

In *Escherichia*, the location of the alpha-haemolysin (*hly*) operon varies from strain to strain. Those strains isolated from the intestines of infected animals possess plasmid-located genes whereas those from human urinary tract infections have the *hly* genes on the chromosome (Müller *et al.*, 1983). Unlike the chromosomally located genes, those found on plasmids are frequently flanked by insertion sequences (Hess *et al.*, 1986; Knapp *et al.*, 1985).

Expression of haemolysin in *E. coli* requires the four genes of the *hly* operon, *hlyCABD*. The *hlyCA* gene products are required for haemolysin synthesis and the *hlyBD* gene products are involved in secretion. The operon is transcribed from a promoter located upstream of the *hylC* gene and two transcripts result (Fig. 8.16). This is because the synthesis and secretion functions can be uncoupled by transcription termination at a rho-independent terminator located between *hlyA* and *hlyB* (Koronakis *et al.*, 1988). Furthermore, the *hlyCA* and the *hlyCABD* mRNAs display differential stability, with the *hlyCA* message being twice as stable as that of *hlyCABD* (Welch & Pellett, 1988). In those *hly* operons originating on transmissible plasmids, high-level expression of the operon depends upon *cis*-acting upstream sequences, including a 600 bp region called *hlyR* (Koronaskis *et al.*, 1989; Vogel *et al.*, 1988). This region affects both

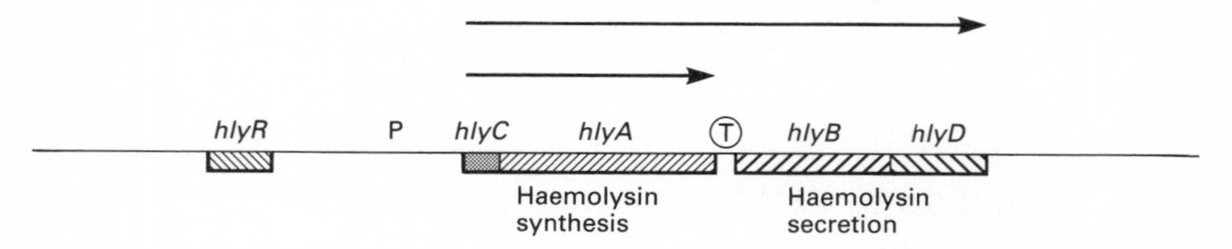

Fig. 8.16 Genetic organization of the *Escherichia coli* haemolysin operon. The *hylR* regulatory sequence is located upstream of the *hly* operon promoter (P). This consists of four structural genes, *hlyCABD*. *hlyC* and *hlyA* are required for haemolysin synthesis while *hlyB* and *hlyD* are required for haemolysin secretion. Transcription of the operon can be interrupted at a rho-independent terminator (T) located between *hlyA* and *hlyB*.

transcription initiation and the antitermination function required for synthesis of the full-length operon transcript. An analogous expression-enhancement system has been found upstream of chromosomally located *hly* operons (Cross *et al.*, 1990).

A gene coding for a transcriptional activator, HlyT, has been identified genetically and shown to be required for haemolysin gene expression (Bailey *et al.*, 1992). The *hlyT* gene is allelic with the *rfaH* (or *sfrB*) regulatory gene which activates expression of genes coding for the lipopolysaccharide core for attachment of the O-antigen of *E. coli* and plasmid-linked genes involved in sex pilus synthesis (Bailey *et al.*, 1992; Pradel & Schnaitman, 1991). The HlyT protein may act as a transcription antiterminator. Data from work on its role in pilus gene regulation (HlyT in the guise of SfrB) supports this hypothesis (Beutin *et al.*, 1981) and it has been shown genetically that HlyT (in the guise of RfaH) probably interacts with Rho factor (Farewell *et al.*, 1991). Other data indicate that the primary (if not the only function) of HlyT is in controlling transcription initiation. Furthermore, the transcription termination site within *hly* is rho-independent so interactions between HlyT and Rho are unlikely to be relevant there (Bailey *et al.*, 1992).

Another unlinked gene, *hha*, also contributes to haemolysin expression, but only in the absence of *hlyR*. The *hha* gene codes for an 8.6 kDa protein, which shows homology to the YmoA histone-like protein of *Yersinia enterocolitica*, and negatively regulates *hly* gene expression (Appendix I). Hha also has effects on the expression of other, unrelated genes carried on recombinant plasmids (Nieto *et al.*, 1991). Whether or not *hha* contributes to the normal regulation of *hly* expression is a matter of contention (Bailey *et al.*, 1992).

DNA supercoiling and virulence gene regulation

The need to reset expression of the genetic material rapidly during environmental adaptation has led to searches for global regulators that are environmentally sensitive. DNA supercoiling appears to fulfil these criteria. It has the potential to reset the expression of every gene in the cell and its level is subject to quantitative variation in response to changes in a range of environmental parameters (temperature, osmolarity, anaerobiosis, etc.) which are likely to be experienced during infection. This has led to the proposal that variable supercoiling of DNA could assist not only environmental adaptation and survival, but also the pathogenic process itself (Dorman, 1991). If this proposal has any merit, then as a minimum, some virulence genes ought to have supercoiling-sensitive promoters.

DNA topology and invasion gene expression in *Shigella flexneri*

Mutations in the *hns* gene in *Shigella flexneri* (*virR* mutations; Maurelli & Sansonetti, 1988b) alter the global level of DNA supercoiling in that organism and deregulate the expression of virulence plasmid-encoded invasion genes (Dorman *et al.*, 1990). Mutations in the *topA* gene of *S. flexneri* also result in loss of thermal regulation, indicating that global and/or local DNA supercoiling levels contribute to the environmental control of virulence gene expression in this organism (Ní Bhriain & Dorman, 1993). Mutations in *hns* alter the expression of several other virulence factors in enteric bacteria (see earlier). Similarly, mutations in genes coding for other proteins with the actual or potential ability to alter the topology of DNA (such as AlgP, H1c, Hha, IHF, YmoA, etc.) alter virulence factor expression (see earlier and Chapter 7).

Invasive phenotype and DNA supercoiling in *Salmonella typhimurium*

Expression of the invasiveness phenotype in *Salmonella typhimurium* is subject to regulation by many environmental parameters known to alter DNA supercoiling. Transcription of the chromosomally encoded *invA* gene is induced eightfold by osmotic stress and this induction requires DNA gyrase. This is consistent with *invA* induction being promoted by increased DNA supercoiling. Increases in the level of DNA supercoiling can also be achieved by inactivation of DNA topoisomerase I. However, in a *topA* mutant, expression of *invA* is poorly expressed. It is possible that this counterintuitive finding could be due to a local requirement for DNA topoisomerase I which is distinct from effects due to the global changes in supercoiling which accompany gyrase inhibition or osmotic insult. *S. typhimurium* strains with deletions of *topA* are less invasive than the wild-type in *in vitro* invasion assays, supporting the notion that this protein (and hence the degree of DNA supercoiling) contributes to the control of the invasive phenotype at some level (Galán & Curtiss, 1990).

Mutations in the *hns* gene of *S. typhimurium* which result in alterations in the global level of DNA supercoiling also attenuate the mouse virulence of this pathogen. This shows that a major component of the bacterial nucleoid which affects the topology of DNA is required *in vivo* for infectivity in the whole animal. However, the contribution of H-NS to virulence is probably indirect. It seems that loss of the protein promotes an increased mutation rate in the bacterium and that it is these second site mutations that are really the attenuating lesions (Harrison, J., Pickard, D., Khan, A., Chatfield, S.N., Dorman, C.J., Dougan, G. & Hormaeche, C., unpublished data).

Context-dependent virulence gene expression in *Vibrio cholerae*

The *acfA* and *acfD* genes of *Vibrio cholerae* are required for expression of the accessory colonization factor (Acf) and their promoters display a context-dependent expression that includes sensitivity to changes in DNA topology. When carried on the chromosome, transcription of these genes is subject to control by the ToxR ToxT regulatory cascade (see earlier). When the genes are cloned in a recombinant plasmid, normal regulation is lost. The ToxR-dependent expression of these genes requires DNA gyrase, since this regulation is lost when the cell is treated with gyrase-inhibiting antibiotics (Parsot & Mekalanos, 1992). The *acfA* and *acfD* genes are transcribed divergently and the start of one open reading frame is only 173 bp away from the start of the other. This could facilitate topological coupling of the promoters through changes in DNA supercoiling generated when these genes are transcribed and translated. Furthermore, the intergenic region is very AT-rich and contains runs of helically phased A residues which might form intrinsically bent DNA. Thus, this structural and genetic arrangement has many of the hallmarks of a DNA sequence whose biological activity is likely to be highly sensitive to changes in topology and that is capable of contributing to such changes.

Alginate synthesis and DNA supercoiling in *Pseudomonas aeruginosa*

The *algD* gene of *Ps. aeruginosa* encodes GDP-mannose dehydrogenase, an essential component of the alginate biosynthetic pathway. Alginate synthesis, leading to the mucoid phenotype, is an important component of the pathology of *Ps. aeruginosa* infection of the cystic lung. Alginate protects the bacteria from the host immune system, from penetration by antibiotics and assists in microcolony formation. Transcription of *algD* is subject to complex control, and is achieved in part by the histone-like protein AlgP (see earlier). When carried on a recombinant plasmid in *E. coli*, *algD* is osmotically inducible and this induction is abolished when the cell is treated with DNA gyrase-inhibiting antibiotics (Berry *et al.*, 1989). Osmotic induction is not seen in a temperature-sensitive *E. coli gyrB* mutant at the non-permissive temperature (DeVault *et al.*, 1990). In addition, dehydration of *Ps. aeruginosa* by ethanol causes induction of *algD*, although the same treatment fails to induce *algD* in *E. coli*, leading to speculation that receptors specific to *Ps. aeruginosa* are required for this signal to be transmitted from the membrane to the genome (DeVault *et al.*, 1990). In *Ps. aeruginosa*, ethanol activation of *algD* is antagonized by treatment with gyrase-inhibiting antibiotics or is reduced when studied in a mutant with a defective gyrase. Furthermore, selection of high-level nalidixic acid resistant derivatives (presumably *gyrA* mutants) of an alginate producing strain results in loss of the mucoid phenotype (DeVault *et al.*, 1990). These data are consistent with a role for DNA supercoiling in the regulation of *algD* expression.

DNA supercoiling and virulence gene expression in *Staphylococcus aureus*

Expression of *eta*, the *Staphylococcus aureus* gene coding for epidermolytic toxin, is under the control of the accessory gene regulator (Agr), which is a member of the histidine protein kinase/response regulator superfamily of signal transduction systems (see earlier). Transcription of *eta* increases dramatically at the end of exponential growth and at the onset of the stationary phase of the growth cycle. This growth phase-dependence is imposed by Agr. Experiments with reporter gene fusions to *eta* reveal that its transcription is osmotically regulated, with the level of expression being related inversely to the level of osmolyte (Sheehan *et al.*, 1992). The *eta* promoter is induced strongly when the cell is treated with the DNA gyrase-inhibiting antibiotic novobiocin and this effect is independent of Agr. Thus, expression of *eta* is subject to regulation at at least two levels: an osmotic-dependent and growth phase-dependent control acting through Agr and an Agr-independent control acting via DNA topology (Sheehan *et al.*, 1992). Thus, DNA supercoiling contributes to the control of virulence factor expression in Gram-positive as well as Gram-negative pathogens.

Pertussis toxin expression and DNA supercoiling

The *ptx* gene of *Bordetella pertussis* codes for pertussis toxin and is environmentally regulated at the level of transcription by the BvgA/BvgS two-component regulatory system (see above). When the promoter for the *ptx* gene (Ptox) is cloned in *E. coli in cis* with the *bvg* regulatory locus, it remains environmentally regulated. When Ptox and the *bvgAS* genes are carried on separate plasmids, transcription from Ptox becomes DNA-context-dependent and can be modulated by inhibiting DNA gyrase with the antibiotic novobiocin (Scarlato *et al.*, 1993).

Ptox can be expressed *in vitro* with *E. coli* RNA polymerase and a total protein extract from *B. pertussis*. Adding DNA gyrase to this mixture inhibits Ptox while adding DNA topoisomerase I overcomes this inhibition. These data are consistent with Ptox function being sensitive to changes in DNA supercoiling. Specifically, an increase in negative supercoiling due to DNA gyrase activity is inhibitory while relaxation of negative supercoiling by DNA topoisomerase I is stimulatory. The *ptx* promoter is an example of a DNA supercoiling-sensitive gene which is also subject to specific control by a dedicated regulatory system, in this case the BvgA/S two-component system.

DNA supercoiling and haemolysin gene expression in *Escherichia coli*

Mutations in the *hha* gene, which codes for a histone-like protein (see above), result in derepression of trancription of the *hly* operon which codes for α-haemolysin. These mutations also alter the level of supercoiling of reporter plasmid DNA. Production of haemolysin expression can be induced by inactivating DNA gyrase with novobiocin, indicating that the *hly* promoter is sensitive to the level of supercoiling. This is supported by data from experiments with *E. coli* strains harbouring mutations in *gyr* genes. Furthermore, growth medium osmolarity, which alters the *in vivo* level of DNA supercoiling, modulates *hly* gene expression (Carmona *et al.*, 1993).

Integration of supercoiling control with specific regulatory networks

Multifactorial control is a feature common to most of the systems described in this section. This is consistent with a role for environmentally controlled DNA supercoiling in setting the background against which more specific controls operate. As with promoters involved in expression of 'housekeeping' genes, the ability of virulence gene promoters to respond to new degrees of DNA supercoiling is constrained by these more specific factors. They will also be limited in their ability to respond by their own primary structure, by the influences of proteins concerned with organizing the architecture of the nucleoid (H-NS, IHF, FIS, HU, etc.) and by the activities of neighbouring genes (see Chapter 6 for a detailed discussion of these issues).

Conclusion

This book has reviewed the mechanisms by which bacterial cells regulate expression of their genetic material in general and how these mechanisms apply in particular to genes coding for virulence determinants. This field of investigation has moved forward very unevenly on many fronts. Consequently, much is known about particular systems while much remains to be done in others. Still others almost certainly remain to be discovered. Despite the gaps in knowledge and the uneven understanding of different systems, it is apparent that the field of gene regulatory studies is entering a period of consolidation. In fact, this is true of prokaryotic cell biology in general. More detail will continue to flow in from the many research groups in the field, and the flow can be expected to increase now that knowledge of the nucleotide sequence of the entire *E. coli* K-12 genome is imminent. Nevertheless, the trend in future will be to fit this new detail into a more and more completely integrated picture of the bacterial cell. As the *E. coli* jigsaw puzzle approaches completion at the level of component identification, new information will emerge on how the whole functions as a unit. This will benefit studies in less well-characterized organisms by giving strong indications about how bacterial cells are organized in general.

Appendix I: Primary Sequences of Selected DNA-Binding Proteins

Prokaryotic histone-like proteins

Primary sequence of H-NS

Escherichia coli

```
        10         20         30         40         50         60
MSEALKILNN IRTLRAQARE CTLETLEEML EKLEVVVNER REEESAAAAE VEERTRKLQQ
        70         80         90        100        110        120
YREMLIADGI DPNELLNSLA AVKSGTKAKR AQRPAKYSYV DENGETKTWT GQGRTPAVIK
       130    137
KAMDEQGKQL DDFLIKQ
```

Proteus vulgaris

```
        10         20         30         40         50         60
MSESLKILNN IRTLRAQARE TSLETLEEML EKLEVVVNER REEEQAMQAE IEERQQKLQK
        70         80         90        100        110        120
YRELLIADGI DPTDLLEAAG ASKSGRAKRA ARPAKYSYVD ENGETKTWTG QGRTLAVIKR
       130   134
AIEEEGKSLE DFLI
```

Salmonella typhimurium

```
        10         20         30         40         50         60
MSEALKILNN IRTLRAQARE CTLETLEEML EKLEVVVNER REEESAAAAE VEERTRKLQQ
        70         80         90        100        110        120
YREMLIADGI DPNELLNSMA AAKSGTKAKR ARRPAKYSYV DENGETKTWT GQGRTPAVIK
       130    137
KAMEEQGKQL EDFLIKE
```

Serratia marcescens

```
        10         20         30         40         50         60
MSERLKILNN IRTLRAQARE CTLETLEEML EKLEVVVNER REEDSQAQAE IEERTRKLQQ
        70         80         90        100        110        120
YREMLIADGI DPNELLNSMA ANKAAGKAKR ARRPAKYQYK DENGELKTWT GQGRTPAVIK
       130  135
KAIEEQGKSL DDFLL
```

Shigella flexneri

```
        10         20         30         40         50         60
MSEALKILNN IRTLRAQARE CTLETLEEML EKLEVVVNER REEESAAAAE VEERTRKLQQ
        70         80         90        100        110        120
YREMLIADGI DPNELLNSLA AVKSGTKAKR AQRPAKYSYV DENGETKTWT GQGRTPAVIK
       130        137
KAMDEQGKQL DDFLIKQ
```

Primary sequence of the factor for inversion stimulation (FIS)

Escherichia coli

```
        10         20         30         40         50         60
MFEQRVNSDV LTVSTVNSQD QVTQKPLRDS VKQALKNYFA QLNGQDVNDL YELVLAEVEQ
        70         80         90         98
PLLDMVMQYT RGNQTRAALM MGINRGTLRK KLKKYGMN
```

Primary sequence of HU

Escherichia coli (HU-β)

```
        10         20         30         40         50         60
MNKSQLIDKI AAGADISKAA AGRALDAIIA SVTESLKEGD DVALVGFGTF AVKERAARTG
        70         80         90
RNPQTGKEIT IAAAKVPSFR AGKALKDAVN
```

Escherichia coli (HU-α)

```
        10         20         30         40         50         60
MNKTQLIDVI AEKAELSKTQ AKAALESTLA AITESLKEGD AVQLVGFGTF KVNHRAERTG
        70         80         90
RNPQTGKEIK IAAANVPAFV SGKALKDAVK
```

Primary sequence of integration host factor (IHF)

Escherichia coli (IHF-α)

```
        10         20         30         40         50         60
MALTKAEMSE YLFDKLGLSK RDAKELVELF FEEIRRALEN GEQVKLSGFG NFDLRDKNQR
        70         80         90        100 101
PGRNNPKTGD KVELEGKYVV PHFKPGQKLK SRVENASPKD E
```

Escherichia coli (IHF-β)

```
        10         20         30         40         50         60
MTKSELIERL ATQQSHIPAK TVEDAVKEML EHMASTLAQG ERIEIRGFGS FSLHYRAPRT
        70         80         90  93
GRNPKTGDKV ELEGKYVPHF KPGKELRDRN IYG
```

Primary sequence of histone-like protein I (HLPI)

Escherichia coli

```
        10         20         30         40         50         60
MKKWLLAAGL GLALATSAQA ADKIAIVNMG SLFQQVAQKT GVSNTLENEF KGRASELQRM
        70         80         90        100        110        120
ETDLQAKMKK LQSMKAGSDR TKLEKDVMAQ RQTFAQKAQA FEQDRARRSN EERGKLVTRI
       130        140        150        160 161
QTAVKSVANS QDIDLVVDAN AVAYNSSDVK DITADVLKQV K
```

Primary sequence of Hcl

Chlamydia trachomatis

```
        10         20         30         40         50         60
MALKDTAKKM TDLLESIQQN LLKAEKGNKA AAQRVRTESI KLEKTAKVYR KESIKAEKMG
        70         80         90        100        110        120
LMKKSKAAAK KAKAAAKKPV RATKTVAKKA CTKRTCATKA KVKPTKKAAP KTKVKTAKKT
  125
RSTKK
```

Primary sequence of AlgP

Pseudomonas aeruginosa

```
        10         20         30         40         50         60
MSANKKPVTT PLHLLQQLSH SLVEHLEDAC KQALVDSEKL LAKLEKQRGK AQEKLHKART
        70         80         90        100        110        120
KLQDAAKAGK TKAQAKARET ISDLEEALDT LKARQADTRT YIVGLKRDVQ ESLKLAQGVG
       130        140        150        160        170        180
KVKEAAGKAL ESRKAKPATK PAAKAAAKPA VKTVAAKPAA KPAAKPAAKP AAKPATKTAA
       190        200        210        220        230        240
GKPAAKPTAK PAAKPAAKPA AKTAAAKPAA KPAAKPVAKP AAKPAAKTAA AKPAAKPAAK
       250        260        270        280        290        300
PVAKPTAKPA AKTAAAKPAA KPAAKPAAKP AAKPVAKSAA AKPAAKPAAK PAAKPAAKPA
       310        320        330        340        350  352
AKPVAAKPAA TKPATAPAAK PAATPSAPAA ASSAASATPA AGSNGAAPTS AS
```

Primary sequence of AlgR3

Pseudomonas aeruginosa

```
        10         20         30         40         50         60
MSANKKPVTT PLHLLQQLSH SLVEHLEGAC KQALVDSEKL LAKLEKQRGK AQEKLHKART
        70         80         90        100        110        120
KLQDAAKAGK TKAQAKARET ISDLEEALDT LKARQADTRT YIVGLKRDVQ ESLKLAQGVG
       130        140        150        160        170        180
KVKEAAGKAL ESRKAKPATK PAAKAAAKPA VKTVAANAAA KPAAKPAAKP AAKTAAAKPA
       190        200        210        220        230        240
AKPAAKPAAK PAAKPAAKTA AAKPAAKPAA KPVAKPAANA AAKTAAAKPA AKPAAKPVAK
       250        260        270        280        290        300
PAAKPAAKTA AAKPAAKPAA KHVAKPAAAK PAAKPAANAA AKPAAKPAAK PVAAKPAAAK
       310        320        330        340
PATAPAANAA ATPSATAAAS SAASATPAAG SNGAAPTSAS
```

Primary sequence of YmoA

Yersinia enterocolitica

```
        10         20         30         40         50         60
MTKTDYLMRL RKCTTIDTLE RVIEKNKYEL SDDELELFYS AADHRLAELT MNKLYDKIPP
      67
TVWQHVK
```

Primary sequence of Hha

Escherichia coli

```
        10         20         30         40         50         60
MVLICRIFRN GGSQVIDYSV VLSMRRKRIL RVYLVRIITT IGRSMSEKPL TKTDYLMRLR
        70         80         90        100        110    116
RCQTIDTLER VIEKNKYELS DNELAVFYSA ADHRLAELTM NKLYDKIPSS VWKFIR
```

Transcription activators

Primary sequence of Crp

Escherichia coli

```
        10         20         30         40         50         60
MVLGKPQTDP TLEWFLSHCH IHKYPSKSTL IHQGEKAETL YYIVKGSVAV LIKDEEGKEM
        70         80         90        100        110        120
ILSYLNQGDF IGELGLFEEG QERSAWVRAK TACEVAEISY KKFRQLIQVN PDILMRLSAQ
       130        140        150        160        170        180
MARRLQVTSE KVGNLAFLDV TGRIAQTLLN LAKQPDAMTH PDGMQIKITR QEIGQIVGCS
       190        200        210
RETVGRILKM LEDQNLISAH GKTIVVYGTR
```

Primary sequence of Fnr

Escherichia coli

```
        10         20         30         40         50         60
MIPEKRIIRR IQSGGCAIHC QDCSISQLSI PFTLNEHELD QLDNIIERKK PIQKGQTLFK
        70         80         90        100        110        120
AGDELKSLYA IRSGTIKSYT ITEQGDEQIT GFHLAGDLVG FDAIGSGHHP SFAQALETSM
       130        140        150        160        170        180
VCEIPFETLD DLSGKMPNLR QQMMRLMSGE IKGDQDMILL LSKKNAEERL AAFIYNLSRR
       190        200        210        220        230        240
FAQRGFSPRE FRLTMTRGDI GNYLGLTVET ISRLLGRFQK SGMLAVKGKY ITIENNDALA
       250
QLAGHTRNVA
```

LysR family members

Primary sequence of SpvR

Salmonella dublin

```
        10         20         30         40         50         60
MDFLINKKLK IFITLMETGS FSIATSVLYI TRTPLSRVIS DLERELKQRL FIRKNGTLIP
        70         80         90        100        110        120
TEFAQTIYRK VKSHYIFLHA LEQEIGPTGK TKQLEIIFDE IYPGSLKNLI ISALTISGQK
       130        140        150        160        170        180
TNIMGRAVNS QIIEELCQTN NCIVISARNY FHRESLVCRT SVEGGVMLFI PKKFFLCGKP
       190        200        210        220        230        240
DINRLAGTPV LFHEGAKNFN LDTIYHFFEQ TLGITNPAFS FDNVDLFSSL YRLQQGLAML
       250        260        270        280        290   295
LIPVRVCRAL GLSTDHALHI KGVALCTSLY YPTKKRETPD YRKAIKLIQQ ELKQS
```

Primary sequence of IrgB

Vibrio cholerae

```
        10         20         30         40         50         60
MQDLSAVKAF HALCQHKSLT AAAKALEQPK STLSRRLAQL EEDLGQSLLM RQGNRLTLTK
        70         80         90        100        110        120
AGEVFAVYSE QLLELANKSQ EALQELNNQV TGELTLVVHP NLIRGWLSQV LDEFMQQHST
       130        140        150        160        170        180
LKIRLLSQFQ HSDEVFEPDL IIWIEHAAPM GYRKERLGYW RYATYASPKY LAHRDKPTHP
       190        200        210        220        230        240
RELIHHPWID FIACRRAELE LHHPEFGSYS LPALESRLQS DNLAMQADAI AKGRGIGLLP
       250        260        270        280        290        298
TWFANGFETA HPGSLIPCVN GWQSQPTEIN CFYPLGRHPL RLRLFIDALR QARPDEWQ
```

AraC family members

Primary sequence of VirF

Shigella flexneri

```
        10         20         30         40         50         60
MMDMGHKNKI DIKVRLHNYI ILYAKRCSMT VSSGNETLTI DEGQIAFIER NININVSIKK
        70         80         90        100        110        120
SDSINPFEII SLDRNLLLSI IRIMEPIYSF QHSYSEEKRG LNKKIFLLSE EEVSIDLFKS
       130        140        150        160        170        180
IKEMPFGKRK IYSLACLLSA VSDEEALYTS ISIASSLSFS DQIRKIVEKN IEKRWRLSDI
       190        200        210        220        230        240
SNNLNLSEIA VRKRLESEKL TFQQILLDIR MHHAAKLLLN SQSYINDVSR LIGISSPSYF
       250        260 262
IRKFNEYYGI TPLLFYLYHK KF
```

Primary sequence of VirF

Yersinia enterocolitica

```
        10         20         30         40         50         60
MASLEIIKLE WATPIFKVVE HSQDGLYILL QGQISWQNSS QTYDLDEGNM LFLRRGSYAV
        70         80         90        100        110        120
RCGTKEPCQL LWIPLPGSFL STFLHRFGSL LSEIRRDNAT PKPLLIFNIS PILSQSIQNL
       130        140        150        160        170        180
CAILERSDFP SVLTQLRIEE LLLLLAFSSQ GALFLSALRH LGNRPEERLQ KFMEENYLQG
       190        200        210        220        230        240
WKLSKFAREF GMGLTTFKEL FGTVYGISPR AWISERRILY AHQLLLNGKM SIVDIAMEAG
       250        260        270 271
FSSQSYFTES YRRRFGCTPS QARLTKIATT G
```

Primary sequence of Rns

Escherichia coli

```
        10         20         30         40         50         60
MDFKYTEEKE TIKINNIMIH KYTVLYTSNC IMDIYSEEEK ITCFSNRLVF LERGVNISVR
        70         80         90        100        110        120
MQKQILSEKP YVAFRLNGDM LRHLKDALMI IYGMSKIDTN ACRSMSRKIM TTEVNKTLLD
       130        140        150        160        170        180
ELKNINSHDN SAFISSLIYL ISKLENNEKI IESIYISSVS FFSDKVRNLI EKDLSRKWTL
       190        200        210        220        230        240
GIIADAFNAS EITIRKRLES ENTNFNQILM QLRMSKAALL LLENSYQISQ ISNMIGISSA
       250        260 264
SYFIRIFNKH YGVTPKQFFT YFGG
```

Primary sequence of ToxT (TcpN)

Vibrio cholerae

```
        10         20         30         40         50         60
MIGKKSFQTN VYRMSKFDTY IFNNLYINDY KMFWIDSGIA KLIDKNCLVS YEINSSSIIL
        70         80         90        100        110        120
LKKNAIQRFS LTSLSDENIN VSVITISDSF IRSLKSYILG DLMIRNLYSE NKDLLLWNCE
       130        140        150        160        170        180
HNDIAVLSEV VNGFREINYS DEFLKVFFSG FFSKVEKKYN SIFITDDLDA MEKISCLVKS
       190        200        210        220        230        240
DITRNWRWAD ICGELRTNRM ILKKELESRG VKFRELINSI RISYSISLMK TGEFIKQIAY
       250        260        270   275
QSGFASVSYF STVFKSTMNV APSEYLFMLT GVAEK
```

Primary sequence of FapR

Escherichia coli

```
        10         20         30         40         50         60
MKLKNIHLYN YVVIYTKNCE IYINKGNEQV YIPPRMVAIF EKNISFNIET IRKGDGVLYE
        70         80         90        100        110        120
SFDMKHELLT SLRRVIEPSV KFAAESYTNK RSFKERIFKV KSCSIVIDLF KRLKDNGSPE
       130        140        150        160        170        180
FTAIYELAFL VSKCENPSMF AISLFSSVAV TFSERIVTLL FSDLTRKWKL SDIAEEMHIS
       190        200        210        220        230        240
EISVRKRLEQ ECLNFNQLIL DVRMNQAAKF IIRSDHQIGM IASLVGYTSV SYFIKTFKEY
       250        260
YGVTPKKFEI GIKENLRCNR
```

Primary sequence of EnvY

Escherichia coli

```
        10         20         30         40         50         60
MQLSSSEPCV VILTEKKVEV SVNNHATFTL PKNYLAAFAC NNNVIELSTL NHVLITHINR
        70         80         90        100        110        120
NRIINDYLLF LNKNLTCVKP WSRLATPVIA CHSTPEVFPL AAHSKQQPSR PCEAELTRAL
       130        140        150        160        170        180
LFTVLSNFLE QSQFIALLMY ILRSSVRDTV CRIIQSDIQH YWNLRIVASS LCLSPSLLKK
       190        200        210        220        230        240
KLKNENTSYS QIVTECRMRY AVQMLLMDNK NITQVAQLCG YSSTSYFISV FKAFYGLTPL
       250  252
NYLAKQRQKV MW
```

Primary sequence of CfaD

Escherichia coli

```
        10         20         30         40         50         60
MDFKYTEEKE MIKINNIMIH KYTVLYTSNC IMDIYSEEEK ITCFSNRLVF LERGVNISVR
        70         80         90        100        110        120
IQKKILSERP YVAFRLNGDI LRHLKNALMI IYGMSKVDTN DCRGMSRKIM TTEVNKTLLD
       130        140        150        160        170        180
ELKNINSHDD SAFISSLIYL ISKIENNEKI IQSIYISSVS FFSDKVRNVI EKDLSRKWTL
       190        200        210        220        230        240
GIIADAFNVS EITIRKRLES ENTNFNQILM QLRMSKAALL LLENSYQISQ ISNMIGISSA
       250        260   265
SYFIRVFNKH YGVTPKQFFT YFKGG
```

Primary sequence of ExsA

Pseudomonas aeruginosa

```
        10         20         30         40         50         60
MPLRSAGHSR YDGKCWGSYN MQGAKSLGRK QITSCHWNIP TFEYRVNKEE GVYVLLEGEL
        70         80         90        100        110        120
TVQDIDSTFC LAPGELLFVR RGSYVVSTKG KDSRILWIPL SAQFLQGFVQ RFGALLSEVE
       130        140        150        160        170        180
RCDEPVPGII AFAATPLLAG CVKGLKELLV HEHPPMLACL KIEELLMLFA FSPQGPLLMS
       190        200        210        220        230        240
VLRQLSNRHV ERLQLFMEKH YLNEWKLSDF SREFGMGLTT FKELFGSVYG VSPRAWISER
       250        260        270        280        290       298
RILYAHQLLL NSDMSIVDIA MEAGFSSQSY FTQSYRRRFG CTPSRSRQGK DECRAKNN
```

Primary sequence of HrpB

Pseudomonas solanacearum

```
        10         20         30         40         50         60
MLGNIYFALA SGLAARERLP EYANAVFAAD FDRAYQLVDH HSSQRGKSDD YAGVLAMADA
        70         80         90        100        110        120
SLLLECDEEA EEGFRLAQRL IRHSDDQLRV VSCRNTGWQA LLRDRYAAAA SCFSRMAEDD
       130        140        150        160        170        180
GATWTQQVEG LIGLALVHHQ LGQQDASDDA LRAAREAADG RSDRGWLATI DLIIYEFAVQ
       190        200        210        220        230        240
AGIRCSNRLL EHAFWQSAEM GATLLANHGG RNGWTPTVSQ GVPMPALIQR RAEYLSLLRR
       250        260        270        280        290        300
MADGDRAAID PLMATLNHSR KLGSRLLMQT KVEVVLAALS GEQYDVAGRV FDQICNRETT
       310        320        330        340        350        360
YGARRWNFDF LYCRAKMAAQ RGDAAGALKF YTTYMQDALR CLRTETVNVR RASAAVPVAS
       370        380        390        400        410        420
RASDDVSARL SAKYRRAYRY IIENIERSDL TTREVAAHIN VTERALQLAF KSAVGMSPSS
       430        440        450        460        470       477
VIRRMRLEGI RSDLLDSERN PSNIIDTASR WGIRSRSALV KGYRKQFNEA PSETIWR
```

Response regulators

Primary sequence of OmpR

Escherichia coli

```
        10         20         30         40         50         60
MQENYKNLVV DDDMRLRALL ERYLTEQGFQ VRSVANAEQM DRLLTRESFH LMVLDLMLPG
        70         80         90        100        110        120
EDGLSICRRL RSQSNPMPII MVTAKGEEVD RIVGLEIGAD DYIPKPFNPR ELLARIRAVL
       130        140        150        160        170        180
RRQANELPGA PSQEEAVIAF GKFKLNLGTR EMFREDEPMP LTSGEFAVLK ALVSHPREPL
       190        200        210        220        230      239
SRDKLMNLAR GREYSAMERS IDVQISRLRR MVEEDPAHPR YIQTVWGLGY VFVPDGSKA
```

Primary sequence of PhoP

Salmonella typhimurium

```
        10         20         30         40         50         60
MMRVLVVEDN ALLRHHLKVQ LQDSGHQVDA AEDAREADYY LNEHLPDIAI VDLGLPDEDG
        70         80         90        100        110        120
LSLIRRWRSS DVSLPVLVLT AREGWQDKVE VLSSGADDYV TKPFHIEEVM ARMQALMRRN
       130        140        150        160        170        180
SGLASQVINI PPFQVDLSRR ELSVNEEVIK LTAFEYTIME TLIRNNGKVV SKDSLMLQLY
       190        200        210        220 224
PDAELRESHT IDVLMGRLRK KIQAQYPHDV ITTVRGQGYL FELR
```

Primary sequence of VirG

Agrobacterium tumefaciens

```
        10         20         30         40         50         60
MIVHPSRENF SSAVNKCSDF RLKGEPLKHV LLVDDDVAMR HLIIEYLTIH AFKVTAVADS
        70         80         90        100        110        120
TQFTRVLSSA TVDVVVVDLN LVREDGLEIV RNLAAKSDIP IIIISGDRLE ETDKVVALEL
       130        140        150        160        170        180
GASDFIAKPF SIREFLARIR VALRVRPNVV RSKDRRSFCF TDWTLNLRQR RLMSEAGGEV
       190        200        210        220        230        240
KLTAGEFNLL LAFLEKPRDV LSREQLLIAS RVRDEEVYDR SIDVLILRLR RKLEADPSSP
       250        260    267
QLIKTARGAG YFFDADVQVS HGGTMAA
```

Primary sequence of BvgA

Bordetella pertussis

```
        10         20         30         40         50         60
MYNKVLIIDD HPVLRFAVRV LMEKEGFEVI GETDNGIDGL KIAREKIPNL VVLDIGIPKL
        70         80         90        100        110        120
DGLEVIARLQ SLGLPLRVLV LTGQPPSLFA RRCLNSGAAG FVCKHENLHE VINAAKAVMA
       130        140        150        160        170        180
GYTYFPSTTL SEMRMGDNAK SDSTLISVLS NRELTVLQLL AQGMSNKDIA DSMFLSNKTV
       190        200        209
STYKTRLLQK LNATSLVELI DLAKRNNLA
```

Primary sequence of NtrC

Klebsiella pneumoniae

```
        10         20         30         40         50         60
MQRGIAWIVD DDSSIRWVLE RALTGAGLSC TTFESGNEVL DALTTKTPDV LLSDIRMPGM
        70         80         90        100        110        120
DGLALLKQIK QRHPMLPVII MTAHSDLDAA VSAYQQGAFD YLPKPFDIDE AVALVDRAIS
       130        140        150        160        170        180
HYQEQQQPRN APINSPTADI IGERPAMQDV FRIIGRLSRS SISVLINGES GTGKELVAHA
       190        200        210        220        230        240
LHRHSPRAKA PFIALNMAAI PKDLIESELF GHEKGAFTGA NTVRQGRFEQ ADGGTLFLDE
       250        260        270        280        290        300
IGDMPLDVQT RLLRVLADGQ FYRVGGYAPV KVDVRIIAAT HQNLELRVQE GKFREDLFHR
       310        320        330        340        350        360
LNVIRVHLPP LRERREDIPR LARHFLQIAA RELGVEAKQL HPETEMALTR LAWPGNVRQL
       370        380        390        400        410        420
ENTCRWLTVM AAGQEVLTQD LPSELFETAI PDNPTQMLPD SWATLLGQWA DRALRSGHQN
       430        440        450        460        469
LLSEAQPEME RTLLTTALRH TQGHKQEAAR LLGWGRNTLT RKLKELGME
```

Primary sequence of AlgB

Pseudomonas aeruginosa

```
        10         20         30         40         50         60
METTSEKQGR ILLVDDESAI LRTFRYCLED EGYSVATASS APQAEALLQR QVFDLCFLDL
        70         80         90        100        110        120
RLGEDNGLDV LAQMRVQAPW MRVVIVTAHS AVDTAVDAMQ AGAVDYLVKP CSPDQLRLAA
       130        140        150        160        170        180
AKQLEVRQLT ARLEALEDEV RRQGDGLESH SPAMAAVLET ARQVAATDAN ILILGESGSG
       190        200        210        220        230        240
KGELARAIHT WSKRAKKPQV TINCPSLTAE LMESELFGHS RGAFTGATES TLGRVSQADG
       250        260        270        280        290        300
GTLFLDEIGD FPLTLQPKLL RFIQDKEYER VGDPVTRRAD VRILAATNRD LGAMVAQGQF
       310        320        330        340        350        360
REDLLYRLNV IVLNLPPLRE RAEDILGLAE RFLARFVKDY GRPARGFSEA AREAMRQYPW
       370        380        390        400        410        420
PGNVRELRNV IERASIICNQ ELVDVDHLGF SAAQSASSAP RIGESLSLED LEKAHITAVM
       430        440        449
ASSATLDQAA KTLGIDASTL YRKRKQYGL
```

Primary sequence of AlgR

Pseudomonas aeruginosa

```
        10         20         30         40         50         60
MNVLIVDDEP LARERLARLV GQLDGYRVLE PSASNGEEAL TLIDSLKPDI VLLDIRMPGL
        70         80         90        100        110        120
DGLQVAARLC EREAPPAVIF CTAHDEFALE AFQVSAVGYL VKPVRSEDLA EALKKASRPN
       130        140        150        160        170        180
RVQLAALTKP PASGGSGPRS HISARTRKGI ELIPLEEVIF FIADHKYVTL RHAQGEVLLD
       190        200        210        220        230        240
EPLKALEDEF GERFVRIHRN ALVARERIER LQRTPLGHFQ LYLKGLDGDA LTVSRRHVAG
     248
VRRLMHQL
```

Primary sequence of ToxR

Vibrio cholerae

```
        10         20         30         40         50         60
MPGLGHNSKE ISMSHIGTKF ILAEKFTFDP LSNTLIDKED SEEIIRLGSN ESRILWLLAQ
        70         80         90        100        110        120
RPNEVISRND LHDFVWREQE FEVDDSSLTQ AISTLRKMLK DSTKSPQYVK TVPKRGYQLI
       130        140        150        160        170        180
ARVETVEEEM ARENEAAHDI SQPESVNEYA ESSSVPSSAT VVNTPQPANV VANKSAPNLG
       190        200        210        220        230        240
NRLFILIAVL LPLAVLLLTN PSQSSFKPLT VVDGVAVNMP NNHPDLSNWL PSIELCVKKY
       250        260        270        280        290  294
NEKHTGGLKP IEVIATGGQN NQLTLNYIHS PEVSGENITL RIVANPNDAI KVCG
```

Primary sequence of AgrA

Staphylococcus aureus

```
        10         20         30         40         50         60
MKIFICEDDP KQRENMVTII KNYIMIEEKP MEIALATDNP YEVLEQAKNM NDIGCYFLDI
        70         80         90        100        110        120
QLSTDINGIK LGSEIRKHDP VGNIIFVTSH SELTYLTFVY KVAAMDFIFK DDPAELRTRI
       130        140        150        160        170        180
IDCLETAHTR LQLLSKDNSV ETIELKRGSN SVYVQYDDIM FFESSTKSHR LIAHLDNRQI
       190        200        210        220        230      238
EFYGNLKELS QLDDRFFRCH NSFVVNRHNI ESIDSKERIV YFKNKEHCYA SVRNVKKI
```

Primary sequence of PilA

Neisseria gonorrhoeae

```
         10         20         30         40         50         60
MFSFFRRKKK QETPALEEAQ VQETAAKVES EVAQIVGNIK EDVESLAESV RAESAVETVS
         70         80         90        100        110        120
KGGAVEQVKE TVAEMPSEAG EAAERVESAK EAVAETVGEA VGQVQEAVAT TEEHKLGWAA
        130        140        150        160        170        180
RLKQGLAKSR DKMAKSLAGV FGGGQIGEDL YEELETVLIT GDMGMEATEY LMKDVRGRVS
        190        200        210        220        230        240
LKGLKDGNEL RGALKEALYD LIKPLEKPLV LPETKEPFVI MLAGINGAGK TTSIGKLAKY
        250        260        270        280        290        300
FQAQGKSVLL AAGDTFRAAA REQLQAWGGR NNVTVISQTT GDSAAVCFDA VQAAKARGID
        310        320        330        340        350        360
IVLADTAGRL PTQLHLMEEI KKVKRVLQKA IPGAPHEIIV VLDANIGQNA VNQVKAFDDA
        370        380        390        400        410     417
LGLTGLIVTK LDGTAKGGIL AALASDRPVP VRYIGVGEGI DDLRPFDARA FVDRLLD
```

Primary sequence of PilR

Pseudomonas aeruginosa

```
         10         20         30         40         50         60
MSRQKALIVD DEPDIRELLE ITLGRMKLDT RSARNVKEAR ELLAREPFDL CLTDMRLPDG
         70         80         90        100        110        120
SGLDLVQYIQ QRHPQTPVAM ITAYGSLDTA IQALKAGAFD FLTKPVDFDF RLRELVATAL
        130        140        150        160        170        180
RLRNPEAEEA PVDNRLLGES PPMRALRNQI GKLARSQAPV YISGESGSGK ELVARLIHEQ
        190        200        210        220        230        240
GPRIERPFVP VNCGAIPSEL MESEFFGHKK GSFTGAIEDK QGLFQAASGG TLFLDEVADL
        250        260        270        280        290        300
PMAMQVKLLR AIQEKAVRAV GGQQEVAVDL RILCATHKDL AAEVGAGRFR QDLYYRLNVI
        310        320        330        340        350        360
ELRVPPLRER REDIPLLAER ILKRLAGDTG LPAARLTGDA QEKLKNYRFP GNVGELENML
        370        380        390        400        410        420
ERAYTLCEDD QIQPHDLRLA DAPGASQEGA ASLSEIDNLE DYLEDIERKL IMQALEETRW
        430        440     446
NRTAAAQRLG LTFRSMRYRL KKLGID
```

Primary sequence of Mry

Streptococcus pyogenes

```
        10         20         30         40         50         60
MYVSKLFTSQ QWRELKLISY VTENADAIGV KDKELSKALN ISMLTLQTCL TNMQFMKEVG
        70         80         90        100        110        120
GITYKNGYIT IWYHQHCGLQ EVYQKALRHS QSFKLLETLF FRDFNSLEEL AEELFVSLST
       130        140        150        160        170        180
LKRLIKKTNA YLTHTFGITI LTSPVQVSGD EHQIRLFYLK YFSEAYKISE WPFGEILNLK
       190        200        210        220        230        240
NCERLLSLMI KEVDVRVNFT LFQHLKILSS VNLIRYYEGY SAVYDNKKTS HRFSQLIQSS
       250        260        270        280        290        300
LETQDLSRLF YLKFGLYLDE TTIAEMFSNH VNDQLEIGYA FDSIKQDSPT GCRKVTNWIH
       310        320        330        340        350        360
LLDELEINLN LSVTNKYEVA VILHNTTVLK EEDITANYLF FDYKKSYLNF YKQEHPHLYK
       370        380        390        400        410        420
AFVAGVEKLM RSEKEPISTE LTNQLIYAFF ITWENSFLKV NQKDEKIRLL VIARRFNSVG
       430        440        450        460        470        480
NFLKKYIGEF FSITNFNELD ALTIDLEEIE KQYDVIVTDV MVGKSDELEI FFFYKMIPEA
       490        500        510        520        530
IIDKLNVFLN ISFADSLPLD KPIKNPLDFH RKELTLPTPP NKLHAPPSTT
```

Primary sequence of Fur

Escherichia coli

```
        10         20         30         40         50         60
MTDNNTALKK AGLKVTLPRL KILEVLQEPD NHHVSAEDLY KRLIDMGEEI GLATVYRVLN
        70         80         90        100        110        120
QFDDAGIVTR HNFEGGKSVF ELTQQHHHDH LICLDCGKVI EFSDDSIEAR QREIAAKHGI
       130        140        148
RLTNHSLYLY GHCAEGDCRE DEHAHEGK
```

Primary sequence of Fur

Pseudomonas aeruginosa

```
        10         20         30         40         50         60
MVENSELRKA GLKVTLPRVK ILQMLDSAEQ RHMSAEDVYK ALMEAGEDVG LATVYRVLTQ
        70         80         90        100        110        120
FEAAGLVVRH NFDGGHAVFE LADSGHHDHM VCVDTGEVIE FMDAEIEKRQ KEIVRERGFE
       130  134
LVDHNLVLYV RKKK
```

Primary sequence of Fur

Vibrio cholerae

```
        10         20         30         40         50         60
MSDNNQALKD AGLKVTLPRL KILEVLQQPE CQHISAEELY KKLIDLSEEI GLATVYRVLN
        70         80         90        100        110        120
QFDDAGIVTR HHFEGGKSVF ELSTQHHHDH LVCLDCGEVI EFSDDVIEQR QKEIAAKYNV
       130        140        150
QLTNHSLYLY GKCGSDGSCK DNPNAHKPKK
```

Primary sequence of Fur

Yersinia pestis

```
        10         20         30         40         50         60
MTDNNKALKN AGLKVTLPRL KILEVLQNPA CHHVSAEDLY KILIDIGEEI GLATVYRCSE
        70         80         90        100        110        120
QFDDAGIVTR HNFEGGKSVF ELTQQHHHDH LICLDCGKVI EFSNESIESL QREIAKQHGI
       130        140        148
KLTNHSLYLY GHCETGNCRE DESAHSKR
```

Primary sequence of DtxR

Corynebacterium diphtheriae

```
        10         20         30         40         50         60
MKDLVDTTEM YLRTIYELEE EGVTPLRARI AERLEQSGPT VSQTVARMER DGLVVVASDR
        70         80         90        100        110        120
SLQMTPTGRT LATAVMRKHR LAERLLTDII GLDINKVHDE ACRWEHVMSD EVERRLVKVL
       130        140        150        160        170        180
KDVSRSPFGN PIPGLDELGV GNSDAAAPGT RVIDAATSMP RKVRIVQINE IFQVETDQFT
       190        200        210        220        226
QLLDADIRVG SEVEIVDRDG HITLSHNGKD VELLDDLAHT IRIEEL
```

Primary sequence of VirB/InvE

Shigella flexneri/Shigella sonnei

```
        10         20         30         40         50         60
MVDLCNDLLS IKEGQKKEFT LHSGNKVSFI KAKIPHKRIQ DLTFVNQKTN VRDQESLTEE
        70         80         90        100        110        120
SLADIIKTIK LQQFFPVIGR EIDGRIEILD GTRRRASAIY AGADLEVLYS KEYISTLDAR
       130        140        150        160        170        180
KLANDIQTAK EHSIRELGIG LNFLKVSGMS YKDIAKKENL SRAKVTRAFQ AASVPQEIIS
       190        200        210        220        230        240
LFPIASELNF NDYKILFNYY KGLEKANESL SSTLPILKEE IKDLDTNLPP DIYKKEILNI
       250        260        270        280        290        300
IKKSKNRKQN PSLKVDSLFI SKDKRTYIKR KENKTNRTLI FTLSKINKTV QREIDEAIRD
       309
IISRHLSSS
```

Primary sequence of LasR

Pseudomonas aeruginosa

```
        10         20         30         40         50         60
MALVDGFLEL ERSSGKLEWS AILQKMASDL GFSKILFGLL PKDSQDYENA FIVGNYPAAW
        70         80         90        100        110        120
REHYDRAGYA RVDPTVSHCT QSVLPIFWEP SIYQTRKQHE FFEEASAAGL VYGLTMPLHG
       130        140        150        160        170        180
ARGELGALSL SVEAENRAEA NRFMESVLPT LWMLKDYALQ SGAGLAFEHP VSKPVVLTSR
       190        200        210        220        230        239
EKEVLQWCAI GKTSWEISVI CNCSEANVNF HMGNIRRKFG VTSRRVAAIM AVNLGLITL
```

Primary sequence of HlyT

Escherichia coli

```
        10         20         30         40         50         60
MQSWYLLYCK RGQLQRAQEH LERQAVNCLA PMITLEKIVR GKRTAVSEPL FPNYLFVEFD
        70         80         90        100        110        120
PEVIHTTTIN ATRGVSHFVR FGASPAIVPS AVIHQLSVYK PKDIVDPATP YPGDKVIITE
       130        140        150        160 162
GAFEGFQAIF TEPDGEARSM LLLNLINKEI KHSVKNTEFR KL
```

References for protein sequences listed in Appendix I

H-NS *E. coli*: Falconi *et al.* (1988); Göransson *et al.* (1990); May *et al.* (1990); Pon *et al.* (1988). H-NS *Ps. vulgaris*: La Teana *et al.* (1989). H-NS *S. marcescans*: La Teana *et al.* (1989). H-NS *S. typhimurium*: Hulton *et al.* (1990); Marsh & Hillyard (1990). H-NS *S. flexneri*: Hromockyj *et al.* (1992). FIS *E. coli*: Koch *et al.* (1988). HU-β *E. coli*: Kano *et al.* (1985); Laine *et al.* (1980); Mende *et al.* (1978). HU-α *E. coli*: Kano *et al.* (1987); Laine *et al.* (1980); Mende *et al.* (1978). IHF-α *E. coli*: Miller (1984). IHF-β *E. coli*: Flamm & Weisberg (1985). HLPI *E. coli*: Hirvas *et al.* (1990). Hc1 *C. trachomatis*: Hackstadt *et al.* (1991). AlgP *Ps. aeruginosa*: Deretic & Konyecsni (1990); Konyescni & Deretic (1990). AlgR3 *Ps. aeruginosa*: Kato *et al.* (1990a). YmoA *Y. enterocolitica*: Cornelis *et al.* (1991). Hha *E. coli*: Nieto *et al.* (1991). Crp *E. coli*: Aiba *et al.* (1982). Fnr *E. coli*: Shaw & Guest (1982). SpvR *S. dublin*: Pullinger *et al.* (1989). IrgB *V. cholerae*: Goldberg *et al.* (1991). VirF *S. flexneri* Sakai *et al.* (1986). VirF *Y. enterocolitica*: Cornelis *et al.* (1989b). Rns *E. coli*: Caron *et al.* (1989). ToxT (TcpN) *V. cholerae*: Higgins *et al.* (1992); Ogierman & Manning (1992). FapR *E. coli*: Klaasen & de Graaf (1990). EnvY *E. coli*: Lundrigan *et al.* (1989). CfaD *E. coli*: Savelkoul *et al.* (1990). ExsA *Ps. aeruginosa*: Frank & Iglewski (1991). HrpB *Ps. solanacearum*: Genin *et al.* (1992). OmpR *E. coli*: Comeau *et al.* (1985). PhoP *S. typhimurium*: Miller *et al.* (1989b). VirG *A. tumefaciens*: Melchers *et al.* (1986). BvgA *B. pertussis*: Aricò *et al.* (1989). NtrC *K. pneumoniae*: Drummond *et al.* (1986). AlgB *Ps. aeruginosa*: Goldberg & Dahnke (1992). AlgR *Ps. aeruginosa*: Deretic *et al.* (1989). ToxR *V. cholerae*: Miller *et al.* (1987). AgrA *S. aureus*: Peng *et al.* (1988). PilA *N. gonorrhoeae*: Taha *et al.* (1988, 1991). PilR *Ps. aeruginosa*: Ishimoto & Lory (1992). Mry *S. pyogenes*: Perez-Casal *et al.* (1991). Fur *E. coli*: Schäffer *et al.* (1985). Fur *Ps. aeruginosa*: Prince *et al.* (1993). Fur *V. cholerae*: Litwin *et al.* (1992). Fur *Y. pestis*: Staggs & Perry (1992). DtxR *C. diphtheriae*: Boyd *et al.* (1990). VirB/InvE *S. flexneri/S. sonnei*: Adler *et al.* (1989); Watanabe *et al.* (1990). LasR *Ps. aeruginosa*: Gambello & Iglewski (1991). HlyT *E. coli*: Bailey *et al.* (1992).

Appendix II: Sequences of Selected Histidine Protein Kinases

Primary sequence of BvgS

Bordetella pertussis

```
        10         20         30         40         50         60
MPAPHRLYPR SLICLAQALL AWALLAWAPA QASQELTLVG KAAVPDVEVA LDGDDWRWLA
        70         80         90        100        110        120
RKRVLTLGVY APDIPPFDVT YGERYEGLTA DYMAIIAHNL GMQAKVLRYP TREQALSALE
       130        140        150        160        170        180
SGQIDLIGTV NGTDGRQQSL RLSVPYAADH PVIVMPIGAR HVPASNLAGQ RLAVDINYLP
       190        200        210        220        230        240
KETLARAYPQ ATLHYFPSSE QALAAVAYGQ ADVFIGDALT TSHLVSQSYF QDVRVVAPAH
       250        260        270        280        290        300
IATGGESFGV RADNTRLLRV VNAVLEAIPP SEHRSLIYRW GLGSSISLDF AHPAYSAREQ
       310        320        330        340        350        360
QWMADHPVVK VAVLNLFAPF TLFRTDEQFE GISAAVLQLL QLRTGLDFEI IGVDTVEELI
       370        380        390        400        410        420
AKLRSGEADM AGALFVNSAR ESFLSFSRPY VRNGMVIVTR QDPDAPVDAD HLDGRTVALV
       430        440        450        460        470        480
RNSAAIPLLQ RRYPQAKVVT ADQPSEAMLM VAQGQADAVV QTQISASYYV NRYFAGKLRI
       490        500        510        520        530        540
ASALDLPPAE IALATTRGQT ELMSILRKAL YSISNDELAS IISRWRGSDG DPRTWYAYRN
       550        560        570        580        590        600
EIYLLIGLGL LSALLFLSWI VYLRRQIRQR KRAERALNDQ LEFMRVLIDG TPNPIYVRDK
       610        620        630        640        650        660
EGRMLLCNDA YLDTFGVTAD AVLGKTIPEA NVVGDPALAR EMHEFLLTRV AAEREPRFED
       670        680        690        700        710        720
RDVTLHGRTR HVYQWTIPYG DSLGELKGII GGWIDITERA ELLRELHDAK ESADAANRAK
       730        740        750        760        770        780
TTFLATMSHE IRTPMNAIIG MLELALLRPT DQEPDRQSIQ VAYDSARSLL ELIGDILDIA
       790        800        810        820        830        840
KIEAGKFDLA PVRTALRVLP EGAIRVFDGL ARQKGIELVL KTDIVGVDDV LIDPLRMKQV
       850        860        870        880        890        900
LSNLVGNAIK FTTEGQVVLA VTARPDGDAA HVQFSVSDTG CGISEADQRQ LFKPFSQVGG
       910        920        930        940        950        960
SAEAGPAPGT GLGLSISRRL VELMGGTLVM RSAPGVGTTV SVDLRLTMVE KSVQAAPPAA
       970        980        990       1000       1010       1020
ATAATPSKPQ VSLRVLVVDD HKPNLMLLRQ QLDYLGQRVI AADSGEAALA LWREHAFDVV
      1030       1040       1050       1060       1070       1080
ITDCNMPGIS GYELARRIRA AEAAPGYGRT RCILFGFTAS AQMDEAQACR AAGMDDCLFK
      1090       1100       1110       1120       1130       1140
PIGVDALRQR LNEAVARAAL PTPPSPQAAA PATDDATPTA FSAESILALT QNDEALIRQL
      1150       1160       1170       1180       1190       1200
LEEVIRTNRA DVDQLQKLHQ QADWPKVSDM AHRLAGGARV VDAKAMIDTV LALEKKAQGQ
      1210       1220       1230     1238
AGPSPEIDGL VRTLAAQSAA LETQLRAWLE QRPHQDQP
```

Primary sequence of VirA (wide host range)

Agrobacterium tumefaciens

```
        10         20         30         40         50         60
MNGRYSPTRQ DFKTGAKPWS ILALIVAAMI FAFMAVASWQ DNATTQAILS QLRSINADSA
        70         80         90        100        110        120
SLQRDVARAH TGTGRNYRPI ISRLGALRKN LEDLKQLFRQ SHIVSESNAA QLLRQLEVSL
       130        140        150        160        170        180
NSADAAVRAF GAQNVRLQDS LASFTRALSS LPGKASTDQT LEKPTELASM MLQFLRQPSP
       190        200        210        220        230        240
AISFEISLEL ERLQKQRGLD EAPVRILARE GPIILSLLPQ VKDLVNMIQT SDTAEIAEML
       250        260        270        280        290        300
QRECLEVYSL KNVEERSARI FLGSASVGLC LYIITLVYRL RKKTDWLARR LDYEELIKEI
       310        320        330        340        350        360
GVCFEGEAAT TSSAQAALRI IQRFFDADTC ALRLVDHDRR WAVETFGAKH PKPVWDDSVL
       370        380        390        400        410        420
REIVSRTKAD ERATVFRIIS SKKIVHLPLE IPGLSILLAH KSTDKLIAVC SLGYQSYRPR
       430        440        450        460        470        480
PCQGEIQLLE LATACLCHYI DVRRKQTECD VLARRLEHAQ RLEAVGTLAG GIAHEFNNIL
       490        500        510        520        530        540
GSILGHAELA QNSVSRTSVT RRYIDYIISS GDRAMLIIDQ ILTLSRKQER MIKPFSVSEL
       550        560        570        580        590        600
VTEIAPLLRM ALPPNIELSF RFDQMQSVIE GSPLELQQVL INICKNASQA MTANGQIDII
       610        620        630        640        650        660
ISQAFLPVKK ILAHGVMPPG DYVLLSISDN GGGIPEAVLP HIFEPFFSTR ARNGGTGLGL
       670        680        690        700        710        720
ASVHGHISAF AGYIDVSSTV GHGTRFDIYL PPSSKEPVNP DSFFGRNKAP RGNGEIVALV
       730        740        750        760        770        780
EPDDLLREAY EDKIAALGYE PVGFRTFNEI RDWISKGNEA DLVMVDQASL PEDQSPNSVD
       790        800        810        820        829
LVLKTSAIII GGNDLKMTLS REDVTRDLYL PKPISSRTMA HAILTKIKT
```

Primary sequence of NtrB

Klebsiella pneumoniae

```
        10         20         30         40         50         60
MATGTLPDAG QILNSLINSI LLVDDDLAVH YANPAAQQLL AQSSRKLFGT PLPELLSYFS
        70         80         90        100        110        120
LNIGLMQESL AAGQGFTDNE VTLVIDGRSH ILSLTAQRLP EGYILLEMAP MDNQRRLSQE
       130        140        150        160        170        180
QLQHAQQIAA RDLVRGLAHE IKNPLGGLRG AAQLLSKALP DPALMEYTKV IIEQADRLRN
       190        200        210        220        230        240
LVDRLLGPQH PGMHVTESIH KVAERVVKLV SMELPDNVKL VRDYDPSLPE LPHDPDQIEQ
       250        260        270        280        290        300
VLLNIVRNAL QALGPEGGEI TLRTRTAFQL TLHGVRYRLA ARIDVEDNGP GIPSHLQDTL
       310        320        330        340        349
FYPMVSGREG GTGLGLSIAR SLIDQHSGKI EFTSWPGHTE FSVYLPIRK
```

Primary sequence of EnvZ

Escherichia coli

```
         10         20         30         40         50         60
MRRLRFSPRS SFARTLLLIV TLLFASLVTT YLVVLNFAIL PSLQQFNKVL AYEVRMLMTD
         70         80         90        100        110        120
KLQLEDGTQL VVPPAFRREI YRELGISLYS NEAAEEAGLR WAQHYEFLSH QMAQQLGGPT
        130        140        150        160        170        180
EVRVEVNKSS PVVWLKTWLS PNIWVRVPLT EIHQGDFSPL FRYTLAIMLL AIGGAWLFIR
        190        200        210        220        230        240
IQNRPLVDLE HAALQVGKGI IPPPLREYGA SEVRSVTRAF NHMAAGVKQL ADDRTLLMAG
        250        260        270        280        290        300
VSHDLRTPLT RIRLATEMMS EQDGYLAESI NKDIEECNAI IEQFIDYLRT GQEMPMEMAD
        310        320        330        340        350        360
LNAVLGEVIA AESGYEREIE TALYPGSIEV KMHPLSIKRA VANMVVNAAR YGNGWIKVSS
        370        380        390        400        410        420
GTEPNRAWFQ VEDDGPGIAP EQRKHLFQPF VRGDSARTIS GTGLGLAIVQ RIVDNHNGML
        430        440        450
ELGTSERGGL SIRAWLPVPV TRAQGTTKEG
```

Primary sequence of PhoQ

Salmonella typhimurium

```
         10         20         30         40         50         60
MNKFARHFLP LSLRVRFLLA TAGVVLVLSL AYGIVALVGY SVSFDKTTFR LLRGESNLFY
         70         80         90        100        110        120
TLAKWENNKI SVELPENLRM NSPTMTLIYD ETGKLLWTQR NIPWLIKSIQ PEWLKTNGFH
        130        140        150        160        170        180
EIETNVDATS TLLSEDHSAQ EKLKEVREDD DDAEMHSVAV NNIYPATARM PQLTIVVVDT
        190        200        210        220        230        240
IPIELKRSYM VWSWFVYVLA ANLLLVIPLL WIAAWWSLRP IEALAREVRE LEDHHREMLN
        250        260        270        280        290        300
PETTRELTSL VRNLNQLLKS ERERYNKYRT TLTDLTHSLK TALAVLQSTL RSLRNEKMSV
        310        320        330        340        350        360
SKAEPVMLEQ ISRISQQIGY YLHRASMRGS GVLLSRELHP VAPLLDNLIS ALNKVYQRKG
        370        380        390        400        410        420
VNISMDISPE ISFVGEQNDF VEVMGNVLDN ACKYCLEFVE ISARQTDDHL HIFVEDDGPG
        430        440        450        460        470        480
IPHSKRSLVF DRGQRADTLR PGQGVGLAVA REITEQYAGQ IIASDSLLGG ARMEVVFGRQ
  487
HPTQKEE
```

Primary sequence of PilB

Neisseria gonorrhoeae

```
        10         20         30         40         50         60
MKHRTFFSLC AKFGCLLALG ACSPKIVDAG TATVPHTLST LKTADNRPAS VYLKKDKPTL
        70         80         90        100        110        120
IKFWASWCPL CLSELGQAEK WAQDAKFSSA NLITVASPGF LHEKKDGEFQ KWYAGLNYPK
       130        140        150        160        170        180
LPVVTDNGGT IAQNLNISVY PSWALIGKDG DVQRIVKGSI NEAQALALIR NPNADLGSLK
       190        200        210        220        230        240
HSPYKPDTQK KDSAIMNTRT IYLAAAASGA WKPISNASTA WLTRYRYANG NTENPSYEDV
       250        260        270        280        290        300
SYRHTGHAET VKVTYDADKL SLDDILQYYF RVVDPTSLNK QGNDTGTQYR SGVYYTDPAE
       310        320        330        340        350        360
KAVIAAALKR EQQKYQLPLV VENEPLKNFY DAEEYHQDYL IKNPNGYCHI DIRKADEPLP
       370        380        390        400        410        420
GKTKAAPQGQ RLRRGQRIKN RVTPNSNAPD RRAIPSDQNS ATEYAFSHEY DHLFKPGIYV
       430        440        450        460        470        480
DVVSGEPLFS SADKYDSGCG WPSFTRPIDA KSVTEHDDFS FNMRRTEVRS RAADSHLGHV
       490        500        510        520 521
FPDGPRDKGG LRYCINGASL KFIPLEQMDA AGYGALKGEV K
```

Primary sequence of AgrB

Staphylococcus aureus

```
        10         20         30         40         50         60
VELLNSYNFV LFVLTQMILM FTIPAIISGI KYSKLDYFFI IVISTLSLFL FKMFDSASLI
        70         80         90        100        110        120
ILTSFIIIMY FVKIKWYSIL LIMSSQIILY CANYMYIVIY AYITKISDSI FVIFPSFFVV
       130        140        150        160        170        180
YVTISILFSY IINRVLKKIS TPYLILNKGF LIVISTILLL TFSLFFFYSQ INSDEAKVIR
       190        200        210        220        230        240
QYSLFYWYHY ILSILTLYSQ FLLKEMKYKR NQEEIETYYE YTLKIEAINN EMRKFRHDYV
       250        260        270        280        290        300
NILTTLSEYI REDDMPGLRD YFNKNIVPMK DNLQMNAIKL NGIENLKVRE IKGLITAKIL
       310        320        330        340        350        360
RAQEMNIPIS IEIPDEVSSI NLNMIDLSRS IGIILDNAIE ASTEIDDPII RVAFIESENS
       370        380        390        400        410        420
VTFIVMNKCA DDIPRIHELF QESFSTKGEG RGLGLSTLKE IADNADNVLL DTIIENGFFY
423
SKS
```

References for protein sequences listed in Appendix II

BvgS, *B. pertussis*: Aricò *et al.* (1991); VirA, *A. tumefaciens*: Leroux *et al.* (1987); NtrB, *K. pneumoniae*: MacFarlane & Merrick (1985); EnvZ, *E. coli*: Comeau *et al.* (1985); PhoQ, *S. typhimurium*: Miller *et al.* (1989b); PilB, *N. gonorrhoeae*: Taha *et al.* (1988); AgrB, *S. aureus*: Kornblum *et al.* (1990).

References

Abraham, J.M., Freitag, C.S., Clements, J.R. & Eisenstein, B.I. (1985) An invertible element of DNA controls phase variation of type I fimbriae of *Escherichia coli*. *Proc. Natl Acad. Sci. USA* **82**, 5724–5727.

Abraham, L.J. & Rood, J.I. (1987) Identification of Tn*4451* and Tn*4452*, chloramphenicol resistance transposons from *Clostridium perfringens*. *J. Bacteriol.* **169**, 1579–1584.

Adams, D.E., Shektman, E.M., Zechiedrich, E.L., Schmid, M.B. & Cozzarelli, N.R. (1992) The role of topoisomerase IV in partitioning bacterial replicons and the structure of catenated intermediates in DNA replication. *Cell* **71**, 277–288.

Adler, B., Sasakawa, C., Tobe, T., Makino, S., Komatsu, K. & Yoshikawa, M. (1989) A dual transcriptional activation system for the 230 kb plasmid genes coding for the virulence-associated antigens of *Shigella flexneri*. *Mol. Microbiol.* **3**, 627–635.

Adzuma, K. & Mizuuchi, K. (1988) Target immunity of Mu transposition reflects a differential distribution of Mu B protein. *Cell* **53**, 257–266.

Aguero, M.E., Aron, L., DeLuca, A.G. & Timmis, K.N. (1984) A plasmid-linked outer membrane protein, TraT, enhances resistance of *Escherichia coli* to phagocytosis. *Infect. Immun.* **46**, 740–746.

Aiba, H. (1985) Transcription of the *Escherichia coli* adenylate cyclase gene is negatively regulated by cAMP–cAMP receptor protein. *J. Biol. Chem.* **260**, 3063–3070.

Aiba, H., Fujimoto, S. & Ozaki, N. (1982) Molecular cloning and nucleotide sequencing of the gene for *E. coli* cAMP receptor protein. *Nucleic Acids Res.* **10**, 1345–1361.

Aiba, H., Nakasai, F., Mizushima, S. & Mizuno, T. (1989) Phosphorylation of a bacterial activator protein, OmpR, by a protein kinase, EnvZ, results in a stimulation of its DNA binding ability. *J. Biochem. (Tokyo)* **106**, 5–7.

Albertini, A.M., Hofer, M., Calos, M.P. & Miller, J.H. (1982) On the formation of spontaneous deletions: the importance of short sequence homologies in the generation of large deletions. *Cell* **29**, 319–328.

Aldea, M., Hernández-Chico, C., de la Campa, A.G., Kushner, S.R. & Vicente, M. (1988) Identification, cloning and expression of *bolA*, an *ftsZ*-dependent morphogene of *Escherichia coli*. *J. Bacteriol.* **170**, 5169–5176.

Aldea, M., Garrido, T., Hernández-Chico, C., Vincente, M. & Kushner, S. R. (1989) Induction of a growth-phase-dependent promoter triggers transcription of *bolA*, an *Escherichia coli* morphogene. *EMBO J.* **8**, 3923–3931.

Aliabadi, Z., Park, Y.K., Slonczewski, J.L. & Foster, J.W. (1988) Novel regulatory loci controlling oxygen- and pH-regulated gene expression in *Salmonella typhimurium*. *J. Bacteriol.* **170**, 842–851.

Allan, I., Kroll, J.S., Dhir, A. & Moxon, E.R. (1988) *Haemophilus influenzae* serotype a: outer membrane protein classification and correlation with DNA polymorphism at the *cap* locus. *Infect. Immun.* **56**, 529–531.

Allet, B. (1979) Mu insertion duplicates a 5 base pair sequence at the host inserted site. *Cell* **16**, 123–129.

Allgood, N.D. & Silhavy, T.J. (1988) Illegitimate recombination in bacteria. In Kucherlapati, R. & Smith, G.R. (eds) *Genetic Recombination*, pp. 309–330. American Society for Microbiology, Washington.

Allison, C., Coleman, N., Jones, P.L. & Hughes, C. (1992a) Ability of *Proteus mirabilis* to invade urothelial cells is coupled to motility and swarming differentiation. *Infect. Immun.* **60**, 4740–4746.

Allison, C., Lai, H.-C. & Hughes, C. (1992b) Coordinate expression of virulence genes during swarm-cell differentiation and population migration of *Proteus mirabilis. Mol. Microbiol.* **6**, 1583–1591.

Almirón, M., Link, A., Furlong, D. & Kolter, R. (1992) A novel DNA-binding protein with regulatory and protective roles in starved *Escherichia coli. Genes Dev.* **6**, 2646–2654.

Alouf, J.E. & Freer, J.H. (1991) *Sourcebook of Bacterial Toxins.* Academic Press, London.

Alton, N.K. & Vapnek, D. (1979) Nucleotide sequence analysis of the chloramphenicol resistance transposon Tn*9*. *Nature* **282**, 864–869.

Amábile-Cuevas, C.F. & Chicurel, M.E. (1992) Bacterial plasmids and gene flux. *Cell* **70**, 189–199.

Amábile-Cuevas, C.F. & Demple, B. (1991) Molecular characterization of the *soxRS* genes of *Escherichia coli*: two genes control a superoxide stress regulon. *Nucleic Acids Res.* **19**, 4479–4484.

Andersen, J., Delihas, N., Ikenaka, K., Green, P.J., Pines, O., Ilercil, O. & Inouye, M. (1987) The isolation and characterization of RNA coded by the *micF* gene in *Escherichia coli. Nucleic Acids Res.* **15**, 2089–2101.

Andersen, J., Forst, A., Zhao, K., Inouye, M. & Delihas, N. (1989) The function of *micF* RNA. *J. Biol. Chem.* **264**, 17961–17970.

Anderson, P. & Bauer, W. (1978) Supercoiling in closed circular DNA: dependence upon ion type and concentration. *Biochemistry* **17**, 594–601.

Anderson, R.P. & Roth, J.R. (1978a) Tandem chromosomal duplications in *Salmonella typhimurium. J. Mol. Biol.* **119**, 147–156.

Anderson, R.P. & Roth, J.R. (1978b) Tandem genetic duplications in *Salmonella typhimurium*: amplification of the histidine operon. *J. Mol. Biol.* **126**, 53–71.

Andersson, D.I. (1992) Involvement of the Arc system in redox regulation of the Cob operon in *Salmonella typhimurium. Mol. Microbiol.* **6**, 1491–1494.

Ansari, A.Z., Chael, M.L. & O'Halloran, T.V. (1992) Allosteric underwinding of DNA is a critical step in positive control of transcription by Hg-MerR. *Nature* **355**, 87–89.

Anzola, J., Luft, B., Gorgone, G., Dattwyler, R.J., Soderberg, C., Lahesmaa, R. & Peltz, G. (1992a) *Borrelia burgdorferi* HSP70 homologue: characterization of an immunoreactive stress protein. *Infect. Immun.* **60**, 3704–3713.

Anzola, J., Luft, B.J., Gorgone, G. & Peltz, G. (1992b) Characterization of a *Borrelia burgdorferi dnaJ* homologue. *Infect. Immun.* **60**, 4965–4968.

Argos, P., Landy, A., Abremski, K., Egan, J.B., Haggard-Ljungqvst, E., Hoess, R.H., Kahn, M.L., Kalionis, B., Narayana, S.V.L., Pierson III, L.S., Sternberg, N. & Leong, J. (1986) The integrase family of site-specific recombinases: regional similarities and global diversity. *EMBO J.* **5**, 433–440.

Aricò, B., Miller, J.F., Roy, C., Stibitz, S., Monack, D., Falkow, S., Gross, R. & Rappuoli, R. (1989) Sequences required for expression of *Bordetella pertussis* virulence factors share homology with prokaryotic signal transduction proteins. *Proc. Natl Acad. Sci. USA* **86**, 6671–6675.

Aricò, B., Scarlato, V., Monack, D.M., Falkow, S. & Rappuoli, R. (1991) Structural and genetic analysis of the *bvg* locus in *Bordetella* species. *Mol. Microbiol.* **5**, 2481–2491.

Armstrong, S. & Parker, C. (1986) Surface proteins of *Bordetella pertussis*: comparison of virulent and avirulent strains and effects of phenotypic modulation. *Infect. Immun.* **54**, 308–314.

Arnold, G.F. & Tessman, I. (1988) Regulation of DNA superhelicity by *rpoB* mutations that suppress defective Rho-mediated transcription termination in *Escherichia coli. J. Bacteriol.* **170**, 4266–4271.

Arnqvist, A., Olsén, A., Pfeifer, J., Russell, D.G. & Normark, S. (1992) The Crl protein activates cryptic genes for curli formation and fibronectin binding in *Escherichia coli. Mol. Microbiol.* **6**, 2443–2452.

Atlung, T., Nielsen, A. & Hansen, F.G. (1989) Isolation, characterization, and nucleotide sequence of *appY*, a regulatory gene for growth-phase dependent gene expression in *Escherichia coli. J. Bacteriol.* **171**, 1683–1691.

Auble, D.T. & deHaseth, P.L. (1988) Promoter recognition by *Escherichia coli* RNA polymerase: influence of DNA structure in the spacer separating the −10 and −35 regions. *J. Mol. Biol.* **202**, 471–482.

Austin, S., Ziese, M. & Sternberg, N. (1981) A novel role for site specific recombination in maintenance of bacterial replicons. *Cell* **25**, 729–736.

Babitzke, P. & Yanofsky, C. (1993) Reconstruction of *Bacillus subtilis trp* attenuation *in vitro* with TRAP, the *trp* RNA-binding attenuation protein. *Proc. Natl Acad. Sci USA* **90**, 133–137.

Bachellier, S., Perrin, D., Hofnung, M. & Gilson, E. (1993) Bacterial interspersed mosaic elements (BIMEs) are present in the genome of *Klebsiella. Mol. Microbiol.* **7**, 537–544.

Bachmann, B.J. (1990) Linkage map of *Escherichia coli* K-12, Edition 8. *Microbiol. Rev.* **54**, 130–197.

Bailey, M.J.A., Koronakis, V., Schmoll, T. & Hughes, C. (1992) *Escherichia coli* HlyT protein, a transcriptional activator of haemolysin synthesis and secretion, is encoded by the *rfaH* (*sfrB*) locus required for expression of sex factor and lipopolysaccharide genes. *Mol. Microbiol.* **6**, 1003–1012.

Baird, G.D., Manning, E.J. & Jones, P.W. (1985) Evidence for related virulence sequences in plasmids of *Salmonella dublin* and *Salmonella typhimurium. J. Gen. Microbiol.* **131**, 1815–1823.

Balke, V.L. & Gralla, J.D. (1987) Changes in linking number of supercoiled DNA accompany growth transitions in *Escherichia coli. J. Bacteriol.* **169**, 4499–4506.

Baracchini, E. & Bremer, H. (1988) Stringent and growth control of rRNA synthesis in *Escherichia coli* are both mediated by ppGpp. *J. Biol. Chem.* **263**, 2597–2602.

Barbour, A. (1989) Antigenic variation in relapsing fever *Borrelia* species: genetic aspects. In Berg, D.E. & Howe, M.M. (eds), *Mobile DNA*, pp. 783–789. American Society for Microbiology, Washington.

Barbour, A.G. (1990a) Antigenic variation in relapsing fever *Borrelia* species. In Iglewski, B.H. & Clark, V.L. (eds), *Molecular Basis of Bacterial Pathogenesis*, pp. 155–176. Academic Press, San Diego.

Barbour, A.G. (1990b) Antigenic variation in a relapsing fever *Borrelia* species. *Annu. Rev. Microbiol.* **44**, 155–171.

Barbour, A.G., Carter, C.J., Burman, N., Freitag, C., Garon, C. & Berstrom, S. (1991a) Tandem insertion sequence-like elements define the expression site for variable antigen genes of *Borrelia hermsii. Infect. Immun.* **59**, 390–397.

Barbour, A.G., Carter, C.J., Burman, N., Kitten, T. & Bergstrom, S. (1991b) Variable antigen genes of the relapsing fever agent *Borrelia hermsii* are activated by promoter addition. *Mol. Microbiol.* **5**, 489–493.

Barbour, A.G. & Garon, C.F. (1987) Linear plasmids of the bacterium *Borrelia burgdorferi* have covalently closed ends. *Science*, **237**, 409–411.

Bardwell, J.C.A. & Craig, E.A. (1987) Eukaryotic Mr 83,000 heat shock protein has a homologue in *Escherichia coli. Proc. Natl Acad. Sci. USA* **84**, 5177–5181.

Baril, C., Richaud, C., Baranton, G. & Saint Girons, I. (1989) Linear chromosome of *Borrelia burgdorferi. Res. Microbiol.* **140**, 507–516.

Barker, R.F., Idler, K.B., Thompson, D.V. & Kemp, J.D. (1983) Nucleotide sequence of the T-DNA region from *Agrobacterium tumefaciens* octopine Ti plasmid pTi5955. *Plant Mol. Biol.* **2**, 335–350.

Barnett, M.J. & Long, S.R. (1990) DNA sequence and translational product of a new nodulation regulatory locus: SyrM has sequence similarity to NodD proteins. *J. Bacteriol.* **172**, 3695–3700.

Barondess, J. & Beckwith, J. (1990) A bacterial virulence determinant encoded by lysogenic coliphage lambda. *Nature* **346**, 871–874.

Barr, G.C., Ní Bhriain, N. & Dorman, C.J. (1992) Identification of two genetically active regions associated with the *osmZ* locus of *Escherichia coli*: role in regulation of *proU* expression and mutagenic effect at *cya*, the structural gene for adenylate cyclase. *J. Bacteriol.* **174**, 998–1006.

Barras, F. & Marinus, M.G. (1989) The great GATC: DNA methylation in *E. coli*. *Trends Genet.* **5**, 139–143.

Barrow, P.A. & Lovell, M.A. (1988) The association between a large molecular mass plasmid and virulence in a strain of *Salmonella pullorum*. *J. Gen. Microbiol.* **134**, 2307–2316.

Barrow, P.A., Simpson, J.M., Lovell, M.A. & Binns, M.M. (1987) Contribution of *Salmonella gallinarum* large plasmid towards virulence in fowl typhoid. *Infect. Immun.* **55**, 388–392.

Barry, C.E. III, Brickman, T.J. & Hackstadt, T. (1993) Hc1-mediated effects on DNA structure: a potential regulator of chlamydial development. *Mol. Microbiol.* **9**, 273–283.

Barry, C.E. III, Hayes, S.F. & Hackstadt, T. (1992) Nucleoid condensation in *Escherichia coli* that express a chlamydial histone homolog. *Science* **256**, 377–379.

Bartkus, J.M. & Leppla, S.H. (1989) Transcriptional regulation of the protective antigen gene of *Bacillus anthracis*. *Infect. Immun.* **57**, 2295–2300.

Batut, J., Daveran-Mingot, M.-L., David, M., Jacobs, J., Garnerone, A.M. & Kahn, D. (1989) *fixK*, a gene homologous with *fnr* and *crp* from *Escherichia coli*, regulates nitrogen fixation genes both positively and negatively in *Rhizobium meliloti*. *EMBO J.* **8**, 1279–1286.

Bavoil, P. (1990) Invasion and intracellular growth of *Chlamydia* species. In Iglewski, B.H. & Clark. V.L. (eds) *Molecular Basis of Bacterial Pathogenesis*, (eds) pp. 273–296. Academic Press, San Diego.

Beckwith, J.R., Signer, E.R. & Epstein, W. (1966) Transposition of the *lac* region of *E. coli*. *Cold Spring Harbor Symp. Quant. Biol.* **31**, 393–401.

Bell, A., Gaston, K., Williams, R., Chapman, K., Kolls, A., Buc, H., Minchin, S., Williams, J. & Busby, S. (1990) Mutations that alter the ability of the *Escherichia coli* cyclic AMP receptor protein to activate transcription. *Nucleic Acids Res.* **18**, 7243–7250.

Belland, R.J. (1991) H-DNA formation by the coding repeat elements of neisserial *opa* genes. *Mol. Microbiol.* **5**, 2351–2360.

Bender, J. & Kleckner, N. (1986) Genetic evidence that Tn*10* transposes by a nonreplicative mechanism. *Cell* **45**, 801–815.

Benjamin, H.W., Matzuk, M.M., Krasnow, M.A. & Cozzarelli, N.R. (1985) Recombination site selection by Tn*3* resolvase: topological tests of a tracking mechanism. *Cell* **40**, 147–158.

Berg, D.E. (1989) Transposon Tn5. In Berg, D.E. & Howe, M.M. (eds) *Mobile DNA*, pp. 185–210. American Society for Microbiology, Washington.

Berg, C.M. & Berg, D.E. (1987) Uses of transposable elements and maps of known insertions. In Neidhardt, F.C., Ingraham, J.L., Low, K.B., Magasanik, B., Schaechter, M. & Umbarger, H.E. (eds) Escherichia coli *and* Salmonella typhimurium *Cellular and Molecular Biology*, pp. 1071–1109. American Society for Microbiology, Washington.

Berg, C.M., Berg, D.E. & Groisman, E. (1989) Transposable elements and the genetic engineering of bacteria. In Berg, D.E. & Howe, M.M. (eds) *Mobile DNA*, pp. 879–925. American Society for Microbiology, Washington.

Berg, D.E. & Berg, C.M. (1983) The prokaryotic transposable element Tn5. *Bio/Technology* **1**, 417–422.

Bergman, T., Håkansson, S., Forsberg, Å., Norlander, L., Macellaro, A., Bäckman, A., Bölin, I. & Wolf-Watz, H. (1991) Analysis of the V antigen *lcrGVH-yopBD* operon of *Yersinia pseudotuberculosis*: evidence for a regulatory role of LcrH and LcrV. *J. Bacteriol.* **173**, 1607–1616.

Bergström, S., Bundoc, V.G. & Barbour, A.G. (1989) Molecular analysis of linear plasmid-encoded major surface proteins, OspA and OspB, of the Lyme disease spirochaete *Borrelia burgdorferi*. *Mol. Microbiol.* **3**, 479–486.

Bernardini, M.L., Fontaine, A. & Sansonetti, P.J. (1990) The two-component regulatory system OmpR–EnvZ controls the virulence of *Shigella flexneri*. *J. Bacteriol.* **172**, 6274–6281.

Berry, A., DeVault, J.D. & Chakrabarty, A.M. (1989) High osmolarity is a signal for enhanced *algD* transcription in mucoid and nonmucoid *Pseudomonas aeruginosa* strains. *J. Bacteriol.* **171**, 2312–2317.

Betley, M.J. & Mekalanos, J.J. (1985) Staphylococcal enterotoxin A is encoded by phage. *Science* **229**, 185–187.

Betley, M.J., Miller, V.L. & Mekalanos, J.J. (1986) Genetics of bacterial enterotoxins. *Annu. Rev. Microbiol.* **40**, 577–605.

Better, M. & Helinski, D.R. (1983) Isolation and characterization of the *recA* gene of *Rhizobium meliloti*. *J. Bacteriol.* **155**, 311–316.

Beutin, L., Bode, L., Özel, M. & Stephan, R. (1990) Enterohemolysin production is associated with a temperate bacteriophage in *Escherichia coli* serogroup O26 strains. *J. Bacteriol.* **172**, 6469–6475.

Beutin, L., Manning, P.A., Achtman, M. & Willetts, N. (1981) *sfrA* and *sfrB* products of *Escherichia coli* K-12 are transcriptional control factors. *J. Bacteriol.* **145**, 840–844.

Beutin, L., Montenegro, M.A., Orskov, I., Orskov, F., Prada, J., Zimmermann, S. & Stephan, T. (1989) Close association of verotoxin (Shiga-like toxin) production with enterohemolysin production in strains of *Escherichia coli*. *J. Clin. Microbiol.* **27**, 2559–2564.

Bevan, M.W. & Chilton, M.-D. (1982) T-DNA of the Agrobacterium Ti and Ri plasmids. *Annu. Rev. Genet.* **16**, 357–384.

Bhat, K.S., Gibbs, C.P., Barrera, O., Morrison, S.G., Jähnig, F., Stern, A., Kupsch, E.-M., Meyer, T. & Swanson, J. (1991) The opacity proteins of *Neisseria gonorrhoeae* strain MS11 are encoded by a family of 11 complete genes. *Mol. Microbiol.* **5**, 1889–1901.

Bhugra, B. & Dybvig, K. (1992) High-frequency rearrangements in the chromosome of *Mycoplasma pulmonis* correlate with phenotypic switching. *Mol. Microbiol.* **6**, 1149–1154

Biel, S.W. & Berg, D.E. (1984) Mechanism of IS*1* transposition in *E. coli*: choice between simple insertion and cointegration. *Genetics* **108**, 319–330.

Binns, M.M., Davies, D.L. & Hardy, K.G. (1979) Cloned fragments of the plasmid ColV,I-K94 specifying virulence and serum resistance. *Nature* **279**, 778–781.

Biot, T. & Cornelis, G.R. (1988) The replication, partition and *yop* regulation of the pYV plasmids are highly conserved in *Yersinia enterocolitica* and *Y. pseudotuberculosis*. *J. Gen. Microbiol.* **134**, 1525–1534.

Blakely, G., Colloms, S., May, G., Burke, M. & Sherratt, D. (1991) *Escherichia coli* XerC recombinase is required for chromosomal segregation at cell division. *New Biol.* **3**, 789–798.

Blander, M.A., Sandler, S.J., Armengod, M.-E., Ream, L.W. & Clark, A.J. (1984) Molecular analysis of the *recF* gene of *Escherichia coli*. *Proc. Natl Acad. Sci. USA* **81**, 4622–4626.

Bliska, J.B. & Cozzarelli, N.R. (1987) Use of site-specific recombination as a probe of DNA structure and metabolism *in vivo*. *J. Mol. Biol.* **194**, 205–218.

Bliska, J.B. & Falkow, S. (1993) The role of host tyrosine phosphorylation in bacterial pathogenesis. *Trends Genet.* **9**, 85–89.

Blomfield, I.C., Calie, P.J., Eberhardt, K.J., McClain, M.S. & Eisenstein, B.J. (1993) Lrp stimulates phase variation of type 1 fimbriation in *Escherichia coli* K-12. *J. Bacteriol.* **175**, 27–36.

Blumenthals, I.I., Kelly, R.M., Gorziglia, M., Kaufman, J.B. & Shiloach, J. (1987) Development of a defined medium and two-step culturing method for improved exotoxin A yields from *Pseudomonas aeruginosa. Appl. Environ. Microbiol.* **53**, 2013–2020.

Blyn, L.B., Braaten, B.A. & Low, D.A. (1990) Regulation of *pap* pilin phase variation by a mechanism involving differential Dam methylation states. *EMBO J.* **9**, 4045–4054.

Bohannon, D.E., Connell, N., Keener, J., Tormo, A., Espinosa-Urgel, M., Zambrano, M.M. & Kolter, R. (1991) Stationary phase-inducible 'gearbox' promoters: differential effects of *katF* mutations and role of σ^{70}. *J. Bacteriol.* **173**, 4482–4492.

Bolin, I., Norlander, L. & Wolf-Watz, H. (1982) Temperature-inducible outer membrane protein of *Yersinia pseudotuberculosis* and *Yersinia enterocolitica* is associated with the virulence plasmid. *Infect. Immun.* **37**, 506–512.

Bolivar, F., Rodriguez, R.L., Betlach, M.C. & Boyer, H.W. (1977a) Construction and characterization of new cloning vehicles. I. Ampicillin-resistant derivatives of the plasmid pMB9. *Gene* **2**, 75–93.

Bolivar, F., Rodriguez, R.L., Greene, P.J., Betlach, M.V., Heynecker, H.L., Boyer, H.W., Crosa, J.H. & Falkow, S. (1977b) Construction and characterization of new cloning vehicles. II. A multipurpose cloning system. *Gene* **2**, 95–113.

Bölker, M. & Kahmann, R. (1989) The *Escherichia coli* regulatory protein OxyR discriminates between methylated and unmethylated states of the phage Mu *mom* promoter. *EMBO J.* **8**, 2403–2410.

Borowiec, J. & Gralla, J. (1985) Supercoiling response of the *lac* p^s promoter *in vitro. J. Mol. Biol.* **184**, 587–598.

Borowiec, J.A. & Gralla, J.D. (1987) All three elements of the *lac* p^s promoter mediate its transcriptional response to DNA supercoiling. *J. Mol. Biol.* **195**, 89–97.

Borowiec, J.A., Zhang, L., Sasse-Dwight, S. & Gralla, J.D. (1987) DNA supercoiling promotes loop formation of a bent repression loop in *lac* DNA. *J. Mol. Biol.* **196**, 101–111.

Borukhov, S., Polyakov, A., Nikiforov, V. & Goldfarb, A. (1992) GreA protein: a transcription elongation factor from *Escherichia coli. Proc. Natl Acad. Sci. USA.* **89**, 8899–8902.

Botsford, J.L. & Harman, J.G. (1992) Cyclic AMP in prokaryotes. *Microbiol. Rev.* **56**, 100–122.

Boulnois, G.J., Roberts, I.S., Hodge, R., Hardy, K.R., Jann, K.B. & Timmis, K.N. (1987) Analysis of the K1 capsule biosynthesis genes of *Escherichia coli*: definition of three functional regions for capsule production. *Mol. Gen. Genet.* **208**, 242–246.

Bouthier de la Tour, C., Portemer, C., Huber, R., Forterre, P. & Dugeut, M. (1991) Reverse gyrase in thermophilic eubacteria. *J. Bacteriol.* **173**, 3921–3923.

Boyd, J., Oza, M.N. & Murphy, J.R. (1990) Molecular cloning and DNA sequence analysis of a diphtheria *tox* iron-dependent regulatory element (*dtxR*) from *Corynebacterium diphtheriae. Proc. Natl Acad. Sci. USA* **87**, 5968–5972.

Braaten, B.A., Blyn, L.B., Skinner, B.S. & Low, D.A. (1991) Evidence for a methylation-blocking factor (*mbf*) locus involved in *pap* pilus expression and phase variation in *Escherichia coli. J. Bacteriol.* **173**, 1789–1800.

Braaten, B.A., Platko, J.V., van der Woude, M.W., Simons, B.H., de Graaf, F.K., Calvo, J.M. & Low, D.A. (1992) Leucine-responsive regulatory protein controls the expression of both *pap* and *fan* pili operons in *Escherichia coli. Proc. Natl Acad. Sci. USA* **89**, 4250–4254.

Bracco, L., Kotlarz, D., Kolb, A., Diekmann, S. & Buc, H. (1989) Synthetic curved DNA sequences can act as transcriptional activators in *Escherichia coli*. *EMBO J.* **8**, 4289–4296.

Bramhill, D. & Kornberg, A. (1988) Duplex opening by dnaA protein at novel sequences in initiation of replication at the origin of the *E. coli* chromosome. *Cell* **52**, 743–755.

Brewer, B.J. (1988) When polymerases collide: replication and the transcriptional organization of the *E. coli* chromosome. *Cell* **53**, 679–686.

Brewer, B.J. (1990) Replication and the transcriptional organization of the *Escherichia coli* chromosome. In Drlica, K. & Riley, M. (eds) *The Bacterial Chromosome*, pp. 61–83. American Society for Microbiology, Washington.

Bringel, F., Van Alstine, G.L. & Scott, J.R. (1991) A host factor absent from *Lactococcus lactis* MG1363 is required for conjugative transposition. *Mol. Microbiol.* **5**, 2983–2993.

Brooks, G.F., Olinger, L., Lammel, C.J., Bhat, K.S., Calvello, C.A., Palmer, M.L., Knapp, J.S. & Stephens, R.S. (1991) Prevalence of gene sequences coding for hypervariable regions of Opa (protein II) in *Neisseria gonorrhoeae*. *Mol. Microbiol.* **5**, 3063–3072.

Broyles, S.S. & Pettijohn, D.E. (1986) Interaction of the *Escherichia coli* HU protein with DNA: evidence for formation of nucleosome-like structure with circular double-stranded DNA. *J. Mol. Biol.* **187**, 47–60.

Brubaker, R.R. (1983) The Vwa$^+$ virulence factor of *Yersinia*: the molecular basis of the attendant nutritional requirement for Ca^{2+}. *Rev. Infect. Dis.* **5**, S748-S758.

Bruckner, R.C. & Cox, M.M. (1989) The histonelike H-protein of *Escherichia coli* is ribosomal protein-S3. *Nucleic Acids Res.* **17**, 3145–3161.

Bruckner, R., Dick, T. & Matzura, H. (1987) Dependence of expression of an inducible *Staphylococcus aureus cat* gene on the translation of its leader sequence. *Mol. Gen. Genet.* **207**, 486–491.

Brumlik, M.J. & Storey, D.G. (1992) Zinc and iron regulate translation of the gene encoding *Pseudomonas aeruginosa* elastase. *Mol. Microbiol.* **6**, 337–344.

Brunelle, A. & Schleif, R. (1989) Determining residue-base interactions between AraC protein and *araI* DNA. *J. Mol. Biol.* **209**, 607–622.

Buchanan-Wollaston, V., Passiatore, J.E. & Cannon, F. (1987) The *mob* and *oriT* mobilization functions of a bacterial plasmid promote its transfer to plants. *Nature* **328**, 172–175.

Buchmeier, N.A. & Heffron, F. (1990) Induction of *Salmonella* stress proteins upon infection of macrophages. *Science* **248**, 730–732.

Buchmeier, N.A., Lipps, C.J., So, M.Y. & Heffron, F. (1993) Recombination-deficient mutants of *Salmonella typhimurium* are avirulent and sensitive to the oxidative burst of macrophages. *Mol. Microbiol.* **7**, 933–936.

Bundoc, V.G. & Barbour, A.G. (1989) Clonal polymorphisms of outer membrane protein OspB of *Borrelia burgdorferi*. *Infect. Immun.* **57**, 2733–2741.

Burke, K.A. & Wilcox, G. (1987) The *araC* gene of *Citrobacter freundii*. *Gene* **61**, 243–252.

Busby, S.J.W. (1986) Positive regulation in gene expression. *Soc. Gen. Microbiol. Symp.* **39**, 51–77.

Bushman, W., Thompson, J.F., Vargas, L. & Landy, A. (1985) Control of directionality in lambda site-specific recombination. *Science* **230**, 906–911.

Bushman, W., Yin, S., Thio, L.L. & Landy, A. (1984) Determinants of directionality in lambda site-specific recombination. *Cell* **39**, 699–706.

Butterton, J.R., Stoebner, J.A., Payne, S.M. & Calderwood, S.B. (1992) Cloning, sequencing, and transcriptional regulation of *viuA*, the gene encoding the ferric vibriobactin receptor of *Vibrio cholerae*. *J. Bacteriol.* **174**, 3729–3738.

Buxton, R.S. & Drury, L.S. (1984) Identification of the *dye* gene product, mutational loss of which alters envelope protein composition and also affects sex factor F expression in *Escherichia coli*. *Mol. Gen. Genet.* **194**, 241–247.

Cairns, J., Overbaugh, J. & Miller, S. (1988) The origin of mutants. *Nature* **335**, 142–145.

Calderwood, S.B., Auclair, F., Donohue-Rolfe, A., Keusch, G.T. & Mekalanos, J.J. (1987) Nucleotide sequence of the shiga-like toxin genes of *Escherichia coli*. *Proc. Natl Acad. Sci. USA* **84**, 4364–4368.

Calderwood, S.B. & Mekalanos, J.J. (1987) Iron regulation of shiga-like toxin expression in *Escherichia coli* is mediated by the *fur* locus. *J. Bacteriol.* **169**, 4759–4764.

Calderwood, S.B. & Mekalanos, J.J. (1988) Confirmation of the Fur operator site by insertion of a synthetic oligonucleotide into an operon fusion plasmid. *J. Bacteriol.* **170**, 1015–1017.

Caldwell, A.L. & Gulig, P.A. (1991) The *Salmonella typhimurium* virulence plasmid encodes a positive regulator of a plasmid-encoded virulence gene. *J. Bacteriol.* **173**, 7176–7185.

Campbell, A. (1981) Some general questions about movable elements and their implications. *Cold Spring Harbor Symp. Quant. Biol.* **45**, 1-9.

Cangelosi, G.A., Best, E.A., Martinetti, G. & Nester, E.W. (1991) Genetic analysis of *Agrobacterium*. *Methods Enzymol.* **204**, 384–397.

Capage, M. & Hill, C.W. (1979) Preferential unequal recombination of the *glyS* region of the *Escherichia coli* chromosome. *J. Mol. Biol.* **127**, 73–87.

Caparon, M.G., Geist, R.T., Perez-Casal, J. & Scott, J.R. (1992) Environmental regulation of virulence in group A streptococci: transcription of the gene encoding M protein is stimulated by carbon dioxide. *J. Bacteriol.* **174**, 5693–5701.

Caparon, M.G. & Scott, J.R. (1987) Identification of a gene that regulates expression of M protein, the major virulence determinant of group A streptococci. *Proc. Natl Acad. Sci. USA* **84**, 8677–8681.

Caparon, M.G. & Scott, J.R. (1989) Excision and insertion of the conjugative transposon Tn*916* involves a novel recombination mechanism. *Cell* **59**, 1027–1034.

Caperon, M.G. & Scott, J.R. (1991) Genetic manipulation of pathogenic streptococci. *Methods Enzymol.* **204**, 556–586.

Carmona, M., Balsalobre, C., Munoa, F., Mourino, M., Jubete, Y., De La Cruz, F. & Juarez, A. (1993) *Escherichia coli hha* mutants, DNA supercoiling and expression of the haemolysin genes from the recombinant plasmid pANN202–312. *Mol. Microbiol.* **9**, 1011–1018.

Carniel, E., Mazigh, D. & Mollaret, H.H. (1987) Expression of iron-regulated proteins in *Yersinia* species and their relation to virulence. *Infect. Immun.* **55**, 277–280.

Caron, J., Coffield, L.M. & Scott, J.R. (1989) A plasmid-encoded regulatory gene, *rns*, required for expression of the CS1 and CS2 adhesins of enterotoxigenic *Escherichia coli*. *Proc. Natl Acad. Sci. USA* **86**, 963–967.

Case, C. C., Bukau, B., Granett, S., Villarejo, M.R. & Boos, W. (1986) Contrasting mechanisms of *envZ* control of *mal* and *pho* regulon genes in *Escherichia coli*. *J. Bacteriol.* **166**, 706–712.

Cashel, M. & Rudd, K.E. (1987) The stringent response. In Neidhardt, F.C., Ingraham, J.L., Low, K.B., Magasanik, B., Schaechter, M. & Umbarger, H.E. (eds) Escherichia coli *and* Salmonella typhimurium *Cellular and Molecular Biology*, pp. 1410–1438. American Society for Microbiology, Washington.

Cass, L.G. & Wilcox, G. (1986) Mutations in the *araC* regulatory gene of *Escherichia coli* B/r that affect repressor and activator functions of AraC protein. *J. Bacteriol.* **166**, 892–900.

Cassel, D. & Pfeuffer T., (1978) Mechanism of cholera toxin action: covalent modification of the guanyl nucleotide binding protein of the adenylate cyclase system. *Proc. Natl Acad. Sci. USA* **75**, 2669–2673.

Cavalli-Sforza, L., Lederberg, J. & Lederberg, E. (1953) An infective factor controlling sex compatibility in *Bacterium coli*. *J. Gen. Microbiol.* **8**, 89–103.

Cesareni, G., Helmer-Citterich, M. & Castagnoli, L. (1991) Control of ColE1 plasmid replication by antisense RNA. *Trends Genet.* **7**, 230–235.

Chaconas, G. & Surette, M.G. (1988) Mechanism of Mu DNA transposition. *BioEssays* **9**, 205–208.

Chan, P.T., Ohmori, H., Tomizawa, J. & Lebowitz, J. (1985) Nucleotide sequence and gene organization of ColE1 DNA. *J. Biol. Chem.* **260**, 8925–8935.

Chandler, M., Allet, B., Gallay, E., Boy de la Tour, E. & Caro, L. (1977) Involvement of IS*1* in the dissociation of the r-determinant and RTF components of the plasmid R100.1. *Mol. Gen. Genet.* **153**, 289–295.

Chandler, M. & Fayet, O. (1993) Translational frameshifting in the control of transposition in bacteria. *Mol. Microbiol.* **7**, 497–503.

Chang, A.C.Y. & Cohen, S.N. (1978) Construction and characterization of amplifiable multicopy DNA cloning vehicles derived from the P15A cryptic miniplasmid. *J. Bacteriol.* **134**, 1141–1156.

Chang, M., Hadero, A. & Crawford, I.P. (1989) Sequence of the *Pseudomonas aeruginosa trpI* activator gene and relatedness of *trpI* to other procaryotic regulatory genes. *J. Bacteriol.* **171**, 172–183.

Charlebois, R.L. & Doolittle, W.F. (1989) Transposable elements and genome structure in Halobacteria. In Berg, D.E. & Howe, M.M. (eds) *Mobile DNA*, pp. 297–307. American Society for Microbiology, Washington.

Chater, K.F. & Hopwood, D.A. (1989) Diversity of bacterial genetics. In Chater, K.F. & Hopwood, D.A. (eds) *Genetics of Bacterial Diversity*, pp. 23–52. Academic Press, London.

Chatfield, S.N., Dorman, C.J., Hayward, C. & Dougan, G. (1991) Role of *ompR*-dependent genes in *Salmonella typhimurium* virulence: mutants deficient in both OmpC and OmpF are attenuated in vivo. *Infect. Immun.* **59**, 449–452.

Chehade, H. & Braun, V. (1988) Iron-regulated synthesis and uptake of colicin V. *FEMS Microbiol Lett.* **52**, 177–182.

Chen, D., Bowater, R., Dorman, C.J. & Lilley, D.M.L. (1992) Activity of a plasmid-borne *leu500* promoter depends on the transcription and translation of an adjacent gene. *Proc. Natl Acad. Sci. USA* **89**, 8784–8788.

Cheung, A.L., Koomey, J.M., Butler, C.A., Projan, S.J. & Fischetti, V.A. (1992) Reglation of exoprotein expression in *Staphylococcus aureus* by a locus (*sar*) distinct from *agr*. *Proc. Natl Acad. Sci. USA* **89**, 6462–6466.

Chiang, R.C., Cavicchioli, R. & Gunsalus, R.P. (1992) Identification and characterization of *narQ*, a second nitrate sensor for nitrate-dependent gene regulation in *Escherichia coli*. *Mol. Microbiol.* **6**, 1913–1923.

Christiansen, G., Pedersen, L.B., Koehler, J.E., Lundemose, A.G. & Birkelund, S. (1993) Interaction between the *Chlamydia trachomatis* histone H1-like protein (Hc1) and DNA. *J. Bacteriol.* **175**, 1785–1795.

Christie, P.J., Korman, R.Z., Zahler, S.A., Adsit, J.C. & Dunny, G.M. (1987) Two conjugation systems with the *Streptococcus faecalis* plasmid pCF-10: identification of a conjugative transposon that transfers between *S. faecalis* and *Bacillus subtilis*. *J. Bacteriol.* **169**, 1212–1220.

Christman, M.F., Storz, G. & Ames, B.N. (1989) OxyR, a positive regulator of hydrogen peroxide-inducible genes in *Escherichia coli* and *Salmonella typhimurium*, is homologous to a family of bacterial regulatory proteins. *Proc. Natl Acad. Sci. USA* **86**, 3484–3488.

Chuba, P., Leon, M., Banerjee, A. & Palchaudhuri, S. (1989) Cloning and sequence of plasmid determinant *iss*, coding for increased serum survival and surface exclusion, which has homology with lambda DNA. *Mol. Gen. Genet.* **216**, 287–292.

Ciampi, M.S., Schmid, M.B. & Roth, J.R. (1982) Transposon Tn*10* provides a promoter for transcription of adjacent genes. *Proc. Natl Acad. Sci. USA* **79**, 5016–5020.

Clark, A.J. (1973) Recombination deficient mutants of *E. coli* and other bacteria. *Annu. Rev. Genet.* **7**, 67–86.

Clark, V.L. (1990) Environmental modulation of gene expression in Gram-negative pathogens. In Inglewski, B.H. & Clark, V.L. (eds) *Molecular Basis of Bacterial Pathogenesis*, pp. 111–135. Academic Press, San Diego.

Clark-Curtiss, J.E. & Docherty, M.A. (1989) A species-specific repetitive sequence in *Mycobacterium leprae* DNA. *J. Infect. Dis.* **159**, 7–15.

Claverie-Martin, F. & Magasanik, B. (1991) Role of integration host factor in the regulation of the *glnHp2* promoter of *Escherichia coli*. *Proc. Natl Acad. Sci. USA* **88**, 1631–1635.

Clerget, M. (1991) Site-specific recombination promoted by a short DNA segment of plasmid R1 and by a homologous segment in the terminus region of the *Escherichia coli* chromosome. *New Biol.* **3**, 780–788.

Clewell, D.B., An, F.Y., White, B.A. & Gawron-Burke, C. (1985) *Streptococcus faecalis* sex pheromone (cAM373) also produced by *Staphylococcus aureus* and identification of a conjugative transposon (Tn*918*). *J. Bacteriol.* **162**, 1212–1220.

Clewell, D.B. & Gawron-Burke, C. (1986) Conjugative transposons and the dissemination of antibiotic resistance in streptococci. *Annu. Rev. Microbiol.* **40**, 635–659.

Clewell, D.B., Senghas, E., Jones, J.M., Flannagan, S.E., Yamamoto, M. & Gawron-Burke, C. (1988) Transposition in *Streptococcus*: structural and genetic properties of the conjugative transposon Tn*916*. *Symp. Soc. Gen. Microbiol.* **43**, 43–58.

Clewell, D.B. & Weaver, K.E. (1989) Sex pheromones and plasmid transfer in *Enterococcus faecalis*. *Plasmid* **21**, 175–184.

Coleman, D.C. (1990) Concomitant control of expression of *Staphylococcus aureus* β-toxin, enterotoxin A and fibrinolysin mediated by serotype F lysogenic converting bacteriophages In Rappuoli, R., Alouf, J.E., Falmagne, P., Fehrenbach, F.J., Freer, J., Gross, R., Jeliaszewicz, J., Montecuccco, C., Tomasi, M., Wadström, T. & Witholt, B. (eds), *Bacterial Protein Toxins*, pp. 333–340. Gustav Fischer Verlag, Stuttgart.

Coleman, D.C., Arbuthnott, J.P., Pomeroy, H.M. & Birkbeck, T.H. (1986) Cloning and expression in *Escherichia coli* and *Staphylococcus aureus* of the beta-lysin determinant from *Staphylococcus aureus*: evidence that bacteriophage conversion of beta-lysin activity is caused by insertional inactivation of the beta-lysin determinant. *Microb. Pathogen.* **1**, 549–564.

Coleman, D.C., Knights, J., Russell, R., Shanley, D., Birkbeck, T.H., Dougan, G. & Charles, I.G. (1991) Insertional inactivation of the *Staphylococcus aureus* β-toxin by bacteriophage ϕ13 occurs by a site- and orientation-specific integration of the ϕ13 genome. *Mol. Microbiol.* **5**, 933–939.

Coleman, D.C., Sullivan, D.J., Russell, R.J., Arbuthnott, J.P., Carey, B.F. & Pomeroy, H.M. (1989) *Staphylococcus aureus* bacteriophages mediating the simultaneous lysogenic conversion of β-lysin, staphylokinase and enterotoxin A: molecular mechanisms of triple conversion. *J. Gen. Microbiol.* **135**, 1679–1697.

Collado-Vides, J., Magasanik, B. & Gralla, J.D. (1991) Control site location and transcriptional regulation in *Escherichia coli*. *Microbiol. Rev.* **55**, 371–394.

Collins, C.M & Falkow, S. (1988) Genetic analysis of an *Escherichia coli* urease locus: evidence of DNA rearrangement. *J. Bacteriol.* **170**, 1041–1045.

Collis, C.M. & Hall, R.M. (1992a) Gene cassettes from the insert region of integrons are excised as covalently closed circles. *Mol. Microbiol.* **6**, 2875–2885.

Collis, C.M. & Hall, R.M. (1992b) Site-specific deletion and rearrangement of integron insert genes catalysed by the integron DNA integrase. *J. Bacteriol.* **174**, 1574–1585.

Colloms, S.D., Sykora, P., Szatmari, G. & Sherratt, D.J. (1990) Recombination at ColE1 *cer* requires the *Escherichia coli xerC* gene product, a member of the lambda integrase family of site-specific recombinases. *J. Bacteriol.* **172**, 6973–6980.

Colman, S.D., Hu, P.C. & Bott, K.F. (1990) Prevalence of novel repeat sequences in and around the P1 operon in the genome of *Mycoplasma pneumoniae*. *Gene* **87**, 91–96.

Colonna, B., Nicoletti, M., Visca, P., Casalino, M., Valenti, P. & Maimone, F. (1985) Composite IS*1* elements encoding hydroxamate-mediated iron uptake in F1*me* plasmids from epidemic *Salmonella* spp. *J. Bacteriol.* **162**, 307–316.

Colonna-Romano, S., Arnold, W., Schlüter, A., Boistard, P., Pühler, A. & Priefer, U.B. (1990) An Fnr-like protein encoded in *Rhizobium leguminosarum* biovar *viciae* shows structural and functional homology to *Rhizobium meliloti* FixK. *Mol. Gen. Genet.* **223**, 138–147.

Comeau, D.E., Ikenaka, K., Tsung, K. & Imouye, M. (1985) Primary characterization of the protein products of the *Escherichia coli ompB* locus: structure and regulation of synthesis of the OmpR and EnvZ proteins. *J. Bacteriol.* **164**, 578–584.

Compan, I. & Touati, D. (1993) Interaction of six global transcription regulators in expression of manganese superoxide dismutase in *Escherichia coli* K-12. *J. Bacteriol.* **175**, 1687–1696.

Connell, T.D., Black, W.J., Kawula, T.H., Barritt, D.S., Dempsey, J.A., Kverneland, K. Jr., Stephenson, A., Schepart, B.S., Murphy, G.L. & Cannon, J.G. (1988) Recombination among protein II genes of *Neisseria gonorrhoeae* generates new coding sequences and increases structural variability in the protein II family. *Mol. Microbiol.* **2**, 227–236.

Connell, T.D., Shaffer, D. & Cannon, J.G. (1990) Characterization of the repertoire of hypervariable regions in the protein II (*opa*) gene family of *Neisseria gonorrhoeae*. *Mol. Microbiol.* **4**, 439–449.

Cooley, M.B., D'Souza, M.R. & Kado, C.I. (1991) The *virC* and *virD* operons of the *Agrobacterium* Ti plasmid are regulated by the *ros* chromosomal gene: analysis of the cloned *ros* gene. *J. Bacteriol.* **173**, 2608–2616.

Cornelis, G.R. (1992) Yersiniae, finely tuned pathogens. In Hormaeche, C.E., Penn, C.W. & Smyth, C.J. (eds) *Molecular Biology of Bacterial Infection: Current Status and Future Perspectives*, pp. 231–265. Cambridge University Press.

Cornelis, G.R., Biot, T., Lambert de Rouvroit, C., Michiels, T., Mulder, B., Sluiters, C., Sory, M.-P., Van Bouchaute, M. & Vanooteghem, J.-C. (1989a) The *Yersinia yop* regulon. *Mol. Microbiol.* **3**, 1455–1459.

Cornelis, G.R, Sluiters, C., Delor, I., Geib, D., Kaniga, K., Lambert de Rouvroit, C., Sory, M.-P., Vanooteghem, J.-C. & Michiels, T. (1991) *ymoA*, a *Yersinia enterocolitica* chromosomal gene modulating the expression of virulence functions. *Mol. Microbiol.* **5**, 1023–1034.

Cornelis, G., Sluiters, C., Lambert de Rouvroit, C. & Michiels, T. (1989b) Homology between VirF, the transcriptional activator of the *Yersinia* virulence regulon, and AraC, the *Escherichia coli* arabinose operon regulator. *J. Bacteriol.* **171**, 254–262.

Correia, F.F., Inouye, S. & Inouye, M. (1986) A 26 base pair repetitive sequence specific for *Neisseria meningitidis* genomic DNA. *J. Bacteriol.* **167**, 1009–1015.

Cossart, P., Vincente, M.F., Mengaud, J., Baquero, F., Perez-Diaz, J.C. & Berche, P. (1989) Listeriolysin O is essential for virulence of *Listeria monocytogenes*: direct evidence obtained by gene complementation. *Infect. Immun.* **57**, 3629–3636.

Cotter, P.A., Chepuri, V., Gennis, R.B. & Gunsalus, R.P. (1990) Cytochrome *o* (*cyoABCDE*) and *d* (*cydAB*) oxidase gene expression in *Escherichia coli* is regulated by oxygen, pH, and the *fnr* gene product. *J. Bacteriol.* **172**, 6333–6338.

Courvalin, P. & Carlier, C. (1986) Transposable multiple drug resistance in *Streptococcus pneumoniae*. *Mol. Gen. Genet.* **205**, 291–297.

Courvalin, P. & Carlier, C. (1987) Tn*1545*, a conjugative shuttle transposon. *Mol. Gen. Genet.* **206**, 259–264.

Cowart, R.E. & Foster, B.G. (1985) Differential effects of iron on the growth of *Listeria monocytogenes*: minimum requirements and mechanism of acquisition. *J. Infect. Dis.* **151**, 721–730.

Cowing, D.W., Bardwell, J.C.A., Craig, E.A., Woolford, C., Hendrix, R.W. & Gross, C.A. (1985) Consensus sequence for *Escherichia coli* heat shock promoters. *Proc. Natl Acad. Sci. USA* **82**, 2679–2683.

Cox, M.M. & Lehman, I.R. (1987) Enzymes of general recombination. *Annu. Rev. Biochem.* **56**, 229–262.

Coy, M. & Neilands, J.B. (1991) Structural dynamics and functional domains of the Fur protein. *Biochemistry* **30**, 8201–8210.

Cozzarelli, N.R. (1980a) DNA gyrase and the supercoiling of DNA. *Science* **207**, 953–960.

Cozzarelli, N.R. (1980b) DNA topoisomerases. *Cell* **22**, 327–328.

Cozzarelli, N.R., Boles, T.C. & White, J. H. (1990) Primer on the topology and geometry of DNA supercoiling. In Cozzarelli, N.R & Wang, J.C. (eds), *DNA Topology and its Biological Effects*, pp. 139–184. Cold Spring Harbor Laboratory Press, Cold Spring Harbor, New York.

Craig, N. L. (1985) Site-specific inversion: enhancers, recombination proteins and mechanism. *Cell* **41**, 649–650.

Craig, N.L. (1988) The mechanism of conservative site-specific recombination. *Annu. Rev. Genet.* **22**, 77–105.

Craig, N.L. (1989) Transposon Tn7. In Berg, D.E. & Howe, M.M. (eds) *Mobile DNA*, pp. 211–225. American Society for Microbiology, Washington.

Craig, N.L. & Kleckner, N. (1987) Transposition and site-specific recombination. In Neidhardt, F.C., Ingraham, J.L., Low, K.B., Magasanik, M., Schaechter, M. & Umbarger, H.E. (eds) Escherichia coli *and* Salmonella typhimurium *Cellular and Molecular Biology*, pp. 1054–1070. American Society for Microbiology, Washington.

Craigie, R., Mizuuchi, M., Adzuma, K. & Mizuuchi, K. (1988) Mechanism of the DNA strand transfer step in transposition of Mu DNA. *Symp. Soc. Gen. Microbiol.* **43**, 131–148.

Cravioto, A., Scotland, S.M. & Rowe, B. (1982) Haemagglutination activity and colonization factor antigens I and II in enterotoxigenic and non-enterotoxigenic strains of *Escherichia coli* isolated from humans. *Infect. Immun.* **351**, 89–97.

Crosa, J.H. (1989) Genetics and molecular biology of siderophore-mediated iron transport in bacteria. *Microbiol. Rev.* **53**, 517–530.

Crosa, L.M., Wolf, M.K., Actis, L.A., Sanders-Loehr, J. & Crosa, J.H. (1988) New aerobactin mediated iron uptake system in a septicemia-causing strain of *Enterobacter cloacae. J. Bacteriol.* **170**, 5539–5544.

Cross, M.A., Koronakis, V., Stanley, P.L.D. & Hughes, C. (1990) HlyB-dependent secretion of haemolysin by uropathogenic *Escherichia coli* requires conserved sequences flanking the chromosomal *hly* determinant. *J. Bacteriol.* **172**, 1217–1224.

Crutz, A.-M., Steinmetz, M., Aymerich, S., Richter, R. & Le Coq, D. (1990) Induction of levansucrase in *Bacillus subtilis*: an antitermination mechanism negatively controlled by the phosphotransferase system. *J. Bacteriol.* **172**, 1043–1050.

Cryz, S.J., Russell, L.M. & Holmes, R.K. (1983) Regulation of toxinogenesis in *Corynebacterium diphtheriae*: mutations in the bacterial genome that alter the effects of iron on toxin production. *J. Bacteriol.* **154**, 245–252.

Csonka, L.N. (1989) Physiological and genetic responses of bacteria to osmotic stress. *Microbiol Rev.* **53**, 121–147.

Csonka, L.N. & Hanson, A.D. (1991) Prokaryotic osmoregulation: genetics and physiology. *Annu. Rev. Microbiol.* **45**, 569–606.

Curtiss, R. III, & Kelly, S.M. (1987) *Salmonella typhimurium* deletion mutants lacking adenylate cyclase and cyclic AMP receptor protein are avirulent and immunogenic. *Infect. Immun.* **55**, 3035–3043.

Dabiri, G.A., Sanger, J.M., Portnoy, D.A. & Southwick, F.S. (1990) *Listeria monocytogenes* moves rapidly through the host cytoplasm by inducing directional actin assembly. *Proc. Natl Acad. Sci. USA* **87**, 6068–6072.

Dagberg, B. & Uhlin, B.E. (1992) Regulation of virulence-associated genes in enteroinvasive *Escherichia coli. J. Bacteriol.* **174**, 7606–7612.

Das, A. (1992) How the lambda *N* gene product suppresses transcription termination: communication of RNA polymerase with regulatory proteins mediated by signals in nascent RNA. *J. Bacteriol.* **174**, 6711–6716.

Dattananda, C.S., Rajkumari, K. & Gowrishankar, J. (1991) Multiple mechanisms contribute to osmotic inducibility of *proU* operon expression in *Escherichia coli*: demonstration of two osmoresponsive promoters and of a negative regulatory element within the first structural gene. *J. Bacteriol.* **173**, 7481–7490.

Dayn, A., Malkhosyan, S., Duzhy, D., Lyamichev, V., Panchenko, Y. & Mirkin, S. (1991) Formation of $(dA\text{-}dT)_n$ cruciforms in *Escherichia coli* cells under different environmental conditions. *J. Bacteriol.* **173**, 2658–2664.

Debarbouille, M., Arnaud, M., Fouet, A., Klier, A. & Rapoport, G. (1990) The *sacT* gene regulating the *sacPA* operon in *Bacillus subtilis* shares strong homology with transcriptional antiterminators. *J. Bacteriol.* **172**, 3966–3973.

De Boer, P.A.J., Cook, W.R. & Rothfield, L.I. (1990) Bacterial cell cycle. *Annu. Rev. Genet.* **24**, 249–274.

De Graaf, F.K. (1988) Fimbrial structures of enterotoxigenic *E. coli. Antonie van Leewenhoek* **54**, 395–404.

De Grandis, S., Ginsberg, J., Toone, M., Climie, S., Friesen, J. & Brunton, J. (1987) Nucleotide sequence and promoter mapping of the *Escherichia coli* Shiga-like toxin operon of bacteriophage H-19B. *J. Bacteriol.* **169**, 4313–4319.

Delius, H. & Worcel, A. (1973) Electron microscopic studies on the folded chromosome of *Escherichia coli. Cold Spring Harbor Symp. Quant. Biol.* **38**, 53–58.

Delius, H. & Worcel, A. (1974) Electron microscopic visualization of the folded chromosome of *E. coli. J. Mol. Biol.* **71**, 107–109.

De Lorenzo, V., Herrero, M., Giovannini, F. & Neilands, J.B. (1988) Fur (ferric uptake regulon) protein and CAP (catabolite activator protein) modulate transcription of *fur* gene in *Escherichia coli. Eur. J. Biochem.* **173**, 537–546.

De Lorenzo, V., Wee, S., Herrero, M. & Neilands, J.B. (1987) Operator sequences of the aerobactin operon of plasmid ColV-K30 binding the ferric uptake regulation (*fur*) repressor. *J. Bacteriol.* **169**, 2624–2730.

Delver, E.P., Kotova, V.U., Zavilgelsky, G.B. & Belogurov, A.A. (1991) Nucleotide sequence of the gene (*ard*) encoding the antirestriction protein of plasmid ColIb-P9. *J. Bacteriol.* **173**, 5887–5892.

Demple, B. & Amábile-Cuevas, F. (1991) Redox redux: the control of oxidative stress response. *Cell* **67**, 837–839.

Demple, B., Johnson, A. & Fung, D. (1986) Exonuclease III and endonuclease IV remove 3′ blocks from DNA synthesis in H_2O_2-damaged *Escherichia coli. Proc. Natl Acad. Sci. USA* **83**, 7731–7735.

Deretic, V., Dikshit, R., Konyecsni, W.M., Chakrabarty, A.M. & Misra, T.K. (1989) The *algR* gene, which regulates mucoidy in *Pseudomonas aeruginosa*, belongs to a class of environmentally responsive genes. *J. Bacteriol.* **171**, 1278–1283.

Deretic, V., Hibler, N.S. & Holt, S.C. (1992a) Immunocytochemical analysis of AlgP (Hp1), a histonelike element participating in control of mucoidy in *Pseudomonas aeruginosa. J. Bacteriol.* **174**, 824–831.

Deretic, V. & Konyecsni, W.M. (1989) Control of mucoidy in *Pseudomonas aeruginosa*: transcription regulation of *algR* and identification of the second regulatory gene, *algQ. J. Bacteriol.* **171**, 3680–3688.

Deretic, V. & Konyecsni, W.M. (1990) A prokaryotic regulatory factor with a histone

H1-like carboxy-terminal domain: clonal variation of repeats within *algP*, a gene involved in regulation of mucoidy in *Pseudomonas aeruginosa*. *J. Bacteriol.* **172**, 5544–5554.

Deretic, V., Leveau, J.H.J., Mohr, C.D. & Hibler, N.S. (1992b) *In vitro* phosphorylation of AlgR, a regulator of mucoidy in *Pseudomonas aeruginosa*, by a histidine protein kinase and effects of small phospho-donor molecules. *Mol. Microbiol.* **6**, 2761–2767.

Deretic, V., Mohr, C.D. & Martin, D.W. (1991) Mucoid *Pseudomonas aeruginosa* in cystic fibrosis: signal transduction and histone-like elements in the regulation of bacterial virulence. *Mol. Microbiol.* **5**, 1577–1583.

Dersch, P., Schmidt, K. & Bremer, E. (1993) Synthesis of the *Escherichia coli* K-12 nucleoid-associated DNA-binding protein HNS is subjected to growth-phase control and autoregulation. *Mol. Microbiol.* **8**, 875–889.

DeVault, J.D., Hendrickson, W., Kato, J. & Chakrabarty, A.M. (1991) Environmentally regulated *algD* promoter is responsive to the cAMP receptor protein in *Escherichia coli*. *Mol. Microbiol.* **5**, 2503–2509.

DeVault, J.D., Kimbara, K. & Chakrabarty, A.M. (1990) Pulmonary dehydration and infection in cystic fibrosis: evidence that ethanol activates alginate gene expression and induction of mucoidy in *Pseudomonas aeruginosa*. *Mol. Microbiol.* **4**, 737–745.

DeVos, G.F. &, Finan, T.M., Signer, E.R. & Walker, G.C. (1984) Host-dependent transposon Tn5-mediated streptomycin resistance. *J. Bacteriol.* **159**, 395–399.

Díaz-Guerra, L., Moreno, F. & San Millán, J.L. (1989) *appR* gene product activates transcription of microcin C7 plasmid genes. *J. Bacteriol.* **171**, 2906–2908.

DiGate, R.J. & Marians, K.J. (1988) Identification of a potent decatenating enzyme from *Escherichia coli*. *J. Biol. Chem.* **263**, 13366–13373.

DiGate, R.J. & Marians, K.J. (1989) Molecular cloning and DNA sequence analysis of *Escherichia coli topB*, the gene encoding topoisomerase III. *J. Biol. Chem.* **264**, 17924–17930.

Dimri, G.P., Rudd, K.E., Morgan, M.K., Bayat, H. & Ames, G.F. (1992) Physical mapping of repetitive extragenic palindromic sequences in *Escherichia coli* and phylogenetic distribution among *Escherichia coli* strains and other enteric bacteria. *J. Bacteriol.* **174**, 4583–4593.

DiNardo, S., Voekel, K.A., Sternglanz, R., Reynolds, A.E. & Wright, A. (1982) *Escherichia coli* DNA topoisomerase I mutants have compensatory mutations in DNA gyrase genes. *Cell* **31**, 43–51.

DiRita, V.J. (1992) Coordinate expression of virulence genes by ToxR in *Vibrio cholerae*. *Mol. Microbiol.* **6**, 451–458.

DiRita, V.J. & Mekalanos, J.J. (1991) Periplasmic interaction between two membrane regulatory proteins, ToxR and ToxS, results in signal transduction and transcriptional activation. *Cell* **64**, 29–37.

DiRita, V.J., Parsot, C., Jander, G. & Mekalanos, J.J. (1991) Regulatory cascade controls virulence in *Vibrio cholerae*. *Proc. Natl Acad. Sci. USA* **88**, 5403–5407.

DiRita, V.J., Peterson, K.M. & Mekalanos, J.J. (1990) Regulation of cholera toxin synthesis. In Iglewski, B.H. & Clark, V.L. (eds) *Molecular Basis of Bacterial Pathogenesis*, pp. 355–376. Academic Press, San Diego.

Dixon, R.A., Henderson, N.C. & Austin, S. (1988) DNA supercoiling and aerobic regulation of transcription from the *Klebsiella pneumoniae nifLA* promoter. *Nucleic Acids Res.* **16**, 9933–9946.

Donachie, W.D. & Robinson, A.C. (1987) Cell division: parameter values and the process. In Neidhardt, F.C., Ingraham, J.L., Low, K.B., Magasanik, B., Schaechter, M. & Umbarger, H.E. (eds), Escherichia coli *and* Salmonella typhimurium *Cellular and Molecular Biology*, pp. 1578–1593. American Society for Microbiology, Washington.

Donnenberg, M.S. & Kaper, J.B. (1992) Enteropathogenic *Escherichia coli*. *Infect. Immun.* **60**, 3953–3961.

Donnenberg, M.S., Yu, J. & Kaper, J.B. (1993) A second chromosomal gene necessary for intimate attachment of enteropathogenic *Escherichia coli* to epithelial cells. *J. Bacteriol.* **175**, 4670–4680.

Dorman, C.J. (1991) DNA supercoiling and environmental regulation of gene expression in pathogenic bacteria. *Infect. Immun.* **59**, 745–749.

Dorman, C.J. (1992) The VirF protein from *Shigella flexneri* is a member of the AraC transcription factor superfamily and is highly homologous to Rns, a positive regulator of virulence genes in enterotoxigenic *Escherichia coli*. *Mol. Microbiol.* **6**, 1575.

Dorman, C.J. & Higgins, C.F. (1987) Fimbrial phase variation in *Escherichia coli:* dependence on integration host factor and homologies with other site-specific recombinases. *J. Bacteriol.* **169**, 3840–3843.

Dorman, C.J., Barr, G.C., Ní Bhriain N. & Higgins, C.F. (1988) DNA supercoiling and the anaerobic and growth phase regulation of *tonB* gene expression. *J. Bacteriol.* **170**, 2816–2826.

Dorman, C.J., Chatfield, S., Higgins, C.F., Hayward, C. & Dougan, G. (1989a) Characterization of porin and *ompR* mutants of a virulent strain of *Salmonella typhimurium*: *ompR* mutants are attenuated in vivo. *Infect. Immun.* **57**, 2136–2140.

Dorman, C.J., Lynch, A.S., Ní Bhriain, N. & Higgins C.F. (1989b) DNA supercoiling in *Escherichia coli*: *topA* mutations can be suppressed by amplifications involving the *tolC* locus. *Mol. Microbiol.* **3**, 531–540.

Dorman, C.J. & Ní Bhriain, N. (1992a) Global regulation of gene expression during environmental adaptation: implications for bacterial pathogenes. In Hormaeche, C.E., Penn, C.W. & Smyth, C.J. (eds) *Molecular Biology of Bacterial Infection: Current Status and Future Perspectives*, pp. 193–230. Cambridge University Press.

Dorman, C.J. & Ní Bhriain, N. (1992b) Thermal regulation of *fimA*, the *Escherichia coli* gene coding for the type 1 fimbrial subunit. *FEMS Microbiol Lett.* **99**, 125–130.

Dorman, C.J., Ní Bhriain N. & Higgins, C.F. (1990) DNA supercoiling and environmental regulation of virulence gene expression in *Shigella flexneri*. *Nature* **344**, 789–792.

Douglas, C.J., Staneloni, R.J., Rubin, R.A. & Nester, E.W. (1985) Identification and genetic analysis of an *Agrobacterium tumefaciens* chromosomal virulence region. *J. Bacteriol.* **161**, 850–860.

Dramsi, S., Kocks, C., Forestier, C. & Cossart, P. (1993) Internalin-mediated invasion of epithelial cells by *Listeria monocytogenes* is regulated by the bacterial growth state, temperature and the pleiotropic activator, *prfA*. *Mol. Microbiol.* **9**, 931–941.

Dri, A.-M., Moreau, P.L. & Rouvière-Yaniv, J. (1992) Role of histone-like proteins OsmZ and HU in homologous recombination. *Gene* **120**, 11–16.

Drlica, K. (1984) Biology of bacterial deoxyribonucleic acid topoisomerases. *Microbiol. Rev.* **48**, 273–289.

Drlica, K. (1987) The nucleoid. In Neidhardt, F.C., Ingraham, J.L., Low, K.B., Magasanik, B., Schaecter, M. & Umbarger, H.E. (eds), Escherichia coli *and* Salmonella typhimurium *Cellular and Molecular Biology*, pp. 91–103. American Society for Microbiology, Washington.

Drlica, K. (1990) Bacterial topoisomerases and the control of DNA supercoiling. *Trends Genet.* **6**, 433–437

Drlica, K. & Riley, M. (1990) A historical introduction to the bacterial chromosome. In Drlica, K. & Riley, M. (eds), *The Bacterial Chromosome*, pp. 3–13. American Society for Microbiology, Washington.

Drlica, K. & Rouvière-Yaniv, J. (1987) Histonelike proteins of bacteria. *Microbiol. Rev.* **51**, 301–319.

Drummond, M., Whitty, P. & Whootton, J. (1986) Sequence and domain relationships of *ntrC* and *nifA* from *Klebsiella pneumoniae*: homologies to other regulatory proteins. *EMBO J.* **5**, 441–447.

Dubnau, D. (1985) Induction of *ermC* requires translation of the leader peptide. *EMBO J.* **4**, 533–537.

Dubos, R.J. & Geiger, J.W. (1946) Preparation and properties of shiga toxin and toxoid. *J. Exp. Med.* **84**, 143–156.

Dunn, T.M., Hahn, S., Ogden, S. & Schleif, R.F. (1984) An operator at – 280 base pairs that is required for repression of *araBAD* operon promoter: addition of DNA helical turns between the operator and promoter cyclically hinders repression. *Proc. Natl Acad. Sci. USA* **81**, 5017–5020.

Dunny, G.M. (1990) Genetic functions and cell–cell interactions in the pheromone-inducible plasmid transfer system of *Enterococcus faecalis*. *Mol. Microbiol.* **4**, 689–696.

Dunny, G.M., Craig, R.A., Garron, R.L. & Clewell, D.B. (1979) Plasmid transfer in *Streptococcus faecalis*: production of multiple sex pheromones by recipients. *Plasmid* **2**, 454–465.

Dürrenberger, M., Bjornst, M.-A., Uetz, T., Hobot, J. & Kellenberger, E. (1988) Intracellular location of the histonelike protein HU in *Escherichia coli*. *J. Bacteriol.* **170**, 4757–4768.

Duvall, E.J., Ambulos, N.P. & Lovett, P.S. (1987) Drug-free induction of a chloramphenicol acetyltransferase gene in *Bacillus subtilis* by stalling ribosomes in a regulatory leader. *J. Bacteriol.* **169**, 4235–4241.

Dynan, W.S. (1989) Understanding the molecular mechanism by which methylation influences gene expression. *Trends Genet* **5**, 35–36.

Edlund, T. & Normark, S. (1981) Recombination between short DNA homologies causes tandem duplications. *Nature* **292**, 269–271.

Edwards, M.S., Kasper, D.L., Jennings, H.J., Baker, C.J. & Nicholson-Weller, A. (1982) Capsule sialic acid prevents activation of the alternative complement pathway by type III, group B streptococci. *J. Immunol.* **128**, 1278–1283.

Eisenstein, B.I., Sweet, D., Vaughn, V. & Friedman, D.I. (1987) Integration host factor is required for the DNA inversion that controls phase variation in *Escherichia coli*. *Proc. Natl Acad. Sci. USA* **84**, 6506–6510.

Eraso, J.M. & Weinstock, G.M. (1992) Anaerobic control of colicin E1 production. *J. Bacteriol.* **174**, 5101–5109.

Erickson, J.W. & Gross, C.A. (1989) Identification of the σ^E subunit of *Escherichia coli* RNA polymerase: a second alternative σ factor involved in high-temperature gene expression. *Genes Dev.* **3**, 1462–1471.

Ernst, R.K., Dombroski, D.M. & Merrick, J.M. (1990) Anaerobiosis, type 1 fimbriae and growth phase are factors that affect invasion of HEp-2 cells by *Salmonella typhimurium*. *Infect. Immun.* **58**, 2014–2016.

Espinosa-Urgel, M. & Tormo, A. (1993) σ^s-dependent promoters in *Escherichia coli* are located in DNA regions with intrinsic curvature. *Nucleic Acids Res.* **21**, 3667–3670.

Evans, D.G. & Evans, D.J. Jr (1978) New surface-associated heat-labile colonization factor antigen (CFA/II) produced by enterotoxigenic *Escherichia coli* of serogroups O6 and O8. *Infect. Immun.* **21**, 636–647.

Evans, D.G., Silver, R.P., Evans, D.J. Jr, Chase, D.G. & Gorbach, S.L. (1975) Plasmid-controlled colonization factor associated with virulence in *Escherichia coli* enterotoxigenic for humans. *Infect. Immun.* **12**, 656–667.

Expert, D. & Toussaint, A. (1985) Bacteriocin-resistant mutants of *Erwinia chrysanthemi*: possible involvement of iron acquisition on pathogenicity. *J. Bacteriol.* **163**, 221–227.

Falconi, M., Gualtieri, M.T., La Teana, A., Losso, M.A. & Pon, C.L. (1988) Proteins from the prokaryotic nucleoid: primary and quaternary structure of the 15 kD *Escherichia coli* DNA binding protein H-NS. *Mol Microbiol.* **2**, 323–329.

Falkow, S. (1991) Bacterial entry into eukaryotic cells. *Cell* **65**, 1099–1102.

Fang, F.C., Krause, M., Roudier, C., Fierer, J. & Guiney, D.G. (1991) Growth regulation of a *Salmonella* plasmid gene essential for virulence. *J. Bacteriol.* **173**, 6783–6789.

Fang, F.C., Libby, S.J., Buchmeier, N.A., Loewen, P.C., Switala, J., Harwood, J. & Guiney, D.G. (1992) The alternative σ factor KatF (RpoS) regulates *Salmonella* virulence. *Proc. Natl Acad. Sci. USA* **89**, 11978–11982.

Farabaugh, P.J., Schmeissner, U., Hofer, M. & Miller, J.H. (1978) Genetic studies of the *lac* repressor. VII. On the molecular nature of spontaneous hotspots in the *lacI* gene of *Escherichia coll. J. Mol. Biol.* **126**, 847–863.

Farewell, A., Brazas, R., Davie, E., Mason, J. & Rothfield, L.I. (1991) Suppression of the abnormal phenotype of *Salmonella typhimurium rfaH* mutants by mutations in the gene for transcription factor rho. *J. Bacteriol.* **173**, 5188–5193.

Farr, S.B. & Kogoma, T. (1991) Oxidative stress responses in *Escherichia coli* and *Salmonella typhimurium. Microbiol. Rev.* **55**, 561–585.

Fasano, A., Baudry, B., Pumplin, D.W., Wasserman, S.S., Tall, B.D., Ketley, J.M. & Kaper, J.B. (1991) *Vibrio cholerae* produces a second enterotoxin, which affects intestinal tight junctions. *Proc. Natl Acad. Sci. USA* **88**, 5242–5246.

Fassler, J.S., Arnold, G.F. & Tessman, I. (1986) Reduced superhelicity of plasmid DNA produced by the *rho-15* mutation in *Escherichia coli. Mol. Gen. Genet.* **204**, 424–429.

Ferber, D.M. & Brubaker, R.R. (1981) Plasmids in *Yersinia pestis. Infect. Immun.* **31**, 839–841.

Ferdows, M.S. & Barbour, A.G. (1989) Megabase-sized linear DNA in the bacterium *Borrelia burgdorferi*, the Lyme disease agent. *Proc. Natl. Acad. Sci. USA* **86**, 5969–5973.

Fields, P.I., Groisman, E.A. & Heffron, F. (1989) A *Salmonella* locus that controls resistance to microbiocidal proteins from phagocytic cells. *Science* **243**, 1059–1062.

Figueroa, N. & Bossi, L. (1988) Transcription induces gyration of the DNA template in *Escherichia coli. Proc. Natl Acad. Sci. USA* **85**, 9416–9420.

Figueroa, N., Wills, N. & Bossi, L. (1991) Common sequence determinants of the response of a prokaryotic promoter to DNA bending and supercoiling. *EMBO J.* **10**, 941–949.

Finkel, S.E. & Johnson, R.C. (1992) The FIS protein: it's not just for DNA inversion anymore. *Mol. Microbiol.* **6**, 3257–3265. Erratum: **7**, 1023.

Finlay, B.B. & Falkow, S. (1989) *Salmonella* as an intracellular parasite. *Mol. Microbiol.* **3**, 1833–1841.

Finlay, B.B. & Falkow, S. (1990) *Salmonella* interactions with polarized human intestinal Caco-2 epithelial cells. *J. Infect. Dis.* **162**, 1096–1106.

Finnegan, J. & Sherratt, D. (1982) Plasmid ColE1 conjugal mobility: the nature of *bom*, a region required for conjugal mobility. *Mol. Gen. Genet.* **185**, 344–351.

Fischetti, V.A. (1991) Streptococcal M protein. *Sci. Am.* **264**, 32–39.

Fisher, L. M. (1984) DNA supercoiling and gene expression. *Nature* **307**, 686–687.

Fitzgerald, G.F. & Clewell, D.B. (1985) A conjugative transposon (Tn*919*) in *Streptococcus sanguis. Infect. Immun.* **47**, 415–420.

Flamm, E.L. & Weisberg, R.A. (1985) Primary structure of the *hip* gene of *Escherichia coli* and of its product, the beta subunit of integration host factor. *J. Mol. Biol.* **183**, 117–128.

Flores, M., Gonzáles, V., Brom, S., Martínez, E., Pinero, D., Romero, D., Dávila, G. & Palacios, R. (1987) Reiterated DNA sequences in *Rhizobium* and *Agrobacterium* spp. *J. Bacteriol.* **169**, 5782–5788.

Folk, W.R. & Berg, P. (1971) Duplication of the structural gene for glycyl transfer RNA synthetase in *Escherichia coli. J. Mol. Biol.* **58**, 595–610.

Folkhard, W., Leonard, K.R., Malsey, S., Marvin, D.A., Dubochet, J., Engel, A., Achtman, M. & Helmuth, R. (1979) X-ray diffraction and electron microscope studies on the structure of bacterial F pili. *J. Mol. Biol.* **130**, 145–160.

Ford, S., Cooper, R.A., Evans, R.E., Hider, R.C. & Williams, P.H. (1988) Domain preference in iron removal from human transferrin by bacterial siderophores aerobactin and enterochelin. *Eur. J. Biochem.* **178**, 477–481.

Forsberg, Å., Viitanen, A.-M., Skurnik, M. & Wolf-Watz, H. (1991) The surface-located YopN protein is involved in calcium signal transduction in *Yersinia pseudotuberculosis. Mol. Microbiol.* **5**, 977–986.

Forst, S., Delgado, J. & Inouye, M. (1989) Phosphorylation of OmpR by the osmosensor EnvZ modulates the expression of the *ompF* and *ompC* genes in *Escherichia coli. Proc. Natl Acad. Sci. USA* **86**, 6052–6056.

Foster, J.W. & Aliabadi, Z. (1989) pH-regulated gene expression in *Salmonella typhimurium*: genetic analysis of *aniG* and cloning of the *earA* regulator. *Mol. Microbiol.* **3**, 1605–1615.

Foster, J.W. & Hall, H.K (1990) Adaptive acidification tolerance response of *Salmonella typhimurium. J. Bacteriol.* **172**, 771–778.

Foster, J.W. & Hall, H.K. (1991) Inducible pH homeostasis and the acid tolerance response of *Salmonella typhimurium. J. Bacteriol.* **173**, 5129–5135.

Foster, J.W. & Hall, H.K. (1992) Effect of *Salmonella typhimurium* ferric uptake regulator (*fur*) mutations on iron- and pH-regulated protein synthesis. *J. Bacteriol.* **174**, 4317–4323.

Foster, P.L. (1992) Directed mutation: between unicorns and goats. *J. Bacteriol.* **174**, 1711–1716.

Foster, T.J. (1983) Plasmid-determined resistance to antimicrobial drugs and toxic metal ions in bacteria. *Microbiol. Rev.* **47**, 361–409.

Foxall, P.A., Drasar, B.S. & Duggleby, C.J. (1990) Evidence for a DNA inversion system in *Bordetella pertussis. FEMS Microbiol. Lett.* **69**, 1–6.

Francis, C.L., Starnbach, M.N. & Falkow, S. (1992) Morphological and cytoskeletal changes in epithelial cells occur immediately upon interaction with *Salmonella typhimurium* grown under low-oxygen conditions. *Mol. Microbiol.* **6**, 3077–3087.

François, V., Louarn, J., Rebollo, J.-E. & Louarn, J.-M. (1990) Replication termination, nondivisible zones and structure of the *Escherichia coli* chromosome. In Drlica, K. & Riley, M. (eds) *The Bacterial Chromosome*, pp. 351–359. American Society for Microbiology, Washington.

Frank, D.W. & Iglewski, B.H. (1988) Kinetics of *toxA* and *regA* mRNA accumulation in *Pseudomonas aeruginosa. J. Bacteriol.* **170**, 4477–4483.

Frank, D.W. & Iglewski, B.H. (1991) Cloning and sequence analysis of a *trans*-regulatory locus required for exoenzyme S synthesis in *Pseudomonas aeruginosa. J. Bacteriol.* **173**, 6460–6468.

Freer, J.H. (1988) Toxins as virulence factors of Gram positive pathogenic bacteria of veterinary importance. In Roth, J.A. (ed) *Virulence Mechanisms of Bacterial Pathogens*, pp. 264–288. American Society for Microbiology, Washington.

Freundlich, M., Ramani, N., Mathew, E., Sirko, A. & Tsui, P. (1992) The role of integration host factor in gene expression in *Escherichia coli. Mol. Microbiol.* **6**, 2557–2563.

Frey, J. & Bagdasarian, M. (1989) The molecular biology of IncQ plasmids. In Thomas, C.M. (ed) *Promiscuous Plasmids of Gram negative Bacteria*, pp. 79–94. Academic Press, New York.

Friedman, D.I. (1988a) Integration Host Factor: a protein for all reasons. *Cell* **55**, 545–554.

Friedman, D.I. (1988b) Regulation of phage gene expression by termination and antitermination of transcription. In Calendar, R. (ed) *The Bacteriophages*, Vol. II, pp. 263–319, Plenum Press, New York.

Friedrich, M.J., Kinsey, N.E., Vila, J. & Kadner, R.J. (1993) Nucleotide sequence of a 13.9 kb segment of the 90 kb virulence plasmid of *Salmonella typhimurium*: the presence of fimbrial biosynthetic genes. *Mol. Microbiol.* **8**, 543–558.

Frosch, M., Weisgerber, C. & Meyer, T.F. (1989) Molecular characterization and expression in *Escherichia coli* of the gene complex encoding the polysaccharide capsule of *Neiserria meningitidis* group B. *Proc. Natl Acad. Sci. USA* **86**, 1669–1673.

Fujita, N. & Ishihama, A. (1987) Heat shock induction of RNA polymerase sigma-32 synthesis in *Escherichia coli*: transcriptional control and a multiple promoter system. *Mol. Gen. Genet.* **210**, 10–15.

Fulks, K.A., Marrs, C.F., Stevens, S.P. & Green, M.R. (1990) Sequence analysis of the inversion region containing the pilin genes of *Moraxella bovis*. *J. Bacteriol.* **172**, 310–316.

Fuller, R.S. & Kornberg, A. (1983) Purified dnaA protein in initiation of replication at the *Escherichia coli* chromosomal origin of replication. *Proc. Natl Acad. Sci. USA* **80**, 5817–5821.

Funnel, B.E. (1988) Participation of *Escherichia coli* integration host factor in the P1 plasmid partition system. *Proc. Natl Acad. Sci USA* **85**, 6657–6661.

Gaillard, J.-L., Berche, P., Frehel, C., Gouin, E. & Cossart, P. (1991) Entry of *L. monocytogenes* into cells is mediated by internalin, a repeat protein reminiscent of surface antigens from Gram-positive cocci. *Cell* **65**, 1127–1141.

Galán, J.E. & Curtiss, R., III (1990) Expression of *Salmonella typhimurium* genes required for invasion is regulated by changes in DNA supercoiling. *Infect. Immun.* **58**, 1879–1885.

Galas, D.J. & Chandler, M. (1982) Structure and stability of Tn9-mediated cointegrates: evidence for two pathways of transposition. *J. Mol. Biol.* **154**, 245–272.

Galas, D.J. & Chandler, M. (1989) Bacterial insertion sequences. In Berg, D.E. & Howe, M.M., (eds) *Mobile DNA*, pp. 109–162. American Society for Microbiology, Washington D.C.

Gallegos, M.-T., Michán, C. & Ramos, J.L. (1993) The XylS/AraC family of regulators. *Nucleic Acids Res.* **21**, 807–810.

Galli, D. & Wirth, R. (1991) Comparative analysis of *Enterococcus faecalis* sex pheromone plasmids identifies a single homologous DNA region which codes for aggregation substance. *J. Bacteriol.* **173**, 3029–3033.

Galli, D., Wirth, R. & Wanner, G. (1989) Identification of aggregation substances of *Enterococcus faecalis* after induction by sex pheromones. *Arch. Microbiol.* **151**, 486–490.

Galloway, D.R. (1991) *Pseudomonas aeruginosa* elastase and elastolysis revisited: recent developments. *Mol. Microbiol.* **5**, 2315–2321.

Gamas, P., Chandler, M.G., Prentki, P. & Galas, D.J. (1987) *Escherichia coli* integration host factor binds specifically to the insertion sequence IS*1* and to its major insertion hot-spot on pBR322. *J. Mol. Biol.* **195**, 261–272.

Gamas, P., Galas, D. & Chandler, M. (1985) DNA sequence at the end of IS*1* required for transposition. *Nature* **317**, 458–460.

Gambello, M.J. & Iglewski, B.H. (1991) Cloning and characterization of the *Pseudomonas aeruginosa lasR* gene, a transcriptional activator of elastase expression. *J. Bacteriol.* **173**, 3000–3009.

Gamer, J., Bujard, H. & Bukau, B. (1992) Physical interaction between heat shock proteins DnaK, DnaJ, and GrpE and the bacterial heat shock transcription factor σ^{32}. *Cell* **69**, 833–842.

Garrett, S. & Silhavy, T.J. (1987) Isolation of mutations in the α operon of *Escherichia coli* that suppress the transcriptional defect conferred by a mutation in the porin regulatory gene *envZ*. *J. Bacteriol.* **169**, 1379–1385.

Gaskill, M.E. & Khan, S.A. (1988) Regulation of enterotoxin B gene in *Staphylococcus aureus*. *J. Biol. Chem.* **263**, 6276–6280.

Gellert, M., Mizuuchi, M., O'Dea, M.H. & Nash, H. (1976) DNA gyrase: an enzyme that introduces superhelical turns into DNA. *Proc. Natl Acad. Sci. USA* **73**, 3872–3876.

Gellert, M. & Nash, H. (1987) Communication between segments of DNA during site-specific recombination. *Nature* **325**, 401–404.

Gelvin, S.B. & Habeck, L.L. (1990) *vir* genes influence conjugal transfer of the Ti plasmid of *Agrobacterium tumefaciens*. *J. Bacteriol.* **172**, 1600–1608.

Genin, S., Gough, C.L., Zischek, C. & Boucher, C.A. (1992) Evidence that the *hrpb* gene encodes a positive regulator of pathogenicity genes from *Pseudomonas solanacearum*. *Mol. Microbiol.* **6**, 3065–3076.

Georgopoulos, C. (1989) The *E. coli* dnaA initiation protein: a protein for all seasons. *Trends Genet.* **5**, 319–321.

Georgopoulos, C., Ang, D., Maddock, A., Raina, S., Lipinska, B. & Zylicz, M. (1990) Heat shock response of *Escherichia coli*. In Drlica, K. & Riley, M. (eds), *The Bacterial Chromosome*, pp. 405–419. American Society for Microbiology, Washington.

Gerdes, J.C. & Romig, W.R. (1975) Complete and defective bacteriophages of classical *Vibrio cholerae*: relationship to the kappa type bacteriophage. *J. Virol.* **15**, 1231–1238.

Gerdes, K., Thisted, T. & Martinussen, J. (1990) Mechanism of post-segregational killing by the *hok/sok* system of plasmid R1: *sok* antisense RNA regulates formation of a *hok* mRNA species correlated with killing of plasmid-free cells. *Mol. Microbiol.* **4**, 1807–1818.

Ghai, J. & Das, A. (1989) The *virD* operon of *Agrobacterium tumefaciens* Ti plasmid encodes a DNA relaxing enzyme. *Proc. Natl Acad. Sci. USA* **86**, 3109–3113.

Gibbs, C.P., Reimann, B.-Y., Schultz, E., Kaufmann, A., Haas, R. & Meyer, T.F. (1989) Reassortment of pilin genes in *Neisseria gonorrhoeae* occurs by two distinct mechanisms. *Nature* **338**, 651–652.

Gibson, M.M., Ellis, E.M., Graeme-Cook, K.A. & Higgins, C.F. (1987) OmpR and EnvZ are pleiotropic regulatory proteins: positive regulation of the tripeptide permease (*tppB*) of *Salmonella typhimurium*. *Mol. Gen. Genet.* **207**, 120–129.

Gilbert, I., Villegas, V. & Barbe, J. (1989) Expression of heat labile enterotoxin is under cyclic AMP control in *Escherichia coli*. *Curr. Microbiol.* **20**, 83–90.

Gill, D.M. & Meren, R. (1978) ADP-ribosylation of membrane proteins catalysed by cholera toxin: basis of the activation of adenylate cyclase. *Proc. Natl Acad. Sci. USA* **75**, 3050–3054.

Gilson, E., Perrin, D. & Hofnung, M. (1990) DNA polymerase I and a protein complex bind specifically to *E. coli* palindromic unit highly repetitive DNA: implications for bacterial chromosome organization. *Nucleic Acids Res.* **18**, 3941–3952.

Gilson, E., Saurin, W., Perrin, D., Bachellier, S. & Hofnung, M. (1991) Palindromic units are part of a new bacterial interspersed mosaic element (BIME). *Nucleic Acids Res.* **19**, 1375–1383.

Gilson, L., Mahanty, H.K. & Kolter, R. (1987) Four plasmid genes are required for colicin V synthesis, export and immunity. *J. Bacteriol.* **169**, 2466–2470.

Giron, J.A., Ho, A.S. & Schoolnik, G.K. (1991) An inducible bundle-forming pilus of enteropathogenic *Escherichia coli*. *Science* **254**, 710–713.

Glare, E.M., Paton, J.C., Premier, R.R., Lawrence, A.J. & Nisbet, I.T. (1990) Analysis of a repetitive DNA sequence from *Bordetella pertussis* and its application to the diagnosis of Pertussis using the polymerase reaction. *J. Clin. Microbiol.* **28**, 1982–1987.

Glasgow, A.C., Hughes, K.T. & Simon, M.I. (1989) Bacterial DNA inversion systems. In Berg, D.E. & Howe, M.M. (eds), *Mobile DNA*, pp. 637–659. American Society for Microbiology, Washington.

Gold, L. (1988) Posttranscriptional regulatory mechanisms in *Escherichia coli*. *Annu. Rev. Biochem.* **57**, 199–233.

Goldberg, I. & Mekalanos, J.J. (1986a) Cloning of the *Vibrio cholerae recA* gene and construction of a *Vibrio cholerae recA* mutant. *J. Bacteriol.* **165**, 715–722.

Goldberg, I. & Mekalanos, J.J. (1986b) Effect of a *recA* mutation on cholera toxin gene amplification and deletion events. *J. Bacteriol.* **165**, 723–731.

Goldberg, J.B. & Dahnke, T. (1992) *Pseudomonas aeruginosa* AlgB, which modulates the expression of alginate, is a member of the NtrC subclass of prokaryotic regulators. *Mol. Microbiol.* **6**, 59–66.

Goldberg, M.B., Boyko, S.A., Butterton, J.R., Stoebner, J.A., Payne, S.M. & Calderwood, S.B. (1992) Characterization of a *Vibrio cholerae* virulence factor homologous to the family of TonB-dependent proteins. *Mol. Microbiol.* **6**, 2407–2418.

Goldberg, M.B., Boyko, S.A. & Calderwood, S.B. (1990a) Transcriptional regulation by iron of a *Vibrio cholerae* virulence gene and homology of the gene to the *Escherichia coli* Fur system. *J. Bacteriol.* **172**, 6863–6870.

Goldberg, M.B., Boyko, S.A. & Calderwood, S.B. (1991) Positive transcriptional regulation of an iron-regulated virulence gene in *Vibrio cholerae. Proc. Natl Acad. Sci. USA* **88**, 1125–1129.

Goldberg, M.B., DiRita, V.J. & Calderwood, S.B. (1990b) Identification of an iron-regulated virulence determinant in *Vibrio cholerae*, using Tn*phoA* mutagenesis. *Infect. Immun.* **58**, 55–60.

Goldstein, E. & Drlica, K. (1984) Regulation of bacterial DNA supercoiling: plasmid linking numbers vary with growth temperature. *Proc. Natl Acad. Sci. USA* **81**, 4046–4050.

Goldstein, J., Pollitt, N.S. & Inouye, M. (1990) Major cold shock protein of *Escherichia coli. Proc. Natl Acad. Sci. USA* **87**, 283–287.

Gomez-Eichelmann, M.C., Levy-Mustri, A. & Ramirez-Santos, J. (1991) Presence of 5-methylcytosine in CC(A/T)GG sequences (Dcm methylation) in DNAs from different bacteria. *J. Bacteriol.* **173**, 7692–7694.

Göransson, M., Forsman, K. & Uhlin, B.E. (1988) Functional and structural homology among regulatory cistrons of pili-adhesin determinants in *Escherichia coli. Mol. Gen. Genet.* **212**, 412–417.

Göransson, M., Sondén, B., Nilsson, P., Dagberg, B., Forsman, K., Emanuelsson, K. & Uhlin, B.E. (1990) Transcriptional silencing and thermoregulation of gene expression in *Escherichia coli. Nature* **344**, 682–685.

Goshorn, S.C. & Schlievert, P.M. (1988) Nucleotide sequence of the streptococcal pyrogenic exotoxin type C. *Infect. Immun.* **56**, 2518–2520.

Gosink, K.K., Ross, W., Leirmo, S., Osuna, R., Finkel, S.E., Johnson, R.C. & Gourse, R.L. (1993) DNA binding and bending are necessary but not sufficient for FIS-dependent activation of *rrnB* P1. *J. Bacteriol.* **175**, 1580–1589.

Gouesbet, G., Abaibou, H., Wu, L.F., Mandrand-Berthelot, M.-A. & Blanco, C. (1993) Osmotic repression of anaerobic metabolic systems in *Escherichia coli. J. Bacteriol.* **175**, 214–221.

Gould, S.J. (1989) *Wonderful Life: The Burgess Shale and the Nature of History*. Penguin Books, London.

Gowrishankar, J. (1989) Nucleotide sequence of the osmoregulatory *proU* operon of *Escherichia coli. J. Bacteriol.* **171**, 1923–1931.

Graeme-Cook, K.A., May, G., Bremer, E. & Higgins, C.F. (1989) Osmotic regulation of porin expression: role for DNA supercoiling. *Mol. Microbiol.* **3**, 1287–1294.

Greenberg, J.T., Chou, J.H., Monach, P.A. & Demple, B. (1991) Activation of oxidative stress genes by mutations at the *soxQ/cfxB/marA* locus of *Escherichia coli. J. Bacteriol.* **173**, 4433–4439.

Greenberg, J.T., Monach, P., Chou, J.H., Josephy, P.D. & Demple, B. (1990) Positive control of a global antioxidant defence regulon activated by superoxide-generating agents in *Escherichia coli. Proc. Natl Acad. Sci. USA* **87**, 6181–6185.

Griffith, J. D. (1976) Visualization of prokaryotic DNA in a regularly condensed chromatin-like fibre. *Proc. Natl Acad. Sci. USA* **73**, 563–567.

Grimsley, N., Hohn, B., Ramos, C., Kado, C.I. & Rogowsky, P. (1989) DNA transfer from *Agrobacterium* to *Zea mays* or *Brassica* by agroinfection is dependent on bacterial virulence functions. *Mol. Gen. Genet.* **217**, 309–316.

Groman, N.B. (1984) Conversion by corynephages and its role in the natural history of diphtheria. *J. Hyg.* **93**, 405–417.

Gross, R. (1993) Signal transduction and virulence regulation in human and animal pathogens. *FEMS Microbiol. Rev.* **104**, 301–326.

Gross, R., Aricò, B. & Rappouli, R. (1989) Families of bacterial signal-transducing proteins. *Mol. Microbiol.* **3**, 1661–1667.

Gross, R. & Rappuoli, R. (1988) Positive regulation of pertussis toxin expression. *Proc. Natl Acad. Sci. USA* **85**, 3913–3917.

Grosskinsky, C.M., Jacobs W.R., Jr, Clark-Curtiss, J.E. & Bloom, B.R. (1989) Genetic relationships among *Mycobacterium leprae, Mycobacterium tuberculosis* and candidate leprosy vaccine strains determined by DNA hybridization: identification of an *M. leprae*-specific repetitive sequence. *Infect. Immun.* **57**, 1535–1541.

Grossman, A.D., Taylor, W.E., Burton, Z.F., Burgess, R.R. & Gross, C.A. (1985) Stringent response in *E. coli* induces expression of heat shock proteins. *J. Mol. Biol.* **186**, 357–365.

Gruss, A., Ross, H.F. & Novick, R.P. (1987) Functional analysis of a palindromic sequence required for normal replication of several staphylococcal plasmids. *Proc. Natl Acad. Sci. USA* **84**, 2165–2169.

Guiney, D.G., Chikami, G., Deiss, C. & Yacobson, E. (1985) The origin of plasmid DNA transfer during bacterial conjugation. In Helinski, D.R., Cohen, S.N., Clewell, R.B., Jackson, D.A. & Hollaender, A. (eds) *Plasmids in Bacteria*, pp. 521–534. Plenum Press, New York.

Gulig, P.A., Caldwell, A.L. & Chiodo, V.A. (1992) Identification, genetic analysis and DNA sequence of a 7.8 kb virulence region of the *Salmonella typhimurium* virulence plasmid. *Mol. Microbiol.* **6**, 1395–1411.

Gulig, P.A. & Chiodo, V.A. (1990) Genetic and DNA sequence analysis of the *Salmonella typhimurium* virulence plasmid gene encoding the 28,000 molecular weight protein. *Infect. Immun.* **58**, 2651–2658.

Gulig, P.A. & Curtiss R. III (1987) Plasmid-associated virulence of *Salmonella typhimurium. Infect. Immun.* **55**, 2891–2901.

Gulig, P.A. & Curtiss R. III (1988) Cloning and transposon insertion mutagenesis of virulence genes of the 100 kilobase plasmid of *Salmonella typhimurium. Infect. Immun.* **56**, 3262–3271.

Gulig, P.A., Danbara, H., Guiney, D.G., Lax, A.J., Norel, F. & Rhen, M. (1993) Molecular analysis of *spv* virulence genes of the salmonella virulence plasmids. *Mol. Microbiol.* **7**, 825–830.

Gulig, P.A. & Doyle, T.J. (1993) The *Salmonella typhimurium* virulence plasmid increases the growth rate of Salmonellae in mice. *Infect. Immun.* **61**, 504–511.

Gyles, C.L., Falkow, S. & Rollins, L. (1978) In vivo transfer of *Escherichia coli* enterotoxin plasmid possessing genes for drug resistance. *Am. J. Vet. Res.* **39**, 1438–1441.

Gyles, C.L., Palchaudhuri, S. & Maas, W.K. (1977) Naturally-occurring plasmid carrying genes for enterotoxin production and drug resistance. *Science* **198**, 198–199.

Haanes, E.J., Heath, D.G. & Cleary, P.P. (1992) Architecture of the *vir* regulons of group A Streptococci parallels opacity factor phenotype and M protein class. *J. Bacteriol.* **174**, 4967–4976.

Haas, R. & Meyer, T.F. (1986) The repertoire of silent pilus genes in *Neisseria gonorrhoeae*: evidence for gene conversion. *Cell* **44**, 107–115.

Haas, R., Schwartz, H. & Meyer, T.F. (1987) Release of soluble pilin antigen coupled with gene conversion in *Neisseria gonorrhoeae. Proc. Natl Acad. Sci. USA* **84**, 9079–9083.

Hackett, J., Wyk, P., Reeves, P. & Mathan, V. (1987) Mediation of serum resistance in *Salmonella typhimurium* by an 11 kilodalton polypeptide encoded by the cryptic plasmid. *J. Infect. Dis.* **155**, 540–549.

Hackstadt, T., Baehr, W. & Ying, Y. (1991) *Chlamydia trachomatis* developmentally regulated protein is homologous to eukaryotic histone H1. *Proc. Natl Acad. Sci. USA* **88**, 3937–3941.

Hahn, S., Henderson, W. & Schleif, R. (1986) Transcription of *Escherichia coli ara in vitro*: the cyclic AMP receptor protein requirement for p_{BAD} induction that depends on the presence and orientation of the *araO*$_2$ site. *J. Mol. Biol.* **188**, 355–367.

Hale, T.L. (1991) Genetic basis of virulence in *Shigella* species. *Microbiol. Rev.* **55**, 206–224.

Hale, T.L., Sansonetti, P.J., Schad, P.A., Austin, S. & Formal, S.B. (1983) Characterization of virulence plasmid-associated outer membrane proteins in *Shigella flexneri, Shigella sonnei*, and *Escherichia coli. Infect. Immun.* **40**, 340–350.

Hall, R.M., Brookes, D.E. & Stokes, H.W. (1991) Site-specific insertion of genes into integrons: role of the 59-base element and determination of the recombination cross-over point. *Mol. Microbiol.* **5**, 1941–1959.

Hamood, A.N., Pettis, G.S., Parker, C.D. and McIntosh, M.A. (1986) Isolation and characterization of the *Vibrio cholerae recA* gene. *J. Bacteriol.* **167**, 375–378.

Hantke, K. (1981) Regulation of ferric ion transport in *E. coli*. Isolation of a constitutive mutant. *Mol. Gen. Genet.* **182**, 288–292.

Hantke, K. (1987) Ferrous iron transport mutants in *Escherichia coli* K-12. *FEMS Microbiol. Lett* **44**, 53–57.

Haring, V. & Scherzinger, E. (1989) Replication proteins of IncQ plasmids. In Thomas, C.M. (ed.) *Promiscuous Plasmids of Gram-Negative Bacteria*, pp. 95–124. Academic Press, New York.

Haring, V., Scholz, P., Scherzinger, E., Frey, J., Derbyshire, K., Hatfull, G., Willetts, N.S. & Bagdasarian, M. (1985) Protein RepC is involved in copy number control of the broad host range plasmid RSF1010. *Proc. Natl Acad. Sci. USA* **82**, 6090–6094.

Harrington, R.E. (1992) DNA curving and bending in protein-DNA recognition. *Mol. Microbiol.* **6**, 2549–2555.

Hassan, H.M. & Sun, H.-C.H. (1992) Regulatory roles of Fnr, Fur, and Arc in expression of manganese-containing superoxide dismutase in *Escherichia coli. Proc. Natl Acad. Sci. USA* **89**, 3217–3221.

Hasunuma, K. & Sekiguchi, M. (1979) Effect of *dna* mutations on the replication of plasmid pSC101 in *Escherichia coli. J. Bacteriol.* **137**, 1095–1099.

Hatfull, G.F., Salvo, J.J., Falvey, E.E., Rimphanitchayakit, V. & Grindley, N.D.F. (1988) Site-specific recombination by the $\gamma\delta$ resolvase. *Symp. Soc. Gen. Microbiol.* **43**, 149–181.

Hayes, W. (1953) Observations on a transmissible agent determining sexual differentiation in *Bacterium coli. J. Gen. Microbiol.* **8**, 72–88.

Hecht, D.W. & Malamy, M.H. (1989) Tn*4399*, a conjugal mobilizing transposon of *Bacteroides fragilis. J. Bacteriol.* **171**, 3603–3608.

Hedstrom, R.C., Funk, C.R., Pavlovskis, O.R. & Galloway, D.R. (1986) Cloning of a gene involved in regulation of exotoxin A expression in *Pseudomonas aeruginosa. Infect. Immun.* **51**, 37–42.

Heffernan, E.I., Harwood, J., Fierer, J. & Guiney, D. (1992) The *Salmonella typhimurium* virulence plasmid complement resistance gene *rck* is homologous to a family of virulence-related outer membrane protein genes, including *pagC* and *ail. J. Bacteriol.* **174**, 84–91.

Heinemann, J.A. (1991) Genetics of gene transfer between species. *Trends Genet.* **7**, 181–185.

Heinemann, J.A. & Sprague, G.F. (1989) Bacterial conjugative plasmids mobilize DNA transfer between bacteria and yeast. *Nature*, **340**, 205–209.

Helmstetter, C.E. (1987) Timing of synthetic activities in the cell cycle. In Neidhardt, F.C.,

Ingraham, J.L., Low, K.B., Magasanik, B., Schaechter, M. & Umbarger, H.E. (eds), *Escherichia coli and* Salmonella typhimurium *Cellular and Molecular Biology*, pp. 1594–1605. American Society for Microbiology, Washington.

Hemmingsen, S.M., Woolford, C., van der Vies, S.M., Tilly, K., Dennis, D.T., Georgopoulos, C.P., Hendrix, R.W. & Ellis, R.J. (1988) Homologous plant and bacterial proteins chaperone oligomeric protein assembly. *Nature* **333**, 330–334.

Hendrickson, W.G., Kusano, T., Yamaki, H., Balakrishnan, R., King, M., Murchie, J. & Schaechter, M. (1982) Binding of the origin of replication of *Escherichia coli* to the outer membrane. *Cell* **30**, 915–923.

Hendrix, R.W., Roberts, J.W., Stahl, F.W. & Weisberg, R.A. eds (1983) *Lambda II*. Cold Spring Harbor Laboratory Press, Cold Spring Harbor, New York.

Hengge-Aronis, R. (1993) Survival of hunger and stress: the role of *rpoS* in early stationary phase gene regulation in *E. coli*. *Cell* **72**, 165–168.

Hengge-Aronis, R. & Fischer, D. (1992) Identification and molecular analysis of *glgS*, a novel growth-phase-regulated and *rpoS*-dependent gene involved in glycogen synthesis in *Escherichia coli*. *Mol. Microbiol.* **6**, 1877–1886.

Hengge-Aronis, R., Klein, W., Lange, R. Rimmele, M. & Boos, W. (1991) Trehalose synthesis genes are controlled by the putative sigma factor encoded by *rpoS* and are involved in stationary-phase thermotolerance in *Escherichia coli*. *J. Bacteriol.* **173**, 7918–7924.

Henikoff, S., Haughn, G.W., Calvo, J.M. & Wallace, J.C. (1988) A large family of bacterial activator proteins. *Proc. Natl Acad. Sci. USA* **85**, 6602–6606.

Hepburn, A.G., Clarke, L.E., Pearson, L. & White, J. (1983) The role of cytosine methylation in control of nopaline synthase gene expression in a plant tumour. *J. Mol. Biol.* **2**, 315–329.

Hernandez, V.J. & Bremer, H. (1991) *Escherichia coli* ppGpp synthetase II activity requires *spoT*. *J. Biol. Chem.* **266**, 5991–5999.

Hess, J., Wels, W., Vogel, M. & Goebel, W. (1986) Nucleotide sequence of a plasmid-encoded haemolysin determinant and its comparison with a corresponding chromosomal haemolysin sequence. *FEMS Microbiol. Lett.* **34**, 1–11.

Hewitt, C.R.A., Hayball, J.D., Lamb, J.R. & O'Hehir, R.E. (1992) The superantigenic activities of bacterial toxins. In Hormaeche, C.E., Penn, C.W. & Smyth, C.J. (eds) *Molecular Biology of Bacterial Infection: Current Status and Future Perspectives*. pp. 149–172. Cambridge University Press.

Heyde, M. & Portalier, R. (1987) Regulation of outer membrane porin proteins of *Escherichia coli* by pH. *Mol. Gen. Genet.* **208**, 511–517.

Heyde, M. & Portalier, R. (1990) Acid shock proteins of *Escherichia coli*. *FEMS Microbiol. Lett.* **69**, 19–26.

Hidaka, M., Kobayashi, T. & Horiuchi, T. (1991) A newly identified DNA replication terminus site, *TerE*, on the *Escherichia coli* chromosome. *J. Bacteriol.* **173**, 391–393.

Hiestand-Nauer, R. & Iida, S. (1983) Sequence of the site-specific recombinase *cin* and of its substrates serving in the inversion of the C segment of bacteriophage P1. *EMBO J.* **2**, 1733–1740.

Higgins, C.F., Ames, G.F.-L., Barnes, W.M., Clement, J.M. & Hofnung, M. (1982) A novel intercistronic regulatory element of procaryotic operons. *Nature* **298**, 760–762.

Higgins, C.F., Dorman, C.J., Stirling, D.A., Waddell, L., Booth, I.R., May, G. & Bremer, E. (1988a) A physiological role for DNA supercoiling in the osmotic regulation of gene expression in *S. typhimurium* and *E. coli*. *Cell*, **52**, 569–584.

Higgins, C.F., Hiles, I.D., Salmond, G.P.C., Gill, D.R., Downie, J.A., Evans, I.J., Holland, I.B., Gray, L., Buckel, S.D., Bell, A.W. & Hermodson, M.A. (1986) A family of related ATP-binding subunits coupled to many distinct biological processes in bacteria. *Nature* **323**, 448–450.

Higgins, C.F., Hinton, J.C.D., Hulton, C.S.J., Owen-Hughes, T., Pavitt, G.D. & Seirafi, A. (1990) Protein H1: a role for chromatin in the regulation of bacterial gene expression and virulence? *Mol. Microbiol.* **4**, 2007–2012.

Higgins, C.F., McLaren, R.S. & Newbury, S.F. (1988b) Repetitive extragenic palindromic sequences, mRNA stability and gene expression: evolution by gene conversion? A review. *Gene* **72**, 3–14.

Higgins, D.E., Nazareno, E. & DiRita, V.J. (1992) The virulence gene activator ToxT from *Vibrio cholerae* is a member of the AraC family of transcriptional activators. *J. Bacteriol.* **174**, 6974–6980.

Higgins, N.P. (1992) Death and transfiguration among bacteria. *Trends Biochem. Sci.* **17**, 207–211.

Hill, C.W. & Combriato, G. (1973) Genetic duplications induced at very high frequency by ultraviolet irradiation in *Escherichia coli*. *Mol. Gen. Genet.* **127**, 197–214.

Hill, C.W., Grafstrom, R.H., Harnish, B.W. & Hillman, B.S. (1977) Tandem duplications resulting from recombination between ribosomal RNA genes in *Escherichia coli*. *J. Mol. Biol.* **116**, 407–428.

Hill, C.W. & Gray, J.A. (1988) Effects of chromosomal inversions on cell fitness in *Escherichia coli* K-12. *Genetics* **119**, 771–778.

Hill, C.W. & Harnish, B.W. (1981) Inversions between ribosomal RNA genes of *Escherichia coli*. *Proc. Natl Acad. Sci. USA* **78**, 7069–7072.

Hindahl, M.S., Frank, D.W., Hamood, A. & Iglewski, B.H. (1988) Characterization of a gene that regulates toxin A synthesis in *Pseudomonas aeruginosa*. *Nucleic Acids Res.* **16**, 5699, Erratum: **16**, 8752.

Hinnebusch, J. & Barbour, A.G. (1991) Linear plasmids of *Borrelia burgdorferi* have a telomeric structure and sequence similar to those of a eukaryotic virus. *J. Bacteriol.* **173**, 7233–7239.

Hinnebusch, J. & Barbour, A.G. (1992) Linear- and circular-plasmid copy numbers in *Borrelia burgdorferi*. *J. Bacteriol.* **174**, 5251–5257.

Hiraga, S. (1992) Chromosome and plasmid partition in *Escherichia coli*. *Annu. Rev. Biochem.* **61**, 283–306.

Hiraga, S., Niki, H., Ogura, T., Ichinose, C., Mori, H., Ezaki, B. & Jaffé, A. (1989) Chromosome partitioning in *Escherichia coli*: novel mutants producing anucleate cells. *J. Bacteriol.* **171**, 1496–1505.

Hirochika, H., Nakamura, K. & Sakaguchi, K. (1984) A linear DNA plasmid from *Streptomyces rochei* with an inverted terminal repetition of 614 base pairs. *EMBO J.* **3**, 761–766.

Hirst, T.R., Sanchez, J., Kaper, K., Hardy, S.J.S. & Holmgren, J. (1984) Mechanism of toxin secretion by *Vibrio cholerae* investigated in strains harbouring plasmids that encode heat-labile enterotoxins of *Escherichia coli*. *Proc. Natl Acad. Sci. USA* **81**, 7752–7756.

Hirvas, L., Coleman, J. Koski, P. & Vaara, M. (1990) Bacterial 'histone-like protein I' (HLP-I) is an outer membrane constituent? *FEBS Lett.* **262**, 123–126.

Hobot, J.A., Villiger, W., Escaig, J., Maeder, M., Ryter, A. & Kellenberger, E. (1985) Shape and fine structure of nucleoids observed on sections of ultrarapidly frozen and cryosubstituted bacteria. *J. Bacteriol.* **162**, 960–971.

Hodges-Garcia, Y., Hagerman, P.J. & Pettijohn, D.E. (1989) DNA ring closure by protein HU. *J. Biol. Chem.* **264**, 14621–14623.

Hoe, N.P., Minion, F.C. & Goguen, J.D. (1992) Temperature sensing in *Yersinia pestis*: regulation of *yopE* transcription by *lcrF*. *J. Bacteriol.* **174**, 4275–4286.

Hoehn, G.T. & Clark, V.L. (1992a) Isolation and nucleotide sequence of the gene (*aniA*) encoding the major anaerobically induced outer membrane protein of *Neisseria gonorrhoeae*. *Infect. Immun.* **60**, 4695–4703.

Hoehn, G.T. & Clark, V.L. (1992b) The major anaerobically induced outer membrane protein of *Neisseria gonorrhoeae*, Pan1, is a lipoprotein. *Infect. Immun.* **60**, 4704–4708.

Hoess, R.H. & Abremski, K. (1984) Interaction of the bacteriophage P1 recombination Cre with the recombining site *loxP. Proc. Natl Acad. Sci. USA* **81**, 1026–1029.

Hoiseth, S.K. & Stocker, B.A.D. (1981) Aromatic-dependent *Salmonella typhimurium* are non-virulent and are effective live vaccines. *Nature* **241**, 238–239.

Holliday, R. (1964) A mechanism for gene conversion in fungi. *Genet. Res.* **5**, 282–304.

Holliday, R. (1989) A different kind of inheritance. *Sci. Am.* **261**, 40–48.

Hollingshead, S.K., Fischetti, V.A. & Scott, J.R. (1987) Size variation in group A streptococcal protein is generated by homologous recombination between intragenic repeats. *Mol. Gen. Genet.* **207**, 196–203.

Holloway, B. & Low, K.B. (1987) F-prime and R-prime factors. In, Neidhardt, F.C., Ingraham, J.L., Low, K.B., Magasanik, B., Schaechter, M. & Umbarger, H.E. (eds), Escherichia coli *and* Salmonella typhimurium *Cellular and Molecular Biology*, pp. 1145–1153. American Society for Microbiology, Washington.

Holmberg, C. & Rutberg, B. (1991) Expression of the gene encoding glycerol-3-phosphate dehydrogenase (*glpD*) in *Bacillus subtilis* is controlled by antitermination. *Mol. Microbiol.* **5**, 2891–2900.

Honigberg, S.M. & Radding, C.M. (1988) The mechanics of winding and unwinding helices in recombination: torsional stress associated with strand transfer promoted by RecA protein. *Cell* **54**, 525–532.

Honoré, N., Nicolas, M.H. & Cole, S.T. (1986) Inducible cephalosporinase production in clinical isolates of *Enterobacter cloacae* is controlled by a regulatory gene that has been deleted from *Escherichia coli. EMBO J.* **5**, 3709–3714.

Hoopes, B.C. & McClure, W.R. (1985) A cII-dependent promoter is located within the Q gene of bacteriophage λ. *Proc. Natl Acad. Sci. USA* **82**, 3134–3138.

Hoopes, B.C. & McClure, W. (1987) Strategies in regulation of transcription initiation. In Neidhardt, F.C., Ingraham, J.L., Low, K.B., Magasanik, B., Schaechter, M. & Umbarger, H.E. (eds), Escherichia coli *and* Salmonella typhimurium *Cellular and Molecular Biology*, pp. 1231–1240. American Society for Microbiology, Washington.

Hoover, T.A., Vodkin, M.H. & Williams, J.C. (1992) A *Coxiella burnetti* repeated DNA element resembling a bacterial insertion sequence. *J. Bacteriol.* **174**, 5540–5548.

Hoover, T.R., Santero, E., Porter, S. & Kutsu, S. (1990) The integration host factor stimulates interaction of RNA polymerase with NIFA, the transcriptional activator for nitrogen fixation operons. *Cell* **63**, 11–22.

Hooykaas, P.J.J. (1989) Tumorigenicity of *Agrobacterium* on plants. In Hopwood, D.A. & Chater, K.F. (eds), *Genetics of Bacterial Diversity*, pp. 373–391. Academic Press, London.

Hopwood, D.A. & Kieser, T. (1990) The *Streptomyces* genome. In Drlica, K. & Riley, M. (eds), *The Bacterial Chromosome*, pp 147–162. American Society for Microbiology, Washington.

Horikoshi, M., Bertuccioli, C., Takada, R., Wang, J., Yamamoto, T. & Roeder, R.G. (1992) Transcription factor TFIID induces DNA bending upon binding to the TATA element. *Proc. Natl Acad. Sci USA* **89**, 1060–1064.

Horiuchi, H., Takagi, M. & Yano, K. (1984) Relaxation of supercoiled plasmid DNA by oxidative stresses in *Escherichia coli. J. Bacteriol.* **160**, 1017–1021.

Howard, E.A., Zupan, J.R., Citovsky, V. & Zambryski, P.C. (1992) The VirD2 protein of *A. tumefaciens* contains a C-terminal bipartite nuclear localization signal: implications for nuclear uptake of DNA in plant cells. *Cell* **68**, 109–118.

Howe, T.R., LaQuier, F.W. & Barbour, A.G. (1986) Organization of genes encoding two outer membrane proteins in *Borrelia burgdorferi* within a single transcriptional unit. *Infect. Immun.* **54**, 207–212.

Howe, T.R., Mayer, L.W. & Barbour, A.G. (1985) A single recombinant plasmid expressing two major outer surface proteins of the lyme disease spirochaete. *Science* **227**, 645–646.

Hromockyj, A.E., Tucker, S.C. & Maurelli, A.T. (1992) Temperature regulation of *Shigella* virulence: identification of the repressor gene *virR*, an analogue of *hns*, and partial complementation by tyrosyl transfer RNA ($tRNA_1^{Tyr}$). *Mol. Microbiol.* **6**, 2113–2124.

Hsieh, L.-S., Burger, R.M. & Drlica, K. (1991a) Bacterial DNA supercoiling and [ATP]/[ADP] changes associated with a transition to anaerobic growth. *J. Mol. Biol.* **219**, 443–450.

Hsieh, L.-S., Rouvière-Yaniv, J. & Drlica, K. (1991b) Bacterial DNA supercoiling and [ATP]/ [ADP] ratio: changes associated with salt shock. *J. Bacteriol.* **173**, 3914–3917.

Htun, H. & Dahlberg, J.E. (1989) Topology and formation of triple-stranded H-DNA. *Science* **243**, 1571–1576.

Hu, M. & Deonier, R.C. (1981) Comparison of IS1, IS2 and IS3 copy number in *Escherichia coli* strains K12, B and C. *Gene* **16**, 161–170.

Hu, S., Ohtsubo, E., Davidson, N. & Saedler, H. (1975) Electron microscope heteroduplex studies of sequence relations among bacterial plasmids: identification and mapping of the insertion sequences IS*1* and IS*2* in F and R plasmids. *J. Bacteriol.* **122**, 764–775.

Huang, A., Friesen, J. & Brunton, J.L. (1987) Characterization of a bacteriophage that carries the genes for production of Shiga-like toxin 1 in *Escherichia coli. J. Bacteriol.* **169**, 4308–4312.

Huang, L., Tsui, P. & Freundlich, M. (1990) Integration host factor is a negative effector of *in vivo* and *in vitro* expression of *ompC* in *Escherichia coli. J. Bacteriol.* **172**, 5293–5298.

Huang, L., Tsui, P. & Freundlich, M. (1992) Positive and negative control of *ompB* transcription in *Escherichia coli* by cyclic AMP and the cyclic AMP receptor protein. *J. Bacteriol.* **174**, 664–670.

Huber, H.E., Iida, S., Arber, W. & Bickle, T.A. (1985) Site-specific DNA inversion is enhanced by a DNA sequence *in cis. Proc. Natl Acad. Sci. USA* **82**, 3776–3780.

Hübscher, U., Lutz, H. & Kornberg, A. (1980) Novel histone H2A-like protein of *Escherichia coli. Proc. Natl Acad. Sci. USA* **77**, 5097–5101.

Huh, Y.J. & Weiss, A.A. (1991) A 23-kilodalton protein, distinct from BvgA, expressed by virulent *Bordetella pertussis* binds to the promoter region of *vir*-regulated toxin genes. *Infect. Immun.* **59**, 2389–2395.

Hultgren, S.J., Normark, S. & Abraham, S.N. (1991) Chaperone-assisted assembly and molecular architecture of adhesive pili. *Annu. Rev. Microbiol.* **45**, 383–415.

Hulton, C.S.J., Higgins, C.F. & Sharp, P.M. (1991) ERIC sequences: a novel family of repetitive elements in the genomes of *Escherichia coli, Salmonella typhimurium* and other enterobacteria. *Mol. Microbiol.* **5**, 825–834.

Hulton, C.S.J., Seirafi, A., Hinton, J.C.D., Sidebotham, J.M., Waddell, L., Pavitt, G.D., Owen-Hughes, T., Spasskey, A., Buc, H. & Higgins, C.F. (1990) Histonelike protein H1 (H-NS), DNA supercoiling, and gene expression in bacteria. *Cell* **63**, 631–642.

Hussain, K., Elliott, E.J. & Salmond, G.P.C. (1987) The $ParD^-$ mutant of *Escherichia coli* also carries a $gyrA_{am}$ mutation: the complete sequence of *gyrA. Mol. Microbiol.* **1**, 259–273.

Hynes, R.O. (1992) Integrins: versatility, modulation and signalling in cell adhesion. *Cell* **69**, 11–25.

Iandolo, J.J. (1990) The genetics of staphylococcal toxins and virulence factors. In Iglewski, B.H. & Clark, V.L. (eds) *Molecular Basis of Bacterial Pathogenesis*, pp. 399–426. Academic Press, New York.

Igo, M.M., Ninfa, A.J., Stock, J.B. & Silhavy, T.J. (1989) Phosphorylation and dephosphorylation of a bacterial activator by a transmembrane receptor. *Genes Dev.* **3**, 1725–1734.

Iida, S. & Arber, W. (1977) Plaque-forming specialized transducing phage P1: isolation of P1CmSmSu, a precursor of P1Cm. *Mol. Gen. Genet.* **153**, 259–269.

Iida, S., Huber, H., Hierstand-Nauer, R., Meyer, J., Bickle, T.A. & Arber, W. (1984) The bacteriophage P1 site-specific recombinase Cin: recombination events and DNA recognition sequences. *Cold Spring Harbor Symp. Quant. Biol.* **49**, 769–777.

Iida, S., Meyer, J., Kennedy, K. & Arber, W. (1982) A site-specific, conservative recombination system carried by bacteriophage P1. *EMBO J.* **1**, 1445–1453.

Isberg, R.R. & Falkow, S. (1985) A single genetic locus encoded by *Yersinia pseudotuberculosis* permits invasion of cultured mammalian cells by *E. coli* K-12. *Nature* **317**, 262–264.

Isberg, R.R., Lazaar, A.L. & Syvanen, M. (1982) Regulation of Tn5 by the right repeat proteins: control at the level of the transposition reactions? *Cell* **30**, 883–892.

Isberg, R.R. & Leong, J.M. (1990) Multiple β1 chain integrins are receptors for invasin, a protein that promotes bacterial penetration into mammalian cells. *Cell* **60**, 861–871.

Isberg, R.R., Swain, A. & Falkow, S. (1988) Analysis of expression and thermoregulation of the *Yersinia pseudotuberculosis inv* gene with hybrid proteins. *Infect. Immun.* **56**, 2133–2138.

Isberg, R.R. & Syvanen, M. (1982) DNA gyrase is a host factor required for transposition of Tn5. *Cell* **30**, 9–18.

Ishimoto, K.S. & Lory, S. (1989) Formation of pilin in *Pseudomonas aeruginosa* requires the alternative σ factor (RpoN) subunit of RNA polymerase. *Proc. Natl Acad. Sci. USA* **86**, 1954–1957.

Ishimoto, K.S. & Lory, S. (1992) Identification of *pilR*, which encodes a transcriptional activator of the *Pseudomonas aeruginosa* pilin gene. *J. Bacteriol.* **174**, 3514–3521.

Iuchi, S., Chepuri, V., Fu, H.-A., Gennis, R.B. & Lin, E.C.C. (1990a) Requirement for terminal cytochromes in generation of the aerobic signal for the *arc* regulatory system in *Escherichia coli*: study utilizing deletions and *lac* fusions of *cyo* and *cyd*. *J. Bacteriol.* **172**, 6020–6025.

Iuchi, S. & Lin, E.C.C. (1988) *arcA* (*dye*), a global regulatory gene in *Escherichia coli* mediating repression of enzymes in aerobic pathways. *Proc. Natl Acad. Sci. USA* **85**, 1888–1892.

Iuchi, S. & Lin, E.C.C. (1991) Adaptation of *Escherichia coli* to respiratory conditions: regulation of gene expression. *Cell* **66**, 5–7.

Iuchi, S. & Lin, E.C.C. (1992) Purification and phosphorylation of the Arc regulatory components of *Escherichia coli*. *J. Bacteriol.* **174**, 5617–5623.

Iuchi, S., Matsuda, Z., Fujiwara, T. & Lin, E.C.C. (1990b) The *arcB* gene of *Escherichia coli* encodes a sensor-regulator protein for anaerobic repression of the *arc* modulon. *Mol. Microbiol.* **4**, 715–727.

Jackson, M.P., Newland, J.W., Holmes, R.K. & O'Brien, A.D. (1987) Nucleotide sequence analysis of the structural genes for shiga-like toxin I encoded by bacteriophage 933J from *Escherichia coli*. *Microb. Pathogen.* **2**, 147–153.

Jacob, F., Brenner, S. & Cuzin, F. (1963) On the regulation of DNA replication in bacteria. *Cold Spring Harbor Symp. Quant. Biol.* **28**, 329–348.

Jann, K. & Jann, B. (1990) Bacterial capsules. *Curr. Top. Microbiol.* **150**, 19–42.

Janzon, L. & Arvidson, S. (1990) The role of the δ-lysin gene (*hld*) in the regulation of virulence genes by the accessory gene regulator (*agr*) in *Staphylococcus aureus*. *EMBO J.* **9**, 1391–1399.

Janzon, L., Lofdahl, S. & Arvidson, S. (1989) Identification and nucleotide sequence of the delta-lysin gene, *hld*, adjacent to the accessory gene regulator (*agr*) of *Staphylococcus aureus*. *Mol. Gen. Genet.* **219**, 480–485.

Jenkins, D.E., Chaisson, S.A. & Matin, A. (1990) Starvation-induced cross protection against osmotic challenge in *Escherichia coli*. *J. Bacteriol.* **172**, 2779–2781.

Jensen, K.F. & Pedersen, S. (1990) Metabolic growth rate control in *Escherichia coli* may be a consequence of subsaturation of the macromolecular biosynthetic apparatus with substrates and catalytic components. *Microbiol. Rev.* **54**, 89–100.

Jin, S., Komari, T., Gordon, M. & Nester, E. (1987) Genes responsible for the supervirulence of *Agrobacterium tumefaciens* A281. *J. Bacteriol.* **169**, 4417–4425.

Jin, S., Prusti, R.K., Roitsch, T., Ankenbauer, R.G. & Nester, E.W. (1990a) Phosphorylation of the VirG protein of *Agrobacterium tumefaciens* by the autophosphorylated VirA protein: essential role in biological activity of VirG. *J. Bacteriol.* **172**, 4945–4950.

Jin, S., Roitsch, T., Ankenbauer, R., Gordon, M. & Nester, E. (1990b) The VirA protein of *Agrobacterium tumefaciens* is autophosphorylated and is essential for *vir* gene regulation. *J. Bacteriol.* **172**, 525–530.

Jin, S., Roitsch, T., Christie, P. & Nester, E. (1990c) The regulatory VirG protein specifically binds to a *cis*-acting regulatory sequence involved in transcriptional activation of *Agrobacterium tumefaciens* virulence genes. *J. Bacteriol.* **172**, 531–537.

Johnson, K., Charles, I., Dougan, G., Pickard, D., O'Gaora, P., Costa, G., Ali, T., Miller, I. & Hormaeche, C. (1991) The role of a stress-response protein in *Salmonella typhimurium* virulence. *Mol. Microbiol.* **5**, 401–407.

Johnson, L.P., Tomai, M.A. & Schlievert, P.M. (1986a) Bacteriophage involvement in group A streptococcal exotoxin A production. *J. Bacteriol.* **166**, 623–627.

Johnson, R.C., Ball, C.A., Pfeffer, D. & Simon, M.I. (1988) Isolation of the gene encoding the hin recombinational enhancer binding protein. *Proc. Natl Acad. Sci. USA* **85**, 3484–3488.

Johnson, R.C., Bruist, M.B. & Simon, M.I. (1986b) Host protein requirements for *in vitro* site-specific DNA inversion. *Cell* **46**, 531–539.

Johnson, R.C., Glasgow, A.C. & Simon, M.I. (1987) Spatial relationship of the Fis binding sites for Hin recombinational enhancer activity. *Nature* **329**, 462–465.

Johnson, R.C. & Simon, M.I. (1985) Hin-mediated site-specific recombination requires two 26 bp recombination sites and a 60 bp recombinational enhancer. *Cell* **41**, 781–791.

Johnson, R.C., Yin, J.C.P. & Reznikoff, W.S. (1982) Control of Tn5 transposition in *Escherichia coli* is mediated by a protein from the right repeat. *Cell* **30**, 873–882.

Johnston, A.W.B. (1989) The symbiosis between *Rhizobium* and legumes. In Hopwood, D.A. & Chater, K.F. (eds) *The Genetics of Bacterial Diversity*, pp. 393–414. Academic Press, London.

Johnston, B.H. (1988) The S1-sensitive form of $d(C-T)_n - d(A-G)_n$: chemical evidence for a three-stranded structure in plasmids. *Science* **241**, 1800–1804.

Jones, G.W., Rabert, D.K., Svinarich, D.M. & Whitfield, H.J. (1982) Association of adhesive, invasive and virulent phenotypes of *Salmonella typhimurium* with autonomous 60 megadalton plasmids. *Infect. Immun.* **38**, 476–486.

Jones, K.F., Hollingshead, S.K., Scott, J.R. & Fischetti, V.A. (1988) Spontaneous M6 protein size mutants of group A streptococci display variation in antigenic and opsonogenic epitopes. *Proc. Natl Acad. Sci. USA* **85**, 8271–8275.

Jones, P.G., Cashel, M., Glaser, G. & Neidhardt, F.C. (1992a) Function of a relaxed-like state following temperature downshifts in *Escherichia coli*. *J. Bacteriol.* **174**, 3903–3914.

Jones, P.G., Krah, R., Tafuri, S.R. & Wolffe, A.P. (1992b) DNA gyrase, CS7.4, and the cold shock response in *Escherichia coli*. *J. Bacteriol.* **174**, 5798–5802.

Jones, P.G., Van Bogelen, R.A. & Neidhardt, F.C. (1987) Induction of proteins in response to low temperature in *Escherichia coli*. *J. Bacteriol.* **169**, 2092–2095.

Jones, S., Yu, B., Bainton, N.J., Birdsall, M., Bycroft, B.W., Chhabra, S.R., Cox, A.J.R., Golby, P., Reeves, P.J., Stephens, S., Winson, M.K., Salmond, G.P.C., Stewart, G.S.A.B. & Williams, P. (1993) The *lux* autoinducer regulates the production of exoenzyme virulence determinants in *Erwinia carotovora* and *Pseudomonas aeruginosa*. *EMBO J.* **12**, 2477–2482.

Jonsson, A.-B., Nyberg, G. & Normark, S. (1991) Phase variation of gonococcal pili by frameshift mutation in *pilC*, a novel gene for pilus assembly. *EMBO J.* **10**, 477–488.

Jordi, B.J.A.M., Dagberg, B., de Haan, L.A.M., Hamers, A.M., van der Zeijst, B.A.M., Gaastra, W. & Uhlin, B.E. (1992) The positive regulator CfaD overcomes the repression mediated by histonelike protein HNS (H1) in the CFA/I fimbrial operon of *Escherichia coli*. *EMBO J.* **11**, 2627–2632.

Jovanovich, S.B. & Lebowitz, J. (1987) Estimation of the effect of coumermycin A1 on *Salmonella typhimurium* promoters using random operon fusions. *J. Bacteriol.* **169**, 4431–4435.

Jung, J.U., Gutierrez, C., Matin, F., Ardourel, M. & Villarejo, M. (1990) Transcription of *osmB*, a gene encoding an *Escherichia coli* lipoprotein, is regulated by dual signals osmotic stress and stationary phase. *J. Mol. Biol.* **265**, 10574–10581.

Kaasen, I., Falkenberg, P., Styrvold, O.B. & Strøm, A.R. (1992) Molecular cloning and physical mapping of the *otsBA* genes, which encode the osmoregulatory trehalose pathway of *Escherichia coli*: evidence that transcription is activated by KatF (AppR). *J. Bacteriol.* **174**, 889–898.

Kahmann, R., Rudt, F., Koch, C. & Mertens, G. (1985) G inversion in bacteriophage Mu is stimulated by a site within the invertase gene and a host factor. *Cell* **41**, 771–780.

Kalman, L.V. & Gunsalus, R.P. (1990) Nitrate- and molybdenum-independent signal transduction mutations in *narX* that alter regulation of anaerobic respiratory genes in *Escherichia coli*. *J. Bacteriol.* **172**, 7049–7056.

Kamp, P. & Kahmann, R. (1981) The relationship of two invertible segments in bacteriophage Mu and *Salmonella typhimurium* DNA. *Mol. Gen. Genet.* **174**, 564–566.

Kano, Y., Osato, K., Wada, M. & Imamoto, F. (1987) Cloning and sequencing of the *HU-2* gene of *Escherichia coli*. *Mol. Gen. Genet.* **209**, 408–410.

Kano, Y., Yoshino, S., Wada, M., Yokoyama, K., Nobura, M. & Imamoto, F. (1985) Molecular cloning and nucleotide sequence of the *HU-1* gene of *Escherichia coli*. *Mol. Gen. Genet.* **201**, 360–362.

Karpel, R., Alon, T., Glaser, G., Schuldiner, S. & Padan, E. (1991) Expression of a sodium proton antiporter (NhaA) in *Escherichia coli* is induced by Na^+ and Li^+ ions. *J. Biol. Chem.* **266**, 21753–21759.

Kass, E.H., Kendrick, M.I., Tsai, Y.-C. & Parsonnet, J. (1987) Interaction of magnesium ion, oxygen tension and temperature in the production of toxic shock syndrome toxin 1 by *Staphylococcus aureus*. *J. Infect. Dis.* **155**, 812–815.

Kato, J., Misra, T.K. & Chakrabarty, A.M. (1990a) AlgR3, a protein resembling eukaryotic histone H1, regulates alginate synthesis in *Pseudomonas aeruginosa*. *Proc. Natl Acad. Sci. USA* **87**, 2887–2891.

Kato, J.-I., Nishimura, Y., Imamura, R., Niki, H., Hiraga, S. & Suzuki, H. (1990b) New topoisomerase essential for chromosome segregation in *E. coli*. *Cell* **63**, 393–404.

Kato, J.-I., Nishimura, Y. & Suzuki, H. (1989) *Escherichia coli parA* is an allele of the *gyrB* gene. *Mol. Gen. Genet.* **217**, 178–181.

Kato, J.-I., Nishimura, Y., Yamada, M., Suzuki, H. & Hirota, Y. (1988) Gene organization in the region containing a new gene involved in chromosome partition in *Escherichia coli*. *J. Bacteriol.* **170**, 3967–3977.

Kawahara, K., Haraguchi, Y., Tsichimoto, M., Terakado, N. & Danbara, N. (1988) Evidence of correlation between 50-kilobase plasmid of *Salmonella cholerasuis* and its virulence. *Microb. Pathogen.* **4**, 155–163.

Kawula, T.H. & Orndorff, P.E. (1991) Rapid site-specific DNA inversion in *Escherichia coli* mutants lacking the histonelike protein H-NS. *J. Bacteriol.* **173**, 4116–4123.

Keener, S.L., McManee, K.E. & McEntee, K. (1984) Cloning and characterization of *recA* genes from *Proteus vulgaris, Erwinia carotovora, Shigella flexneri* and *Escherichia coli*. *J. Bacteriol.* **160**, 153–160.

Kehoe, M.A., Miller, L., Poirer, T.R., Beachey, E.H., Lee, M. & Harrington, D. (1987) Genetics of type 5 M protein of *Streptococcus pyogenes*. In Ferretti, J.J. & Curtiss, R., III. (eds), *Streptococcal Genetics*, pp. 112–116. American Society for Microbiology, Washington.

Kehoe, M.A., Poirer, T.P., Beachey, E.H. & Timmis, K.N. (1985) Cloning and genetic analysis of serotype 5 M protein determinant of group A streptococci: evidence for multiple copies of the M5 determinant in the *Streptococcus pyogenes* genome. *Infect. Immun.* **48**, 190–197.

Kellenberger, E. (1990) Intracellular organization of the bacterial genome. In Drlica, K. & Riley, M. (eds), *The Bacterial Chromosome*, eds pp. 173–186. American Society for Microbiology, Washington.

Ketley, J.M., Kaper, J.B., Herrington, D.A., Losonsky, G. & Levine, M.M. (1990) Diminished immunogenicity of a recombination-deficient derivative of *Vibrio cholerae* vaccine strain CVD103. *Infect. Immun.* **58**, 1481–1484.

Kier, L.D., Weppleman, R.M. & Ames, B.N. (1977) Regulation of nonspecific acid phosphatase in *Salmonella typhimurium*: *phoN* and *phoP* genes. *J. Bacteriol.* **138**, 155–161.

Kikuchi, A. (1990) Reverse gyrase and other archaebacterial topoisomerases. In Cozzarelli, N.R. & Wang, J.C. (eds) *DNA Topology and its Biological Effects*, eds pp. 285–298. Cold Spring Harbor Laboratory Press, Cold Spring Harbor, New York.

Kikuchi, A. & Asai, K. (1984) Reverse gyrase — a topoisomerase which introduces positive superhelical turns into DNA. *Nature* **309**, 677–681.

Kiley, P.J. & Reznikoff, W.S. (1991) Fnr mutants that activate gene expression in the presence of oxygen. *J. Bacteriol.* **173**, 16–22.

Kimbara, K. & Chakrabarty, A.M. (1989) Control of alginate synthesis in *Pseudomonas aeruginosa*: regulation of the *algR1* gene. *Biochem. Biophys. Res. Commun.* **164**, 601–608.

Kitten, T. & Barbour, A.G. (1990) Juxtaposition of expressed variable antigen genes with a conserved telomere in the bacterium *Borrelia hermsii*. *Proc. Natl Acad. Sci. USA* **87**, 6077–6081.

Klaasen, P. & De Graaf, F.K. (1990) Characterization of FapR, a positive regulator of expression of the 987P operon in enterotoxigenic *Escherichia coli*. *Mol. Microbiol.* **4**, 1779–1783.

Klaasen, P., Woodward, M.J., van Zijderveld, F.G. & De Graaf, F.K. (1990) The 987P gene cluster in enterotoxigenic *Escherichia coli* contains a STpa transposon that activates 987P expression. *Infect. Immun.* **58**, 801–807.

Kleckner, N. (1979) DNA sequence analysis of Tn*10* insertions: origin and role of 9 bp flanking repetitions during Tn*10* translocation. *Cell* **16**, 711–720.

Kleckner, N. (1989) Transposon Tn*10*. In Berg, D.E. & Howe, M.M. (eds) *Mobile DNA*, pp. 227–268. American Society for Microbiology, Washington.

Kleckner, N., Steele, D., Reichardt, K. & Botstein, D. (1979) Specificity of insertion of the translocatable tetracycline resistance element Tn*10*. *Genetics* **92**, 1023–1040.

Klemm, P. (1986) Two regulatory *fim* genes, *fimB* and *fimE*, control the phase variation of type I fimbriae in *Escherichia coli*. *EMBO J.* **5**, 1389–1393.

Knapp, S., Hacker, J., Jarchau, T. & Goebel, W. (1986) Large, unstable inserts in the chromosome affect virulence properties of uropathogenic *Escherichia coli* O6 strain 536. *J. Bacteriol.* **168**, 22–30.

Knapp, S. & Mekalanos, J.J. (1988) Two *trans*-acting regulatory genes (*vir* and *mod*) control antigenic modulation in *Bordetella pertussis*. *J. Bacteriol.* **170**, 5059–5066.

Knapp, S., Then, I., Wels, W., Michel, G., Tschape, H., Hacker, J. & Goebel, W. (1985) Analysis of the flanking regions from different haemolysin determinants of *Escherichia coli*. *Mol. Gen. Genet.* **200**, 385–392.

Koch, C. & Kahmann, R.K. (1986) Purification and properties of the *Escherichia coli* host factor required for inversion of the G segment in bacteriophage Mu. *J. Mol. Biol.* **261**, 15673–15678.

Koch, C., Vandekerckhove, J. & Kahmann, R. (1988) *Escherichia coli* host factor for site-specific DNA inversion: cloning and characterization of the *fis* gene. *Proc. Natl Acad. Sci. USA* **85**, 4237–4241.

Kohara, Y., Akiyama K. & Isono K. (1987) The physical map of the whole *E. coli* chromosome: application of a new strategy for rapid analysis and sorting of a large genomic library. *Cell* **50**, 495–508.

Kolter, R. (1992) Life and death in stationary phase. *ASM News* **58**, 75–79.

Kondo, I. & Fujise, K. (1977) Serotype B Staphylococcal bacteriophage singly converting staphylokinase. *Infect. Immun.* **18**, 266–272.

Kondorosi, E., Pierre, M., Cren, M., Haumann, U., Buiré, M., Hoffmann, B., Schell, J. & Kondorosi, A. (1991) Identification of NolR, a negative transacting factor controlling the *nod* regulon in *Rhizobium meliloti. J. Biol. Chem.* **222**, 885–896.

Konyecsni, W.M. & Deretic, V. (1990) DNA sequence and expression analysis of *algP* and *algQ*, components of the multigene system transcriptionally regulating mucoidy in *Pseudomonas aeruginosa*: *algP* contains multiple direct repeats. *J. Bacteriol.* **172**, 2511–2520.

Koomey, J.M. & Falkow, S. (1987) Cloning of the *recA* gene of *Neisseria gonorrhoeae* and construction of gonococcal *recA* mutants. *J. Bacteriol.* **169**, 790–795.

Koomey, J.M., Gotschlich, E.C., Robbins, K., Bergström, S. & Swanson, J. (1987) Effects of *recA* mutations on pilus antigenic variation and phase variation in *Neisseria gonorrhoeae. Genetics* **117**, 391–398.

Korba, B.E. & Hayes, J.B. (1982) Partially-deficient methylation of cytosine in DNA at CC(A/T)GG sites stimulates genetic recombination of bacteriophage lambda. Cell **28**, 531–541.

Kornberg, A. & Baker, T. (1992) *DNA Replication*, 2nd edn. Freeman, New York.

Kornblum, J., Kreiswirth, B.N., Prrojan, S.J., Ross, H. & Novick, R.P. (1990) Agr: a ploycistronic locus regulating exoprotein synthesis in *Staphylococcus aureus* In, Novick R. P. (ed.) *Molecular Biology of the Staphylococci* pp. 373–402. VCH Publishers, Inc., New York.

Koronakis, V., Cross, M. & Hughes, C. (1988) Expression of the *E. coli* haemolysin secretion gene *hlyB* involves transcript anti-termination within the *hly* operon. *Nucleic Acids Res.* **16**, 4789–4800.

Koronakis, V., Cross, M. & Hughes, C. (1989) Transcription anti-termination in an *Escherichia coli* haemolysin operon is directed progressively by *cis*-acting DNA sequences upstream of the promoter region. *Mol. Microbiol.* **3**, 1397–1404.

Kosturko, L.D., Daub, E. & Murialdo, H. (1989) The interaction of *E. coli* integration host factor and λ *cos* DNA: multiple complex formation and protein-induced bending. *Nucleic Acids Res.* **17**, 317–334.

Kowalski, D. & Eddy, M.J. (1989) The DNA unwinding element: a novel, *cis*-acting component that facilitates opening of the *Escherichia coli* replication origin. *EMBO J.* **8**, 4335–4344.

Kramer, H., Amouyal, M., Nordheim, A. & Müller-Hill, B. (1988) DNA supercoiling changes the spacing requirements of two *lac* operators for DNA loop formation with *lac* repressor. *EMBO J.* **7**, 547–556.

Krasnow M.A. & Cozzarelli, N.R. (1983) Site-specific relaxation and recombination of the Tn*3* resolvase: recognition of the DNA path between oriented sites. *Cell* **32**, 1313–1324.

Krause, H.M. & Higgins, N.P. (1986) Positive and negative regulation of the Mu operator by Mu repressor and *E. coli* integration host factor. *J. Biol. Chem.* **261**, 3744–3752.

Krause, M., Fang, F.C. & Guiney, D.G. (1992) Regulation of plasmid virulence gene expression in *Salmonella dublin* involves an unusual operon structure. *J. Bacteriol.* **174**, 4482–4489.

Krause, M. & Guiney, D.G. (1991) Identification of a multimer resolution system involved in stabilization of the *Salmonella dublin* virulence plasmid pSDL2. *J. Bacteriol.* **173**, 5754–5762.

Krause, M., Roudier, C., Fierer, J., Harwood, J. & Guiney, D. (1991) Molecular analysis of the virulence locus of the *Salmonella dublin* plasmid pSDL2. *Mol. Microbiol.* **5**, 307–316.

Krawiec, S. & Riley, M. (1990) Organization of the bacterial chromosome. *Microbiol. Rev.* **54**, 502–539.

Krinke, M., Mahoney, M. & Wulff, D.L. (1991) The role of the OOP antisense RNA in coliphage λ development. *Mol. Microbiol.* **5**, 1265–1272.

Krinke, L. & Wulff, D.L. (1987) OOP RNA, produced from multicopy plasmids, inhibits λ *c*II gene expression through an RNase III-dependent mechanism. *Genes Dev.* **1**, 1005–1013.

Krohn, M., Pardon, B. & Wagner, R. (1992) Effects of template topology on RNA polymerase pausing during *in vitro* transcription of the *Escherichia coli rrnB* leader region. *Mol. Microbiol.* **6**, 581–589.

Krolewski, J.W., Murphy, E., Novick, R.P. & Rush, M.G. (1981) Site-specificity of the chromosomal insertion of *Staphylococcus aureus* transposon Tn554. *J. Mol. Biol.* **152**, 19–33.

Kroll, J.S., Hopkins, I. & Moxon, E.R. (1988) Capsule loss in *H. influenzae* type b occurs by recombination-mediated disruption of a gene essential for polysaccharide export. *Cell* **53**, 347–356.

Kroll, J.S., Loynds, B., Brophy, L.N. & Moxon, E.R. (1990) The *bex* locus in encapsulated *Haemophilus influenzae*: a chromosomal region involved in capsule polysaccharide export. *Mol. Microbiol.* **4**, 1853–1862.

Kroll, J.S., Loynds, B.M. & Moxon, E.R. (1991) The *Haemophilus influenzae* capsulation gene cluster: a compound transposon. *Mol. Microbiol.* **5**, 1549–1560.

Kroll, J.S. & Moxon, E.R. (1988) Capsulation and gene copy number at the *cap* locus of *Haemophilus influenzae* type b. *J. Bacteriol.* **170**, 859–864.

Kroll, J.S., Zamze, S., Loynds, B. & Moxon, E.R. (1989) Common organization of chromosomal loci for production of different capsular polysaccharides in *Haemophilus influenzae*. *J. Bacteriol.* **171**, 3343–3347.

Krummel, B. & Chamberlain, M.J. (1989) RNA chain initiation by *Escherichia coli* RNA polymerase: structural transitions of the enzyme in the early ternary complexes. *Biochemistry* **28**, 7829–7842.

Krylov, V.N., Bogush, V.G. & Shapiro, J. (1979) Bacteriophages of *Pseudomonas aeruginosa*, the DNA structure of which is similar to phage Mu1 DNA. I. General description, localization of endonuclease sensitive sites in DNA and structure of homoduplexes of phage D3112. *Genetika* **16**, 824–832.

Kuempel, P.L., Henson, J.M., Dircks, L., Tecklenburg, M. & Lim, D.F. (1991) *dif*, a *recA* — independent recombination site in the terminus of the chromosome of *Escherichia coli*. *New Biol.* **3**, 799–811.

Kuempel, P.L., Pelletier, A.J. & Hill, T.M. (1990) *tus* and the terminators: inhibition of replication in *Escherichia coli*. In Drlica, K. & Riley, M. (eds) *The Bacterial Chromosome*, pp 147–162. American Society for Microbiology, Washington.

Kües, U. & Stahl, U. (1989) Replication of plasmids in gram-negative bacteria. *Microbiol. Rev.* **53**, 491–516.

Kusano, T., Steinmetz, D., Hendrickson, W.G., Murchie, J., King, M., Benson, A. & Schaechter, M. (1984) Direct evidence for specific binding of the replication origin of the *Escherichia coli* chromosome to the membrane. *J. Bacteriol.* **158**, 313–316.

Kutsukake, K. & Iino, T. (1980) Inversion of specific DNA segments in flagellar phase variation of *Salmonella* and inversion systems of bacteriophage P1 and Mu. *Proc. Natl Acad. Sci. USA* **77**, 7338–7341.

Kutsukake, K., Ohya, Y. & Iino, T. (1990) Transcriptional analysis of the flagellar regulon of *Salmonella typhimurium*. *J. Bacteriol.* **172**, 741–747.

Lacey, B.W. (1960) Antigenic modulation of *Bordetella pertussis*. *J. Hyg.* **58**, 57–93.

Lacey, R.W. (1980) Evidence for two mechanisms of plasmid transfer in mixed cultures of *S. aureus in vitro*. *J. Gen. Microbiol.* **119**, 423–435.

Laine, B., Kmiecik,D., Sautiere, P., Biserte, G. & Cohen-Solal, M. (1980) Complete amino acid sequences of DNA binding proteins HU-1 and HU-2 from *Escherichia coli*. *Eur. J. Biochem.* **103**, 447–461.

Laird, W. & Groman, N. (1976) Orientation of the *tox* gene in the prophage of corynebacteriophage beta. *J. Virol.* **19**, 228–321.

Lambert de Rouvroit, C., Sluiters, C. & Cornelis, G.R. (1992) Role of the transcriptional activator, VirF, and temperature in the expression of the pYV plasmid genes of *Yersinia enterocolitica*. *Mol. Microbiol.* **6**, 395–409.

Lamond, A.I. (1985) Supercoiling response of a bacterial tRNA gene. *EMBO J.* **4**, 501–507.

Lamond, A.I. & Travers, A.A. (1985) Genetically separable functional elements mediate the optimal expression and stringent regulation of a bacterial tRNA gene. *Cell* **40**, 319–326.

Landick, R. & Yanofsky, C. (1987) Transcription attenuation. In Neidhardt, F.C., Ingraham, J.L., Low, K.B., Magasanik, B., Schaechter, M. & Umbarger, H.E. (eds), Escherichia coli *and* Salmonella typhimurium, *Cellular and Molecular Biology*, pp. 1276–1301. American Society for Microbiology, Washington.

Landoulsi, A., Malki, A., Kern, R., Kohiyama, M. & Hughes, P. (1990) The *E. coli* cell surface specifically prevents the initiation of DNA replication at *oriC* on hemimethylated DNA templates. *Cell* **63**, 1053–1060.

Landy, A. (1989) Dynamic, structural and regulatory aspects of λ site-specific recombination. *Annu. Rev. Biochem.* **58**, 913–949.

Lane, D., de Feyter, R., Kennedy, M., Phua, S.-H. & Semon, D. (1986) D-protein of mini-F plasmid acts as a repressor of transcription and as a site-specific recombinase. *Nature* **14**, 365–372.

Lange, R. & Hengge-Aronis, R. (1991a) Growth phase-regulated expression of *bolA* and morphology of stationary-phase *Escherichia coli* cells are controlled by the novel sigma factor σ^s. *J. Bacteriol.* **173**, 4474–4481.

Lange, R. & Hengge-Aronis, R. (1991b) Identification of a central regulator of stationary-phase gene expression in *Escherichia coli*. *Mol. Microbiol.* **5**, 49–59.

La Teana, A., Bandi, A., Falconi, M., Sprino, R., Pon, C.L. & Gualerzi, C.O. (1991) Identification of a cold shock transcriptional enhancer of the *Escherichia coli* gene encoding nucleoid protein H-NS. *Proc. Natl Acad. Sci. USA* **88**, 10907–10911.

La Teana, A., Falconi, M., Scarlato, V. & Pon, C.L. (1989) Characterization of the structural genes for the DNA-binding protein H-NS in enterobacteriaceae. *FEBS Lett.* **244**, 34–38.

Lathigra, R.B., Young, D.B., Sweetser, D. & Young, R.A. (1988) A gene from *Mycobacterium tuberculosis* which is homologous to the DnaJ heat shock protein of *E. coli*. *Nucleic Acids Res.* **16**, 1636.

Lauble, H., Georgalis, Y. & Heinemann, U. (1989) Studies on the domain structure of the *Salmonella typhimurium* AraC protein. *Eur. J. Biochem.* **185**, 319–325.

Laudenbach, D.E. & Grossman, A.R. (1991) Characterization of sulphur-regulated genes in a cyanobacterium: evidence for function in sulphate transport. *J. Bacteriol.* **173**, 2739–2750.

Lawlor, K.M. & Payne, S.M. (1984) Aerobactin genes in *Shigella* spp. *J. Bacteriol.* **160**, 266–272.

Lax, A.J., Pullinger, G.D., Baird, G.D. & Williamson, C.M. (1990) The virulence plasmid of *Salmonella dublin*: detailed restriction map and analysis by transposon insertion. *J. Gen. Microbiol.* **136**, 1117–1123.

Lee, C.A. & Falkow, S. (1990) The ability of *Salmonella* to enter mammalian cells is affected by bacterial growth state. *Proc. Natl Acad. Sci. USA* **87**, 4304–4308.

Lee, C.A., Jones, B.D. & Falkow, S. (1992) Identification of a *Salmonella typhimurium* invasion locus by selection for hyperinvasive mutants. *Proc. Natl Acad. Sci. USA* **89**, 1847–1851.

Lee, C.-H., Hu, S.-T., Swiatek, P.J., Moseley, S.L., Allen, S.D. & So, M. (1985) Isolation of a novel transposon which carries the *Escherichia coli* enterotoxin STII gene. *J. Bacteriol.* **162**, 615–620.

Lee, C.Y. & Iandolo, J.J. (1986a) Lysogenic conversion of staphylococcal lipase is caused by insertion of the bacteriophage L54a genome into the lipase structural gene. *J. Bacteriol.* **166**, 385–391.

Lee, C.Y. & Iandolo, J.J. (1986b) Integration of staphylococcal page L54a occurs by site-specific recombination: structural analysis of the attachment site. *Proc. Natl Acad. Sci. USA* **83**, 5474–5478.

Lee, D.-H. & Schleif, R.F. (1989) *In vivo* DNA loops in *araCBAD*: size limits and helical repeat. *Proc. Natl Acad. Sci. USA* **86**, 476–480.

Lee, D.K., Horikoshi, M. & Roeder, R.G. (1991) Interaction of TFIID in the minor groove of the TATA element. *Cell* **67**, 1241–1250.

Lee, J. & Goldfarb, A. (1991) *lac* repressor acts by modifying the initial transcribing complex so that it cannot leave the promoter. *Cell* **66**, 793–798.

Leemans, J., Deblaere, R., Willmitzer, L., de Greve, H., Hernalsteens, J.P., van Montague, M. & Schell, J. (1982) Genetic identification of functions of T_L-DNA transcripts in octopine crown galls. *EMBO J.* **1**, 147–152.

Leimeister-Wächter, M., Domann, E. & Chakroborty, T. (1992) The expression of virulence genes in *L. monocytogenes* is thermoregulated. *J. Bacteriol.* **174**, 947–952.

Lejeune, P. & Danchin, A. (1990) Mutations in the *bglY* gene increase the frequency of spontaneous deletions in *Escherichia coli* K-12. *Proc. Natl Acad. Sci. USA* **87**, 360–363.

Leroux, B., Yanofsky, M.F., Winans, S.C., Ward, J.E., Ziegler, S.F. & Nester, E.W. (1987) Characterization of the *virA* locus of *Agrobacterium tumefaciens*: a transcriptional regulator and host range determinant. *EMBO J.* **6**, 849–856.

Lesley, S.A., Jovanovich, S.B., Tse-Dinh, Y.-C. & Burgess, R.B. (1990) Identification of a heat shock promoter in the *topA* gene of *Escherichia coli*. *J. Bacteriol.* **172**, 6871–6874.

Levine, M.M., Ristaino, P., Marley, G., Smyth, C., Knutton, S., Boedeker, E., Black, R., Young, C., Clemens, M.L., Cheney, C. & Patnaik, R. (1984) Coli surface antigens 1 and 3 of colonization factor II-positive enterotoxigenic *Escherichia coli*: morphology, purification and immune responses in humans. *Infect. Immun.* **44**, 409–420.

Liao, S.-M., Wu, T.-H., Chiang, C.H., Susskind, M.M. & McClure, W.R. (1987) Control of gene expression in bacteriophage P22 by a small antisense RNA. I. Characterization *in vitro* of the P_{sar} promoter and the *sar* RNA transcript. *Genes Dev.* **1**, 197–203.

Lichenstein, C. & Brenner, S. (1982) Unique insertion site of Tn7 in the *E. coli* chromosome. *Nature* **297**, 601–603.

Liljestrom, P., Laamanen, I. & Palva, E.T. (1988) Structure and expression of the *ompB* operon, the regulatory locus for the outer membrane porin regulon in *Salmonella typhimurium* LT-2 *J. Mol. Biol.* **201**, 663–673.

Lilley, D.M.J. (1986a) Bacterial chromatin: a new twist to an old story. *Nature* **320**, 14–15.

Lilley, D.M.J. (1986b) The genetic control of DNA supercoiling, and vice versa. *Symp. Soc. Gen. Microbiol.* **39**, 105–126.

Lilley, D.M.J. (1991) When the CAP fits bent DNA. *Nature* **354**, 359–360.

Lin, R., D'Ari, R. & Newman, E.B. (1990) The leucine regulon of *Escherichia coli* K-12: a mutation in *rblA* alters expression of L-leucine-dependent metabolic operons. *J. Bacteriol.* **172**, 4529–4535.

Lin, R., D'Ari, R. & Newman, E.B. (1992) λ p*lac*Mu insertions in genes of the leucine regulon: extension of the regulon to genes not regulated by leucine. *J. Bacteriol.* **174**, 1948–1955.

Lin, R.-J., Capage, M. & Hill, C.W. (1984) A repetitive DNA sequence, *rhs*, responsible for duplications within the *Escherichia coli* K-12 chromosome. *J. Mol. Biol.* **177**, 1–18.

Linderoth, N.A. & Calendar, R.L. (1991) The Psu protein of bacteriophage P4 is an antitermination factor for Rho-dependent transcription termination. *J. Bacteriol.* **173**, 6722–6731.

Lindgren, P.B., Peet, R.C. & Panopoulos, N.J. (1986) Gene cluster of *Pseudomonas syringae* pv. *phaseolicola* controls pathogenicity of bean plants and hypersensitivity on nonhost plants. *J. Bacteriol.* **168**, 512–522.

Lindquist, S., Lindberg, F. & Normark, S. (1989) Binding of the *Citrobacter freundii* AmpR regulator to a single DNA site provides both autoregulation and activation of the inducible *ampC* β-lactamase gene. *J. Bacteriol.* **171**, 3746–3753.

Lipinska, B., Zylicz, M. & Georgopoulos, C. (1990) The HtrA (DegP) protein, essential for *Escherichia coli* survival at high temperatures is an endopeptidase. *J. Bacteriol.* **172**, 1791–1797.

Little, J.W. (1993) LexA cleavage and other self-processing reactions. *J. Bacteriol.* **175**, 4943–4950.

Litwin, C.M., Boyko, S.A. & Calderwood, S.B. (1992) Cloning, sequencing and transcriptional regulation of the *Vibrio cholerae fur* gene. *J. Bacteriol.* **174**, 1897–1903.

Liu, L.F. & Wang, J.C. (1987) Supercoiling of the DNA template during transcription. *Proc. Natl Acad. Sci. USA* **84**, 7024–7027.

Lloyd, R.G. & Buckman, C. (1985) Identification and genetic analysis of *sbcC* mutations in commonly used *recBC sbcB* strains of *Escherichia coli* K-12. *J. Bacteriol.* **164**, 836–844.

Lloyd, R.G. & Sharples, G.J. (1991) Molecular organization and nucleotide sequence of the *recG* locus of *Escherichia coli* K-12. *J. Bacteriol.* **173**, 6837–6843.

Lobell, R.B. & Schleif, R.F. (1990) DNA looping and unlooping by AraC protein. *Science* **250**, 528–532.

Lobell, R.B. & Schleif, R.F. (1991) AraC-DNA looping: orientation and distance-dependent loop breaking by the cyclic AMP receptor protein. *J. Mol. Biol.* **218**, 45–54.

Lockshon, D. & Morris, D.R. (1983) Positively supercoiled plasmid DNA is produced by treatment of *Escherichia coli* with DNA gyrase inhibitors. *Nucleic Acids Res.* **11**, 2999–3017.

Loewen, P.C., von Ossowski, I., Switala, J. & Mulvey, M.R. (1993) KatF (σ^s) synthesis in *Escherichia coli* is subject to posttranscriptional regulation. *J. Bacteriol.* **175**, 2150–2153.

Lombardo, M.-J., Bagga, D. & Miller, C.G. (1991) Mutations in *rpoA* affect expression of anaerobically regulated genes in *Salmonella typhimurium*. *J. Bacteriol.* **173**, 7511–7518.

Lonetto, M., Gribskov, M. & Gross, C.A. (1992) The σ^{70} family: sequence conservation and evolutionary relationships. *J. Bacteriol.* **174**, 3843–3849.

Louarn, J.-M., Bouche, J.-P., Lengendre, F., Louarn, J. & Patte, J. (1985) Characterization and properties of very large inversions of the *E. coli* chromosome along the origin-to-terminus axis. *Mol. Gen. Genet.* **201**, 467–476.

Louarn, J., Bouche, J.P., Patte, J. & Louarn, J.-M. (1984) Genetic inactivation of topoisomerase I suppresses a defect in initiation of chromosome replication in *Escherichia coli*. *Mol. Gen. Genet.* **195**, 170–174.

Loughney, K., Lund, E. & Dahlberg, J.E. (1983) Deletion of an rRNA gene set in *Bacillus subtilis*. *J. Bacteriol.* **154**, 529–532.

Lovett, C.M. & Roberts, J.W. (1985) Purification of a RecA protein analogue from *Bacillus subtilis*. *J. Biol. Chem.* **260**, 3305–3313.

Low, K.B. (1987) Hfr strains of *Escherichia coli* K-12. In Neidhardt, F.C., Ingraham, J.L., Low, K.B., Magasanik, B., Schaechter, M. & Umbarger, H.E. (eds), Escherichia coli *and* Salmonella typhimurium *Cellular and Molecular Biology*, pp. 1134–1137. American Society for Microbiology, Washington.

Lucht, J.M. & Bremer, E. (1991) Characterization of mutations affecting the osmoregulated *proU* promoter of *Escherichia coli* and identification of 5′ sequences required for high-level expression. *J. Bacteriol.* **173**, 801–809.

Lukat, G.S., McCleary, W.R., Stock, A.M. & Stock, J.B. (1992) Phosphorylation of bacterial response regulator proteins by low molecular weight phospho-donors. *Proc. Natl Acad. Sci. USA* **89**, 718–722.

Lundrigan, M.D. & Earhardt, C.F. (1981) Reduction in three iron-regulated outer membrane proteins and protein a by the *Escherichia coli* K-12 *perA* mutation. *J. Bacteriol.* **146**, 804–807.

Lundrigan, M.D. & Earhardt, C.F. (1984) Gene *envY* of *Escherichia coli* K-12 affects thermoregulation of major porin expression. *J. Bacteriol.* **157**, 262–268.

Lundrigan, M.D., Friedrich, M.J. & Kadner, R.J. (1989) Nucleotide sequence of the *Escherichia coli* porin thermoregulatory gene *envY*. *Nucleic Acids Res.* **17**, 800.

Lupski, J.R. & Weinstock, G.M. (1992) Short, interspersed repetitive DNA sequences in prokaryotic genomes. *J. Bacteriol.* **174**, 4525–4529.

Lutkenhaus, J. (1990) Regulation of cell division in *E. coli*. *Trends Genet.* **6**, 22–25.

McAllister, C.F. & Achberger, E.C. (1989) Rotational orientation of upstream curved DNA affects promoter function in *Bacillus subtilis*. *J. Biol. Chem.* **264**, 10451–10456.

McCann, M.P., Fraley, C.D. & Matin, A. (1993) The putative σ factor KatF is regulated post-transcriptionally during carbon starvation. *J. Bacteriol.* **175**, 2143–2149.

McCann, M.P., Kidwell, J.P. & Matin, A. (1991) The putative σ factor KatF has a central role in development of starvation-mediated general resistance in *Escherichia coli*. *J. Bacteriol.* **173**, 4188–4194.

McClain, M.S., Blomfield, I.C. & Eisenstein, B.I. (1991) Roles of *fimB* and *fimE* in site-specific DNA inversion associated with phase variation of Type 1 fimbriae in *Escherichia coli*. *J. Bacteriol.* **173**, 5308–5314.

McCleary, W.R., Stock, J.B. & Ninfa, A. J. (1993) Is acetyl phosphate a global signal in *Escherichia coli*? *J. Bacteriol.* **175**, 2793–2798.

McClellan, J.A., Boublikova, P., Palecek, E. & Lilley, D.M.J. (1990) Superhelical torsion in cellular DNA responds directly to environmental and genetic factors. *Proc. Natl Acad. Sci. USA* **87**, 8373–8377.

McDougall, S. & Neilands, J.B. (1984) Plasmid- and chromosomal-coded aerobactin synthesis in enteric bacteria: insertion sequences flank operon in plasmid-mediated systems. *J. Bacteriol.* **159**, 300–305.

MacFarlane, S.A. & Merrick, M. (1985) The nucleotide sequence of the nitrogen regulation gene *ntrB* and the *glnA-ntrBC* intergenic region of *Klebsiella pneumoniae*. *Nucleic Acids Res.* **13**, 7591–7606.

Machida, Y., Machida, C. & Ohtsubo, E. (1984) Insertion element IS1 encodes two structural genes required for its transposition. *J. Mol. Biol.* **177**, 229–245.

MacInnes, J.I., Kim, J.E., Lian, C.-J. & Soltes, G.A. (1990) *Actinobacillus pleuropneumoniae hylX* gene homology with the *fnr* gene of *Escherichia coli*. *J. Bacteriol.* **172**, 4587–4592.

McIntosh, M.A. & Earhart, C.F. (1976) Effect of iron on the relative abundance of two large polypeptides of *Escherichia coli* outer membrane. *Biochem. Biophys. Res. Commun.* **70**, 315–322.

McMacken, R., Silver, L. & Georgopoulos, C. (1987) DNA replication. In Neidhardt, F.C., Ingraham, J.L., Low, K.B., Magasanik, B., Schaechter, M. & Umbarger, H.E. (eds), Escherichia coli *and* Salmonella typhimurium *Cellular and Molecular Biology*, pp. 564–612. American Society for Microbiology, Washington.

Madiraju, M.V.V.S., Templin, A. & Clark, A.J. (1988) Properties of a mutant *recA*-encoded protein reveal a possible role for *Escherichia coli recF*-encoded protein in genetic recombination. *Proc. Natl Acad. Sci. USA* **85**, 6592–6596.

Magasanik, B. & Neidhardt, F.C. (1987) Regulation of carbon and nitrogen utilization. In Neidhardt, F.C., Ingraham, J.L., Low, K.B., Magasanik, B., Schaecter, M. & Umbarger, H.E. (eds) Escherichia coli *and* Salmonella typhimurium, *Cellular and Molecular Biology*, pp. 1318–1325. American Society for Microbiology, Washington.

Mahadevan, S. & Wright, A. (1987) A bacterial gene involved in transcription antitermination: regulation of a rho-independent terminator in the *bgl* operon of *E. coli*. *Cell* **50**, 485–494.

Mahajan, S.K. (1988) Pathways of homologous recombination in *E. coli*. In Kucherlapati, R. & Smith, G.R. (eds), *Genetic Recombination*, pp. 87–140. American Society for Microbiology, Washington.

Mahajna, J., Oppenheim, A.B., Rattray, A. & Gottesman, M. (1986) Translation initiation of bacteriophage lambda gene *c*II requires integration host factor. *J. Bacteriol.* **165**, 167–174.

Mahan, M.J., Segall, A.M. & Roth, J.R. (1990) Recombination events that rearrange the chromosome: barriers to inversion. In Drlica, K. & Riley, M. (eds), *The Bacterial Chromosome*, pp. 341–349. American Society for Microbiology, Washington.

Makino, S.-I., Sasakawa, C., Uchida, I., Terekado, N. & Yoshikawa, M. (1988) Cloning and CO_2-dependent expression of the genetic region for encapsulation from *Bacillus anthracis*. *Mol. Microbiol.* **2**, 371–376.

Makino, S., Sasakawa, C. & Yoshikawa, M. (1988) Genetic relatedness of the basic replicon of the virulence plasmid in shigellae and enteroinvasive *Escherichia coli*. *Microb. Pathogen.* **5**, 267–274.

Makman, R.S. & Sutherland, E.W. (1965) Adenosine 3′,5′-phosphate in *Escherichia coli*. *J. Biol. Chem.* **240**, 1309–1314.

Makris, J.C., Nordmann, P.L. & Reznikoff, W.S. (1988) Mutational analysis of insertion sequence *50* (IS*50*) and transposon *5* (Tn*5*) ends. *Proc. Natl Acad. Sci. USA* **85**, 2224–2228.

Mamelak, L. & Boyer, H.W. (1970) Genetic control of the secondary modification of deoxyribonucleic acid in *Escherichia coli*. *J Bacteriol.* **104**, 57–62.

Manen, D., Goebel, T. & Caro, L. (1990) The *par* region of pSC101 affects plasmid copy number as well as stability. *Mol. Microbiol.* **4**, 1839–1846.

Marians, K.J. (1992) Prokaryotic DNA replication. *Annu. Rev. Biochem.* **61**, 673–719.

Marinus, M.G. (1987) DNA methylation in *Escherichia coli*. *Annu. Rev. Genet.* **21**, 113–131.

Marolda, C.L., Valvano, M.A., Lawlor, K.M., Payne, S.M. & Crosa, J.H. (1987) Flanking and internal regions of chromosomal genes mediating aerobactin iron uptake in entero-invasive *Escherichia coli* and *Shigella flexneri*. *J. Gen. Microbiol.* **133**, 2269–2278.

Marrs, C.F., Rozsa, F.W., Hackel, M., Stevens, S.P. & Glasgow, A.C. (1990) Identification, cloning and sequencing of *piv*, a new gene involved in inverting the pilin genes of *Moraxella lacunata*. *J. Bacteriol.* **172**, 4370–4377.

Marrs, C.F., Ruehl, W.W., Schoolnik, G.K. & Falkow, S. (1988) Pilin gene phase variation of *Moraxella bovis* is caused by an inversion of the pilin genes. *J. Bacteriol.* **170**, 3032–3039

Marsh, M. & Hillyard, D.R. (1990) Nucleotide sequence of *hns* encoding the DNA-binding protein H-NS of *Salmonella typhimurium*. *Nucleic Acids Res.* **18**, 3397.

Martinez-Cadena, M.G., Guzman-Verduzco, L.M., Steiglitz, H. & Koperstoch-Portnoy, Y.M. (1981) Catabolite repression of *Escherichia coli* heat-stable enterotoxin activity. *J. Bacteriol.* **145**, 722–728.

Maskell, D.J., Szabo, M.J., Butler, P.D., Williams, A.E. & Moxon, E.R. (1991) Molecular analysis of a complex locus from *Haemophilus influenzae* involved in phase-variable lipopolysaccharide biosynthesis. *Mol. Microbiol.* **5**, 1013–1022.

Masters, M., Moir, P.D., Spiegelberg, R., Pringle, J.H. & Vermeulen, C.W. (1985) Is the chromosome of *E. coli* differentiated along its length with respect to gene density or accessibility to transcription? In Schaecter, M., Neidhardt, F., Ingraham, J. & Kjelgaard, N. (eds), *The Molecular Biology of Bacterial Growth*, pp. 335–343. Jones & Bartlett, Boston.

Matsuyama, S. & Mizushima, S. (1987) Novel *rpoA* mutation that interferes with the function of OmpR and EnvZ, positive regulators of the *ompF* and *ompC* genes that code for outer-membrane proteins in *Escherichia coli* K-12. *J. Mol. Biol.* **195**, 847–853.

Maurelli, A.T. (1990) Regulation of virulence genes in shigella. *Mol. Biol. Med.* **6**, 425–432.

Maurelli, A.T. & Sansonetti, P.J. (1988a) Genetic determinants of *Shigella* pathogenicity. *Annu. Rev. Microbiol.* **42**, 127–150.

Maurelli, A.T. & Sansonetti, P.J. (1988b) Identification of a chromosomal gene controlling temperature-regulated expression of *Shigella* virulence. *Proc. Natl Acad. Sci. USA* **85**, 2820–2824.

May, G., Dersch, P., Haardt, M., Middendorf, A. & Bremer, E. (1990) The *osmZ* (*bglY*) gene encodes the DNA binding protein H-NS (H1a), a component of the *Escherichia coli* K12 nucleoid. *Mol. Gen. Genet.* **224**, 81–90.

Mazodier, P., Petter, R. & Thompson, C. (1989) Intergeneric conjugation between *Escherichia coli* and *Streptomyces* species. *J. Bacteriol.* **171**, 3583–3585.

Mehlert, A. & Young, D.B. (1989) Biochemical and antigenic characterization of the *Mycobacterium tuberculosis* 71 kDa antigen, a member of the 70 kDa heat shock protein family. *Mol. Microbiol.* **3**, 125–130.

Meier, J.T., Simon, M.I. & Barbour, A.G. (1985) Antigenic variation is associated with DNA rearrangements in a relapsing fever borrelia. *Cell* **41**, 403–409.

Mekalanos, J.J. (1983) Duplication and amplification of toxin genes in *Vibrio cholerae*. *Cell* **35**, 253–263.

Mekalanos, J.J. (1992) Environmental signals controlling expression of virulence determinants in bacteria. *J. Bacteriol.* **174**, 1–7.

Mekalanos, J.J., Collier, R.J. & Romig, W.R. (1979) Enzymatic activity of cholera toxin. II. Relationships to proteolytic processing, disulphide bond reduction and subunit composition. *J. Biol. Chem.* **254**, 5855–5861.

Mekalanos, J.J., Moseley, S.L., Murphy, J.R. & Falkow, S. (1982) Isolation of enterotoxin structural gene deletion mutations in *Vibrio cholerae* induced by two mutagenic vibriophages. *Proc. Natl Acad. Sci. USA* **79**, 151–155.

Mekalanos, J.J., Peterson, K.M., Finn, T. & Knapp, S. (1988) The pathogenesis and immunology of *Vibrio cholerae* and *Bordetella pertussis*. *Antonie van Leeuwenhoek* **54**, 379–387.

Mekalanos, J.J., Swartz, D.J., Pearson, G.D.N., Harford, N., Groyne, F. & de Wilde, M. (1983) Cholera toxin genes: nucleotide sequence, deletion analysis and vaccine development. *Nature* **306**, 551–557.

Melchers, L., Regensburg-Tuink, T., Bourret, R., Sedee, N., Schilperoort, R. & Hooykaas, P. (1989) Membrane topology and functional analysis of sensory protein VirA of *Agrobacterium tumefaciens*. *EMBO J.* **8**, 1919–1925.

Melchers, L.S., Thompson, D.V., Idler, K.B., Schilperoort, R.A. & Hooykaas, P.J.J. (1986)

Nucleotide sequence of the virulence gene *virG* of the *Agrobacterium tumefaciens*, octopine Ti plasmid: significant homology between *virG* and the regulatory genes *ompR*, *phoB* and *dye* of *E. coli*. *Nucleic Acids Res.* **14**, 9933–9942.

Melton, A.R. & Weiss, A.A. (1989) Environmental regulation of expression of virulence determinants in *Bordetella pertussis*. *J. Bacteriol.* **171**, 6206–6212.

Memelink, J., dePater, B.S., Hoge, J.H.C. & Schilperoot, R.A. (1987) T-DNA hormone biosynthetic genes: phytohormone and gene expression in plants. *Dev. Genet.* **8**, 321–337.

Mende, L., Timm, B. & Subramanian, A.R. (1978) Primary structures of two homologous ribosome-associated DNA-binding proteins of *Escherichia coli*. *FEBS Lett.* **96**, 395–398.

Menendez, M., Kolb, A. & Buc, H. (1987) A new target for CRP action at the *malT* promoter. *EMBO J.* **6**, 4227–4234.

Menzel, R. & Gellert, M. (1983) Regulation of the genes for *E. coli* DNA gyrase: homeostatic control of DNA supercoiling. *Cell* **34**, 105–113.

Menzel, R. & Gellert, M. (1987) Fusions of the *Escherichia coli gyrA* and *gyrB* control regions to the galactokinase gene are inducible by coumermycin treatment. *J. Bacteriol.* **169**, 1272–1278.

Mertens, G., Klippel, A., Fuss, H., Blöcker, H., Frank, R. & Kahmann, R. (1988) Site-specific recombination in bacteriophage Mu: characterization of binding sites for the DNA invertase Gin. *EMBO J.* **7**, 1219–1227.

Meyer, T.F. (1990) Variation of pilin and opacity-associated protein in pathogenic *Neisseria* species. In Iglewski, B.H. & Clark, V.L. (eds) *Molecular Basis of Bacterial Pathogenesis*, pp. 137–153, Academic Press, San Diego.

Meyer, T.F., Billyard, E., Haas, R., Störzbach, S. & So, M. (1984) Pilus gene of *Neisseria gonorrhoeae*: chromosomal organization and DNA sequence. *Proc. Natl Acad. Sci. USA* **81**, 6110–6114.

Meyer, T.F., Frosch, M., Gibbs, C.P., Haas, R., Halter, R., Pohlner, J. & van Putten, J.P.M. (1988) Virulence functions and antigen variation in pathogenic neisseriae. *Antonie van Leeuwenhoek* **54**, 421–430.

Meyer, T.F., Gibbs, C.P. & Haas, R. (1990) Variation and control of protein expression in *Neisseria*. *Annu. Rev. Microbiol.* **44**, 451–477.

Michiels, T., Popoff, M.Y., Durviaux, S., Coynault, C. & Cornelis, G. (1987) A new method for the physical and genetic mapping of large plasmids: application of the localization of the virulence determinants on the 90 kb plasmid of *Salmonella typhimurium*. *Microb. Pathogen.* **3**, 109–116.

Mickelsen, P.A. & Sparling, P.F. (1981) Ability of *Neisseria gonorrhoeae*, *Neisseriae meningitis*, and commensal *Neisseria* species to obtain iron from transferrin and iron compounds. *Infect. Immun.* **33**, 555–564.

Mikesell, P., Ivins, B.E., Ristroph, J.D. & Drier, T.M. (1983) Evidence for plasmid-mediated toxin production in *Bacillus anthracis*. *Infect. Immun.* **39**, 371–376.

Miller, C.A., Beaucage, S.L. & Cohen, S.N. (1990a) Role of DNA superhelicity in partitioning of the pSC101 plasmid. *Cell* **62**, 127–133.

Miller, H.I. (1984) Primary structure of the *himA* gene of *Escherichia coli*: homology with DNA-binding protein HU and association with the phenylalanyl-tRNA synthetase operon. *Cold Spring Harbor Symp. Quant. Biol.* **49**, 691–698.

Miller, J.F., Mekalanos, J.J. & Falkow, S. (1989a) Coordinate regulation and sensory transduction in the control of bacterial virulence. *Science* **243**, 916–922.

Miller, S.I., Kukral, A.M. & Mekalanos, J.J. (1989b) A two-component regulatory system (*phoP phoQ*) controls *Salmonella typhimurium* virulence. *Proc. Natl Acad. Sci. USA* **86**, 5054–5058.

Miller, V.L., Bliska, J.B. & Falkow, S. (1990b) Nucleotide sequence of the *Yersinia enterocolitica ail* gene and characterization of the Ail protein product. *J. Bacteriol.* **172**, 1062–1069.

Miller, V.L., DiRita, V.J. & Mekalanos, J.J. (1989c) Identification of *toxS*, a regulatory gene whose product enhances ToxR-mediated activation of the cholera toxin promoter. *J. Bacteriol.* **171**, 1288–1293.

Miller, V.L. & Falkow, S. (1988) Evidence for two genetic loci from *Yersinia enterocolitica* that can promote invasion of epithelial cells. *Infect. Immun.* **56**, 1242–1248.

Miller, V.L. & Mekalanos, J.J. (1984) Synthesis of cholera toxin is positively regulated at the transcriptional level by *toxR*. *Proc. Natl Acad. Sci. USA* **81**, 3471–3475.

Miller, V.L., Taylor, R.K. & Mekalanos, J.J. (1987) Cholera toxin transcriptional activator ToxR is a transmembrane DNA binding protein. *Cell* **48**, 271–279.

Mirkin, S.M., Lyamichev, V.I., Drumshlyak, K.N., Dobrynin, V.N., Filippov, S.A., & Frank, K.M. (1987) DNA H form requires a homopurine–homopyrimidine mirror repeat. *Nature* **330**, 495–497.

Mizuno, T., Chou, M.Y. & Inouye, M. (1984) A unique mechanism regulating gene expression: translational inhibition by a complementary RNA transcript (micRNA). *Proc. Natl Acad. Sci. USA* **81**, 1966–1970.

Mizuno, T. & Mizushima, S. (1990) Signal transduction and gene regulation through the phosphorylation of two regulatory components: the molecular basis for the osmotic regulation of the porin genes. *Mol. Microbiol.* **4**, 1077–1082.

Mizuno, T., Wurtzel, E.T. & Inouye, M. (1982) Osmoregulation of gene expression. II. DNA sequence of the *envZ* gene of the *ompB* locus of *Escherichia coli* and characterization of gene product. *J. Biol. Chem.* **257**, 13692–13698.

Mizusawa, S. & Gottesman, S. (1983) Protein degradation in *Escherichia coli*: the *lon* gene controls the stability of the SulA protein. *Proc. Natl Acad. Sci. USA* **80**, 358–362.

Mohr, C.D., Hibler, N.S. & Deretic, V. (1991a) AlgR, a response regulator controlling mucoidy in *Pseudomonas aeruginosa*, binds to the FUS sites of the *algD* promoter located unusually far upstream from the mRNA start site. *J. Bacteriol.* **173**, 5136–5143.

Mohr, C.D., Leveau, J.H.J., Kreig, D.P., Hibler, N.S. & Deretic, V. (1992) AlgR-binding sites within the *algD* promoter make up a set of inverted repeats separated by a large intervening segment of DNA. *J. Bacteriol.* **174**, 6624–6633.

Mohr, C.D., Martin, D.W., Konyecsni, W.M., Govan, J.R.W., Lory. S. & Deretic, V. (1990) Role of the far-upstream sites of the *algD* promoter and the *algR* and *rpoN* genes in environmental modulation of mucoidy in *Pseudomonas aeruginosa*. *J. Bacteriol.* **172**, 6576–6580.

Mohr, S.C., Sokolov, N.V.H.A., He, C. & Setlow, P. (1991b) Binding of small acid-soluble spore proteins from *Bacillus subtilis* changes the conformation of DNA from B to A. *Proc. Natl Acad. Sci. USA* **88**, 77–81.

Monack, D., Aricò, B., Rappouli, R. & Falkow, S. (1989) Phase variants of *Bordetella bronchiseptica* arise by spontaneous deletions in the *vir* locus. *Mol. Microbiol.* **3**, 1719–1728.

Morales, V., Bagdasarian, M.M. & Bagdasarian, M. (1990) Promiscuous plasmids of the IncQ group: mode of replication and use for gene cloning in Gram-negative bacteria. In Silver, S., Chakrabarty, A.M., Iglewski, B. & Kaplan, S. (eds) *Pseudomonas: Biotransformations, Pathogenesis and Evolving Biotechnology*, pp. 229–241. American Society for Microbiology, Washington.

Morgan, R.W., Christman, M.F., Jacobson, F.S., Storz, G. & Ames, B.N. (1986) Hydrogen peroxide-inducible proteins in *Salmonella typhimurium* overlap with heat shock and other stress proteins. *Proc. Natl Acad. Sci. USA.* **83**, 8059–8063.

Morisato, D., Way, J.C., Kim, H.-J. & Kleckner, N. (1983) Tn*10* transposase acts preferentially on nearby transposon ends in vivo. *Cell* **32**, 799–807.

Moser, D.R., Ma, D., Moser, C.D. & Campbell, J.L. (1984) *cis*-acting mutations that affect

Rop protein control of plasmid copy number. *Proc. Natl Acad. Sci. USA* **81**, 4465–4469.

Moxon, E.R. & Kroll, J.S. (1990) The role of bacterial polysaccharide capsules as virulence factors. In Jann, K. & Jann, B. (eds) *Current Topics in Microbiology and Immunology*, Vol. 150, *Bacterial Capsules* pp. 65–85. Springer-Verlag, Berlin.

Moxon, E.R. & Maskell, D. (1992) *Haemophilus influenzae* lipopolysaccharide: the biochemistry and biology of a virulence factor. In Hormaeche, C.E., Penn, C.W. & Smyth, C.J. (eds) *Molecular Biology of Bacterial Infection: Current Status and Future Perspectives.* pp. 75–96. Cambridge University Press.

Muesing, M., Tamm, J., Shepherd, H.M. & Polisky, B. (1981) A single base-pair alteration is responsible for the DNA overproduction phenotype of a plasmid copy-number mutant. *Cell* **24**, 235–242.

Mullany, P., Field, A.M., McConnell, M.M., Scotland, S.M., Smith, H.R. & Rowe, B. (1983) Expression of plasmids coding for colonization factor antigen II (CFA/II) and enterotoxin production in *Escherichia coli. J. Gen. Microbiol.* **129**, 3591–3601.

Müller, D., Hughes, C. & Goebel, W. (1983) Relationship between plasmid and chromosomal haemolysin determinants in *Escherichia coli. J. Bacteriol.* **153**, 846–851.

Mulvey, M.R. & Loewen, P.C. (1989) Nucleotide sequence of *katF* of *Escherichia coli* suggests KatF is a novel σ transcription factor. *Nucleic Acids Res.* **17**: 9979–9991.

Mulvey, M.R., Switala, J., Borys, A. & Loewen, P.C. (1990) Regulation of transcription of *katE* and *katF* in *Escherichia coli. J. Bacteriol.* **172**, 6713–6720.

Murphy, E. (1989) Transposable elements in gram-positive bacteria. In Berg, D.E. & Howe, M.M. (eds), *Mobile DNA*, pp. 269–288.

Murphy, E. & Novick, R.P. (1979) Physical mapping of *S. aureus* penicillinase plasmid pI524: characterization of an invertible region. *Mol. Gen. Genet.* **175**, 19–30.

Murphy, G.L., Connell, T.D., Barritt, D.S., Koomey, M. & Cannon, J.G. (1989) Phase variation of gonococcal protein II: regulation of gene expression by slipped strand mispairing of a repetitive DNA sequence. *Cell* **56**, 539–547.

Murray, P.J. & Young, R.A. (1992) Stress and immunological recognition in host–pathogen interactions. *J. Bacteriol.* **174**, 4193–4196.

Musser, J.M., Kroll, J.S., Moxon, E.R. & Selander, R.K. (1988) Evolutionary genetics of the encapsulated strains of *Haemophilus influenzae. Proc. Natl Acad. Sci. USA* **85**, 7758–7762.

Nadal, M., Mirambeaue, G., Forterre, P., Reiter, W.D. & Duguet, M. (1986) Positively supercoiled DNA in a virus particle of an archaebacterium. *Nature* **321**, 256–258.

Nakamura, M., Sato, S., Ohya, T., Suzuki, S. & Ikeda, S. (1985) Possible relationship of a 36-megadalton *Salmonella enteritidis* plasmid to virulence in mice. *Infect. Immun.* **47**, 831–833.

Nakashima, K., Sugiura, A., Momoi, H. & Mizuno, T. (1992) Phosphotransfer signal transduction between two regulatory factors involved in the osmoregulated *kdp* operon in *Escherichia coli. Mol. Microbiol.* **6**, 1777–1784.

Napoli, C., Gold, L. & Singer, B.S. (1981) Translational reinitiation of the rIIb cistron of bacteriophage T4. *J. Mol. Biol.* **149**, 433–449.

Narayanan, C. & Dubnau, D. (1987) Demonstration of erythromycin-dependent stalling of ribosomes on the *ermC* leader transcript. *J. Biol. Chem.* **262**, 1766–1771.

Nash, H. (1990) Bending and supercoiling of DNA at the attachment site of bacteriophage λ. *Trends Biochem. Sci.* **15**, 222–227.

Nash, H. & Granston, A.E. (1991) Similarity between the DNA-binding domains of IHF protein and TFIID protein. *Cell* **67**, 1037–1038.

Nassif, X. & Sansonetti, P.J. (1986) Correlation of the virulence of *Klebsiella pneumoniae* K1 and K2 with the presence of a plasmid encoding aerobactin. *Infect. Immun.* **54**, 603–608.

Neidhardt, F.C. (1987) Multigene systems and regulons. In Neidhardt, F.C., Ingraham, J.L., Low, K.B., Magasanik, B., Schaechter, M. & Umbarger, H.E. (eds) Escherichia coli *and* Salmonella typhimurium *Cellular and Molecular Biology*, pp. 1313–1317. American Society for Microbiology, Washington.

Neidhardt, F.C. & Van Bogelen, R.A. (1987) Heat shock response. In Neidhardt, F.C., Ingraham, J.L., Low, K.B., Magasanik, B., Schaechter, M. & Umbarger, H.E. (eds) Escherichia coli *and* Salmonella typhimurium *Cellular and Molecular Biology*, pp. 1334–1345. American Society for Microbiology, Washington.

Neilands, J.B. (1990) Molecular biology and regulation of iron acquisition by *Escherichia coli* K-12. In Iglewski, B.H. & Clark, V.L. (eds) *Molecular Basis of Bacterial Pathogenesis*, pp. 205–223, Academic Press, San Diego.

Neilands, J.B., Konopka, K., Schwyn, B., Coy, M., Francis, R.T., Paw, B.H. & Bagg, A. (1982) Microbial envelope proteins related to iron. *Annu. Rev. Microbiol.* **36**, 285–309.

Newland, J., Strockbine, N., Miller, S., O'Brien, A. & Holmes, R.K. (1985) Cloning of the shiga-like toxin structural genes of a toxin converting phage of *Escherichia coli*. *Science* **230**, 179–181.

Newman, E.B., D'Ari, R. & Lin, R.T. (1992) The leucine-Lrp regulon in *E. coli*: a global response in search of a raison d'être. *Cell* **68**, 617–619.

Ní Bhriain, N. & Dorman, C.J. (1993) Isolation and characterization of a *topA* mutant of *Shigella flexneri*. *Mol. Microbiol.* **7**, 351–358.

Ní Bhriain, N., Dorman, C.J. & Higgins, C.F. (1989) An overlap between osmotic and anaerobic stress responses: a potential role for DNA supercoiling in the coordinate regulation of gene expression. *Mol. Microbiol.* **3**, 933–942.

Nielsen, A.K., Thorsted, P., Thisted, T., Wagner, E.G.H. & Gerdes, K. (1991) The rifampicin-inducible genes *srnB* from F and *pnd* from R483 are regulated by antisense RNAs and mediate plasmid maintenance by killing of plasmid-free segregants. *Mol. Microbiol.* **5**, 1961–1973.

Nieto, J.M., Carmona, M., Bolland, S., Jubete, Y., de la Cruz, F. & Juárez, A. (1991) The *hha* gene modulates haemolysin expression in *Escherichia coli*. *Mol. Microbiol.* **5**, 1285–1293.

Nikaido, H. & Rosenberg, E.Y. (1983) Porin channels in *Escherichia coli*: studies with liposomes reconstituted from purified proteins. *J. Bacteriol.* **153**, 241–252.

Nikaido, H. & Vaara, M. (1987) Outer membrane. In Neidhardt, F.C., Ingraham, J.L., Low, K.B., Magasanik, B., Schaechter, M. & Umbarger, H.E. (eds) Escherichia coli *and* Salmonella typhimurium *Cellular and Molecular Biology*, pp. 7–22. American Society for Microbiology, Washington.

Niki, H., Jaffé, A., Imamura, R., Ogura, T. & Hiraga, S. (1991) The new gene *mukB* codes for a 177 kda protein with coiled-coil domains involved in chromosome partitioning in *Escherichia coli*. *EMBO J.* **10**, 183–193.

Nilsson, L., Vanet, A., Vijgenboom, E. & Bosch, L. (1990) The role of FIS in *trans* activation of stable RNA operons of *E. coli*. *EMBO J.* **9**, 727–734.

Ninfa, A.J. & Magasanik, B. (1986) Covalent modification of the *glnG* product, NR_I, by the *glnL* product, NR_{II}, regulates the transcription of the *glnALG* operon in *Escherichia coli*. *Proc. Natl Acad. Sci. USA* **83**, 5909–5913.

Ninfa, A.J., Ninfa, E.G., Lupas, A., Stock, A., Magasanik, B. & Stock, J. (1988) Crosstalk between bacterial chemotaxis signal transduction proteins and the regulators of transcription of the Ntr regulon: evidence that nitrogen assimilation and chemotaxis are controlled by a common phosphotransfer mechanism. *Proc. Natl Acad. Sci. USA* **85**, 5492–5496.

Ninnemann, O., Koch, C. & Kahmann, R. (1992) The *E. coli fis* promoter is subject to stringent control and autoregulation. *EMBO J.* **11**, 1075–1083.

Nishiguchi, R., Takanami, M. & Oka, A. (1987) Characterization sequence determination of the replicator origin in the hairy-root-inducing plasmid pRiA4b. *Mol. Gen. Genet.* **206**, 1–8.

Nohno, T., Noji, S., Taniguchi, S. & Saito, T. (1989) The *narX* and *narL* genes encoding the nitrate-sensing regulators of *Escherichia coli* are homologous to a family of prokaryotic two-component regulatory genes. *Nucleic Acids Res.* **17**, 2947–2957.

Nordström, K., Bernarder, R. & Dasgupta, S. (1991) The *Escherichia coli* cell cycle: one cycle or multiple independent processes that are coordinated. *Mol. Microbiol.* **5**, 769–774.

Norel, F., Pisano, M.-R., Nicoli, J. & Popoff, M.Y. (1989a) Nucleotide sequence of the plasmid-borne virulence gene *mfkA* encoding a 28 kDa polypeptide from *Salmonella typhimurium. Res. Microbiol.* (Institut Pasteur/Elsevier, Paris) **140**, 263–265.

Norel, F., Pisano, M.-R., Nicoli, J. & Popoff, M.Y. (1989b) Nucleotide sequence of the plasmid-borne virulence gene *mkfB* from *Salmonella typhimurium. Res. Microbiol.* (Institut Pasteur/Elsevier, Paris) **140**, 455–457.

Norel, F. Pisano, M.-R., Nicoli, J. & Popoff, M.Y. (1989c) A plasmid-borne virulence region (2.8 kb) from *Salmonella typhimurium* contains two open reading frames. *Res. Microbiol.* (Institut Pasteur/Elsevier, Paris) **140**, 627–630.

Nou, X., Skinner, B., Braaten, B., Blyn, L., Hirsch, D. & Low, D. (1993) Regulation of pyelonephritis-associated pili phase-variation in *Escherichia coli*: binding of the PapI and the Lrp regulatory proteins is controlled by DNA methylation. *Mol. Microbiol.* **7**, 545–553

Novick, R.P. (1976) Plasmid–protein relaxation complexes in *Staphylococcus aureus. J. Bacteriol.* **127**, 1177–1187.

Novick, R.P. (1989) Staphylococcal plasmids and their replication. *Annu. Rev. Microbiol.* **43**, 537–566.

Novick, R.P. (1991) Genetic systems in Staphylococci. *Methods Enzymol.* **204**, 587–636.

Nunoshiba, T., Hidalgo, E., Amábile-Cuevas, C.F. & Demple, B. (1992) Two-stage control of an oxidative stress regulon: the *Escherichia coli* SoxR protein triggers redox-inducible expression of the *soxS* regulatory gene. *J. Bacteriol.* **174**, 6054–6060.

Nur, I., LeBlanc, J. & Tully, J.G. (1987) Short, interspersed and repetitive DNA sequences in *Spiroplasma* species. *Plasmid* **17**, 110–116.

Nyman, K., Nakamura, K., Ohtsubo, H. & Ohtsubo, E. (1981) Description of the insertion sequence IS*1* in gram negative bacteria. *Nature* **289**, 609–612.

Obar, R.A., Collins, C.A., Hammarback, J.A., Shpetner, H.S. & Vallee, R.B. (1990) Molecular cloning of the microtubule-associated mechanochemical enzyme dynamin reveals homology with a new family of GTP-binding proteins. *Nature* **347**, 256–261.

O'Brien, A.D., LaVeck, G.D., Thompson, M.R. & Formal, S.B. (1982) Production of a *Shigella dysenteriae* 1-like cytotoxin by *Escherichia coli. J. Infect. Dis.* **146**, 763–769.

O'Byrne, C., Ní Bhriain, N. & Dorman, C.J. (1992) The DNA supercoiling-sensitive expression of the *Salmonella typhimurium his* operon requires the *his* attenuator and is modulated by anaerobiosis and by osmolarity. *Mol. Microbiol.* **6**, 2467–2476.

Ochman, H. & Wilson, A.C. (1987) Evolutionary history of enteric bacteria. In Neidhardt, F.C., Ingraham, J.L., Low, K.B., Magasanik, B., Schaecter, M. & Umbarger, H.E. (eds), Escherichia coli *and* Salmonella typhimurium *Cellular and Molecular Biology*, pp. 1649–1654. American Society for Microbiology, Washington.

Ogden, G.B., Pratt, M.J. & Schaechter, M. (1988) The replicative origin of the *E. coli* chromosome binds to cell membrane only when fully methylated. *Cell* **54**, 127–135.

Ogierman, M.A. & Manning, P.A. (1992) Homology of TcpN, a putative regulatory protein of *Vibrio cholerae*, to the AraC family of transcriptional activators. *Gene* **116**, 93–97.

Ogura, T., Niki, H., Kano, Y., Imamoto, F. & Hiraga, S. (1990a) Maintenance of plasmids in HU and IHF mutants of *Escherichia coli. Mol. Gen. Genet.* **220**, 197–203.

Ogura, T., Niki, H., Mori, H., Morita, M., Hasegawa, M., Ichinose, C. & Hiraga, S. (1990b) Identification and characterization of *gyrB* mutants of *Escherichia coli* that are defective in partitioning of mini-F plasmids. *J. Bacteriol.* **172**, 103–112.

Ohman, D.E., West, M.A., Flynn, J.L. & Goldberg, J.B. (1985) Method for gene replacement in *Pseudomonas aeruginosa* used in construction of *recA* mutants: *recA*-independent instability of alginate production. *J. Bacteriol.* **162**, 1068–1074.

Ohnishi, K., Kutsukake, K., Suzuki, H. & Iino,T. (1992) A novel transcriptional regulation mechanism in the flagellar regulon of *Salmonella typhimurium*: an anti-sigma factor inhibits the activity of the flagellum-specific sigma factor, σ^F. *Mol. Microbiol.* **6**, 3149–3157.

Okada, N., Geist, R.T. & Caparon, M.G. (1993) Positive transcriptional control of *mry* regulates virulence in the group A streptococcus. *Mol. Microbiol.* **7**, 893–903.

Okamoto, K. & Freundlich, M. (1986) Mechanism for the autogenous control of the *crp* operon: transcriptional inhibition by a divergent RNA transcript. *Proc. Natl Acad. Sci. USA* **83**, 5000–5004.

Olsén, A., Arnqvist, A., Sukupolvi, S. & Normark, S. (1993) The RpoS sigma factor relieves H-NS-mediated repression of *csgA*, the subunit gene of fibronectin-binding curli in *Escherichia coli*. *Mol. Microbiol.* **7**, 523–536.

Olsén, A., Jonsson, A. & Normark, S. (1989) Fibronectin binding mediated by a novel class of surface organelles on *Escherichia coli*. *Nature* **338**, 652–655.

Olson, E.R. & Chung, S.T. (1988) Transposon Tn4556 of *Streptomyces fradiae*: sequence of the ends and the target sites. *J. Bacteriol.* **170**, 1955–1957.

Oppenheim, D.S. & Yanofsky, C. (1980) Translational coupling during expression of the tryptophan operon of *Escherichia coli*. *Genetics* **95**, 783–795.

Ortega-Barria, E. & Pereira, M.E.A. (1991) A novel *T. cruzi* heparin-binding protein promotes fibroblast adhesion and penetration of engineered bacteria and trypanosomes into mammalian cells. *Cell* **67**, 411–421.

O'Toole, P. & Foster, T.J. (1986) Molecular cloning and expression of the epidermolytic toxin A gene of *Staphylococcus aureus*. *Microb. Pathogen.* **1**, 583–594.

O'Toole, P. & Foster, T.J. (1987) Nucleotide sequence of the epidermolytic toxin A gene of *Staphylococcus aureus*. *J. Bacteriol.* **169**, 3910–3915.

O'Toole, P., Stenberg, L., Rissler, M. & Lindahl, G. (1992) Two major classes in the M protein family of group A streptococci. *Proc. Natl Acad. Sci. USA* **89**, 8661–8665.

Otridge, J. & Gollnick, P. (1993) MtrB from *Bacillus subtilus* binds specifically to *trp* leader RNA in a tryptophan-dependent manner. *Proc. Natl Acad. Sci. USA* **90**, 128–132.

Ottemann, K.M., DiRita, V.J. & Mekalanos, J.J. (1992) ToxR proteins with substitutions in residues conserved with OmpR fail to activate transcription from the cholera toxin promoter. *J. Bacteriol.* **174**, 6807–6814.

Otto, B.R., Verweij, A.M.J.J. & MacLaren, D.M. (1992) Transferrins and haeme-compounds as iron sources for pathogenic bacteria. *Crit. Rev. Microbiol.* **18**, 217–233.

Oudega, B., Oldenziel-Werner, W.J.M., Klaasen-Boor, P., Rezee, A., Glas, J. & DeGraaf, F.K. (1979) Purification and characterization of cloacin DF13 receptor from *Enterobacter cloacae* and its interaction with cloacin DF13 *in vitro*. *J. Bacteriol.* **138**, 7–16.

Overdier, D.G. & Csonka, L.N. (1992) A transcriptional silencer downstream of the promoter in the osmotically controlled *proU* operon of *Salmonella typhimurium*. *Proc. Natl Acad. Sci. USA* **89**, 3140–3144.

Overdier, D.G., Olson, E.R., Erickson, B.D., Ederer, M.M. & Csonka, L.N. (1989) Nucleotide sequence of the transcriptional control region of the osmotically regulated *proU* operon of *Salmonella typhimurium* and identification of the 5' end of the *proU* mRNA. *J. Bacteriol.* **171**, 4694–4706.

Owen-Hughes, T.A., Pavitt, G.D., Santos, D.S., Sidebotham, J.M., Hulton, C.S.J., Hinton, J.C.D. & Higgins, C.F. (1992) The chromatin-associated protein H-NS interacts with curved DNA to influence DNA topology and gene expression. *Cell* **71**, 1–20.

Palecek, E. (1991) Local supercoil-stabilized DNA structures. *Crit. Rev. Biochem. Mol. Biol.*, **26**, 151–226.

Pappenheimer, A.M., Jr. (1977) Diphtheria toxin. *Annu. Rev. Biochem.* **46**, 69–94.

Parker, L.L. & Hall, B.G. (1990) Characterization and nucleotide sequence of the cryptic *cel* operon of *E. coli* K-12. *Genetics* **124**, 455–471.

Parsot, C. & Mekalanos, J.J. (1991a) Expression of the *Vibrio cholerae* gene encoding aldehyde dehydrogenase is under control of ToxR, the cholera toxin transcriptional activator. *J. Bacteriol.* **173**, 2842–2851.

Parsot, C. & Mekalanos, J.J. (1991b) Expression of ToxR, the transcriptional activator of the virulence factors in *Vibrio cholerae*, is modulated by the heat shock response. *Proc. Natl Acad. Sci. USA* **87**, 9898–9902.

Parsot, C. & Mekalanos, J.J. (1992) Structural analysis of the *acfA* and *acfD* genes of *Vibrio cholerae*: effects of DNA topology and transcriptional activators on expression. *J. Bacteriol.* **174**, 5211–5218.

Passador, L., Cook, J.M., Gambello, M.J., Rust, L. & Iglewski, B.H. (1993) Expression of *Pseudomonas aeruginosa* virulence genes requires cell-to-cell communication. *Science* **260**, 1127–1130.

Pato, M.L. (1989) Bacteriophage Mu. In Berg, D.E. & Howe, M.M. (eds), *Mobile DNA*, pp. 23–52. American Society for Microbiology, Washington.

Paul, K., Ghosh, S.K. & Das, J. (1986) Cloning and expression in *Escherichia coli* of a *recA*-like gene from *Vibrio cholerae*. *Mol. Gen. Genet.* **203**, 58–63.

Payne, S.M. & Lawlor, K.M. (1990) Molecular studies on iron acquisition by non-*Escherichia coli* species. In Iglewski, B.H. & Clark, V.L. (eds) *Molecular Basis of Bacterial Pathogenesis*, pp. 225–248. Academic Press, San Diego.

Pearson, G.D.N. & Mekalanos, J.J. (1982) Molecular cloning of the *Vibrio cholerae* enterotoxin genes in *Escherichia coli* K-12. *Proc. Natl Acad. Sci. USA* **79**, 2976–2980.

Peden, K.W.C. (1983) Revised sequence of the tetracycline resistance gene of pBR322. *Gene* **22**, 277–280.

Penaranda, M.E., Evans, D.G., Murray, B.E. & Evans, J. Jr (1983) ST:LT:CFA/II plasmids in enterotoxigenic *Escherichia coli* belonging to serogroups O6, O8, O85 and O139. *J. Bacteriol.* **154**, 980–983.

Peng, H.-L., Novick, R.P., Kreiswirth, B., Kornblum, J. & Schlievert, P. (1988) Cloning, characterization and sequencing of an accessory gene regulator (*agr*) in *Staphylococcus aureus*. *J. Bacteriol.* **170**, 4365–4372.

Peralta, E.G., Hellmiss, R. & Ream, W. (1986) *Overdrive*, a T-DNA transmission enhancer on the *Agrobacterium tumefaciens* tumour-inducing plasmid. *EMBO J.* **5**, 1137–1142.

Perez-Casal, J., Caparon, M.G. & Scott, J.R. (1991) Mry, a *trans*-acting positive regulator of the M protein gene of *Streptococcus pyogenes* with similarity to the receptor proteins of two-component regulatory systems. *J. Bacteriol.* **173**, 2617–2624.

Perez-Casal, J.F. & Crosa, J.H. (1984) Aerobactin iron uptake sequences in plasmid ColV-K30 are flanked by inverted IS*1*-like elements and replication regions. *J. Bacteriol.* **160**, 256–265.

Perry, R.D. & Brubaker, R.R. (1979) Accumulation of iron by *Yersiniae*. *J. Bacteriol.* **137**, 1290–1298.

Perry, R.D. & Brubaker, R.R. (1987) Transport of Ca^{2+} by *Yersinia pestis*. *J. Bacteriol.* **169**, 4861–4864.

Petes, T.D. & Hill, C.W. (1988) Recombination between repeated genes in microorganisms. *Annu. Rev. Genet.* **22**, 147–168.

Pettijohn, D.E. (1982) Structure and properties of the bacterial nucleoid. *Cell* **30**, 667–669.

Pettijohn, D.E. (1988) Histonelike proteins and bacterial chromosome structure. *J. Biol. Chem.* **263**, 12793–12796.

Pettijohn, D.E. & Pfenninger, O. (1980) Supercoils in prokaryotic DNA restrained in vivo. *Proc. Natl Acad. Sci. USA* **77**, 1331–1335.

Philips, G.J., Prasher, D.C. & Kushner, S.R. (1988) Physical and biochemical characterization of cloned *sbcB* and *xonA* mutations from *Escherichia coli*. *J. Bacteriol.* **170**, 2089–2094.

Philips, S. & Novick, R. (1979) Tn554: a repressible site-specific transposon in *Staphyloccus aureus*. *Nature* **278**, 476–478.

Pickens, R.N., Mazaitis, A.J., Saadi, S. & Maas, W.K. (1984) Characterization of the basic replicons of the chimeric R/Ent plasmid pCG86 and the related Ent plasmid P307. *Plasmid* **12**, 10–18.

Pierre, A. & Paoletti, C. (1983) Purification and characterization of RecA protein from *Salmonella typhimurium*. *J. Biol. Chem.* **258**, 2870–2874.

Pierson, D. & Falkow, S. (1990) Non-pathogenic isolates of *Yersinia enterocolitica* do not contain functional *inv*-homologous sequences. *Infect. Immun.* **58**, 1059–1064.

Piper, K.R., Beck von Bodman, S. & Farrand, S.K. (1993) Conjugation factor of *Agrobacterium tumefaciens* regulates Ti plasmid transfer by autoinduction. *Nature* **362**, 448–450.

Pirhonen, M., Flego, D., Heikinheimo, R. & Palva, E.T. (1993) A small diffusible signal molecule is responsible for the global control of virulence and exoenzyme production in the plant pathogen *Erwinia carotovora*. *EMBO J.* **12**, 2467–2476.

Plamann, L.S. & Stauffer, G.V. (1987) Nucleotide sequence of the *Salmonella typhimurium metR metE* control region. *J. Bacteriol.* **169**, 3932–3937.

Plasterk, R.H.A. (1991) Frameshift control of IS*1* transposition. *Trends Genet.* **7**, 203–204.

Plasterk, R.H.A. (1992) Genetic switches: mechanism and function. *Trends Genet.* **8**, 403–406.

Plasterk, R.H.A., Brinkman, A. & van de Putte, P. (1983) DNA inversions in the chromosome of *E. coli* and in bacteriophage Mu: relationship to other site-specific recombination systems. *Proc. Natl Acad. Sci. USA* **80**, 5355–5358.

Plasterk, R.H.A., Simon, M.I. & Barbour, A.G. (1985) Transposition of structural genes to an expression sequence on a linear plasmid causes antigenic variation in the bacterium *Borrelia hermsii*. *Nature* **318**, 257–263.

Plasterk, R.H.A. & van de Putte, P. (1985) The invertible P-DNA segment in the chromosome of *Escherichia coli*. *EMBO J.* **4**, 237–242.

Plaston, R.R. & Wartell, R.M. (1987) Sequence distributions associated with DNA curvature are found upstream of strong *E. coli* promoters. *Nucleic Acids Res.* **15**, 785–796.

Podbielski, A., Peterson, J.A. & Cleary, P. (1992) Surface protein–CAT reporter fusions demonstrate differential gene expression in the *vir* regulon of *Streptococcus pyogenes*. *Mol. Microbiol.* **6**, 2253–2265.

Polarek, J.W., Williams, G. & Epstein, W. (1992) The products of the *kdpDE* operon are required for expression of the Kdp ATPase of *Escherichia coli*. *J. Bacteriol.* **174**, 2145–2151.

Pon, C.L., Calogero, R.A. & Gualerzi, C.O. (1988) Identification, cloning, nucleotide sequence and chromosomal map location of *hns*, the structural gene for *Escherichia coli* DNA-binding protein H-NS. *Mol. Gen. Genet.* **212**, 199–202.

Pontiggia, A., Negri, A., Beltrame, M. & Bianchi, M.E. (1993) Protein HU binds specifically to kinked DNA. *Mol. Microbiol.* **7**, 343–350.

Poole, K. & Braun, V. (1988) Iron regulation of *Serratia marcescens* haemolysin gene expression. *Infect. Immun.* **56**, 2967–2971.

Portnoy, D.A., Chakraborty, T., Goebel, W. & Cossart, P. (1992) Molecular determinants of *Listeria monocytogenes* pathogenesis. *Infect. Immun.* **60**, 1263–1267.

Portnoy, D.A. & Falkow, S. (1981) Virulence-associated plasmid from *Yersinia enterocolitica* and *Yersinia pestis*. *J. Bacteriol.* **148**, 877–883.

Postle, K. (1990) TonB and the Gram-negative dilemma. *Mol. Microbiol.* **4**, 2019–2025.

Poulsen, L.K., Refn, A., Molin, S. & Andersson, P. (1991) The *gef* gene from *Escherichia coli* is regulated at the level of translation. *Mol. Microbiol.* **5**, 1639–1648.

Pradel, E. & Schnaitman, C.A. (1991) Effect of *rfaH* (*sfrB*) and temperature on expression of *rfa* genes of *Escherichia coli* K-12. *J. Bacteriol.* **173**, 6428–6431.

Prince, R.W., Cox, C.D. & Vasil, M.L. (1993) Coordinate regulation of siderophore and exotoxin A production: molecular cloning and sequencing of the *Pseudomonas aeruginosa fur* gene. *J. Bacteriol.* **175**, 2589–2598.

Prince, R.W., Storey, D.G., Vasil, A.I. & Vasil, M.L. (1991) Regulation of *toxA* and *regA* by the *Escherichia coli fur* gene and identification of a Fur homologue in *Pseudomonas aeruginosa* PA103 and PA01. *Mol. Microbiol.* **5**, 2823–2831.

Projan, S.J. & Archer, G.L. (1989) Mobilization of relaxable *Staphylococcus aureus* plasmid pC221 by the conjugative plasmid pG01 involves three pC221 loci. *J. Bacteriol.* **171**, 1841–1845.

Projan, S.J. & Novick, R.P. (1986) Incompatibility between plasmids with independent copy control. *Mol. Gen. Genet.* **204**, 341–348.

Provence, D.L. & Curtiss III, R. (1992) Role of *crl* in avian pathogenic *Escherichia coli*: a knockout mutation of *crl* does not affect haemaglutination activity, fibronectin binding or curli production. *Infect. Immun.* **60**, 4460–4467.

Pruss, G.J. & Drlica, K. (1986) Topoisomerase I mutants: the gene on pBR322 that encodes resistance to tetracycline affects plasmid DNA supercoiling. *Proc. Natl Acad. Sci. USA* **83**, 8952–8956.

Pruss, G.J. & Drlica, K. (1989) DNA supercoiling and prokaryotic transcription. *Cell* **56**, 521–523.

Pruss, G.J., Manes, S.H. & Drlica, K. (1982) *Escherichia coli* DNA topoisomerase mutants: increased supercoiling is corrected by mutations near gyrase genes. *Cell* **31**, 35–42.

Ptashne, M. (1986) Gene regulation by proteins acting nearby and at a distance. *Nature* **322**, 697–701.

Pulkkinen, W.S. & Miller, S.I. (1991) A *Salmonella typhimurium* virulence protein is similar to a *Yersinia enterocolitica* invasion protein and a bacteriophage lambda outer membrane protein. *J. Bacteriol.* **173**, 86–93.

Pullinger, G.D., Baird, G.D., Williamson, C.M. & Lax, A.J. (1989) Nucleotide sequence of a plasmid gene involved in the virulence of *Salmonellas*. *Nucleic Acids Res.* **17**, 7983.

Quackenbush, R.L. & Falkow, S. (1979) Relationship between colicin V activity in *Escherichia coli*. *Infect. Immun.* **24**, 562–564.

Rahme, L.G., Mindrinos, M.N. & Panopoulos, N.J. (1992) Plant and environmental sensory signals control the expression of *hrp* genes in *Pseudomonas syringae* pv. phaseolicola. *J. Bacteriol.* **174**, 3499–3507.

Rahmouni, A.R. & Wells, R.D. (1989) Stabilization of Z-DNA *in vivo* by localized supercoiling. *Science* **246**, 358–363.

Raji, A., Zabel, D.J., Laufer, C.S. & Depew, R.E. (1985) Genetic analysis of mutations that compensate for loss of *Escherichia coli* DNA topoisomerase I. *J. Bacteriol.* **162**, 1173–1179.

Ramos, J.L., Rojo, F., Zhou, L. & Timmis, K.N. (1990) A family of positive regulators related to the *Pseudomonas putida* TOL plasmid XylS and the *Escherichia coli* AraC activators. *Nucleic Acids Res.* **18**, 2149–2152.

Rayssiguier, C., Thaler, D.S. & Radman, M. (1989) The barrier to recombination between *Escherichia coli* and *Salmonella typhimurium* is disrupted in mismatch–repair mutants. *Nature*, **342**, 396–401.

Razin, A. & Cedar, H. (1991) DNA methylation and gene expression. *Microbiol. Rev.* **55**, 451–458.

Recsei, P., Kreiswirth, B., O'Reilly, M., Schlievert, P., Gruss, A. & Novick, R. (1985) Regulation of exoprotein gene expression by *agr*. *Mol. Gen. Genet.* **202**, 58–61.

Rees, C.E.D. & Wilkins, B.M. (1990) Protein transfer into the recipient cell during bacterial conjugation: studies with F and RP4. *Mol. Microbiol.* **4**, 1199–1205.

Regassa, L.B. & Betley, M.J. (1992) Alkaline pH decreases expression of the accessory gene regulator (*agr*) in *Staphylococcus aureus*. *J. Bacteriol.* **174**, 5095–5100.

Restrepo, B.I., Kitten, T., Carter, C.J., Infante, D. & Barbour, A.G. (1992) Subtelomeric expression regions of *Borrelia hermsii* linear plasmids are highly polymorphic. *Mol. Microbiol.* **6**, 3299–3311.

Reznikoff, W.S. (1992a) Catabolite gene activator protein activation of *lac* transcription. *J. Bacteriol.* **174**, 655–658.

Reznikoff, W.S. (1992b) The lactose operon-controlling elements: a complex paradigm. *Mol. Microbiol.* **6**, 2419–2422.

Rhen, M. & Sukupolvi, S. (1988) The role of the *traT* gene of the *Salmonella typhimurium* virulence plasmid for serum resistance and growth within liver macrophages. *Microb. Pathogen.* **5**, 275–285.

Riccio, A., Bruin, C.B., Rosenberg, M., Gottesman, M., McKenny, K. & Blasi, F. (1985) Regulation of single and multicopy *his* operons in *Escherichia coli*. *J. Bacteriol.* **163**, 1172–1179.

Richardson, S.M.H., Higgins, C.F. & Lilley, D.M.J. (1984) The genetic control of DNA supercoiling in *Salmonella typhimurium*. *EMBO J.* **3**, 1745–1752.

Richardson, S.M.H., Higgins, C.F. & Lilley, D.M.J. (1988) DNA supercoiling and the *leu500* promoter of *Salmonella typhimurium*. *EMBO J.* **7**, 1863–1869.

Richet, E. & Raibaud, O. (1991) Supercoiling is essential for the formation and stability of the initiation complex at the divergent *malEp* and *malKp* promoters. *J. Mol. Biol.* **218**, 529–542.

Richet, E., Vidal-Ingigliardi, D. & Raibaud, O. (1991) A new mechanism for activation of transcription initiation: repositioning of an activator triggered by the binding of a second activator. *Cell*, **66**, 1185–1195.

Rimsky, S. & Spassky, A. (1990) Sequence determinants for H1 binding on *Escherichia coli lac* and *gal* promoters. *Biochemistry* **29**, 3765–3771.

Rioux, C.R., Friedrich, M.J. & Kadner, R.J. (1990) Genes on the 90-kilobase plasmid of *Salmonella typhimurium* confer low-affinity cobalamin transport: relationship to fimbria biosynthesis genes. *J. Bacteriol.* **172**, 6217–6222.

Ritossa, F. (1962) A new puffing pattern induced by temperature and DNP in *Drosophila*. *Experimentia* **18**, 571–573.

Roberts, D.E., Hoopes, B.C., McClure, W.R. & Kleckner, N. (1985) IS*10* transposition is regulated by DNA adenine methylation. *Cell* **43**, 117–130.

Roberts, D. & Kleckner, N. (1988) Tn*10* transposition promotes RecA-dependent induction of a λ prophage. *Proc. Natl Acad. Sci. USA* **85**, 6037–6041.

Roberts, I.S. & Coleman, M.J. (1991) The virulence of *Erwinia amylovora*: molecular genetic perspectives. *J. Gen. Microbiol.* **137**, 1453–1457.

Robertson, B.D. & Meyer, T.F. (1992) Genetic variation in pathogenic bacteria. *Trends Genet.* **8**, 422–427.

Ronson, C.W., Nixon, B.T. & Ausubel, F.M. (1987) Conserved domains in bacterial regulatory proteins that respond to environmental stimuli. *Cell* **49**, 579–581.

Roosendaal, B., Boots, M. & de Graaf, F.K. (1987) Two novel genes, *fanA* and *fanB*, involved in the biogenesis of K99 fimbriae. *Nucleic Acids Res.* **15**, 5937–5984.

Rosa, P.A., Schwan, T. & Hogan, D. (1992) Recombination between genes encoding major outer surface proteins A and B of *Borrelia burgdorferi*. *Mol. Microbiol.* **6**, 3031–3040.

Rosenshine, I. & Finlay, B.B. (1993) Exploitation of host signal transduction pathways and cytoskeletal functions by invasive bacteria. *BioEssays* **15**, 17–24.

Rosqvist, R., Skurnik, M. & Wolf-Watz, H. (1988) Increased virulence of *Yersinia pseudotuberculosis* by two independent mutations. *Nature* **334**, 522–525.

Ross, B.C., Raios, K., Jackson, K. & Dwyer, B. (1992) Molecular cloning of a highly repeated DNA element from *Mycobacterium tuberculosis* and its use as an epidemiological tool. *J. Clin. Microbiol.* **30**, 942–946.

Rothstein, S.J. & Reznikoff, W.S. (1981) The functional differences in the inverted repeats of Tn5 are caused by a single base pair nonhomology. *Cell* **23**, 191–199.

Roudier, C., Fierer, J. & Guiney, D.G. (1992) Characterization of translation termination mutations in the *spv* operon of the *Salmonella* virulence plasmid pSDL2. *J. Bacteriol.* **174**, 6418–6423.

Roudier, C., Krause, M., Fierer, J. & Guiney, D.G. (1990) Correlation between the presence of sequences homologous to the *vir* region of *Salmonella dublin* plasmid pSDL2 and the virulence of twenty-two *Salmonella* serotypes in mice. *Infect. Immun.* **58**, 1180–1185.

Rouvière-Yaniv, J., Bonnefoy, E., Huisman, O. & Almeida, A. (1990) Regulation of HU protein synthesis in *Escherichia coli*. In Drlica, K. & Riley, M. (eds), *The Bacterial Chromosome*, pp. 247–257. American Society for Microbiology, Washington.

Rouvière-Yaniv, J., Germond, J. & Yaniv, M. (1979) *E. coli* DNA binding protein HU forms nucleosome-like structures with circular double-stranded DNA. *Cell* **17**, 265–274.

Rowland, S.-J. & Dyke, K.G.H. (1988) A DNA invertase from *Staphylococcus aureus* is a member of the Hin family of site-specific recombinases. *FEMS Microbiol. Lett.* **50**, 253–258.

Roy, C.R., Miller, J.F. & Falkow, S. (1989) The *bvgA* gene of *Bordetella pertussis* encodes a transcriptional activator required for coordinate regulation of several virulence genes. *J. Bacteriol.* **171**, 6338–6344.

Roy, C.R., Miller, J.F. & Falkow, S. (1990) Autogenous regulation of the *Bordetella pertussis bvgABC* operon. *Proc. Natl Acad. Sci. USA* **87**, 3763–3767.

Rozsa, F.W. & Marrs, C.F. (1991) Interesting sequence differences between the pilin gene inversion regions of *Moraxella lacunata* ATCC 17956 and *Moraxella bovis* Epp63. *J. Bacteriol.* **173**, 4000–4006.

Rudd, K.E. (1992) Alignment of *E. coli* DNA sequences to a revised, integrated genomic restriction map. In Miller, J.H. (ed) *A Short Course in Bacterial Genetics: A Laboratory Manual and Handbook for* Escherichia coli *and Related Bacteria*, pp. 2.3–2.43. Cold Spring Harbor Laboratory Press, Cold Spring Harbor.

Rudd, K.E. & Menzel, R. (1987) *his* operons of *Escherichia coli* and *Salmonella typhimurium* are regulated by DNA supercoiling. *Proc. Natl Acad. Sci. USA* **84**, 517–521.

Ruhfel, R.E., Manias, D.A. & Dunny, G.M. (1993) Cloning and characterisation of a region of the *Enterococcus faecalis* conjugative plasmid, pCF10, encoding a sex pheromone-binding function. *J. Bacteriol.* **175**, 5253–5259.

Saadi, S., Maas, W.K., Hill, D.F. & Bergquist, P.L. (1987) Nucleotide sequence analysis of RepFIC, a basic replicon present in IncFI plasmids P307 and F, and its relation to the RepA replicon of IncFII plasmids. *J. Bacteriol.* **169**, 1836–1846.

Sadowsky, A.B., Davidson, A., Lin, R.-J. & Hill, C.W. (1989) *rhs* gene family of *Escherichia coli* K-12. *J. Bacteriol.* **171**, 636–642.

Sadowsky, P. (1986) Site-specific recombinases: changing partners and doing the twist. *J. Bacteriol.* **165**, 341–347.

Saito, T., Duly, D. & Williams, R.P.J. (1991) The histidines of the iron-uptake regulation protein, Fur. *Eur. J. Biochem.* **197**, 39–42.

Saito, T. & Williams, R.P.J. (1991) The binding of the ferric uptake regulation protein to a DNA fragment. *Eur. J. Biochem.* **197**, 43–47.

Sak, B.D., Eisenstark, A. & Touti, D. (1989) Exonuclease III and the catalase hydroperoxidase II in *Escherichia coli* are both regulated by the *katF* gene product. *Proc. Natl Acad. Sci. USA* **86**, 3271–3275.

Sakai, T., Sasakawa, C., Makino, S. & Yoshikawa, M. (1986) DNA sequence and product analysis of the *virF* locus responsible for Congo red binding and cell invasion in *Shigella flexneri* 2a. *Infect. Immun.* **54**, 395–402.

Sako, T., Sawaki, S., Sakurai, T., Ito, S., Yoshizawa, Y. & Kondo, I. (1983) Cloning and expression of the staphylokinase gene of *Staphylococcus aureus* in *Escherichia coli. Mol. Gen. Genet.* **90**, 271–277.

Sanders, D.A., Gillece-Castro, B.L., Burlingame, A.L. & Koshland, D.E. (1992) Phosphorylation site of NtrC, a protein phosphatase whose covalent intermediate activates transcription. *J. Bacteriol.* **174**, 5117–5122.

Sanderson, K.E. & Roth, J.R. (1988) Linkage map of *Salmonella typhimurium* Edition VII. *Microbiol. Rev.* **52**, 485–532.

Sansonetti, P.J. (1992) Molecular and cellular biology of epithelial invasion by *Shigella flexneri* and other enteroinvasive pathogens. In Hormaeche, C.E., Penn, C.W. & Smyth, C.J. (eds) *Molecular Biology of Bacterial Infection: Current Status and Future Perspectives*, pp. 47–60. Cambridge University Press.

Sansonetti, P.J., d'Hauteville, H., Ecobichon, C. & Pourcel, C. (1983) Molecular comparison of virulence plasmids in *Shigella* and enteroinvasive *Escherichia coli. Ann. Microbiol.* (Paris) **134A**, 295–318.

Sansonetti, P.J., Kopecko, D.J. & Formal, S.B. (1981) *Shigella sonnei* plasmids: evidence that a large plasmid is necessary for virulence. *Infect. Immun.* **34**, 75–83.

Sansonetti, P.J., Kopecko, D.J. & Formal, S.B. (1982) Involvement of a plasmid in the invasive ability of *Shigella flexneri. Infect. Immun.* **35**, 852–860.

Sansonetti, P.J., Ryter, A., Clerc, P., Maurelli, A.T. & Mourier, J. (1986) Multiplication of *Shigella flexneri* within HeLa cells: lysis of the phagocytic vacuole and plasmid mediated contact hemolysis. *Infect. Immun.* **51**, 461–469.

Santero, E., Hoover, T.R., North, A.K., Berger, D.K., Porter, S.C. & Kustu, S. (1992) Role of integration host factor in stimulating transcription from the σ^{54}-dependent *nifH* promoter. *J. Mol. Biol.* **227**, 602–620.

Sanzey, B. (1979) Modulation of gene expression by drugs affecting deoxyribonucleic acid gyrase. *J. Bacteriol.* **138**, 40–47.

Sapienza, C. & Doolittle, W. F. (1982) Unusual physical organization of the *Halobacterium* genome. *Nature* **295**, 384–389.

Sarma, V. & Reeves, P. (1977) Genetic locus (*ompB*) affecting a major outer membrane protein in *Escherichia coli* K-12. *J. Bacteriol.* **132**, 23–27.

Sasakawa, C., Uno, Y. & Yoshikawa, M. (1987) *lon-sulA* regulatory function affects the efficiency of transposition of Tn5 from l *b*221 *c*1857 Pam Oam to the chromosome. *Biochem. Biophys. Res. Commun.* **142**, 879–884.

Saunders, J.R. (1989) The molecular basis of antigenic variation in pathogenic *Neisseria*. In Hopwood, D.A. & Chater, K.F. (eds) *Genetics of Bacterial Diversity*, pp. 267–285. Academic Press, London.

Savelkoul, P.H.M., Willshaw, G.A., McConnell, M.M., Smith, H.R., Hamers, A.M., van der Zeijst, B.A.M. & Gaastra, W. (1990) Expression of CFA/I fimbriae is positively regulated. *Microb. Pathogen.* **8**, 91–99.

Savic, D.J., Romac, S.P. & Ehrlich, S.D. (1983) Inversion in the lactose region of the *E. coli* K-12: inversion termini map within IS*3* elements α*3*β*3* and β5α5. *J. Bacteriol.* **155**, 943–946.

Sawers, R. G. (1991) Identification and molecular characterization of a transcriptional regulator from *Pseudomonas aeruginosa* PAO1 exhibiting structural and functional similarity to the Fnr protein of *Escherichia coli. Mol. Microbiol.* **5**, 1469–1481.

Scarlato, V., Aricò, B., Prugnola, A. & Rapouli, R. (1991) Sequential activation and environmental regulation of virulence genes in *Bordetella pertussis*. *EMBO J.* **10**, 3971–3975.

Scarlato, V., Aricò, B. & Rappuoli, R. (1993) DNA topology affects transcriptional regulation of the pertussis toxin gene of *Bordetella pertussis* in *Escherichia coli* and *in vitro*. *J. Bacteriol.* **175**, 4764–4771.

Scarlato, V., Prugnola, A., Aricò, B. & Rappuoli, R. (1990) Positive transcriptional feedback at the *bvg* locus controls expression of virulence factors in *Bordetella pertussis*. *Proc. Natl Acad. Sci. USA* **87**, 6753–6757.

Schäffer, S., Hantke, K. & Braun, V. (1985) Nucleotide sequence of the iron regulatory gene *fur*. *Mol. Gen. Genet.* **200**, 110–113.

Schell, M.A. & Sukordhaman, M. (1989) Evidence that the transcription activator encoded by the *Pseudomonas putida nahR* gene is evolutionarily related to the transcription activators encoded by the *Rhizobium nodD* genes. *J. Bacteriol.* **171**, 1952–1959.

Schellhorn, H.E. & Stones, V.L. (1992) Regulation of *katF* and *katE* in *Escherichia coli* K-12 by weak acids. *J. Bacteriol.* **174**, 4769–4776.

Schiemann, D.A. & Shope, S.R. (1991) Anaerobic growth of *Salmonella typhimurium* results in increased uptake by henle 407 epithelial and mouse peritoneal cells *in vitro* and repression of a major outer membrane protein. *Infect. Immun.* **59**, 437–440.

Schlagman, S., Hattman, S., May, M.S. & Berger, L. (1976) *In vivo* methylation of *Escherichia coli* K-12 *mec*$^+$ deoxyribonucleic acid-cytosine methylase protects against in vitro cleavage by the RII restriction endonuclease (R.*Eco*RII). *J. Bacteriol.* **126**, 990–996.

Schleif, R. (1992) DNA looping. *Annu. Rev. Biochem.* **61**, 199–223.

Schmid, M.B. (1990) More than just 'histonelike' proteins. *Cell* **63**, 451–453.

Schmid, M.B. & Roth, J.R. (1987) Gene location affects expression level in *Salmonella typhimurium*. *J. Bacteriol.* **169**, 2872–2875.

Schmitt, M.P. & Holmes, R.K. (1991) Iron-dependent regulation of diphtheria toxin and siderophore expression by the cloned *Corynebacterium diphtheriae* repressor gene *dtxR* in *C. diphtheriae* C7 strains. *Infect. Immun.* **59**, 1899–1904.

Schmitt, M.P. & Holmes, R.K. (1993) Analysis of diphtheria toxin repressor–operator interactions and characterisation of a mutant repressor with decreased binding activity for divalent metals. *Mol. Microbiol.* **9**, 173–181.

Schneider, D. & Parker, C. (1982) Effect of pyridines on phenotypic properties of *Bordetella pertussis*. *Infect. Immun.* **38**, 548–553.

Schnetz, K. & Rak, B. (1988) Regulation of the *bgl* operon of *Escherichia coli* by transcription antitermination. *EMBO J.* **7**, 3271–3277.

Schnetz, K. & Rak, B. (1990) Beta-glucoside permease represses the *bgl* operon of *Escherichia coli* by phosphorylation of the antiterminator protein and also interacts with glucose-specific enzyme III, the key element in catabolite control. *Proc. Natl Acad. Sci. USA* **87**, 5074–5078.

Scholz, P., Haring, V., Scherzinger, E., Lurz, R., Bagdasarian, M.M., Schuster, H. & Bagdasarian, M. (1984) Replication determinants of the broad host range plasmid RSF1010 In Helinski, D.R., Cohen, S.N., Clewell, D.B., Jackson, D.A. & Hollaender, A. (eds) *Plasmids in Bacteria*, pp. 243–259. Plenum Press, New York.

Scholz, P., Haring, V., Wittmann-Liebold, B., Ashman, K., Bagdasarian, M. & Scherzinger, E. (1989) Complete nucleotide sequence and gene organization of the broad-host-range plasmid RSF1010. *Gene* **75**, 271–288.

Schultz, S.C., Shields, G.C. & Steitz, T.A. (1991) Crystal structure of a CAP-DNA complex: the DNA is bent by 90°. *Science* **253**, 1001–1007.

Schumperli, D., McKenny, R., Sobieski, D.A. & Rosenberg, M. (1982) Translational coupling at an intercistronic boundary of the *Escherichia coli* galactose operon. *Cell* **30**, 865–871.

Schwan, T.G. & Burgdorfer, W. (1987) Antigenic changes of *Borrelia burgdorferi* as a result of *in vitro* cultivation. *J. Inf. Dis.* **156**, 852–853.

Schwartz, M. (1987) The maltose regulon. In Neidhardt, F.C., Ingraham, J.L., Low, K.B., Magasanik, B., Schaecter, M. & Umbarger, H.E. (eds) Escherichia coli *and* Salmonella typhimurium *Cellular and Molecular Biology*, pp. 91–103. American Society for Microbiology, Washington.

Scott, J.R. (1990) The M protein of Group A *Streptococcus*: evolution and regulation. In Iglewski, B.H. & Clark, V.L. (eds) *Molecular Basis of Bacterial Pathogenesis*, pp. 177–203, Academic Press, San Diego.

Scott, J.R. (1992) Sex and the single circle: conjugative transposition. *J. Bacteriol.* **174**, 6005–6010.

Scott, J.R., Kirchman, P.A. & Caparon, M.G. (1988) An intermediate in transposition of the conjugative transposon Tn*916*. *Proc. Natl Acad. Sci. USA* **85**, 4809–4813.

Seifert, H.S., Ajioka, R.S., Marchal, C., Sparling, P.F. & So, M. (1988) DNA transformation leads to pilin antigenic variation in *Neisseria gonorrhoeae. Nature* **336**, 392–395.

Segall, A., Mahan, M.J. & Roth, J.R. (1988) Rearrangement of the bacterial chromosome: forbidden inversions. *Science* **241**, 1314–1318.

Sekine, Y. & Ohtsubo, E. (1989) Frameshifting is required for production of the transposase encoded by insertion sequence *1*. *Proc. Natl Acad. Sci. USA* **86**, 4609–4613.

Selvaraj, G. & Iyer, V.N. (1984) Transposon Tn*5* specifies streptomycin resistance in *Rhizobium* sp. *J. Bacteriol.* **158**, 580–589.

Shand, R.F., Blurn, P.H., Mueller, R.D., Riggs, D.L. & Artz, S.W. (1989) Correlation between histidine operon expression and guanosine 5′-diphosphate 3′-diphosphate levels during amino acid downshift in stringent and relaxed strains of *Salmonella typhimurium. J. Bacteriol.* **171**, 737–743.

Shaw, D.J. & Guest, J.R. (1982) Nucleotide sequence of the *fnr* gene and primary structure of the Fnr protein of *Escherichia coli. Nucleic Acids Res.* **10**, 6119–6130.

Sheehan, B.J., Foster, T.J., Dorman, C.J., Park, S. & Stewart, G.S.A.B. (1992) Osmotic and growth-phase dependent regulation of the *eta* gene of *Staphylococcus aureus*: a role for DNA supercoiling. *Mol. Gen. Genet.* **232**, 49–57.

Shellman, V.L. & Pettijohn, D.E. (1991) Introduction of proteins into living bacterial cells: distribution of labelled HU protein in *Escherichia coli. J. Bacteriol.* **173**, 3047–3059.

Sherratt, D. (1989) Tn*3* and related transposable elements: site-specific recombination and transposition. In Berg, D.E. & Howe, M.M. (eds), *Mobile DNA*, pp. 163–184. American Society for Microbiology, Washington.

Shi, X., Waasdorp, B.C. & Bennett, G.N. (1993) Modulation of acid-induced amino acid decarboxylase gene expression by *hns* in *Escherichia coli. J. Bacteriol.* **175**, 1182–1186.

Shimamura, T., Watanabe, S. & Sasaki, S. (1985) Enhancement of enterotoxin production by carbon dioxide in *Vibrio cholerae. Infect. Immun.* **49**, 455–456.

Shyamala, V., Schneider, E. & Ames, G.F.-L. (1990) Tandem chromosomal duplications: a role of REP sequences in the recombination event at the join-point. *EMBO J.* **9**, 939–946.

Siegle, D.A. & Kolter, R. (1992) Life after log. *J. Bacteriol.* **174**, 345–348.

Sigel, S.P., Stoebner, J.A. & Payne, S.M. (1985) Iron-vibriobactin transport system is not required for virulence of *Vibrio cholerae. Infect. Immun.* **47**, 360–362.

Sikkema, D.J. & Brubaker, R.R. (1989) Outer membrane peptides of *Yersinia pestis* mediating siderophore-independent assimilation of iron. *Biol. Metals* **2**, 174–184.

Sikorski, R.S., Michaud, W., Levin, H.L., Boeke, J.D. & Hieter, P. (1990) Trans-kingdom promiscuity. *Nature*, **345**, 581–582.

Silva, R.M., Saadi, S. & Maas, W.K. (1988) A basic replicon of virulence associated plasmids of *Shigella* spp. and enteroinvasive *Escherichia coli* is homologous with a basic replicon in plasmids of IncF groups. *Infect. Immun.* **56**, 836–842.

Silver, R.P., Vann, W.F. & Aaronson, W. (1984) Genetic and molecular analysis of *Escherichia coli* K1 antigen genes. *J. Bacteriol.* **157**, 568–575.

Silver, R.P. & Vimr, E.R. (1990) Polysialic acid capsule of *Escherichia coli* K1. In Iglewski, B.H. & Clark, V.L. (eds) *Molecular Basis of Bacterial Pathogenesis*, pp. 39–60. Academic Press, New York.

Silver, S. & Walderhaug, M. (1992) Gene regulation of plasmid- and chromosome-determined inorganic ion transport in bacteria. *Microbiol. Rev.* **56**, 195–228.

Silverman, M., Martin, M. & Engebracht, J. (1989) Regulation of luminescence in marine bacteria. In Hopwood, D.A. & Chater K.F. (eds) *Genetics of Bacterial Diversity* pp. 71–86. Academic Press, London.

Simons, R.W., Hoopes, B., McClure, W. & Kleckner, N. (1983) Three promoters near the ends of IS*10*: p-IN, p-OUT and p-III. *Cell* **34**, 673–682.

Simons, R.W. & Kleckner, N. (1983) Translational control of IS*10* transposition. *Cell* **34**, 683–691.

Simons, R.W. & Kleckner, N. (1988) Biological regulation by antisense RNA in prokaryotes. *Annu. Rev. Genet.* **22**, 567–600.

Sinden, R.R. & Pettijohn, D.E. (1981) Chromosomes in living *Escherichia coli* cells are segregated into domains of supercoiling. *Proc. Natl Acad. Sci. USA* **78**, 224–228.

Skurnik, M. & Toivanen, P. (1992) LcrF is the temperature-regulated activator of the *yadA* gene of *Yersinia enterocolitica* and *Yersinia pseudotuberculosis*. *J. Bacteriol.* **174**, 2047–2051.

Slauch, J.M., Russo, F.D. & Silhavy, T.J. (1991) Suppressor mutations in *rpoA* suggest that OmpR controls transcription by direct interaction with the α subunit of RNA polymerase *J. Bacteriol.* **173**, 7501–7510.

Slauch, J.M. & Silhavy, T.J. (1989) Genetic analysis of the switch that controls porin gene expression in *Escherichia coli* K-12. *J. Mol. Biol.* **210**, 281–292.

Slauch, J.M. & Silhavy, T.J. (1991) *cis*-acting *ompF* mutations that result in OmpR-dependent constitutive expression. *J. Bacteriol.* **173**, 4039–4048.

Slonczewski, J.L. (1992) pH-regulated genes in enteric bacteria. *ASM News* **58**, 140–144.

Small, P.L.C. & Falkow, S. (1992) A genetic analysis of acid resistance in *Shigella flexneri:* the requirement for a *katF* homologue. *Abstr. 92nd Annu. Meet. Am. Soc. Microbiol.* (1992), abstr. B-74, p. 38.

Smeltzer, M.S., Hart, M.E. & Iandolo, J.J. (1993) Phenotypic characterization of extracellular virulence factors in *Staphylococcus aureus*. *Infect. Immun.* **61**, 919–925.

Smith, G.R. (1981) DNA supercoiling: another level for regulating gene expression. *Cell* **24**, 599–600.

Smith, G.R. (1988) Homologous recombination in procaryotes. *Microbiol. Rev.* **52**, 1–28.

Smith, G.R. & Stahl, F.W. (1985) Homologous recombination promoted by Chi sites and RecBC enzyme of *Escherichia coli*. *BioEssays* **2**, 244–249.

Smith, H. (1990) Pathogenicity and the microbe *in vivo*. *J. Gen. Microbiol.* **136**, 377–393.

Smith, H.O. & Danner, D.B. (1981) Genetic transformation. *Annu. Rev. Biochem.* **50**, 41–68.

Smith, H.W. & Huggins, M.B. (1978) The effect of plasmid-determined and other characteristics on the survival of *Escherichia coli* in the alimentary tract of human beings. *J. Gen. Microbiol.* **109**, 375–379.

Smith, I., Dubnau, E., Gaur, N., Lewandoski, M., Weir, J., Cabane, K. & Nair, G. (1990) Sporulation: a comprehensive stress response which leads to differentiation. In Drlica, K. & Riley, M. (eds) *The Bacterial Chromosome*, pp. 389–403. American Society for Microbiology, Washington.

Smith, J.C., Liechty, M.C., Rasmussen, J.L. & Macrina, F.L. (1985) Genetics of clindomycin resistance in *Bacteroides*. In Helinski, D.R., Cohen, S.N., Clewell, R.B., Jackson, D.A. & Hollaender, A. (eds) *Plasmids in Bacteria*, pp. 555–570. Plenum, New York.

Smyth, C.J. (1982) Two mannose resistant haemagglutinins on enterotoxigenic *Escherichia coli* of serotype O6:K15:H16 or H^- isolated from traveller's and infantile diarrhoea. *J. Gen. Microbiol.* **128**, 2081–2096.

Smyth, C.J., Boylan, M., Matthews, H.M. & Coleman, D.C. (1991) Fimbriae of human enterotoxigenic *Escherichia coli* and control of their expression. In Ron, E.Z. & Rottem, S. (eds) *Microbial Surface Components and Toxins in Relation to Pathogenesis*, pp. 37–53. Plenum Press, New York.

Smyth, C.J. & Smith, S.G.J. (1992) Bacterial fimbriae: variation and regulatory mechanisms. In Hormaeche, C.E., Penn, C.W. & Smyth, C.J. (eds) *Molecular Biology of Bacterial Infection: Current Status and Future Prospects*, pp. 267–297. Cambridge University Press.

So, M., Heffron, F. & McCarthy, B.J. (1979) The *E. coli* gene encoding heat stable toxin is a bacterial transposon flanked by inverted repeats of IS*1*. *Nature* **277**, 453–456.

Sodeinde, O.A. & Goguen, J.D. (1988) Genetic analysis of the 9.5 kilobase virulence plasmid of *Yersinia pestis*. *Infect. Immun.* **56**, 2743–2748.

Sodeinde, O.A. & Goguen, J.D. (1989) Nucleotide sequence of the plasminogen activator gene of *Yersinia pestis*: relationship to *ompT* of *Escherichia coli* and gene E of *Salmonella typhimurium*. *Infect. Immun.* **57**, 1517–1523.

Sodeinde, O.A., Sample, A.K., Brubaker, R.R. & Goguen, J.D. (1988) Plasminogen activator/coagulase gene of *Yersinia pestis* is responsible for degradation of plasmid-encoded outer membrane proteins. *Infect. Immun.* **56**, 2749–2752.

Sodeinde, O.A., Subramanyam, Y.V.B.K., Stark, K., Quan, T., Bao, Y. & Goguen, J.D. (1992) A surface protease and the invasive character of plague. *Science* **258**, 1004–1007.

Sohel, I., Puente, J.L., Murray, W.J., Vuopio-Varkila, J. & Schoolnik, G.K. (1993) Cloning and characterization of the bundle-forming pilin gene of enteropathogenic *Escherichia coli* and its distribution in *Salmonella* serotypes. *Mol. Microbiol.* **7**, 563–575.

Sokolovic, Z. & Goebel, W. (1989) Synthesis of lysteriolysin in *Listeria monocytogenes* under heat shock conditions. *Infect. Immun.* **57**, 295–298.

Spassky, A., Rimsky, S., Garreau, H. & Buc, H. (1984) H1a, an *E. coli* DNA-binding protein which accumulates in stationary phase, strongly compacts DNA *in vitro*. *Nucleic Acids Res.* **12**, 5321–5340.

Spears, P.A., Schauer, D. & Orndorff, P.E. (1986) Metastable regulation of type 1 piliation in *Escherichia coli* and isolation of a phenotypically stable mutant. *J. Bacteriol.* **168**, 179–185.

Spiro, S. & Guest, J.R. (1988) Inactivation of the Fnr protein of *Escherichia coli* by targeted mutagenesis in the N-terminal region. *Mol. Microbiol.* **2**, 701–707.

Spiro, S. & Guest, J.R. (1990) FNR and its role in oxygen-regulated gene expression in *Escherichia coli*. *FEMS Microbiol. Rev.* **75**, 399–428.

Spiro, S., Roberts, R.E. & Guest, J.R. (1989) Fnr-dependent repression of the *ndh* gene of *Escherichia coli* and metal ion requirement for Fnr-regulated gene expression. *Mol. Microbiol.* **3**, 601–608.

Sporecke, I., Castro, D. & Mekalanos, J.J. (1984) Genetic mapping of *Vibrio cholerae* enterotoxin structural genes. *J. Bacteriol.* **157**, 253–261.

Squires, C. & Squires, C.L. (1992) The Clp proteins: proteolysis regulators or molecular chaperones. *J. Bacteriol.* **174**, 1081–1085.

Srivenugopal, K.S., Lockshon, D. & Morris, D.R. (1984) *Escherichia coli* DNA topoisomerase III: purification and characterization of a new type I enzyme. *Biochemistry* **23**, 1899–1906.

Stachel, S.E., Timmerman, B. & Zambryski, P. (1986) Generation of single-stranded T-DNA molecules during the initial stages of T-DNA transfer from *Agrobacterium tumefaciens* to plant cells. *Nature* **322**, 706–712.

Stachel, S.E. & Zambryski, P.C. (1986) *virA* and *virG* control the plant-induced activation of the T-DNA transfer process of *Agrobacterium tumefaciens*. *Cell* **46**, 325–333.

Staggs, T.M. & Perry, R.D. (1991) Identification and cloning of a *fur* regulatory gene in *Yersinia pestis*. *J. Bacteriol.* **173**, 417–425.

Staggs, T.M. & Perry, R.D. (1992) Fur regulation in *Yersinia* species. *Mol. Microbiol.* **6**, 2507–2516.

Stark, W.M., Boocock, M.R. & Sherratt, D.J. (1989a) Site-specific recombination by Tn*3* resolvase. *Trends Genet.* **5**, 304–309.

Stark, W.M., Boocock, M.R. & Sherratt, D.J. (1992) Catalysis by site-specific recombinases. *Trends Genet.* **8**, 432–439.

Stark, W.M., Sherratt, D.J. & Boocock, M.R. (1989b) Site-specific recombination by Tn*3* resolvase: topological changes in the forward and reverse reactions. *Cell* **58**, 779–790.

Starnbach, M.N. & Lory, S. (1992) The *fliA* (*rpoF*) gene of *Pseudomonas aeruginosa* encodes an alternative sigma factor required for flagellin synthesis. *Mol. Microbiol.* **6**, 459–469.

Steck, T.R. & Kado, C.I. (1990) Virulence genes promote conjugative transfer of the Ti plasmid between *Agrobacterium* strains. *J. Bacteriol.* **172**, 2191–2193.

Stephens, J.C., Artz, S.W. & Ames, B.N. (1975) Guanosine-5′-diphosphate-3′-diphosphate (ppGpp): positive effector for histidine operon transcription and general signal for amino acid deficiency. *Proc. Natl Acad. Sci. USA* **72**, 4389–4393.

Stern, A., Brown, M., Nickel, P. & Meyer, T.F. (1986) Opacity genes in *Neisseria gonorrhoeae*: control of phase variation and antigenic variation. *Cell* **47**, 61–71.

Stern, A. & Meyer, T.F. (1987) Common mechanism controlling phase variation and antigenic variation in pathogenic *Neisseriae*. *Mol. Microbiol.* **1**, 5–12.

Stern, A., Nickel, P., Meyer, T.F. & So, M. (1984a) Opacity determinants of *Neisseria gonorrhoeae*: gene expression and chromosomal linkage to the gonococcal pilus gene. *Cell* **37**, 447–456.

Stern, M.J., Ames, G.F.-L., Smith, N.H., Robinson, E.C. & Higgins, C.F. (1984b) Repetitive extragenic palindromic sequence: a major component of the bacterial genome. *Cell* **37**, 1015–1026.

Stibitz, S. & Yang, M.-S. (1991) Subcellular localization and immunological detection of proteins encoded by the *vir* locus of *Bordetella pertussis*. *J. Bacteriol.* **173**, 4288–4296.

Stirling, C.J., Colloms, S.D., Collins, J.F., Szatmari, G. & Sherratt, D.J. (1989a) *xerB*, an *Escherichia coli* gene required for plasmid ColE1 site-specific recombination, is identical to *pepA*, encoding aminopeptidase A, a protein with substantial similarity to bovine lens leucine aminopeptidase. *EMBO J.* **8**, 1623–1627.

Stirling, C.J., Stewart, G. & Sherratt, D.J. (1988a) Multicopy plasmid stability in *Escherichia coli* requires host-encoded functions that lead to plasmid site-specific recombination. *Mol. Gen Genet.* **214**, 80–84.

Stirling, C.J., Szatmari, G., Stewart, G., Smith, M.C.M. & Sherratt, D.J. (1988b) The arginine repressor is essential for plasmid-stabilizing site-specific recombination at the ColE1 *cer* locus. *EMBO J.* **7**, 4389–4395.

Stirling, D.A., Hulton, C.S.J., Waddell, L., Park, S.F., Stewart, G.S.A.B., Booth, I.R. & Higgins, C.F. (1989b) Molecular characterization of the *proU* loci of *Salmonella typhimurium* and *Escherichia coli* encoding osmoregulated glycine betaine transport systems. *Mol. Microbiol.* **3**, 1025–1038.

Stock, J.B., Ninfa, A.J. & Stock, A.M. (1989) Protein phosphorylation and regulation of adaptive responses in bacteria. *Microbiol. Rev.* **53**, 450–490.

Stoebner, J.A. & Payne, S.M. (1988) Iron-regulated haemolysin production and utilization of haeme and haemoglobin by *Vibrio cholerae*. *Infect. Immun.* **56**, 2891–2895.

Stojiljkovic, I. & Hantke, K. (1992) Hemin uptake system of *Yersinia enterocolitica*: similarities with other TonB-dependent systems in Gram-negative bacteria. *EMBO J.* **11**, 4359–4367.

Stokes, H.W. & Hall, R.M. (1989) A novel family of potential mobile DNA elements encoding site-specific gene integration functions: integrons. *Mol. Microbiol.* **3**, 1669–1683.

Stoorvogel, J., van Bussel, M.J.A.W.M., Tommasse, J. & van de Klundert, J.A.M. (1991a) Molecular characterization of an *Enterobacter cloacae* outer membrane protein (OmpX). *J. Bacteriol.* **173**, 156–160.

Stoorvogel, J., van Bussel, M.J.A.W.M., Tommasse, J. & van de Klundert, J.A.M. (1991b) Biological characterization of an *Enterobacter cloacae* outer membrane protein (OmpX). *J. Bacteriol.* **173**, 161–167.

Storey, D.G., Frank, D.W., Farinha, M.A., Kropinski, A.M. & Iglewski, B.H. (1990) Multiple promoters control the regulation of the *Pseudomonas aeruginosa regA* gene. *Mol. Microbiol.* **4**, 499–503.

Storz, G., Tartaglia, A. & Ames, B.N. (1990) Transcriptional regulator of oxidative stress-inducible genes: direct activation by oxidation. *Science* **248**, 189–194.

Stragier, P. & Patte, J.-C. (1983) Regulation of diaminopimelate decarboxylase synthesis in *Escherichia coli*. III. Nucleotide sequence and regulation of the *lysR* gene. *J. Mol. Biol.* **168**, 333–350.

Strancy, S.B. & Crothers, D.M. (1987) Lac repressor is a transient gene-activating protein. *Cell* **51**, 699–707.

Straney, D.C., Straney, S.M. & Crothers, D.M. (1989) Synergy between *Escherichia coli* CAP protein and RNA polymerase in the *lac* promoter open complex. *J. Mol. Biol.* **206**, 41–57.

Strauch, K.L. & Beckwith, J. (1988) An *Escherichia coli* mutation preventing degradation of abnormal periplasmic proteins. *Proc. Natl Acad. Sci. USA* **85**, 1576–1580.

Straus, D.B., Walter, W.A. & Gross, C.A. (1987) The heat shock response of *E. coli* is regulated by changes in the concentration of σ^{32}. *Nature* **329**, 348–351.

Stringer, S.L., Hong, S.-T., Giutoli, D. & Stringer, R.J. (1991) Repeated DNA in *Pneumocystis carinii*. *J. Clin. Microbiol.* **29**, 1194–1201.

Su, W., Porter, S., Kutsu, S. & Echols, H. (1990) DNA looping and enhancer activity: association between NtrC activator and RNA polymerase at the bacterial *glnA* promoter. *Proc. Natl Acad. Sci. USA* **87**, 5504–5508.

Sugiura, A., Nakashima, K., Tanaka, K. & Mizuno, T. (1992) Clarification of the structural and functional features of the osmoregulated *kdp* operon of *Escherichia coli*. *Mol. Microbiol.* **6**, 1769–1776.

Sukupolvi, S., Vuorio, R., Qi, S.-Y., O'Connor, D. & Rhen, M. (1990) Characterization of the *traT* gene and mutants that increase outer membrane permeability from the *Salmonella typhimurium* virulence plasmid. *Mol. Microbiol.* **4**, 49–57.

Summers, D.K. & Sherratt, D.J. (1984) Multimerization of high copy number plasmids causes instability: ColE1 encodes a determinant for plasmid monomerization and stability. *Cell* **36**, 1097–1103.

Sung, Y.-C. & Fuchs, J.A. (1992) The *Escherichia coli* K-12 *cyn* operon is positively regulated by a member of the *lysR* family. *J. Bacteriol.* **174**, 3645–3650.

Surette, M.G., Buch, S.J. & Chaconas, G. (1987) Transpososomes: stable protein-DNA complexes involved in the in vitro transposition of bacteriophage Mu DNA. *Cell* **49**, 253–262.

Sutcliffe, J.G. (1979) Complete sequence of the *Escherichia coli* plasmid pBR322. *Cold Spring Harbor Symp. Quant. Biol.* **43**, 77–90.

Sutherland, L., Cairney, J., Elmore, M.J., Booth, I.R. & Higgins, C.F. (1986) Osmotic regulation of transcription: induction of the *proU* betaine transport gene is dependent on accumulation of intracellular potassium. *J. Bacteriol.* **168**, 805–814.

Symonds, N. (1988) Action at a distance. *Nature* **333**, 18–19.

Symonds, N., Toussaint, A., van de Putte, P. & Howe, M., eds. (1987) *Phage Mu*. Cold Spring Harbor Laboratory Press, Cold Spring Harbor, New York.

Szabo, M., Maskell, D., Butler, P., Love, J. & Moxon, R. (1992) Use of chromosomal gene fusions to investigate the role of repetitive DNA in regulation of genes involved in lipopolysaccharide biosynthesis in *Haemophilus influenzae*. *J. Bacteriol.* **174**, 7245–7252.

Tabata, S., Hooykaas, P.J. & Oka, A. (1989) Sequence determination and characterization of the replicator region in the tumor-inducing plasmid pTiB6S3. *J. Bacteriol.* **171**, 1665–1672.

Taha, M.K., Dupuy, B., Saurin, W., So. & Marchal, C. (1991) Control of expression in *Neisseria gonorrhoeae* as an original system in the family of two-component regulators. *Mol. Microbiol.* **5**, 137–148.

Taha, M.K., So, M., Seifert, H.S., Billyard, E. & Marchal, C. (1988) Pilin expression in *Neisseria gonorrhoeae* is under both positive and negative transcriptional control. *EMBO J.* **7**, 4367–4378.

Tai, S.S. & Holmes, R.K. (1988) Iron regulation of the cloned diphtheria toxin promoter in *Escherichia coli*. *Infect. Immun.* **56**, 2430–2436.

Taira, S. & Rhen, M. (1989) Molecular organization of genes constituting the virulence determinant on the *Salmonella typhimurium* 96 kilobase pair plasmid. *FEBS Lett.* **257**, 274–278.

Taira, S., Riikonen, P., Saarilahti, H., Sukupolvi, S. & Rhen, M. (1991) The *mkaC* virulence gene of the *Salmonella* serovar Typhimurium 96 kb plasmid encodes a transcriptional activator. *Mol. Gen. Genet.* **228**, 381–384.

Tanabe, H., Goldstein, J., Yang, M. & Inouye, M. (1992) Identification of the promoter region of the *Escherichia coli* major cold shock gene, *cspA*. *J. Bacteriol.* **174**, 3867–3873.

Tanaka, K., Takayanagi, Y., Fujita, N., Ishihama, A. & Takahashi, H. (1993) Heterogeneity of the principal σ factor in *Escherichia coli*: the *rpoS* gene product, σ^{38}, is a second principal σ factor of RNA polymerase in stationary phase *Escherichia coli*. *Proc. Natl Acad. Sci. USA* **90**, 3511–3515.

Tanimoto, K., An, F.Y. & Clewell, D.B. (1993) Characterisation of the *traC* determinant of the *Enterococcus faecalis* haemolysin-bacteriocin plasmid pAD1: binding of sex pheromone. *J. Bacteriol.* **175**, 5260–5264.

Tao, K., Fujita, N. & Ishihama, A. (1993) Involvement of the RNA polymerase α subunit C-terminal region in co-operative interaction and transcriptional activation with OxyR protein. *Mol. Microbiol.* **7**, 859–864.

Tao, X., Boyd, J. & Murphy, J.R. (1992) Specific binding of the diphtheria *tox* regulatory element DtxR to the *tox* operator requires divalent heavy metal ions and a 9-base-pair interrupted palindromic sequence. *Proc. Natl Acad. Sci. USA* **89**, 5897–5901.

Tao, X. & Murphy, J.R. (1992) Binding of the metalloregulatory protein DtxR to the diphtheria *tox* operator requires a divalent metal ion and protects the palindromic sequence from DNase I digestion. *J. Biol. Chem.* **267**, 21761–21764.

Tardat, B. & Touati, D. (1993) Iron and oxygen regulation of *Escherichia coli* MnSOD expression: competition between the global regulators Fur and ArcA for binding to DNA. *Mol. Microbiol.* **9**, 53–63.

Tartaglia, L.A., Storz, G. & Ames, B.N. (1989) Identification and molecular analysis of OxyR-regulated promoters important for the bacterial adaptation to oxidative stress. *J. Mol. Biol.* **210**, 709–719.

Taylor, A.F. (1992) Movement and resolution of Holliday junctions by enzymes from *E. coli*. *Cell* **69**, 1063–1065.

Taylor, A.F., Schultz, D.W., Ponticelli, A.S. & Smith, G.R. (1985) RecBC enzyme nicking at Chi sites during DNA unwinding: location and orientation dependence of the cutting. *Cell* **41**, 153–163.

Taylor, R.K. (1989) Genetic studies of enterotoxin and other potential virulence factors of

Vibrio cholerae. In Hopwood, D.A. & Chater, K.F. (eds) *Genetics of Bacterial Diversity*, pp. 309–329. Academic Press, London.

Taylor-Robinson, D. (1990) The *Mycoplasmatales*. In Parker, T.M. & Duerden, B.I. (eds) *Topley and Wilson's Principles of Bacteriology, Virology and Immunity*, Vol. 2, pp. 663–681. Edward Arnold, London.

Tennent, J.M., Hultgren, S., Marklund, B.-I., Forsman, K., Göransson, M., Uhlin, B.E. & Normark, S. (1990) Genetics of adhesin expression in *Escherichia coli.* In Iglewski, B.H. & Clark, V.L. (eds) *Molecular Basis of Bacterial Pathogenesis*, pp. 79–110. Academic Press, San Diego.

Thisted, T. & Gerdes, K. (1992) Mechanism of post-segregational killing by the *hok/sok* system of plasmid R1: Sok antisense RNA regulates *hok* gene expression indirectly through the overlapping *mok* gene. *J. Mol. Biol.* **223**, 41–54.

Thliveris, A.T., Ennis, D.G., Lewis, L.K. & Mount, D.W. (1990) SOS functions. In Drlica, K. & Riley, M. (eds) *The Bacterial Chromosome*, pp. 381–387. American Society for Microbiology, Washington.

Thomas, C.D., Balson, D. & Shaw, W.V. (1988) Identification of the tyrosine residue involved in bond formation between replication origin and initiator protein of plasmid pC221. *Biochem. Soc. Trans.* **16**, 758–759.

Thomas, C.M. & Smith, C.A. (1987) Incompatibility group P plasmids: genetics, evolution and use in genetic manipulation. *Annu. Rev. Microbiol.* **41**, 77–101.

Thompson, J.F. & Landy, A. (1988) Empirical estimation of protein-induced bending angles: application to λ site-specific recombination complexes. *Nucleic Acids Res.* **16**, 9687–9705.

Thompson, J.F. & Landy, A. (1989) Regulation of bacteriophage lambda site-specific recombination. In Berg, D.E. & Howe, M.M. (eds) *Mobile DNA*, pp. 1–22. American Society for Microbiology, Washington.

Thompson, J.F., Moitoso de Vargas, L., Koch, K., Kahman, R. & Landy, A. (1987) Cellular factors couple recombination with growth phase: characterization of a new component in the λ site-specific recombination pathway. *Cell* **50**, 901–908.

Thöny, B., Hwang, D.S., Fradkin, L. & Kornberg, A. (1991) *iciA*, an *Escherichia coli* gene encoding a specific inhibitor of chromosomal initiation of replication *in vitro*. *Proc. Natl Acad. Sci. USA* **88**, 4066–4070.

Tilly, K., Hauser, R., Campbell, J. & Ostheimer, G.J. (1993) Isolation of *dnaJ*, *dnaK*, and *grpE* homologues from *Borrelia burgdorferi* and complementation of *Escherichia coli* mutants. *Mol. Microbiol.* **7**, 359–369.

Timmons M.S., Lieb, M. & Deonier, R.C. (1986) Recombination between IS5 elements: requirement for homology and recombination functions. *Genetics* **113**, 797–810.

Tinge, S.A. & Curtiss III, R. (1990) Isolation of the replication and partitioning regions of the *Salmonella typhimurium* virulence plasmid and stabilization of heterologous replicons. *J. Bacteriol.* **172**, 5266–5277.

Tissieres, A., Mitchell, H.K. & Tracy, U.M. (1974) Protein synthesis in salivary glands of *Drosophila melanogaster*: relation to chromosome puffs. *J. Mol. Biol.* **84**, 389–398.

Tlsty, T.D., Albertini, A.M. & Miller, J.H. (1984) Gene amplification in the *lac* region of *E. coli Cell* **37**, 217–224.

Tobe, T., Nagai, S., Okada, N., Adler, B., Yoshikawa, M. & Sasakawa, C. (1991) Temperature-regulated expression of invasion genes in *Shigella flexneri* is controlled through the transcriptional activation of the *virB* gene on the large plasmid. *Mol. Microbiol.* **5**, 887–893.

Tobe, T., Sasakawa, C., Okada, N., Honma, Y. & Yoshikawa, M. (1992) *vacB*, a novel chromosomal gene required for the expression of virulence genes on the large plasmid of *Shigella flexneri*. *J. Bacteriol.* **174**, 6359–6367.

Tobin, J.F. & Schleif, R.F. (1987) Positive regulation of the *Escherichia coli* L-rhamnose operon is mediated by the products of tandemly repeated regulatory genes. *J. Mol. Biol.* **196**, 789–799.

Tominaga, A., Ikemizu, S. & Enomoto, M. (1991) Site-specific recombinase genes in three *Shigella* subgroups and nucleotide sequences of a *pinB* gene and an invertible B segment from *Shigella boydii*. *J. Bacteriol.* **173**, 4079–4087.

Toro, N., Datta, A., Carmi, O.A., Young, C., Prusti, R.K. & Nester, E.W. (1989) The *Agrobacterium tumefaciens virC1* gene product binds to overdrive, a T-DNA transfer enhancer. *J. Bacteriol.* **171**, 6845–6849.

Torres-Cabassa, A.S. & Gottesman, S. (1987) Capsule synthesis in *Escherichia coli* K-12 is regulated by proteolysis. *J. Bacteriol.* **169**, 981–989.

Totten, P.A., Lara, J.C. & Lory, S. (1990) The *rpoN* gene product of *Pseudomonas aeruginosa* is required for expression of diverse genes, including the flagellin gene. *J. Bacteriol.* **172**, 389–396.

Toussaint, A. & Résibois, A. (1983) Phage Mu: transposition as a lifestyle. In Shapiro, J.A. (ed.), *Mobile Genetic Elements*, pp. 105–158. Academic Press, New York.

Travers, A.A. (1984) Conserved regions of coordinately regulated *E. coli* promoters. *Nucleic Acids Res.* **12**, 2605–2618.

Travers, A. (1989) Curves with a function. *Nature* **341**, 184–185.

Trieu-Cuot, P., Carlier, C., Martin, P. & Courvalin, P. (1987) Plasmid transfer by conjugation from *Escherichia coli* to Gram-positive bacteria. *FEMS Microbiol. Lett.* **48**, 289–294.

Trieu-Cuot, P., Carlier, C. & Courvalin, P. (1988) Conjugative plasmid transfer from *Enterococcus faecalis* to *Escherichia coli*. *J. Bacteriol.* **170**, 4388–4391.

Trifonov, E.N. (1985) Curved DNA. *CRC Crit. Rev. Biochem.* **19**, 89–106.

Tsaneva, I.R. & Weiss, B. (1990) *soxR*, a locus governing a superoxide response regulon in *Escherichia coli* K-12. *J. Bacteriol.* **172**, 4197–4205.

Tsao, Y.-P., Wu, H.-Y. & Liu, L.F. (1989) Transcription-driven supercoiling of DNA: direct biochemical evidence from *in vitro* studies. *Cell* **56**, 111–118.

Tse-Dinh, Y.-C. (1985) Regulation of the *Escherichia coli* Topoisomerase I gene by DNA supercoiling. *Nucleic Acids Res.* **13**, 4751–4763.

Tse-Dinh, Y.-C. & Beran, R.K. (1988) Multiple promoters for transcription of the *Escherichia coli* DNA topoisomerase I gene and their regulation by DNA supercoiling. *J. Mol. Biol.* **202**, 735–742.

Tsui, P., Helu, V. & Freundlich, M. (1988) Altered osmoregulation of *ompF* in integration host factor mutants of *Escherichia coli*. *J. Bacteriol.* **170**, 4950–4953.

Uchida, I., Sekizaki, T., Hashimoto, R. & Terakado, N. (1985) Association of the encapsulation of *Bacillus anthracis* with a 60 megadalton plasmid. *J. Gen. Microbiol.* **131**, 363–367.

Ueguchi, C. & Ito, K. (1992) Multicopy suppression: an approach to understanding intracellular functioning of the protein export system. *J. Bacteriol.* **174**, 1454–1461.

Ueguchi, C. & Mizuno, T. (1993) The *Escherichia coli* nucleoid protein H-NS functions directly as a transcriptional repressor. *EMBO J.* **12**, 1039–1046.

Vale, R.D. & Goldstein, S.B. (1990) One motor, many tails: an expanding repertoire of force-generating enzymes. *Cell* **60**, 883–885.

Valone, S.E., Chikami, G.K. & Miller, V.L. (1993) Stress induction of the virulence proteins (SpvA, -B, and -C) from native plasmid pSDL2 of *Salmonella dublin*. *Infect. Immun.* **61**, 705–713.

Valvano, M.A. & Crosa, J.H. (1984) Molecular cloning, expression and regulation in *Escherichia coli* K1. *Infect. Immun.* **46**, 159–167.

Van Bogelen, R.A., Kelley, P.M. & Neidhardt, F.C. (1987) Differential induction of heat shock, SOS, and oxidative stress regulons and accumulation of nucleotides in *Escherichia coli*. *J. Bacteriol.* **169**, 26–32.

Vandenbosch, J.L., Kurlandsky, D.R., Urdangaray, R. & Jones, G.W. (1989a) Evidence of coordinate regulation of virulence in *Salmonella typhimurium* involving the *rsk* element of the 95 kilobase plasmid. *Infect. Immun.* **57**, 2566–2568.

Vandenbosch, J.L., Rabert, D.K., Kurlandsky, D.R. & Jones, G.W. (1989b) Sequence analysis of *rsk*, a portion of the 95 kilobase plasmid of *Salmonella typhimurium* associated with resistance to the bactericidal activity of serum. *Infect. Immun.* **57**, 850–857.

Vandenesch, F., Kornblum, J. & Novick, R.P. (1991) A temporal signal, independent of *agr*, is required for *hla* but not *spa* transcription in *Staphylococcus aureus*. *J. Bacteriol.* **173**, 6313–6320.

Van de Putte, P. & Goosen, N. (1992) DNA inversions in phages and bacteria. *Trends Genet.* **8**, 457–462.

Van der Woude, M.W., Braaten, B.A. & Low, D.A. (1992) Evidence for global regulatory control of pilus expression in *Escherichia coli* by Lrp and DNA methylation: model building based on analysis of *pap*. *Mol. Microbiol.* **6**, 2429–2435.

Van Haaren, M.J.J., Sedee, N.J.A., Schilperoot, R.A. & Hooykaas, P.J.J. (1987) Overdrive is a T-region enhancer which stimulates T-strand production in *Agrobacterium tumefaciens*. *Nucleic Acids Res.* **15**, 8983–8997.

Vanooteghem, J.-C. & Cornelis, G.R. (1990) Structural and functional similarities between the replication region of the *Yersinia* virulence plasmid and the RepFIIA replicons. *J. Bacteriol.* **172**, 3600–3608.

Varshavsky, A.J., Nedospasov, A., Bakayeva, V.V., Bakayeva, T.G. & Georgiev, G. (1977) Histonelike proteins in the purified *Escherichia coli* deoxyribonucleoprotein. *Nucleic Acids Res.* **4**, 2725–2745.

Vega-Palas, M.A., Flores, E. & Herro, A. (1992) NtcA, a global nitrogen regulator from the cyanobacterium *Synechococcus* that belongs to the Crp family of bacterial regulators. *Mol. Microbiol.* **6**, 1853–1859.

Veluthambi, K., Jayaswal, R.K. & Gelvin, S.B. (1987) Virulence genes *A*, *G*, and *D* mediate the double stranded border cleavage of T-DNA from the *Agrobacterium* Ti plasmid. *Proc. Natl Acad. Sci. USA* **84**, 1881–1885.

Versalovic, J., Koeuth, T. & Lupski, J.R. (1991) Distribution of repetitive DNA sequences in eubacteria and application to fingerprinting of bacterial genomes. *Nucleic Acids Res.* **19**, 6823–6831.

Vijayakumar, M.N., Priebe, S.D. & Guild, W.R. (1986) Structure of a conjugative element in *Streptococcus pneumoniae*. *J. Bacteriol.* **166**, 978–984.

Vimr, E.R., Aaronson, W. & Silver, R.P. (1989) Genetic analysis of chromosomal mutations in the polysialic acid gene cluster of *Escherichia coli* K1. *J. Bacteriol.* **171**, 1106–1117.

Vinella, D., Jaffé, A., D'Ari, R., Kohiyama, M. & Hughes, P. (1992) Chromosome partitioning in *Escherichia coli* in the absence of Dam-directed methylation. *J. Bacteriol.* **174**, 2388–2390.

Vinograd, J., Lebowitz, J., Radloff, R., Watson, R. & Laipis, P. (1965) The twisted circular form of polyoma viral DNA. *Proc. Natl Acad. Sci. USA* **53**, 1104–1111.

Virji, M. & Heckels, J.E. (1986) The effect of protein II and pili on the interaction of *Neisseria gonorrhoeae* with human polymorphonuclear leucocytes. *J. Gen. Microbiol.* **132**, 503–512.

Virji, M., Kayhty, H., Ferguson, D.J.P., Alexandrescu, C., Heckels, J.E. & Moxon, E.R. (1991) The role of pili in the interactions of pathogenic *Neisseria* with cultured human endothelial cells. *Mol. Microbiol.* **5**, 1831–1841.

Vodkin, M.H. & Williams, J.C. (1988) A heat shock operon in *Coxiella burnetti* produces a major antigen homologous to a protein in both mycobacteria and *Escherichia coli*. *J. Bacteriol.* **170**, 1227–1234.

Vogel, M., Hess, J., Then, I., Juarez, A. & Goebel, W. (1988) Characterization of a sequence (*hlyR*) which enhances synthesis of haemolysin in *Escherichia coli*. *Mol. Gen. Genet.* **212**, 76–84.

Von Ossowski, I., Mulvey, M.R., Leco, P.A., Borys, A. & Loewen, P.C. (1991) Nucleotide sequence of *Escherichia coli katE*, which encodes catalase HPII. *J. Bacteriol.* **173**, 514–520.

Walderhaug, M.O., Polarek, J.W., Voelkner, P., Daniel, J.M., Hesse, J.E., Altendorf, K. & Epstein, W. (1992) KdpD and KdpE, proteins that control expression of the *kdpABC* operon, are members of the two-component sensor-effector class of regulators. *J. Bacteriol.* **174**, 2152–2159.

Walker, G.C. (1987) The SOS response of *Escherichia coli*. In Neidhardt, F.C., Ingraham, J.L., Low, K.B., Magasanik, B., Schaechter, M. & Umbarger, H.E. (eds) Escherichia coli *and* Salmonella typhimurium *Cellular and Molecular Biology*, pp. 1346–1357. American Society for Microbiology, Washington.

Wall, D., Delaney, J.M., Fayet, O., Lipinska, B., Yamamoto, T. & Georgopoulos, C. (1992) *arc*-dependent thermal regulation and extragenic suppression of the *Escherichia coli* cytochrome *d* operon. *J. Bacteriol.* **174**, 6554–6562.

Wandersman, C.F., Moreno, F. & Schwartz, M. (1980) Pleiotropic mutations rendering *Escherichia coli* resistant to bacteriophage TP1. *J. Bacteriol.* **143**, 1374–1383.

Wang, J.C. (1985) DNA topoisomerases. *Annu. Rev. Biochem.* **54**, 665–697.

Wang, J.C. (1987) Recent studies of DNA topoisomerases. *Biochim. Biophys. Acta* **909**, 1–9.

Wang, J.C., Caron, P.R. & Kim, R.A. (1990) The role of DNA topoisomerases in recombination and genome stability: a double-edged sword? *Cell* **62**, 403–406.

Wang, J.C. & Giaever, G.N. (1988) Action at a distance along a DNA. *Science* **240**, 300–304.

Wang, J.C. & Liu, L.F. (1990) DNA replication: topological aspects and the roles of DNA topoisomerases. In Cozzarelli, N.R. & Wang, J.C. (eds), *DNA Topology and its Biological Effects*, pp. 321–340. Cold Spring Harbor Laboratory Press, Cold Spring Harbor, New York.

Wang, J.-Y. & Syvanen, M. (1992) DNA twist as a transcriptional sensor for environmental changes. *Mol. Microbiol.* **6**, 1861–1866.

Wang, K., Herrera-Estrella, L., Van Montague, M. & Zambryski, P. (1984) Right 25 bp terminus sequence of the nopaline T-DNA is essential for and determines direction of DNA transfer from *Agrobacterium* to the plant genome. *Cell* **38**, 455–462.

Wang, K., Stachel, S.E., Timmerman, B., Van Montague, M. & Zambryski, P.C. (1987) Site-specific nick in the T-DNA border sequence as a result of *Agrobacterium vir* gene expression. *Science* **235**, 587–591.

Wang, Q. & Calvo, J.M. (1993) Lrp, a major regulatory protein in *Escherichia coli*, bends DNA and can organize the assembly of a higher-order nucleoprotein structure. *EMBO J.* **12**, 2495–2501.

Wanner, B.L. (1983) Overlapping and separate controls on the phosphate regulon in *Escherichia coli* K-12. *J. Mol. Biol.* **166**, 283–308.

Wanner, B.L. (1992) Is cross regulation by phosphorylation of two-component response regulator proteins important in bacteria? *J. Bacteriol.* **174**, 2053–2058.

Wanner, B.L. & Wilmes-Reisenberg, M.R. (1992) Involvement of phosphotransacetylase, acetate kinase and acetyl phosphate synthesis in control of the phosphate regulon in *Escherichia coli*. *J. Bacteriol.* **174**, 2124–2130.

Wanner, G., Formanek, H., Galli, D. & Wirth, R. (1989) Localization of aggregation substances of *Enterococcus faecalis* after induction by sex pheromones. *Arch. Microbiol.* **151**, 491–497.

Watanabe, E., Inamoto, S., Lee, M.-H., Kim, S.U., Ogua, T., Mori, H., Hiraga, S., Yamasaki, M. & Nagai, K. (1989) Purification and characterization of the *sopB* gene product which is responsible for stable maintenance of mini-F plasmid. *Mol. Gen. Genet.* **218**, 431–436.

Watanabe, H. (1988) Genetics of virulence of *Shigella* species. *Microbiol. Sci.* **5**, 307–310.

Watanabe, H., Arakawa, E., Ito, K.-I., Kato, J.-I. & Nakamura, A. (1990) Genetic analysis of an invasion region by use of a Tn*3-lac* transposon and identification of a second positive regulator gene, *invE*, for cell invasion of *Shigella sonnei*: significant homology of InvE with ParB of plasmid P1. *J. Bacteriol.* **172**, 619–629.

Waters, V.L. & Crosa, J.H. (1988) Divergence of the aerobactin iron uptake system encoded by plasmids pColV-K30 in *Escherichia coli* and pSMN1 in *Aerobacter aerogenes* 62–1. *J. Bacteriol.* **170**, 5153–5160.

Waters, V.L. & Crosa, J.H. (1991) Colicin V virulence plasmids. *Microbiol Rev.* **55**, 437–450.

Watson, N., Dunyak, D.S., Rosey, E.L., Slonczewski, J.L. & Olson, E.R. (1992) Identification of elements involved in transcriptional regulation of the *Escherichia coli cad* operon by external pH. *J. Bacteriol.* **174**, 530–540.

Webster, C., Kempsell, K., Booth, I. & Busby, S. (1987) Organization of the regulatory region of the *Escherichia coli* melibiose operon. *Gene* **83**, 253–263.

Webster, R.E. (1991) The *tol* gene products and the import of macromolecules into *Escherichia coli. Mol. Microbiol.* **5**, 1005–1011.

Weeks, C.R. & Ferretti, J.J. (1986) Nucleotide sequence of the type A streptococcal exotoxin (erythrogenic toxin) gene from *Streptococcus pyogenes* bacteriophage T12. *Infect.. Immun.* **52**, 144–150.

Weiner, L., Brissette, J.L. & Model, P. (1991) Stress-induced expression of the *Escherichia coli* phage shock protein operon is dependent on σ^{54} and modulated by positive and negative feedback mechanism. *Genes Dev.* **5**, 1912–1923.

Weinreich, M.D. & Reznikoff, W.S. (1992) FIS plays a role in Tn*5* and IS*50* transposition. *J. Bacteriol.* **174**, 4530–4537.

Weinstock, G.M. (1987) General recombination in *Escherichia coli.* In Neidhardt, F.C., Ingraham, J.L., Low, K.B., Magasanik, B., Schaechter, M. & Umbarger, H.E. (eds) *Escherichia coli and* Salmonella typhimurium *Cellular and Molecular Biology*, pp. 1034–1043.

Weisbeek, P.J., Bitter, W., Leong, J., Koster, M. & Marugg, J.D. (1990) Genetics of iron uptake in plant growth-promoting *Pseudomonas putida* WCS358. In Silver, S., Chakrabarty, A.M., Iglewski, B. & Kaplan, S. (eds) Pseudomonas: *Biotransformations, Pathogenesis and Evolving Biotechnology*, pp. 64–73. American Society for Microbiology, Washington.

Weiser, J.N., Love, J.M. & Moxon, E.R. (1989) The molecular mechanism of phase variation of *H. influenzae* lipopolysaccharide. *Cell* **59**, 657–665.

Weiss, D.S., Batut, J., Klose, K.E., Kener, J. & Kutsu, S. (1991) The phosphorylated form of the enhancer-binding protein NtrC has an ATPase activity that is essential for activation of transcription. *Cell* **67**, 155–167.

Weiss, V. & Magasanik, B. (1988) Phosphorylation of nitrogen regulator 1 (NR1) of *Escherichia coli. Proc. Natl Acad. Sci. USA* **85**, 8919–8923.

Wek, R.C. & Hatfield, G.W. (1986) Nucleotide sequence and *in vivo* expression of *ilvY* and *ilvC* genes in *Escherichia coli* K-12. *J. Biol. Chem.* **261**, 643–663.

Welch, R.A. & Pellett, S. (1988) Transcriptional organization of the *Escherichia coli* haemolysin genes. *J. Bacteriol.* **170**, 1622–1630.

Wells, R.D. (1988) Unusual DNA structures. *J. Mol. Biol.* **263**, 1095–1098.

Wenzel, R. & Herrmann, R. (1988) Repetitive DNA sequences in *Mycoplasma pneumoniae. Nucleic Acids Res.* **16**, 8337–8350.

Weppleman, R., Kier, L.D. & Ames, B.N. (1977) Properties of two phosphatases and a cyclic phosphodiesterase of *Salmonella typhimurium. J. Bacteriol.* **130**, 411–419.

West, S.C., Countryman, J.K. & Howard-Flanders, P. (1983) Purification and properties of the RecA protein from *Proteus mirabilis*. *J. Biol. Chem.* **258**, 4648–4654.

West, S.E.H. & Sparling, P.F. (1985) Response of *Neisseria gonorrhoeae* to iron limitation: alterations in expression of membrane proteins without apparent siderophore production. *Infect. Immun.* **47**, 388–394.

Whitehall, S., Austin, S. & Dixon, R. (1992) DNA supercoiling response of the σ^{54}-dependent *Klebsiella pneumoniae nifL* promoter *in vitro*. *J. Mol. Biol.* **225**, 591–607.

Whitson, P.A., Hsieh, W.-T., Wells, R.D. & Matthews, K.S. (1987) Supercoiling facilitates *lac* operator–repressor–pseudooperator interactions. *J. Biol. Chem.* **262**, 4943–4946.

Whoriskey, S.K., Nghiem, V.-H., Leong, P.-M., Masson, J.-M. & Miller, J.H. (1987) Genetic rearrangements and gene amplification in *Escherichia coli*: DNA sequences at the junctures of amplified gene fusions. *Genes Dev.* **1**, 227–237.

Wick, M.J., Frank, D.W., Storey, D.G. & Iglewski, B.H. (1990) Identification of *regB*, a gene required for optimal exotoxin A yields in *Pseudomonas aeruginosa*. *Mol. Microbiol.* **4**, 489–497.

Widom, R.L., Jarvis, E.D., LaFauci, G. & Rudner, R. (1988) Instability of rRNA operons in *Bacillus subtilis*. *J. Bacteriol.* **170**, 605–610.

Willems, R., Paul, A., van der Heide, H.G.J., ter Avest, A.R. & Mooi, F.R. (1990) Fimbrial phase variation in *Bordetella pertussis*: a novel mechanism for transcriptional regulation. *EMBO J.* **9**, 2803–2809.

Willetts, N. & Skurray, R. (1987) Structure and function of the F factor and mechanism of conjugation. In Neidhardt, F.C., Ingraham, J.L., Low, K.B., Magasanik, B., Schaechter, M. & Umbarger, H.E. (eds), Escherichia coli *and* Salmonella typhimurium *Cellular and Molecular Biology*, pp. 1110–1133. American Society for Microbiology, Washington.

Willetts, N.S. & Wilkins, B. (1984) Processing of plasmid DNA during bacterial conjugation. *Microbiol. Rev.* **48**, 24–41.

Williams, D.R. & Thomas C.M. (1992) Active partitioning of bacterial plasmids. *J. Gen. Microbiol.* **138**, 1–16.

Williams, P., Bainton, N.J., Swift, S., Chhabra, S.R., Winson, M.K., Stewart, G.S.A.B., Salmond, G.P.C. & Bycroft, B.W. (1992) Small molecule-mediated density-dependent control of gene expression in prokaryotes: bioluminescence and the biosynthesis of carbapenem antibiotics. *FEMS Microbiol. Lett.* **100**, 161–168.

Williams, P., Brown, M.R.W. & Lambert, P.A. (1984) Effect of iron deprivation on the production of siderophores and outer membrane proteins in *Klebsiella aerogenes*. *J. Gen. Microbiol.* **130**, 2357–2365.

Williams, P.H. (1979) Novel iron uptake system specified by ColV plasmids: an important component in the virulence of invasive strains of *Escherichia coli*. *Infect. Immun.* **26**, 925–932.

Williams, P.H. & Roberts, M. (1989) Iron scavenging in the pathogenesis of *Escherichia coli*. In Hopwood, D.A. & Chater, K.F. (eds), *The Genetics of Bacterial Diversity*, pp. 331–350. Academic Press, London.

Williamson, C.M., Baird, G.D. & Manning, E.J. (1988) A common virulence region on plasmids from eleven serotypes of *Salmonella*. *J. Gen. Microbiol.* **134**, 975–982.

Willimsky, G., Bang, H., Fischer, G. & Marahiel, M.A. (1992) Characterization of *cspB*, a *Bacillus subtilis* inducible cold shock gene affecting cell viability at low temperatures. *J. Bacteriol.* **174**, 6326–6335.

Willins, D.A., Ryan, C.W., Platko, J.V. & Calvo, J.M. (1991) Characterization of Lrp, an *Escherichia coli* regulatory protein that mediates a global response to leucine. *J. Mol. Biol.* **266**, 10768–10774.

Winans, S.C. (1990) Transcriptional induction of an *Agrobacterium* regulatory gene at tandem promoters by plant-released phenolic compounds, phosphate starvation and acidic growth media. *Zj. Bacteriol.* **172**, 2433–2438.

Winans, S.C. (1992) Two-way chemical signalling in *Agrobacterium*-plant interactions. *Microbiol. Rev.* **56**, 12–31.

Winans, S.C., Ebert, P.R., Stachel, S.E., Gordon, M.P. & Nester, E.W. (1986) A gene for *Agrobacterium* virulence is homologous to a family of positive regulatory loci. *Proc. Natl Acad. Sci. USA* **83**, 8278–8282.

Winans, S.C., Kerstetter, R.A. & Nester, E.W. (1988) Transcriptional regulation of the *virA* and *virG* genes of *Agrobacterium tumefaciens. J. Bacteriol.* **170**, 4047–4054.

Winans, S.C., Kerstetter, R.A., Ward, J.E. & Nester, E.W. (1989) A protein required for transcriptional regulation of *Agrobacterium* virulence genes spans the cytoplasmic membrane. *J. Bacteriol.* **171**, 1616–1622.

Winkler, M.E. (1987) Biosynthesis of histidine. In Neidhardt, F.C, Ingraham, J.L., Low, K.B., Magasanik, B., Schaecter, M. & Umbarger, H.E. (eds), Escherichia coli *and* Salmonella typhimurium *Cellular and Molecular Biology*, pp. 395–411. American Society for Microbiology, Washington.

Winkler, M.E., Roth, D.J. & Hartman, P.E. (1978) Promoter- and attenuator-related regulation of the *Salmonella typhimurium* histidine operon. *J. Bacteriol.* **133**, 830–843.

Wlodarczyk, M. & Nowicka, B. (1988) Preliminary evidence for the linear nature of *Thiobacillus versutus* pTAV2 plasmid. *FEMS Microbiol. Lett* **55**, 125–128.

Wong, P.-Z., Projan, S.J., Henriquez, V. & Novick, R.P. (1992) Specificity of origin recognition by replication initiator protein in plasmids of the pT181 family is determined by a six amino acid residue element. *J. Mol. Biol.* **223**, 145–158.

Worcel, A. & Burgi, E. (1972) Structure of the folded chromosome of *E. coli. J. Mol. Biol.* **71**, 127–147.

Wozniak, D.J., Cram, D.C., Daniels, C.J. & Galloway, D.R. (1987) Nucleotide sequence and characterization of *toxR*: a gene involved in exotoxin A regulation in *Pseudomonas aeruginosa. Nucleic Acids Res.* **15**, 2123–2135.

Wozniak, D.J. & Ohman, D.E. (1993) Involvement of the alginate *algT* gene and integration host factor in the regulation of the *Pseudomonas aeruginosa algB* gene. *J. Bacteriol.* **175**, 4145–4153.

Wren, B.W. (1992) Bacterial enterotoxin interactions. In Hormaeche, C.E., Penn, C.W. & Smyth, C.J. (eds) *Molecular Biology of Bacterial Infection: Current Status and Future Perspectives*, pp. 127–147. Cambridge University Press.

Wright, K. (1990) Bad news bacteria. *Science*, **249**, 22–24.

Wu, H.-Y., Shyy, S., Wang, J.C. & Liu, L.F. (1988) Transcription generates positively and negatively supercoiled domains in the template. *Cell* **53**, 433–440.

Wu, J. & Weiss, B. (1991) Two divergently transcribed genes, *soxR* and *soxS*, control a superoxide response regulon in *Escherichia coli. J. Bacteriol.* **173**, 2864–2871.

Wu, J. & Weiss, B. (1992) Two-stage induction of the *soxRS* (superoxide response) regulon of *Escherichia coli. J. Bacteriol.* **174**, 3915–3920.

Wu, T.-H., Liao, S.-M., McClure, W. & Susskind, M.M. (1987) Control of gene expression in bacteriophage P22 by a small antisense RNA. II. Characterization of mutants defective in repression. *Genes Dev.* **1**, 204–212.

Wurtzel, E.T., Chou, M.-Y. & Inouye, M. (1982) Osmoregulation of gene expression. I. DNA sequence of the *ompR* gene of the *ompB* operon of *Escherichia coli* and characterization of its gene product. *J. Biol. Chem.* **257**, 13685–13691.

Xiao, H., Kalman, M., Ikehara, K., Zemel, S., Glaser, G. & Cashel, M. (1991) Residual 3′,5′-bispyrophosphate synthetic activity of *relA* null mutants can be eliminated by *spoT* null mutations. *J. Biol. Chem.* **266**, 5980–5990.

Yager, T.D. & von Hippel, P.H. (1987) Transcript elongation and termination in *Escherichia coli.* In Neidhardt, F.C., Ingraham, J.L., Low, K.B., Magasanik, B., Schaechter, M. & Umbarger, H.E. (eds) Escherichia coli *and* Salmonella typhimurium, *Cellular and Molecular Biology*, pp. 1241–1275. American Society for Microbiology, Washington.

Yamada, H., Muramatsu, S. & Mizuno, T. (1990) An *Escherichia coli* protein that preferentially binds to sharply curved DNA. *J. Biochem.* **108**, 420–425.

Yamada, H., Yoshida, T., Tanaka, K.-I., Sasakawa, C. & Mizuno, T. (1991) Molecular analysis of the *Escherichia coli hns* gene encoding a DNA-binding protein, which preferentially recognizes curved DNA sequences. *Mol. Gen. Genet.* **230**, 332–336.

Yamamoto, N. & Droffner, M.L. (1985) Mechanisms determining aerobic or anaerobic growth in the facultative anaerobe *Salmonella typhimurium. Proc. Natl Acad. Sci. USA* **82**, 2077–2081.

Yamamoto, T., Gojobori, T. & Yokota, T. (1987) Evolutionary origin of pathogenic determinants in enterotoxigenic *Escherichia coli* and *Vibrio cholerae* O1. *J. Bacteriol.* **169**, 1352–1357.

Yamamoto, T. & Yokota, T. (1981) *Escherichia coli* heat-labile enterotoxin genes are flanked by repeated deoxyribonucleic acid sequences. *J. Bacteriol.* **145**, 850–860.

Yanagida, M. & Sternglanz, R. (1990) Genetics of topoisomerases. In Cozzarelli, N.R. & Wang, J.C. (eds), *DNA Topology and its Biological Effects*, pp. 299–320. Cold Spring Harbor Laboratory Press, Cold Spring Harbor, New York.

Yang, C. & Konisky, J. (1984) Colicin V-treated *Escherichia coli* does not generate membrane potential. *J. Bacteriol.* **158**, 757–759.

Yang, Y. & Ames, G. F.-L. (1989) DNA gyrase binds to the family of prokaryotic repetitive extragenic palindromic sequences. *Proc. Natl Acad. Sci. USA* **85**, 8850–8854.

Yang, Y. & Ames, G. (1990) The family of repetitive extragenic palindromic sequences: interaction with DNA gyrase and histonelike protein HU. In Drlica, K. & Riley, M. (eds), *The Bacterial Chromosome*, pp. 211–225. American Society for Microbiology, Washington.

Yanofsky, M.F., Porter, S.G., Young, C., Albright, L.M., Gordon, M.P. & Nester, E.W. (1986) The *virD* operon of *Agrobacterium tumefaciens* encodes a site-specific endonuclease. *Cell* **47**, 471–477.

Yim, H.H. & Villarejo, M. (1992) *osmY*, a new hyperosmotically inducible gene, encodes a periplasmic protein in *Escherichia coli. J. Bacteriol.* **174**, 3637–3644.

Yin, J.C.P., Krebs, M.P. & Reznikoff, W.S. (1988) The effect of *dam* methylation on Tn5 transposition. *J. Mol. Biol.* **199**, 35–45.

Yin, J.C.P. & Reznikoff, W.S. (1987) *dnaA*, an essential host gene, and Tn5 transposition. *J. Bacteriol.* **169**, 4637–4645.

Yogev, D., Rosengarten, R., Watson-McKown, R. & Wise, K.S. (1991) Molecular basis of *Mycoplasma* surface antigenic variation: a novel set of divergent genes undergo spontaneous mutation of periodic coding regions and 5' regulatory sequences. *EMBO J.* **10**, 4069–4079.

Yoshikawa, M., Sasakawa, C., Makino, S., Okada, N., Lett, M.-C., Sakai, T., Yamada, M., Komatsu, K., Kamata, K., Kurata, T. & Sata, T. (1988) Molecular genetic approaches to the pathogenesis of bacillary dysentery. *Microbiol. Sci.* **5**, 333–339.

Young, D.B., Mehlert, A., Bal, V., Mendez-Samperio, P., Ivanyi, J. & Lamb, J.R. (1988) Stress proteins and the immune response to mycobacteria: antigens as virulence factors? *Antonie van Leeuwenhoek* **54**, 431–439.

Young, V.B., Miller, V.L., Falkow, S. & Schoolnik, G.K. (1990) Sequence, localization and function of the invasin protein of *Yersinia enterocolitica. Mol. Microbiol.* **4**, 1119–1128.

Yu, C.-E. & Ferretti, J.J. (1991) Frequency of the erythrogenic toxin B and C genes (*speB* and *speC*) among clinical isolates of Group A streptococci. *Infect. Immun.* **59**, 211–215.

Zabeau, M., Friedman, S., Van Montagu, M. & Schell, J. (1980) The *ral* gene of phage λ. Identification of a nonessential gene that modulates restriction and modification in *E. coli. Mol. Gen. Genet.* **179**, 63–74.

Zagaglia, C., Casalino, M., Colonna, B., Conti, C., Calconi, A. & Nicoletti, M. (1991) Virulence plasmids of enteroinvasive *Escherichia coli* and *Shigella flexneri* integrate into a specific site on the host chromosome: integration greatly reduces expression of plasmid-carried virulence genes. *Infect. Immun.* **59**, 792–799.

Zambryski, P. (1988) Basic processes underlying *Agrobacterium*-mediated DNA transfer to plant cells. *Annu. Rev. Genet.* **22**, 1–30.

Zhang, L. & Kerr, A. (1991) A diffusible compound can enhance conjugal transfer of the Ti plasmid in *Agrobacterium tumefaciens. J. Bacteriol.* **173**, 1867–1872.

Zhang, L., Murphy, P.J., Kerr, A. & Tate, M.E. (1993) *Agrobacterium* conjugation and gene regulation by *N*-acyl-L-homoserine lactones. *Nature* **362**, 446–448.

Zhang, Q.Y., DeRyckere, D., Lauer, P. & Koomey, M. (1992) Gene conversion in *Neisseria gonorrhoeae*: evidence for its role in pilus antigenic variation. *Proc. Natl Acad. Sci. USA* **89**, 5366–5370.

Zhou, Y.N., Kusukawa, J., Erickson, J.W., Gross, C.A. & Yura, T. (1988) Isolation and characterization of *Escherichia coli* mutants that lack the heat shock sigma factor (σ^{32}). *J. Bacteriol.* **170**, 3640–3649.

Zieg, J. & Simon, M.I. (1980) Analysis of the nucleotide sequence of an invertible controlling element. *Proc. Natl Acad. Sci. USA* **77**, 4196–4200.

Zimmermann, A., Reimmann, C., Galimand, M. & Haas, D. (1991) Anaerobic growth and cynanide synthesis of *Pseudomonas aeruginosa* depend on *anr*, a regulatory gene homologous with *fnr* of *Escherichia coli. Mol. Microbiol.* **5**, 1483–1490.

Zou, C., Fujita, N., Igarashi, K. & Ishihama, A. (1992) Mapping the cAMP receptor protein contact site on the α subunit of *Escherichia coli* RNA polymerase. *Mol. Microbiol.* **6**, 2599–2605.

Zuerner, R.L. & Bolin, C.A. (1988) Repetitive sequence element cloned from *Leptospira interrogans* serovar hardjo type hardjo-bovis provides a sensitive probe for bovine leptospirosis. *J. Clin. Microbiol.* **26**, 2495–2500.

Zyskind, J.W. (1990) Priming and growth rate regulation: questions concerning the initiation of DNA replication in *Escherichia coli.* In *The Bacterial Chromosome*, Drlica, K. & Riley, M. (eds), pp. 269–278. American Society for Microbiology, Washington.

Zyskind, J.W., Svitil, A.L., Stine, W.B., Biery, M.C. & Smith, D.W. (1992) RecA protein of *Escherichia coli* and chromosome partitioning. *Mol. Microbiol.* **6**, 2525–2537.

Index